MW00760484

Electron Correlation in Molecules and Condensed Phases

PHYSICS OF SOLIDS AND LIQUIDS

Current Volumes in the Series

AMORPHOUS SOLIDS AND THE LIQUID STATE
Edited by Norman H. March, Robert A. Street, and Mario P. Tosi

CHEMICAL BONDS OUTSIDE METAL SURFACES
Norman H. March

CRYSTALLINE SEMICONDUCTING MATERIALS AND DEVICES
Edited by Paul N. Butcher, Norman H. March, and Mario P. Tosi

ELECTRON CORRELATION IN MOLECULES AND CONDENSED PHASES
N. H. March

EXCITATION ENERGY TRANSFER PROCESSES IN CONDENSED MATTER: Theory and Applications
Jai Singh

FRACTALS
Jens Feder

INTERACTION OF ATOMS AND MOLECULES WITH SOLID SURFACES
Edited by V. Bortolani, N. H. March, and M. P. Tosi

LOCAL DENSITY THEORY OF POLARIZABILITY
Gerald D. Mahan and K. R. Subbaswamy

MANY-PARTICLE PHYSICS, Second Edition
Gerald D. Mahan

ORDER AND CHAOS IN NONLINEAR PHYSICAL SYSTEMS
Edited by Stig Lundqvist, Norman H. March, and Mario P. Tosi

PHYSICS OF LOW-DIMENSIONAL SEMICONDUCTOR STRUCTURES
Edited by Paul Butcher, Norman H. March, and Mario P. Tosi

QUANTUM TRANSPORT IN SEMICONDUCTORS
Edited by David K. Ferry and Carlo Jacoboni

Electron Correlation in Molecules and Condensed Phases

N. H. March
University of Oxford
Oxford, England

Plenum Press • New York and London

Library of Congress Cataloging-in-Publication Data

March, Norman H. (Norman Henry), 1927-
Electron correlation in molecules and condensed phases / N.H. March.
p. cm. -- (Physics of solids and liquids)
Includes bibliographical references and index.
ISBN 0-306-44844-0
1. Electron configuration. 2. Condensed matter. 3. Molecules. I. Title. II. Series.
IN PROCESS
530.4'11--dc20 95-4195
CIP

ISBN 0-306-44844-0

A Division of Plenum Publishing Corporation
233 Spring Street, New York, N. Y. 10013

10 9 8 7 6 5 4 3 2 1

Printed in the United States of America

Preface

This book had its origins in lectures presented at EPFL, Lausanne, during two separate visits (the most recent being to IRRMA). The author is most grateful to Professors A. Baldereschi, R. Car, and A. Quattropani for making these visits possible, and for the splendidly stimulating environment provided. Professors S. Baroni and R. Resta also influenced considerably the presentation of material by constructive help and comments.

Most importantly, Chapters 4 and 5 were originally prepared for a review article by Professor G. Senatore, then at Pavia and now in Trieste, and myself for Reviews of Modern Physics (1994). In the course of this collaboration, he has taught me a great deal, especially about quantum Monte Carlo procedures, and Chapter 5 is based directly on this review article. Also in Chapter 4, my original draft on Gutzwiller's method has been transformed by his deeper understanding; again this is reflected directly in Chapter 4; especially in the earlier sections.

In addition to the above background, it is relevant here to point out that, as a backcloth for the present, largely "state of the art," account, there are two highly relevant earlier books: *The Many-body Problem in Quantum Mechanics* with W. H. Young and S. Sampanthar (Dover reprint: 1995) and *Collective Effects in Solids and Liquids* with M. Parrinello. These still provide excellent background for the present Volume, as does the review with J. C. Stoddart (1968) on *Localization of Electrons in Condensed Matter.* More recent places for the reader to acquire further relevant background are: (i) G. D. Mahan, *Many-Particle Physics,* in this same series and (ii) S. Wilson, *Electron Correlation in Molecules.* As well as many-body techniques, a good knowledge of general condensed matter background is assumed; for example as in the recent book on *Introductory Solid State Physics* by H. P. Myers (1990).

While I am, as set out above, indebted to many workers, I must, of course, accept sole responsibility for any errors that remain.

My book, though about concepts and theory, was written with experimental physicists, materials scientists and physical and inorganic chemists much in mind, especially those who are keen to relate their experimental programs to basic theoretical questions that remain. The book should also prove useful for graduate students in these areas, when supplemented by the background reading set out above.

Finally I must acknowledge help from numerous authors relating to the figures in the present work. Especially I must name Drs. B. Farid, F. Siringo, V. Cheranovskii, and G. D. Mahan. Dr. Farid generously made available to me many of the diagrams in Chapter 6 and in some Appendices. Dr. Siringo helped with the diagrams in Chapter 8 and thereby influenced the presentation of material there. Drs. Cheranovskii and Mahan permitted me to adopt their presentations of material on model Hamiltonians and the so-called GW approximations, respectively. It is also a pleasure to thank Professor M. P. Tosi and Drs. B. Farid and A. Holas for much fruitful collaboration and invaluable advice on many-electron theory over many years. The author acknowledges here material reprinted with permission from *Phys. Rev.* **B44**, 13356–13373 (1991), *Phys. Rev.* **A48**, 3561–3566 (1993), and *Phys. Rev. Lett.* **68**, 121–124 (1992): ©American Physical Society. Also material is reprinted with permission from *Rev. Mod. Phys.* **66**, 445–479 (1994): ©American Institute of Physics.

N. H. March

Contents

Appendixes

Electron Correlation in Molecules and Condensed Phases

Chapter 1

Outline

This volume will treat those aspects of electron correlation in atoms, molecules, and solids about which fundamental progress has proved possible. The main tools to be employed here are the first- and second-order density matrices (see Section 2.2), including the important observable electron density $\rho(r)$ for the analytical development presented. Closely related tools, e.g., the Greens function, are used in discussing quantum computer simulations.

Chapter 2 deals with some aspects of electron correlation in atoms, with particular reference to the role of electron correlation energy. Progress in understanding the asymptotic forms of the density matrices will be summarized and related to the known asymptotic behavior of the electron density, which is dominated by the ionization potential. The relation between N-electron and $(N-1)$-electron systems will be stressed, with particular reference to their respective low-order density matrices.

Then attention shifts to the general theory of the inhomogeneous electron gas, equivalently known as density functional theory. Here one naturally enough builds from the homogeneous electron fluid, i.e., the so-called jellium model of a metal, in which electrons interacting Coulombically move within a nonresponsive uniform positive neutralizing background charge. This is treated in Chapter 3, leading into a summary of density functional theory. Since at least three recent books on this field exist (Parr and Yang, 1989; March, 1991; Kryachko and Ludena, 1991), this discussion will be kept relatively short. It was felt, however, to be of interest to focus on the spin density formulation, with a review of the important progress made on local moment formation in metals, as well as on stability conditions for cooperative magnetism, both problems being closely linked with the correlation energy in narrow d-band transition metals.

One point to be stressed at the outset is that, while density functional theory reveals the way in which electron correlation enters the calculation of single-particle density, as well as the important chemical potential of the inhomogeneous electron gas, there is no means within the formal framework of the theory to calculate the required functionals describing correlation. Besides local density approximations derived from the jellium model, a brief summary will be given of the quantum chemical prescription of Colle and Salvetti (1975).

Chapter 4 contrasts the multicenter problem of molecules and solids with the atomic case and stresses the qualitative aspects of electron correlation by summarizing the idea behind the Coulson–Fischer wave function for the ground state of the H_2 molecule. This idea is then taken up in relation to the work of Gutzwiller on strongly correlated electrons in narrow energy bands. In this treatment, electron interactions are treated by means of the Hubbard U which, by definition, is the energy required to place two electrons with antiparallel spin on the same site. As in the Coulson–Fischer treatment, one reduces the weight of the ionic configurations that are obtained by expanding out a single Slater determinant of Bloch wave functions.

To this stage, the main theoretical development occurs largely by means of analytical methods. However, important progress in the calculation of correlation energy in multicenter problems has resulted from the development of quantum computer simulation, pioneered in general terms by Kalos and Anderson and applied to jellium by Ceperley and Alder (1980). The parallel development in calculating correlation energies in small molecules will also be described. At the time of writing, this computer simulation approach is the most accurate way of calculating electronic correlation energies in multicenter problems. Plasmons, quasiparticles, and the so-called G–W approximation in many-body theory are then treated.

With a view to emphasizing a qualitative consequence of electron correlation in multicenter problems, a substantial chapter is concerned with metal–insulator transitions in the presence of both electron–ion and electron–electron interactions with particular emphasis on chemically bonded systems. After a brief discussion of the exact solution of the Hubbard Hamiltonian for linear polyenes and its comparison with the Gutzwiller results, a phenomenological theory of the metal–insulator transition at absolute zero is presented. Brief reference is then made to the elevated temperature studies of Rice *et al.* in relation to heavy fermion problems. Because of the current excitement concerning superconductivity, the relevance of the Hubbard Hamiltonian in providing a quantitative basis for the development of Pauling's ideas on the resonating valence-bond theory of metals is briefly summarized, together with a review of Luttinger liquids in low dimensional electronic assemblies.

Electron correlation in disordered systems is then considered at some length, with specific applications to liquid metals. The volume concludes with a chapter on the two-dimensional magnetically induced Wigner solid, in particular its phase diagram. This latter is determined by the equilibrium between an electron solid and the Laughlin (1983) electron liquid, therefore leading to a brief consideration of integral and fractional quantum Hall effects as well as a short section on anyons.

Chapter 2

Electron Density, Density Matrices and Atomic Properties

Though the focus of the volume is on correlation effects in the multicentre problems of molecules and condensed phases, it will be helpful to start by considering the single-centre problem of atoms. A brief discussion of gross trends of atomic ground-state energy with increasing number of electrons serves already to highlight the importance of the electron density in many-body theory—a recurring theme through this volume. It also prompts the generalization of electron density to its off-diagonal form, the first-order density matrix. This latter quantity determines the kinetic energy, of an atom say, but one needs the second-order density matrix to yield the total ground-state energy (see Eq. (2.21) below). Collective effects are then briefly considered; and the Chapter concludes with reference to (i) d-dimensional theory and (ii) the coupled cluster method.

2.1. GROSS TRENDS OF GROUND-STATE ENERGY IN ATOMS WITH INCREASING NUMBER OF ELECTRONS

Consider non-interacting electrons, in closed-shell configurations, moving in the bare Coulomb field of a nucleus of charge Ze. The Bohr formula for the one-electron levels is

$$\varepsilon_n = -\frac{Z^2}{2n^2}\frac{e^2}{a_0} : n = 1, 2, \ldots \tag{2.1}$$

where a_0 is the Bohr radius $\hbar^2/me^2$. If N closed shells are filled, then since a closed shell of principal quantum number n contains $2n^2$ electrons, the energy per shell is $-Z^2e^2/a_0$ and the total energy E (sum of one-electron energies in this simple model) is

$$E = -Z^2 N \frac{e^2}{e_0}. \tag{2.2}$$

A neutral atom, to which our attention will be confined at this stage, must evidently satisfy

$$Z = \sum_1^N 2n^2 = \frac{N(N+1)(2N+1)}{3} \tag{2.3}$$

For simplicity, consider the statistical limit of many electrons when Eq. (2.3) takes the approximate form

$$N = (\tfrac{3}{2})^{1/3} Z^{1/3} - \tfrac{1}{2} + O(Z^{-1}) \tag{2.4}$$

and therefore, from Eq. (2.2),

$$E = -(\tfrac{3}{2})^{1/3} Z^{7/3}(e^2/a_0) + \tfrac{1}{2} Z^2(e^2/a_0) + \ldots \tag{2.5}$$

This bare Coulomb model predicts, in the nonrelativistic framework used throughout this volume, that the total energy of heavy neutral atoms increases as $Z^{7/3}$.

To confirm this result for the Z-dependence and to correct the coefficient $(\frac{3}{2})^{1/3}$ for self-consistent field effects, one uses the theory of Thomas (1926) and Fermi (1928), the forerunner of the density functional theory to be discussed in Chapter 3. Here one assumes the electrons to move a self-consistent common potential energy $V(\mathbf{r})$, and if $p_f(\mathbf{r})$ is the maximum momentum (Fermi momentum) of electrons at position $\mathbf{r}$ then the energy equation for the fastest electron reads, in its classical form,

$$\mu = \frac{p^2(\mathbf{r})}{2m} + V(\mathbf{r}) \tag{2.6}$$

Here one must note that (a) it has been anticipated that in this statistical theory the total energy of the fastest electron is to be identified with the chemical potential μ, and (b) the condition of equilibrium in the electron cloud is that while the kinetic and potential energy contributions on the right-hand side of Eq. (2.6) evidently vary from point to point in the charge cloud, the left-hand side must be a constant; that is, it must take the same value at every point $\mathbf{r}$ in the charge distribution. For otherwise the electrons could redistribute in space to lower the total ground-state energy of the atom. If one now combines Eq. (2.6) with the local electron gas relation that the number of electrons per unit volume at $\mathbf{r}$, say $\rho(\mathbf{r})$, is related to the Fermi momentum $p_f(\mathbf{r})$ by

$$\rho(\mathbf{r}) = \frac{8\pi}{3h^3} p_f^3(\mathbf{r}), \tag{2.7}$$

then the self-consistent potential energy in the atom of nuclear charge Ze can be established, once and for all, by solving Poisson's equation relating $\rho(\mathbf{r})$ and $V(\mathbf{r})$.

The result for the total energy of the neutral atom is then found to be (Milne, 1927; see, for example, March, 1975)

$$E_{\mathrm{TF}}(Z) = -0.77 Z^{7/3}(e^2/a_0). \tag{2.8}$$

the screening of the bare nucleus by the electrons reducing the coefficient $(3/2)^{1/3} = 1.14$ in Eq. (2.5) to the self-consistent field value of 0.77 in Eq. (2.8). Though Eq. (2.8) is the correct law for atomic binding energies as Z tends to infinity in the nonrelativistic Schrödinger theory, it turns out that even for the uranium atom, with $Z = 92$, there are important quantitative corrections to this limiting law [Eq. (2.8)] of the Thomas–Fermi statistical theory. Thus, if one returns to the bare Coulomb result [Eq. (2.5)], there is a correction term, which is $O(1/Z^{1/3})$ of the leading term. This $Z^{-1/3}$ is the correct expansion parameter for the neutral atom energy, and indeed Scott (1952) has argued that the correction $\frac{1}{2}Z^2e^2/a_0$ in the bare Coulomb field carries over directly into the self-consistent field result, Eq. (2.8). The reason for this is that the Thomas–Fermi theory is in error primarily through its treatment of the K shell, and just in this shell the self-consistent field effects are minimal (see also the discussion in Schwinger, 1985). A further correction, due to Scott, is the addition of exchange energy. He accomplished this by calculating the so-called Dirac–Slater exchange energy $-c_e \int \rho^{4/3}\, dr$; $c_e = (3/4)(3/\pi)^{1/3}e^2$ (cf. Chapter 3 below), but with the Thomas–Fermi approximation to $\rho(\mathbf{r})$. He obtained

$$A = -c_e \int \{\rho_{\mathrm{TF}}(\mathbf{r})\}^{4/3} d\mathbf{r} = -0.22 Z^{5/3}(e^2/a_0) \tag{2.9}$$

As a numerical example, for Uranium, with $Z = 92$, the term $\frac{1}{2}Z^2(e^2/a_0)$ is 14% of the leading Thomas–Fermi term [Eq. (2.8)], while the exchange energy is down by a factor of 10 on the $\frac{1}{2}Z^2$ term. But still these corrections represent huge energies in chemically significant terms.

No treatment of correlation is as well based as those for kinetic and exchange energy discussed above, as a "universal" function of Z. But Clementi (1963; see also Appendix 4.1) has studied numerically the way the correlation energy of atoms varies with Z, and again a clear trend emerges. The variation with Z is roughly linear (see his Figs. 1 and 2), the order of magnitude of the correlation energy per electron being 0.1 Ry (cf. March and Bader, 1980).

Following this brief introduction to atomic energies, which appealed to the electron density $\rho(\mathbf{r})$ in Eqs. (2.7) and (2.9) we turn immediately to discuss density matrices.

2.2. DEFINITION OF DENSITY MATRICES

2.2.1. First-Order Density Matrices and Particle Density

The general spin-dependent (σ) first-order density matrix $\gamma_\sigma(\mathbf{r}_1'\sigma_1', \mathbf{r}_1\sigma_1)$ is defined from the normalized many-body wave function $\Psi(\mathbf{r}_1\sigma_1, \ldots, \mathbf{r}_N\sigma_N)$ by

$$\gamma_\sigma(\mathbf{r}_1'\sigma_1', \mathbf{r}_1\sigma_1) = N \sum_{\sigma_2 \dots \sigma_N} \int \Psi^*(\mathbf{r}_1'\sigma_1', \mathbf{r}_2\sigma_2, \dots, \mathbf{r}_N\sigma_N)$$
$$\times \Psi(\mathbf{r}_1\sigma_1, \mathbf{r}_2\sigma_2, \dots, \mathbf{r}_N\sigma_N)d\mathbf{r}_2 d\mathbf{r}_3 \dots d\mathbf{r}_N. \tag{2.10}$$

The widely useful spinless form $\gamma(\mathbf{r}_1, \mathbf{r}_1)$ of Eq. (2.10) is simply

$$\gamma(\mathbf{r}_1', \mathbf{r}_1) = \sum_{\sigma_1} \gamma_\sigma(\mathbf{r}_1'\sigma_1, \mathbf{r}_1\sigma_1). \tag{2.11}$$

The normalization factor N in Eq. (2.10) has been chosen such that the diagonal element $\gamma(\mathbf{r}_1,\mathbf{r}_1)$ of Eq. (2.11) is the electron density $\rho(\mathbf{r}_1)$, which evidently must integrate to N through the whole of space. In Section 2.2.5 an explicit calculation of γ for a plane-wave determinant will be referred to, but before applying Eq. (2.11) let us introduce the second-order forms (see also March et al., 1995).

2.2.2. Second-Order Density Matrices and Pair Function

In a similar manner, one defines the second-order matrix (Löwdin, 1955) $\Gamma(\mathbf{r}_1'\sigma_1', \mathbf{r}_2'\sigma_2'; \mathbf{r}_1\sigma_1, \mathbf{r}_2\sigma_2)$ as

$$\Gamma_\sigma(\mathbf{r}_1'\sigma_1', \mathbf{r}_2'\sigma_2'; \mathbf{r}_1\sigma_1, \mathbf{r}_2\sigma_2)$$
$$= \frac{N(N-1)}{2} \sum_{\sigma_3 \dots \sigma_N} \int \Psi^*(\mathbf{r}_1'\sigma_1', \mathbf{r}_2'\sigma_2', \mathbf{r}_3\sigma_3, \dots, \mathbf{r}_N\sigma_N)$$
$$\times \Psi(\mathbf{r}_1\sigma_1, \mathbf{r}_2\sigma_2, \dots, \mathbf{r}_N\sigma_N)d\mathbf{r}_3 \dots d\mathbf{r}_N \tag{2.12}$$

and the corresponding spinless form as

$$\Gamma(\mathbf{r}_1', \mathbf{r}_2'; \mathbf{r}_1, \mathbf{r}_2) = \sum_{\sigma_1 \dots \sigma_2} \Gamma_\sigma(\mathbf{r}_1'\sigma_1, \mathbf{r}_2'\sigma_2; \mathbf{r}_1\sigma_1, \mathbf{r}_2\sigma_2). \tag{2.13}$$

Just as with the first-order density matrix, the diagonal element of Eq. (2.13) has immediate physical significance, as it yields the number of distinct pairs multiplied by the probability density of simultaneously finding one particle at $\mathbf{r}_1$and another at $\mathbf{r}_2$. It frequently proves useful to rewrite $\Gamma(\mathbf{r}_1\mathbf{r}_2; \mathbf{r}_1, \mathbf{r}_2)$ as a constant times $g(\mathbf{r}_1, \mathbf{r}_2)$, where the constant is chosen such that $g \to 1$ for large separation $|\mathbf{r}_1 - \mathbf{r}_2|$. The quantity $g(\mathbf{r}_1, \mathbf{r}_2)$ will thus be referred to as the *pair function.*

2.2.3. Relation between First- and Second-Order Matrices, and Total Energy

It is clear from the definitions of Γ_σ and γ_σ that the first contains the second, the explicit relation following immediately from Eqs. (2.12) and (2.10) as

$$\gamma_\sigma(\mathbf{r}_1'\sigma_1', \mathbf{r}_1\sigma_1) = \frac{2}{N-1} \sum_{\sigma_2} \int \Gamma_s(\mathbf{r}_1'\sigma_1', \mathbf{r}_2\sigma_2; \mathbf{r}_1\sigma_1, \mathbf{r}_2\sigma_2)d\mathbf{r}_2. \tag{2.14}$$

Similarly, for the spinless forms,

$$\gamma(\mathbf{r}_1', \mathbf{r}_1) = \frac{2}{N-1} \int G(\mathbf{r}_1', \mathbf{r}_2; \mathbf{r}_1, \mathbf{r}_2) d\mathbf{r}_2. \tag{2.15}$$

The total energy E of the assembly will now be written in terms of the density matrices. To do so, one notes first that the total energy of a state defined by a total wave function Ψ is

$$E = \sum_{\text{spin}} \int \Upsilon^* H \Psi d\mathbf{r}_1 \ldots d\mathbf{r}_N. \tag{2.16}$$

One will assume a Hamiltonian H of the form

$$H = \sum_{i=1}^{N} (-) \frac{\hbar^2}{2m} \nabla_i^2 + \sum_{i<j} v(\mathbf{r}_i, \mathbf{r}_j), \tag{2.17}$$

or if there are internal degrees of freedom, most usually spin, that affect the particle interaction and if the particles are also subject to some external potential, then the generalization will be taken as

$$H = \sum_{i=1}^{N} U(\mathbf{r}_i\sigma_i) + \tfrac{1}{2} \sum_{i=1}^{N} \sum_{j=1}^{N}{}' v(\mathbf{r}_i\sigma_i, \mathbf{r}_j\sigma_j). \tag{2.18}$$

One can now readily evaluate the energy E. To be quite explicit, let us neglect spin dependence in H and also write the one-particle term $\sum_{i=1}^{N} U(\mathbf{r}_i)$ explicitly as

$$\frac{-\hbar^2}{2m} \nabla^2 + U_1(\mathbf{r}_i). \tag{2.19}$$

Then it is clear that its expectation value can be expressed in terms of $\gamma(\mathbf{r}', \mathbf{r})$. Similarly the average of the two-particle contribution

$$\sum_{i<j} v(\mathbf{r}_i, \mathbf{r}_j) \tag{2.20}$$

depends only on the diagonal element $\Gamma(\mathbf{r}_1, \mathbf{r}_2; \mathbf{r}_1, \mathbf{r}_2)$. The final form that results after a quite straightforward calculation is

$$E = \frac{-\hbar^2}{2m} [\nabla^2 \gamma(\mathbf{r}', \mathbf{r})]_{\mathbf{r}'=\mathbf{r}} d\mathbf{r} + \int U_1(\mathbf{r})\rho(\mathbf{r}) d\mathbf{r}$$
$$+ \int v(\mathbf{r}_1, \mathbf{r}_2)\Gamma(\mathbf{r}_1, \mathbf{r}_2; \mathbf{r}_1, \mathbf{r}_2) d\mathbf{r}_1 d\mathbf{r}_2. \tag{2.21}$$

2.2.4. Dirac Density Matrix

The density matrices, as defined above, are to be calculated using the exact many-electron wave functions. However, when the total wave function is a determinant of single-particle states, the density matrices take a particularly simple form (Dirac, 1930). Thus if the total wave function Ψ is

$$\Psi = \frac{1}{(N!)^{1/2}} \det \phi_i(\mathbf{r}_j, \sigma_j), \tag{2.22}$$

then it is readily shown that

$$\gamma_\sigma(\mathbf{r}'\sigma', \mathbf{r}\sigma) = \sum_1^N \phi_i^*(\mathbf{r}'\sigma')\,\phi_i(\mathbf{r}\sigma), \tag{2.23}$$

where the ϕ_i are the one-particle states entering the determinant. Similarly, the second-order density matrix is

$$\Gamma_\sigma(\mathbf{r}_1'\sigma_1', \mathbf{r}_2'\sigma_2'; \mathbf{r}_1\sigma_1, \mathbf{r}_2\sigma_2)$$

$$= \frac{1}{2}\begin{vmatrix} \gamma_\sigma(\mathbf{r}_1'\sigma_1', \mathbf{r}_1\sigma_1) & \gamma_\sigma(\mathbf{r}_1'\sigma_1', \mathbf{r}_2\sigma_2) \\ \gamma_\sigma(\mathbf{r}_2'\sigma_2', \mathbf{r}_1\sigma_1) & \gamma_\sigma(\mathbf{r}_2'\sigma_2', \mathbf{r}_2\sigma_2) \end{vmatrix} \tag{2.24}$$

When one considers levels occupied by two electrons with opposed spins, then the states $\phi(\mathbf{r}_j\sigma_j)$ are built up as simple products of $\frac{1}{2}N$ space orbitals $\phi_i(\mathbf{r}_j)$ with the spin functions α and β. It then follows that

$$\gamma(\mathbf{r}_1, \mathbf{r}_2) = 2\sum_1^{1/2N} \phi_i^*(\mathbf{r}_1)\phi_i(\mathbf{r}_2) \tag{2.25}$$

and

$$2\Gamma(\mathbf{r}_1, \mathbf{r}_2; \mathbf{r}_1, \mathbf{r}_2) = \rho(\mathbf{r}_1)\,\rho(\mathbf{r}_2) - \tfrac{1}{2}[\gamma(\mathbf{r}_1, \mathbf{r}_2)]^2. \tag{2.26}$$

Eqs. (2.23) and (2.25) define the Dirac density matrix—the latter, of course, being the spinless form. Furthermore, with a single determinant, it can be seen from Eq. (2.26) that the Dirac density matrix also determines the second-order density matrix. This is not generally true. Combining the results in Eqs. (2.25) and (2.26) with the energy expression, Eq. (2.21), it is clear that one could obtain the best possible orbitals ϕ_i, in the sense of the variational method, by minimization. This is the Hartree–Fock theory.

In many circumstances, one may ask for the one-electron orbitals that are not quite optimal but satisfy a one-electron wave equation

$$\left[\frac{-\hbar^2}{2m}\nabla^2 + V(\mathbf{r})\right]\phi_i(\mathbf{r}) = \varepsilon_i\phi_i(\mathbf{r}), \tag{2.27}$$

where $V(\mathbf{r})$ is a common potential energy in which all the electrons move and which should eventually be calculated by a self-consistent procedure. This will be illustrated in Chapter 3 in some detail; for the present, let us regard $V(\mathbf{r})$ as given.

In this simplified case, the problem is reduced to finding Dirac's density matrix as a functional of $V(\mathbf{r})$. Unfortunately, no closed solution of this problem has, as yet, been found for an arbitrary potential energy $V(\mathbf{r})$. However, for the most important case, when the lowest single-particle states are occupied (that is, for the ground state), a complete perturbation solution was given by March and Murray (1960, 1961), and their results are summarized in March *et al.* (1995). Since this expansion is based on plane waves, it will be helpful at this point to obtain explicit forms for the density matrices in this free-electron case.

2.2.5. Plane-Wave Density Matrices

As a simple, but nevertheless important, example of the calculation of the density matrices γ and Γ in the approximation of a single determinant, let us consider the ground state for the case of free particles. Thus the single-electron wave functions are the eigenfunctions of momentum, normalized in volume Ω:

$$\frac{1}{\Omega^{1/2}}\exp(i\mathbf{k}\cdot\mathbf{r}),$$

and when these are substituted into Eq. (2.25), one finds that

$$\gamma(\mathbf{r}', \mathbf{r}) = \frac{2}{\Omega}\sum_{k<k_f}\exp(-i\mathbf{k}\cdot\mathbf{r}')\exp(i\mathbf{k}\cdot\mathbf{r}). \tag{2.28}$$

with k_f the Fermi wave number. Passing from a summation to an integration in the usual fashion, one has

$$\gamma(\mathbf{r}', \mathbf{r}) = \frac{2}{8\pi^3}\int_{k<k_f}\exp(i\mathbf{k}\cdot(\mathbf{r}-\mathbf{r}'))d\mathbf{k}. \tag{2.29}$$

The integration over angles simply means replacing the plane wave $\exp(i\mathbf{k}\cdot\mathbf{r})$ by the s-wave part, $(\sin kr)/kr$, of its expansion in spherical waves, and hence

$$\begin{aligned}\gamma(\mathbf{r}', \mathbf{r}) &= \frac{1}{\pi^2}\int_0^{k_f} k\,\frac{\sin k|\mathbf{r}'-\mathbf{r}|}{|\mathbf{r}'-\mathbf{r}|}\,dk \\ &= \frac{k_f^3}{\pi^2}\,\frac{j_1(k_f|\mathbf{r}'-\mathbf{r}|)}{k_f|\mathbf{r}'-\mathbf{r}|},\end{aligned} \tag{2.30}$$

where $j_1(x)$ is the first-order spherical Bessel function defined explicitly by

$$j_1(x) = \frac{\sin x - x\cos x}{x^2}. \tag{2.31}$$

Taking the limit $\mathbf{r}'\to\mathbf{r}$ in Eq. (2.30) and noting that, for small x, $j_1(x) \sim x/3$, one immediately obtains

$$\gamma(\mathbf{r}, \mathbf{r}) = \frac{k_f^3}{3\pi^2}, \tag{2.32}$$

which is simply the constant density, ρ_0, say, of the free-electron gas.

The second-order density matrix is readily obtained from Eq. (2.26), and, in particular, its spinless diagonal element $\Gamma(\mathbf{r}_1, \mathbf{r}_2; \mathbf{r}_1, \mathbf{r}_2)$ is given from Eqs. (2.26) and (2.30) as

$$\Gamma(\mathbf{r}_1, \mathbf{r}_2; \mathbf{r}_1, \mathbf{r}_2) = \frac{k_f^6}{18\pi^4} - \frac{k_f^6}{4\pi^4}\left[\frac{j_1(k_f|\mathbf{r}_1 - \mathbf{r}_2|)}{k_f|\mathbf{r}_1 - \mathbf{r}_2|}\right]^2. \tag{2.33}$$

This yields the pair correlation function $g(r)$ as

$$g(r) = 1 - \frac{9}{2}\left[\frac{j_1(k_f r)}{k_f r}\right]^2. \tag{2.34}$$

These explicit results for $\gamma(\mathbf{r}', \mathbf{r})$ and $g(r)$ are central to the discussion of interacting electrons in Chapter 3, as well as to the plane-wave perturbation theory of March and Murray (1961; see also March et al., 1995). A model, in which interacting density matrices are calculated, is set up in Appendix 2.1.

2.3. ASYMPTOTIC FORM OF FIRST-ORDER DENSITY MATRIX AND IONIZATION POTENTIAL

As has already been emphasized, one of the important tools to use in describing electron correlation is the first-order density matrix $\gamma(\mathbf{r}', \mathbf{r})$, defined in Eq. (2.11). The ground-state electron density $\rho(\mathbf{r})$ ($\equiv\gamma(\mathbf{r}, \mathbf{r})$) is evidently given in terms of the many-electron wave function by

$$\rho(\mathbf{r}_1) = N \int \Psi_N^*(\mathbf{r}_1, \mathbf{r}_2, \ldots, \mathbf{r}_N)\, \Psi_N(\mathbf{r}_1, \mathbf{r}_2, \ldots, \mathbf{r}_N)\, d\mathbf{r}_2 d\mathbf{r}_3 \ldots d\mathbf{r}_N \tag{2.35}$$

This is obviously the way to define the electron density at $\mathbf{r}_1$; the choice of electron 1 becomes irrelevant when one remembers that (a) electrons are indistinguishable and (b) $\rho(\mathbf{r}_1)$ must normalize to the total number N:

$$\int \rho(\mathbf{r})\, d\mathbf{r} = N \tag{2.36}$$

2.3.1 Natural Orbitals

At this stage, one specializes again to the case of a neutral atom. As March and Pucci (1982) have pointed out, the natural orbital expansion of the first-order density matrix, introduced by Löwdin (1955), namely

$$\gamma(\mathbf{r}',\mathbf{r}) = \sum_i n_i \chi_i(\mathbf{r}')\chi_i(\mathbf{r}), \tag{2.37}$$

where χ_i are the natural orbitals and n_i their occupation numbers, is of such a form that, at large distances from the nucleus of the atom, the natural orbitals decay with the same exponential factor, and it turns out that for large $\mathbf{r}'$ but for all $\mathbf{r}$ the density matrix factorizes to read

$$\gamma(\mathbf{r}',\mathbf{r}) = \rho^{1/2}(\mathbf{r}')\, f(\mathbf{r}). \tag{2.38}$$

The author (March, 1983) has proposed a one-body Schrödinger equation for the calculation of $f(\mathbf{r})$, in which the eigenvalue is the ionization potential I of the neutral atom. This then reads

$$\nabla^2 f + \frac{2m}{\hbar^2}[-I - V_{\text{eff}}(\mathbf{r})]f = 0, \tag{2.39}$$

where $V_{\text{eff}}(\mathbf{r})$ tends to zero as the distance $\mathbf{r}$ from the nucleus tends to infinity. The results in Eqs. (2.38) and (2.39) can, in fact, be put on a rigorous footing (Dawson and March, 1983) by utilizing an alternative expansion to the natural orbital expansion (2.37). This takes the form

$$\gamma(1',1) = \sum_j \gamma_j(1', 1) = \sum_j g_j(1')\, g_j(1), \tag{2.40}$$

where 1 now stands for both space coordinate $\mathbf{r}$ and spin. In contrast to the expansion in Eq. (2.37), where the χ_i form a complete orthonormal set, the g_i are not orthogonal, but their definition follows naturally from introducing a complete set of Fock states into the definition of the first-order density matrix in terms of the field operator ψ,

$$\gamma(1',1) = \sum_i < \Psi_N^0 | \Psi^\dagger(1') | \Psi_{N-1}^i > < \Psi_{N-1}^i | \Psi(1) | \Psi_N^0 >, \tag{2.41}$$

where Ψ_N^0 is again the ground-state wave function of the N-electron system, while Ψ_{N-1}^i are the exact and complete set of wave functions for the (N−1)-electron ion. In contrast to the property of the natural orbitals above, namely that they all decay with the same exponential factor, from Eq. (2.39) it follows that in an atom the lowest term in the series dominates when $\mathbf{r}$ is large. Thus Eq. (2.38) is recovered, for $g_0(\mathbf{r})$ differs from $f(\mathbf{r})$ only by a normalization factor. The functions $f(\mathbf{r})$ or $g_0(\mathbf{r})$ can be described as measuring the overlap of the N- and (N−1)-electron ground states, and then it will occasion no surprise that the ionization potential I, the difference between the exact ground-state energies of these two states, enters the one-body Eq. (2.39).

2.3.2. Effective One-Body Potential

Of course, all the many-body effects must therefore be built into the effective potential $V_{\text{eff}}(\mathbf{r})$; but as shown by Dawson and March (1983), a rather simple screened potential given by Berrondo *et al.* (1979) already leads to an ionization potential transcending the Hartree–Fock Koopmans value. The conclusion is therefore that correlation (in the sense that it affects the ionization potential), and the factorizable asymptotic form of the first-order density matrix to Eq. (2.35) can be built into atomic theory by means of an effective one-body potential. Of course, Eq. (2.37) warns us that, as one comes away from the asymptotic region, it will become increasingly important to include the overlap of the N-electron ground-state wave function with the excited states of the ($N-1$)-electron system.

In the context of this introductory discussion of correlation effects in the one-center example of neutral atoms, one notes that Eq. (2.39) has automatically built into it the correct asymptotic decay of the electron density $\rho(\mathbf{r})$ as a constant $x \exp[-2(2I)^{1/2}r]$ far from the nucleus (cf. Hoffmann-Ostenhof *et al.*, 1977). The possibility that electron correlation could introduce qualitative effects in the sense of collective oscillations in the electron cloud in atoms will be touched on in the next section where reference is made to the theory of plasmons in inhomogeneous electron gases. The work of Lundqvist (1983) and Wendin (1976) in this general area is, of course, of considerable interest for atomic theory, as is the study of Kirzhnits *et al.* (1975) and of Dellafiore and Matera (1990).

2.4. COLLECTIVE EFFECTS

Following the discussion of the gross trend of energy with increasing atomic number, brief reference will be made to what has been done on collective effects induced by Coulomb correlations in atoms before turning to the main focus of this volume, electron correlation in multicenter problems: that is, molecules and solids.

Collective effects in atoms have been studied in two ways. In the first, either hydrodynamic theory or some local approximate dielectric theory, neither of which takes into account the shell structure and corresponding single-particle spectrum, has been used. These methods are capable, at most, of giving gross trends in dynamical properties, such as a universal photoabsorption curve (cf. Lundqvist, 1983). The second route instead uses a fully quantal description based on the one-electron excitation spectrum and corresponding wave functions. The papers in a Nobel Symposium (Lindgren and Lundqvist, 1980) may be consulted for an extensive set of methods and results of the application of many-body techniques in atomic theory (see also Lindgren and Morrison, 1982).

Because of links with Chapter 3 below, let us mention here the density functional work of Zangwill and Soven (1980), who have calculated the photo-absorption spectrum of Ne, Ar, Kr, Xe, Ba, Cl, and Ce. Most of these atoms have also been treated by Amusia *et al.* (1974) and by Wendin (1982). These workers employed the RPAE,* and in a few cases transcended this approximation by including relaxation effects.

*Random phase approximation (see Appendix 3.4) with exchange.

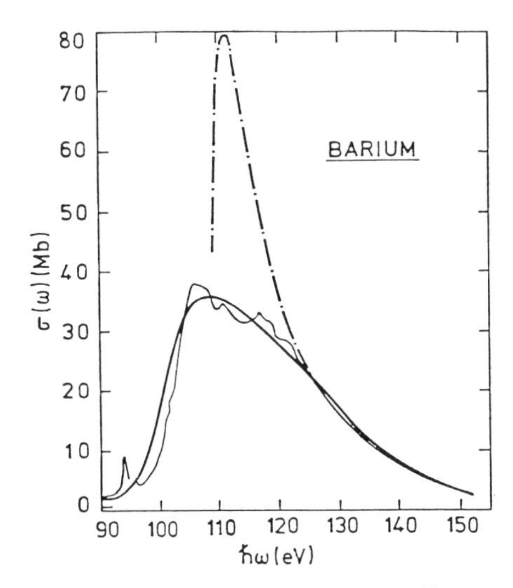

FIGURE 2.1. The photoabsorption cross section $\sigma(\omega)$ for the $4d^{10}$ shell in Ba. Solid line, TDLDA (Time Dependent Local Density Approximation); dashed-dot line, RPAE. (From Lundquist, 1983.)

One example, reviewed by Lundqvist (1983), will be mentioned here. This concerns the $4d^{10}$ absorption spectrum in Ba, which is plotted in Fig. 2.1. The RPAE result gives the peak position rather accurately but fails to account for the line shape. The local density functional scheme (cf. Chapter 3 below) has also been applied, and it can be seen from Fig. 2.1 that it agrees very well with the experimental curve except for the detailed structure near threshold and at the absorption peak.

It seems probable that if relaxation effects were included, then the RPAE result would be brought into agreement with the experimental curve to the same sort of accuracy as the local density theory. The surprising feature is that the density functional scheme, as Lundqvist emphasizes, does not include relaxation effects, yet this scheme gives results comparable to those obtained if relaxation effects are included in the RPAE scheme, and it seems superior to the RPAE results.

2.5. VISUALIZATION OF ELECTRON CORRELATION IN He AND H⁻

Rehmus *et al.* (1978a) have presented an illuminating analysis of the spatial correlation of atomic electrons in doubly excited helium. For any two-electron wave function whose angular dependence is given in terms of the spherical harmonics of the individual electrons and/or θ_{12}, where θ_{12} is the interelectronic angle, they generate a transformation which reduces $|\Psi(r_1, r_2, \theta_1, \theta_2, \theta_{12})|^2$ to $|\Psi(r_1, r_2, \theta_{12})|^2$.

This method has been applied by Rehmus *et al.* (1978b) to study the conditional probability density $d(\mathbf{r}_2,\theta_{12}|\mathbf{r}_1)$ for finding one electron at a distance r_2 with angle θ_{12} between the vectors $\mathbf{r}_1$ and $\mathbf{r}_2$, when the other electron is at a distance r_1 from the nucleus. With this probability density *d*, one can compare wave functions of different qualities, see what roles long-range and short-range correlation play in various states, exhibit the relative importance of angular and radial correlation, and compare correlation in different atomic systems.

Results taken from Rehmus *et al.* (1978b) are shown in Fig. 2.2 for the H^- ion, which is perhaps the most dramatic example in which electron correlation determines the properties of an atomic species. In Fig. 2.2a, an electron is fixed at its expectation distance from the nucleus, while in Fig. 2.2b the electron is at its most probable distance. The most striking characteristic of Fig. 2.2 is the extreme spatial diffuseness of H^- (to compare with corresponding pictures for He, the reader can consult Rehmus *et al.*, 1978b). There is considerable similarity with

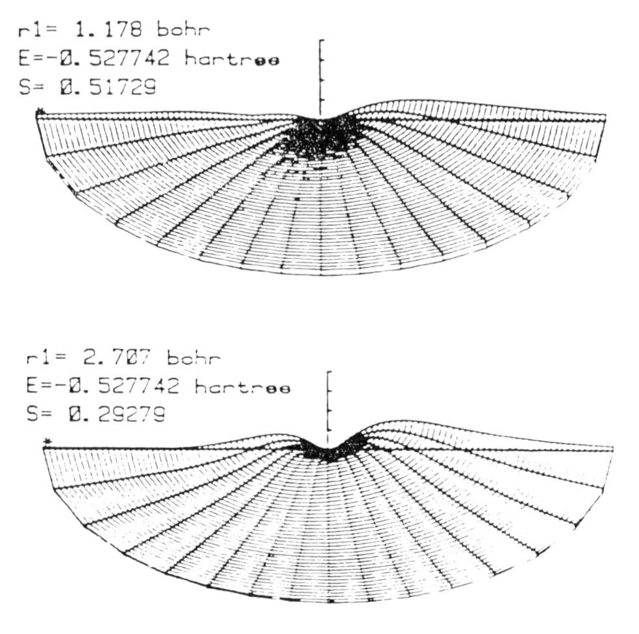

FIGURE 2.2. Conditional probability densities for the H^- ion out to a maximal radial distance of 5 bohr. (a) fixes r_1 at its average value $< r_1 >$; (b) fixes r_1 at its most probable value. The value *S* is proportional to the probability that an electron actually attains the specified distance from the nucleus. The density on the vertical axis is in units of 0.1. (From Rehmus *et al.*, 1978b.)

the earlier description of correlation in H^- given by Chandrasekhar (1947), apart from the fact that his discussion does not quite do justice to the extent of the spread-out in H^- noted above.

Similar visualization of electron correlation in a series of helium S states is to be found in the later study of Rehmus and Berry (1979). Their work, as well as that described above, generalizes previous quantitative descriptions of electron correlation in terms of joint radial distributions for pairs of electrons (Dickens and Linnett, 1957; Fano, 1976), from differences in charge densities (Coulson and Neilson, 1961), from angular distributions (Banyard and Ellis, 1975; Banyard, 1994), or from moments of various distances and angles between electrons. Other notable early work in this general area is that of Wulfman and Kumei (1973) and of Herrick and Sinanoglu (1975), together, of course, with the book by Sinanoglu and Brueckner (1970). A soluble model in which two electrons with constant interaction move in an external oscillator potential is discussed in Appendix 2.5.

2.6. *d*-DIMENSIONAL THEORY

At this point, reference should be made to the promise afforded by generalizing atomic and molecular theory to d dimensions. Progress has come along two routes:

1. Generalization of the Coulomb interaction by taking the solution of Poisson's equation in d dimensions
2. Direct generalization of $1/r_{ij}$ to d dimensions.

As to 1, it has so far been developed predominantly for heavy atoms and ions. Here, as emphasized in Section 2.1, correlation energy enters the neutral atom binding energy at $O(Z^n)$: $n \sim 1 - 1 \cdot 3$ in three dimensions. So far attention has focused on generalizing the scaling result $Z^{7/3}f(N/Z)$ for the total energy $E(Z, N)$ of heavy atomic ions. This generalization, according to March (1985), is

$$E_d(Z, N) = Z^{(4+4d-d^2)/d(4-d)} f_d(N/Z) \tag{2.42}$$

from atomic number Z and N electrons, with $N/Z \leq 1$.

From the standpoint of the present discussion, route (2), applied to light atoms, is the most promising for estimating correlation energy. Thus, the pioneering work of Herrick and Stillinger (1966) on two-electron ions, shown to relate to the density function theory of heavy atomic ions through the study of Senatore and March (1985), has been developed in a series of papers by Herschbach and his co-workers (see Herschbach, 1986, and references therein). Roughly speaking, for two-electron systems, the $d \to \infty$ limit is exactly soluble (in its physical content, it bears some resemblance to the Wigner electron crystal limit of the homogeneous electron assembly treated in Chapter 3). It turns out that some progress in this large

d limit is provided by a separable semiclassical model, which includes correlation effects and offers clear structural predictions for two-electron atoms. This limit, in fact, provides a reasonable approximation to the wave function near the most probable region of configuration space. Systematic corrections to this limiting model can be calculated as a series in $1/d$. However the convergence properties of this series are not very favorable, and moreover it seems likely that calculation of more than a few terms for all but the smallest systems will be extremely difficult.

Doren and Herschbach (1987) emphasize that one can also expand about the $d = 1$ case. The ground-state energy of one- or two-electron atoms turns out to be singular at $d = 1$, but a scaling argument reduces this problem to a limiting Hamiltonian with only two degrees of freedom, in which all Coulombic potentials effectively reduce to δ functions. Doren and Herschbach (1987) assert that, since the singularity at $d = 1$ dominates the energy at nearby dimensions, this limit forms the basis for an expansion in $(d - 1)/d$, which is reasonably accurate at $d = 3$. By combining results from this equation with the $1/d$ expansion about the $d \to \infty$ limit, one can obtain estimates of the energy of a two-electron atom with accuracy considerably better than that of either series alone.

To summarize the above in a quantitative manner, when $Z = 2$ the results are as follows:

a. *Large d series*

$$E(d) = -(2/d)^2(0.6844423 + 1.514398d^{-1} + 2.215546d^{-2} + \ldots) \quad (2.43)$$

b. *(d − 1)/d series*

$$E(d) = -(2/\{d-1\})^2[0.788847 - 0.07911(\{d-1\}/d) + \ldots] \quad (2.44)$$

Doren and Herschbach (1987) construct hybrid series from these; the reader is referred to their paper for the details of this (see also Appendix 2.2).

With this same interaction, March and Cizek (1987) have made a start toward generalizing the important $1/Z$ expansion, introduced by Hylleraas (1930) and developed especially by Layzer (1959), to d dimensions for heavy atoms, but this problem needs full investigation in the future. Their result for the dimensionality dependence of the total energy for closed shells in a bare Coulomb field for large atomic number is

$$E(Z, N, d) = \frac{-2}{(d-1)(d-2)} Z^{3-(2/d)} \left[\frac{Nd!}{4Z}\right]^{1-(2/d)} \frac{1}{(d-2)!}, \quad (2.45)$$

reducing to the familiar result for $d = 3$. This suggests for the total energy scaling property of heavy atoms in d dimensions $E(Z, N, d) = Z^{3-(2/d)} f_d(N/Z)$, but it must be stressed that this has not been proved to date.

2.7. COUPLED-CLUSTER WAVE FUNCTION

To conclude this Chapter, it is worthy of note that, at the time of writing, there is considerable interest in the coupled-cluster method, introduced into the many-body area initially in nuclear physics (Coester and Kummel, 1960). The method was first applied to quantum-chemical problems by Cizek (1966).

The many-electron wave function Ψ_0 in this method takes the form (see also Bartlett, 1981, 1989):

$$\Psi_0 = \exp(S)\Phi_0. \tag{2.46}$$

A derivation of the equations of the coupled-cluster method, given by Fulde and coworkers, using field-theoretic techniques, is summarized in Appendix 2.3 (see also the somewhat related techniques in Appendix 4.1). There the nature of the operator S, and possible choices of the model state wave function Φ_0, appearing in Eq. (2.46) are considered.

Subsequently, Jankowski and Malinowski (1993) have studied the four-electron systems Be, B^+, and C^{2+} by a variant of the coupled-cluster method, while Nooijen and Snijders (1993) have considered a hybrid of Green-function and coupled-cluster methods (see also Chapter 5). The studies of Bartlett and co-workers are also specially noteworthy here (Watts and Bartlett, 1994a,b; Perera *et al.*, 1994; Szalay and Bartlett, 1994).

In addition to the sophisticated methods mentioned in this and the previous section for estimating correlation energies, Appendices 4.2 and 4.3 summarize results of the simpler density functional approach (see Chapter 3) for atoms and molecules respectively. Chapter 5 reports refined quantal computer simulation of correlation energies.

Chapter 3

Homogeneous and Inhomogeneous Electron Assemblies

Let us turn now from the discussion of correlation in atoms to the treatment of electron correlation in a model appropriate to simple metals. It will be useful to focus first on the electron assembly formed by the conduction electrons in a metal like Na or K. In these cases, there is a whole body of evidence which demonstrates the weakness of the electron–ion interaction. Therefore the jellium model, introduced initially by Sommerfeld long ago, in which the ionic lattice is smeared out into a uniform neutralizing background of positive charge where the correlated electronic motion takes place, affords a valuable starting point.

Electron correlation can be treated quantitatively in this jellium model (see also Section 5.4); it will be helpful first to consider two limiting cases, very high and very low densities, before briefly summarizing results on the range of metallic densities appropriate to simple metals such as Na and K.

Though, at first sight, the problem looks totally different from the molecular case (see especially Section 4.1), there is in fact a close parallel in the sense that in the high density limit a delocalized picture (cf. molecular orbitals) is correct while in the low density limit, electron localization (cf. valence bond theory) is again induced by strong Coulomb repulsion between electrons.

3.1. HIGH DENSITY THEORY

The jellium model is characterized by one physical quantity, the number of electrons per unit volume $\rho_0 = N/\Omega$ for N electrons in total volume Ω. It is often convenient to work with the mean interelectronic spacing $r_s a_0$ defined by

$$\rho_0 = \frac{3}{4\pi(r_s a_0)^3}, \tag{3.1}$$

where r_s is then a pure number and a_0 is the Bohr radius. Simple *sp* metals have values of r_s for their conduction electrons ranging from approximately 2 (for Al) to 5.5 (for Cs).

However, it is very helpful to consider the limiting cases of $r_s \ll 1$ and r_s very large; in this section the former, high density, case is treated. In the Hartree–Fock limit $r_s \to 0$, it is well known (cf. March *et al.*, 1995) that the first- and second-order density matrices are, respectively, [Eqs. (2.30) and (2.33)]

$$\gamma_{\text{H-Fock}}(\mathbf{r}'\mathbf{r}) = \frac{k_f^3}{\pi^2}\,\frac{j_1(k_f|\mathbf{r}' - \mathbf{r}|)}{k_f|\mathbf{r}' - \mathbf{r}|},$$

where

$$j_1(x) = \frac{\sin x - x\cos x}{x^2} \tag{3.2}$$

and

$$\Gamma_{\text{H-Fock}}(\mathbf{r}_1\mathbf{r}_2) = \rho(\mathbf{r}_1)\rho(\mathbf{r}_2) - \tfrac{1}{2}[\gamma(\mathbf{r}_1\mathbf{r}_2)]^2, \tag{3.3}$$

which as written is quite general for Hartree–Fock theory. In jellium, the inhomogeneous density $\rho(\mathbf{r})$ becomes a constant ρ_0. Γ in Eq. (3.3) then depends on $|\mathbf{r}_1 - \mathbf{r}_2| \equiv r$: the pair correlation function as $r_s \to 0$ follows as [Eq. (2.34)]

$$\frac{\Gamma(r)}{\rho_0^2} = g(r) = 1 - \frac{9}{2}\left[\frac{j_1(k_f r)}{k_f r}\right]^2. \tag{3.4}$$

This so-called Fermi hole has $g(0) = 1/2$, and when one allows antiparallel spins to correlate through their Coulomb repulsion by increasing r_s from zero, then $g(0, r_s) < \frac{1}{2}$, $(r_s > 0)$.

Evidently, in this extreme high density limit the kinetic energy per electron T/N is given by the usual Fermi gas form as $\frac{3}{5}$ of the Fermi energy, or $2.21/r_s^2$ Ry. Likewise, the potential energy per electron U/N is readily written down as

$$\frac{U}{N} = \tfrac{1}{2}e^2\rho_0 \int \frac{[g(r) - 1]}{r}\,dr, \tag{3.5}$$

which is a general result, expressing the fact that U/N is the potential energy of the electron one chooses to sit on at the origin in the field of the hole, carrying a deficit of one electron, of density $\rho_0[g(r) - 1]$. Using $\int_0^\infty dx[j_1(x)]^2/x = \frac{1}{4}$, if one inserts Eq. (3.4) into Eq. (3.5) one finds U/N as the exchange energy per electron, yielding the high density limit for the total energy per electron E/N as

$$\frac{E}{N} = \left[\frac{2.21}{r_s^2} - \frac{0.916}{r_s}\right] \text{Rydbergs} \tag{3.6}$$

Correlation Energy

In the high density limit $r_s \ll 1$, the pioneering work of Gell-Mann and Brueckner (1957) using many-body perturbation theory established the correlation energy correction to the Hartree–Fock energy Eq. (3.6) as

$$\frac{E_{\text{correlation}}}{N} = A \ln r_s + C + \dots . \tag{3.7}$$

with $A = (2/\pi^2)(1 - \ln 2)$ and $C = 0.096$, the energy units again being rydbergs. While Eq. (3.7) is not useful for the range of the real metallic densities, it is, of course, of great interest as a first-principles calculation of correlation energy. Subsequent progress in many body Brillouin-Wigner-Feenberg perturbation theory has also proved possible (Kulic, 1970, 1972).

From the virial theorem, which for jellium reads (March, 1958; Argyres, 1967)

$$2T + U = -r_s \frac{dE}{dr_s}, \tag{3.8}$$

kinetic and potential correlation energies can evidently be constructed from Eq. (3.7). An obvious property of "switching on" electron–electron interactions away from the limit $r_s \to 0$ is to promote electrons outside the Fermi sphere of radius k_f, leaving holes inside (cf. Fig. 4.2). This creation of electron–hole pairs obviously increases the kinetic energy. As will be seen below, this increase in kinetic energy becomes so great in the low density limit $r_s \to \infty$ that the "coefficient" K in the kinetic energy K/r_s^2 [cf. Eq. (3.6)] tends to infinity as $r_s \to \infty$—in fact, as $r_s^{1/2}$—and the Fermi sphere approximation hence becomes invalid.

3.2. LOW DENSITY WIGNER ELECTRON CRYSTAL

Turning then to the extreme low density limit $r_s \to \infty$, Wigner (1934, 1938) pointed out that the delocalized picture of Section 3.1 breaks down completely, and when the potential energy became large compared with the kinetic energy, then electrons avoid each other maximally. He stressed that this situation would be achieved by electrons becoming localized on the sites of a lattice. He argued that one must find the stable lattice by minimizing the Madelung energy, and of the lattices so far examined the body-centered-cubic (*bcc*) lattice has the lowest Madelung term. This yields, for the electron *bcc* Wigner crystal, the energy

$$\lim_{r_s \to \infty} \frac{E}{N} = -\frac{1.792}{r_s} \tag{3.9}$$

showing, by comparison with the Hartree–Fock plane wave result of Eq. (3.6), that this latter approximation is no longer of any physical utility, the energy being too high by a factor of about 2. As the repulsive coupling between electrons is relaxed—that is, r_s is reduced below the range of validity of Eq. (3.9)—the electrons vibrate about the *bcc* lattice sites. Since the *bcc* lattice has a Wigner–Seitz cell of high symmetry, it is a useful first approximation to neglect the (multipole) fields of the other cells in considering the vibration of an electron in its own cell. Then the potential energy $V(r)$ in which this electron vibrates is created solely by the uniform positive background in its own cell, which in the spherical approximation gives

$$V(r) = \frac{e^2 r^2}{2r_s^3} + \text{const}, \tag{3.10}$$

with ground-state isotropic harmonic oscillator wave function as

$$\psi = \left[\frac{\alpha}{\pi}\right]^{3/4} \exp(-\tfrac{1}{2}\alpha r^2),\ \alpha = (r_s)^{-3/2}. \tag{3.11}$$

This Wigner oscillator orbital leads to a kinetic energy per electron in the low density limit as

$$\lim_{r_s \to \infty} \frac{T}{N} = \tfrac{3}{2}(r_s)^{-3/2}, \tag{3.12}$$

which verifies the claim made above that creation of particle–hole pairs becomes so prolific as $r_s \to \infty$ that if one writes T/N as K/r_s^2 [cf. Eq. (3.6)], then $K \propto (r_s)^{1/2}$ as $r_s \to \infty$, and the Fermi sphere picture breaks down completely. The calculation yielding Eq. (3.12) is an Einstein-type model, whereas one should, of course, treat the vibrational modes of the *bcc* Wigner crystal by collective phonon theory. The coefficient 3 in Eq. (3.12) is then reduced to 2.66 (Carr, 1961; Coldwell-Horsfall and Maradudin, 1963).

The Fermi distribution turns out to be profoundly changed by the creation of particle–hole pairs. One way to demonstrate this is to take the Fourier transform of the Gaussian orbital [Eq. (3.11)]. Naturally, this then yields a Gaussian momentum distribution totally different from the Fermi gas step function. At most, then, remnants of a Fermi surface remain in the low-density limit. In one dimension, this statement is made quantitative by Holas and March (1991). The Wigner electron crystal will be taken up again in Section 3.8 after discussing density functional theory.

Having summarized the results that can be obtained by largely analytical methods in the high and low density limits, let us now discuss the intermediate density regime, which remains somewhat intractable unless extensive numerical studies are employed (see Section 5.4 below).

3.3. INTERMEDIATE DENSITIES: FORM OF PAIR CORRELATION FUNCTION $g(r)$ AND STRUCTURE FACTOR $S(k)$

Having discussed the correlation energy in the jellium model in extreme high and low density limits, let us consider next the influence of Coulomb correlations between electrons on the pair correlation function $g(r)$ and its Fourier transform, the important structure factor $S(k)$, defined precisely by

$$S(k) - 1 = \rho_0 \int [g(r)-1] \exp(i\mathbf{k}\cdot\mathbf{r})d\mathbf{r}, \tag{3.13}$$

ρ_0 being as usual the mean electron density. The Fermi hole form [Eq. (3.4)] for $g(r)$ yields, when inserted in Eq. (3.13), the structure factor $S(k)$ in the extreme high density limit as

$$\left.\begin{aligned} S(k)_{\text{Fermi hole}} &= \frac{3}{4}\left[\frac{k}{k_f}\right] - \frac{1}{16}\left[\frac{k}{k_f}\right]^3, & k < 2k_f \\ &= 1 & k > 2k_f \end{aligned}\right\}. \tag{3.14}$$

It will be seen below that one very clear-cut effect of the long-range Coulomb interactions on the structure factor of the electron fluid is to suppress the term proportional to k in Eq. (3.14) at small k and to introduce a k^2 term. This result is closely related to the organized plasma oscillations whose importance in the treatment of correlations in the electron liquid was first recognized in the pioneering work of Bohm and Pines (1953).

Equation (3.14) focuses not only on small k, but also on k equal to $2k_f$, the diameter of the Fermi sphere. This, as will be seen later from an experimental study in Be metal using synchrotron radiation, is certainly an important region of k. Using the asymptotic form

$$\frac{j_1(k_f r)}{k_f r} \sim \frac{\cos k_f r}{(k_f r)^2}, \tag{3.15}$$

it is seen that the Fermi hole $g(r) \sim \cos^2(k_f r)/r^4$ at large r. Using $\cos 2k_f r = 2\cos^2 k_f r - 1$, this is $\sim(1 + \cos 2k_f r)/r^4$, the first term coming from the small k term $\propto k$ and the second from the nonanalytic behavior at $2k_f$. As mentioned above, the collective plasmon mode removes the $1/r^4$ term, but since there is a sharp Fermi surface in the presence of the electron–electron interactions (cf. Daniel and Vosko, 1960), one expects a term $\sim \cos 2k_f r/r^4$ to remain (see also Holas and March, 1987), though its coefficient depends on the magnitude of the discontinuity in the occupation number $n(k)$ at the Fermi surface, which will be discussed further below.

Bohm and Pines (1953) incorporated collective effects directly into the theory, electing to do so dynamically by introducing the plasma frequency

$$\omega_p = \left[\frac{4\pi\rho_0 e^2}{m}\right]^{1/2}. \tag{3.16}$$

Since the present section concerns the static pair correlation function $g(r)$, it is relevant to note that an alternative point of view would be to argue that, once the plasma oscillations are accounted for, the residual electron–electron interaction is short-range. Thus

$$\frac{e^2}{r_{ij}} \to \frac{e^2}{r_{ij}} \exp\left[\frac{-r_{ij}}{l}\right], \tag{3.17}$$

where, in order of magnitude, the characteristic screening length l is the Fermi velocity v_f multiplied by the period $2\pi/\omega_p$ of the plasma oscillations. This discussion is equivalent to the introduction of the Thomas–Fermi screening length or the corresponding wave number $q = l^{-1}$ through

$$q^2 = \frac{4k_f}{\pi a_0}, \tag{3.18}$$

where k_f as usual is the Fermi wave number.

3.3.1. Quantitative Modification of *S*(*k*) at Small *k* Due to Collective Plasmon Modes

The next question to be addressed is the quantitative way in which the collective modes, the plasmons, affect $g(r)$ at large r in the electron fluid (see also Appendices 3.4 and 3.5). One can use an analogy with classical atomic liquids to see how the result arises. In such liquids, the important quantity to work with is the Ornstein–Zernike direct correlation function $c(r)$. Defining the total correlation function $h(r) = g(r)-1$, Ornstein and Zernike write (see also March and Tosi, 1985)

$$h(r) = c(r) + \rho_0 \int h(|\mathbf{r}-\mathbf{r}'|)\, c(r')\, d\mathbf{r}', \tag{3.19}$$

which separates h into a direct part $c(r)$ and an indirect part represented by the convolution of h and c. In a classical liquid, while $h(r)$ is represented by introducing a potential of mean force U into the Boltzmann factor through

$$h(r) = \exp\left[\frac{-U}{k_B T}\right] - 1, \tag{3.20}$$

the direct correlation function $c(r)$ at large r is correctly approximated by $\exp\left[\frac{-\phi(r)}{k_B T}\right] -1 \sim -\phi(r)/k_B T$, where $\phi(r)$ is the pair interaction between the particles.

Evidently, in the ground state of the electron fluid we have the direct interaction $\phi(r)$ as e^2/r, but $k_B T$ must be replaced by an approprite characteristic energy, which in the light of the above discussion, is the zero-point energy of the plasma oscillations $\frac{1}{2}\hbar\,\omega_p$. Thus, at large r in the electron fluid,

$$c(r) \sim -\frac{e^2/r}{\frac{1}{2}\hbar\,\omega_p}. \tag{3.21}$$

To translate this into $h(r) = g(r) - 1$ at large r, it is simplest to remove the convolution in Eq. (3.19) by Fourier transform to obtain

$$S(k) = \frac{1}{1 - c(k)}. \tag{3.22}$$

But the large r form of $c(r)$ above determines the small k behavior of $c(k)$ as

$$c(k) = -\frac{4\pi e^2 \rho_0}{k^2 \frac{1}{2} \hbar\omega_p}, \tag{3.23}$$

which clearly yields, since $c(k) \to \infty$ as $k \to 0$,

$$S(k) \simeq -\frac{1}{c(k)} = \frac{\frac{1}{2}\hbar\,\omega_p k^2}{4\pi e^2 \rho_0}. \tag{3.24}$$

This result shows that $S(k) \propto k^2$ as $k \to 0$, the size of this small k quadratic term being fixed by the zero-point energy of the collective plasma mode. This is quite different from the Fermi hole behavior of Eq. (3.14), obtained by Fourier transforming Eq. (3.4).

In concluding this discussion of the large r form of $g(r)$, it is to be noted that the result in Eq. (3.24) for $S(k)$ at small k follows from the Bijl–Jastrow wave function (cf. Section 5.4), which is a Slater determinant D of plane waves multiplied by a product of two-particle wave functions, the details being given by March (1981; see also Gaskell, 1962). Precisely the same coefficient of k^2 is found as obtained above by rather intuitive arguments.

Having discussed the large r behavior of $g(r)$ in some detail, we turn now to the short-range form of $g(r)$ in the electron fluid regime.

3.3.2. Short-Range Behavior of $g(r)$

To motivate the argument below it is useful to refer to Kato's theorem. This is easily understood by appeal to the example of the hydrogen atom's ground state density, namely

$$\rho(r) = \frac{1}{\pi a_0^3} \exp\left[\frac{-2r}{a_0}\right]. \tag{3.25}$$

Kato's theorem is a generalization to the correlated electron problem of the result, which follows immediately from Eq. (3.25):

$$\left.\frac{\partial \rho(r)}{\partial r}\right|_{r=0} = -\frac{2}{a_0}\rho(0). \tag{3.26}$$

This is to be viewed as a consequence of the $-e^2/r$ attraction between electron and nucleus at short distances. The negative slope in Eq. (3.26)—i.e., the decrease in $\rho(r)$ from $\rho(0)$, its value at the nucleus—is due to the attractive interaction. The atomic scattering factor $f(k) = \int \rho(\mathbf{r})\, e^{i\mathbf{k}\cdot\mathbf{r}} d\mathbf{r}$ falls off as k^{-4} at large k, with a coefficient proportional to $\rho(0)$, from Kato's theorem.

Turning to the pair correlation function $g(r)$ in the uniform interacting assembly, one notes, following Kimball (1973), a result, of the same kind as Eq. (3.26) for $g(r)$, namely

$$\left.\frac{\partial g(r)}{\partial r}\right|_{r=0} = \frac{g(0)}{a_0}. \tag{3.27}$$

The sign change is due to the repulsive interaction, or $g(r)$ rising from $g(0)$, while the factor of 2 has been removed because one has a reduced mass $\mu = (1/2)m$ in discussing the relative motion of two electrons at very close interelectronic separation. The argument of Kimball goes as follows: At small separation, write a wave function $\psi(r)$ for the relative motion of two electrons at separation r as the solution of

$$\left[\frac{-\hbar^2}{2\mu}\frac{\partial^2}{\partial r^2} - \frac{\hbar^2}{\mu r}\frac{\partial}{\partial r} + \frac{e^2}{r}\right]\psi(r) = \varepsilon\psi(r) \tag{3.28}$$

where ε is, in fact, a complicated, but bounded operator. One now writes, following the above analogy,

$$\psi(r) = a + br + \ldots \qquad \text{(small } r\text{)}, \tag{3.29}$$

and one can relate a and b by substituting in the Schrödinger equation, Eq. (3.28) for the relative motion and equating terms in $1/r$. The result is

$$-(\hbar^2/\mu)\, b + e^2 a = 0. \tag{3.30}$$

Since $g(r) \propto \psi^*\psi$, one immediately obtains the desired result, Eq. (3.27). Of course, this equation only relates the linear term in $g(r)$ at small r to $g(0)$, which is itself a function of r_s to be calculated explicitly in a complete theory.

But, just as for the atomic scattering factor $f(k)$ referred to above, $S(k) - 1$

$\sim k^{-4}$ at large k, which, by comparison with the Fermi hole form in Eq. (3.24), is a direct consequence of electron correlation. Precisely one has

$$g(0) = \frac{3\pi a_0}{8k_f} \operatorname*{Limit}_{k\to\infty} k^4[1 - S(k)] \tag{3.31}$$

and hence, at large k, $S(k)$ approaches unity from below.

The value of $g(0, r_s)$ is also, it turns out, related to the form of the momentum distribution $n(k)$ at large k (Kimball, 1975; March, 1975b), with $n(k) \propto k^{-8} g(0, r_s)$.

Various treatments exist giving approximations for $g(r)$ for all r as a function of r_s, the most celebrated being the work of Singwi *et al.* (1968). A full review of this area has been given by Singwi and Tosi (1981). The work of Dawson and March (1985) and the application of their method by Schinner (1987) are also relevant in this context.

3.3.3. Pair Function in Wigner Crystal Regime

In the Wigner crystal regime, and within the approximate framework of localized electrons behaving like Einstein oscillators, it is a straightforward matter to construct $g(r)$. Results for different values of r_s are shown in Fig. 3.1, taken from the work of March and Young (1995).

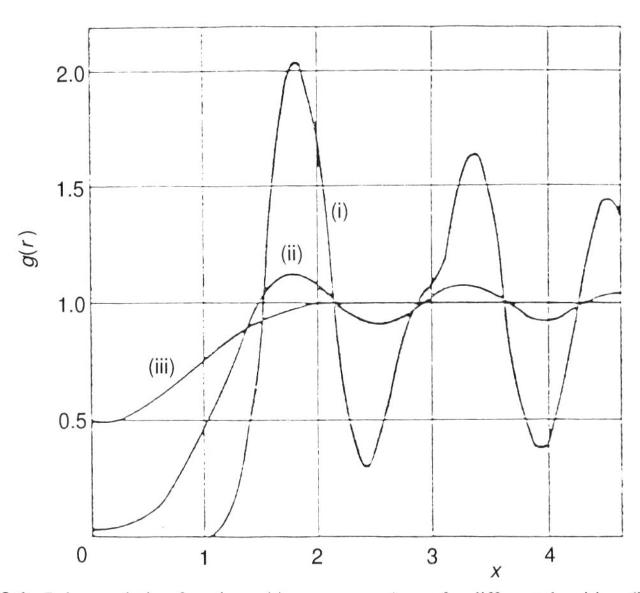

FIGURE 3.1. Pair correlation function $g(r)$ versus $x = r/r_s\alpha_0$, for different densities: (i) Wigner form for $r_s = 100$, (ii) Wigner form for $r_s = 4$, (iii) Fermi hole, correct in the limit $r_s \to 0$.

3.4. DIFFRACTION AND OTHER EXPERIMENTAL EVIDENCE FOR SHORT-RANGE ORDER IN ELECTRON ASSEMBLY IN SIMPLE METALS

Since the review of Singwi and Tosi (1981) fully covers the important calculations of the correlation hole through the integration of the dynamical structure factor $S(k, \omega)$, the van Hove generalization of $S(k)$ over frequency, which connects correlation effects with the so-called local field corrections in the static case, this section on jellium will be concluded with some brief comments on experimental evidence for short-range order in the conduction electron assembly in simple metals.

The first point to be emphasized, following Egelstaff *et al.* (1974; see also Ashcroft and Straus, 1976; Trigger, 1976; Chihara, 1987) is that in a molten metal such as Na or K, one can use the description of simple liquid metals as two-component systems, built of the ions Na^+ and K^+ and electrons. Naturally, one then has three partial structure factors: the ionic structure factor $S_{ii}(k) = S(k)$, the electron–ion structure factor $S_{ie}(k)$, and the electron–electron structure factor $S_{ee}(k)$.

For molten Na, with very weak electron–ion interaction, Cusack *et al.* (1976) have used a pseudopotential method to estimate the change in the jellium structure factor (for the appropriate r_s for Na) because of this electron–ion interaction. Their main conclusion is that this correction to the jellium pair function is so small that, at least in principle, the combination of three diffraction experiments (X-ray, electron, and neutron), as set out by Egelstaff *et al.*, could lead to a direct determination of the electron pair function in jellium. Though some progress has been made with this program of experimental work (Dobson, 1976; Johnson *et al.*, 1994) it is too early to be certain that quantitative results for $S_{ee}(k)$ can be thus obtained. We believe that this area remains of considerable interest for the future; the reader is referred to the reviews by Tamaki (1987) and March (1987).

The second area to be referred to here is the use of synchrotron radiation on solids. Results have been obtained on Be by Platzman and Eisenberger (1974) that demonstrate the presence of an electron liquid with appreciable short-range order.

Of course, even in the lowest-density metal, Cs, with $r_s \sim 5.5a_0$, one is far from the jellium limit ($r_s \sim$ 80–100 a_0 as discussed in detail in Section 5.3) of Wigner crystallization, which must be sought elsewhere, for three-dimensional examples, though in two-dimensional geometry classical electron crystals have been observed (e.g., on the surface of liquid helium).

Having discussed the theory of electron correlation in jellium, let us now turn to the basic theory of the inhomogeneous electron gas, alternatively called density functional theory. Though a general theory can be set up, it has, so far, been given practically useful content by employing local density theory [see Eq. (3.37) below] judiciously, sometimes with direct inclusion of density gradient corrections. Let us turn then to discussion of this theory and its generalization to magnetic systems with nonzero spin density in the next section.

3.5. WEAKLY INHOMOGENEOUS ELECTRON ASSEMBLIES

One approach to incorporating electron correlation, of course approximately, is afforded by this density functional theory. The theory has developed from the pioneering work of Thomas (1926), Fermi (1928), and Dirac (1930), a later paper by Slater (1951) being also very important in developing the form of density functional theory presented below. The step that was lacking, namely the proof that the ground-state energy of a many-electron system is indeed a unique functional of the electron density, was taken by Hohenberg and Kohn (1964), who supplied the formal proof for a nondegenerate ground state. For an atomic ion, for example, of atomic number Z and N electrons, $N = \int \rho(\mathbf{r})d\mathbf{r}$ and Z is determined also by $\rho(\mathbf{r})$ through Eq. (3.26). But Z and N determine H and hence, the ground-state energy E.

3.5.1. General Euler Equation

It has proved helpful in developing the theory to separate the energy functional into a number of parts, closely paralleling the energy principle of the original Thomas–Fermi theory (cf. March, 1975a).

Thus, one writes the total energy $E[\rho]$ as a sum of a single-particle kinetic energy functional $T_s[\rho]$, classical potential energy terms, and then a contribution $E_{xc}[\rho]$ from exchange and correlation which, are usefully considered together. Explicitly, we write

$$E[\rho] = T_s[\rho] + \int \rho V_N d\mathbf{r} + \tfrac{1}{2}\int \frac{\rho(\mathbf{r})'\rho(\mathbf{r})d\mathbf{r}d\mathbf{r}'}{|\mathbf{r}' - \mathbf{r}|} + E_{xc}[\rho], \tag{3.32}$$

where V_N is the total potential energy of the nuclear framework ($-Ze^2/r$ in an atom). ($4\pi\varepsilon_0$ of SI units is set equal to unity throughout.)

Now one minimizes this total energy with respect to the ground-state electron density ρ, subject only to the condition that the electron density must satisfy the normalization condition

$$\int \rho d\mathbf{r} = N. \tag{3.33}$$

Introducing a Lagrange multiplier μ, which has the significance of the chemical potential of the electron cloud of the atom, molecule, or solid being considered, the Euler equation of the variational problem reads

$$\mu = \frac{\delta T_s}{\delta \rho(\mathbf{r})} + V_N(\mathbf{r}) + V_e(\mathbf{r}) + \frac{\delta E_{xc}}{\delta \rho(\mathbf{r})} \tag{3.34}$$

which evidently expresses the constant chemical potential μ throughout the charge distribution as a sum of three contributions, each of which varies from point to point, arising from kinetic, classical potential, and exchange plus correlation contributions. In this form, as Kohn and Sham (1965) emphasized, thereby formally completing the treatment of Slater (1951), one can interpret the problem as posed in terms of equivalent single-particle-equations, which then bypasses the fact that even the single-particle kinetic energy functional is still not known in closed form. Thus, by solving single-particle Schrödinger equations, with a total one-body potential energy given by

$$V(\mathbf{r}) = V_N(\mathbf{r}) + V_e(\mathbf{r}) + \frac{\delta E_{xc}}{\delta \rho(\mathbf{r})}, \tag{3.35}$$

one can avoid any approximation in the single-particle kinetic energy. Furthermore, the above argument constitutes a proof that, in principle, the exact many-electron problem of calculating the ground-state electron density $\rho(r)$ can be reduced to a one-body problem. Naturally, the many-electron effects in the correlation energy functional are now subsumed in the one-body potential energy $V(\mathbf{r})$ through the functional derivative $\delta E_{xc}/\delta\rho(\mathbf{r})$. Of course, exact knowledge of this quantity would require exact solution of the many-electron problem, which is currently not feasible. Extraction of the potential $V(\mathbf{r})$ directly from the ground-state density has proved possible for the Be atom, however (Hunter and March, 1989; Nagy and March, 1989).

3.5.2. Local Density Approximations to Exchange and Correlation

As the simplest example of the use of Eq. (3.35), let us mention the so-called Dirac–Slater exchange potential. Here one neglects correlation and approximates the exchange energy density by its value in a uniform electron gas, with its local density inserted at the point in question. This leads to the result that the total exchange energy A, say, in Dirac–Slater approximation is given by

$$A = -c_e \int \rho(\mathbf{r})^{4/3}\, d\mathbf{r}: \qquad c_e = \frac{3e^2}{4}\left[\frac{3}{\pi}\right]^{1/3} \tag{3.36}$$

Taking the functional derivative required by Eq. (3.35) leads immediately to the Dirac–Slater exchange potential

$$V_{\text{Dirac–Slater}}(\mathbf{r}) = -\frac{4}{3}\, c_e \rho(\mathbf{r})^{1/3}. \tag{3.37}$$

This local density approximation has proved very valuable; one can also take the above discussion of the correlation energy density in jellium (cf. Section 3) into a local density theory, and this has also had wide utility, as discussed in the review by Callaway and March (1984).

It turns out that in the few examples which can be solved for the exchange energy, the Dirac–Slater form is a very useful approximation (see, however, Overhauser, 1985). For example, Miglio *et al.* (1981) have shown that in the infinite barrier model of a metal surface, in which the electron density varies strongly at the surface, the exchange theory discussed above nevertheless remains a remarkably useful approximation.

In Section 5.4, the QMC study of Ceperley and Alder (1980) is discussed. Here the concern is with potentially useful formulas for fitting their ground-state energy data.

As noted by Herman and March (1984), the theoretical formula given by Coldwell-Horsfall and Maradudin (1960; Carr *et al.*, 1961) for the ground state energy, from the theory of the vibrating *bcc* Wigner electron crystal (see also Section 3.2), is

$$\frac{E(r_s)}{N} = -\frac{1.79186}{r_s} + \frac{2.65}{r_s^{3/2}} - \frac{0.73}{r_s^2}\ \text{Ry}. \tag{3.38}$$

Without any adjustment at all, this reproduces the Ceperley–Alder energies at $r_s = 100$, 130, and 200 to within 0.003, 0.002, and 0.006 mRy, respectively. The corresponding computer simulation uncertainties are 0.003, 0.002, and 0.001 mRy.

The ground-state energies for the paramagnetic (P) and ferromagnetic (F) metallic phases of jellium have been calculated by Ceperley and Alder (1980) for $1.0 \le r_s \le 10$ and $2.0 \le r_s \le 100$, respectively, at selected values of r_s. These results have been carefully fitted by Vosko *et al.* (1980; see also Painter, 1981) and by Perdew and Zunger (1981). In spite of their drastically different analytical forms, the VWN and PZ interpolation formulas both provide good overall fits to the Ceperley–Alder data over the entire range of r_s. Looked at in greater detail, however, the PZ fit is closer than the VMN fit for $r_s \le 20$, while for $r_s \ge 50$ the PZ fitting errors, though quite small, are about five times larger than their VMN counterparts overall, and in some cases are of opposite sign.

In summary then, formulas exist that allow the correlation energy per electron in jellium to be reproduced almost to the accuracy of the QMC calculations. These form the basis for most approximations (local density) to the correlation energy of an inhomogeneous electron liquid; namely,

$$E_{xc}[\rho] \simeq \int \varepsilon^0_{xc}[\rho(\mathbf{r})]d\mathbf{r} \tag{3.39}$$

where ε^0_{xc} is the jellium model exchange and correlation energy density as a function of the constant density ρ_0.

However, in some situations the above theory, based as it is on using uniform electron gas relations locally, is too crude. This seems to be the case in situations of highly directional bonding, in particular in the diamond lattice semiconductors [see, for instance, Kane (1971, 1972)], which are classic examples of covalent bonding. It seems also true for the *d* electrons in transition metals (cf. Friedel and Sayers, 1977). Furthermore, with the same type of emphasis (namely, strong directional bonding), Goodgame and Goddard (1981) have argued that one cannot treat transition metal dimers very successfully by this means (see, however, Delley *et al.*, 1983; Bernholc and Holzwarth, 1983).

3.6. NONLOCAL CONTRIBUTIONS IN DENSITY FUNCTIONAL THEORY

The density functional Euler equation [Eq. (3.34)], involving the single-particle kinetic energy functional $T_s[\rho]$, can be bypassed if one is willing, following the proposal of Kohn and Sham (1965), which formally completes the work of Slater (1951), to solve one-electron equations with the one-body potential energy given in [Eq. (3.35)].

However, the functional $T_s[\rho]$ remains of outstanding interest for basic theory, and the simplest available procedure is to add density gradient corrections to the Thomas–Fermi kinetic energy $c_k \int \rho^{5/3} d\mathbf{r}$ (see, for example, March, 1975a). Following the early work of von Weizsäcker, Kirznits (1957) gave the correction to the slowly varying density and potential result as

$$T_s = c_k \int \rho^{5/3} d\mathbf{r} + \frac{\hbar^2}{72m} \int \frac{(\nabla\rho)^2}{\rho} d\mathbf{r} + \ldots . \tag{3.40}$$

Higher-order terms were given by Hodges (1973), and both the $(\nabla\rho)^2/\rho$ term given in Eq. (3.40) as well as the Hodges fourth-order gradient correction have been evaluated numerically by Wang *et al.* (1976) using Hartree–Fock quality densities for atoms. This is to be contrasted with calculating the density from the Euler equation based on Eq. (3.40), which leads to disappointing results because of the absence of any shell structure in the radial electron density. Although the results of the numerical studies of Wang *et al.* were somewhat encouraging, for chemically significant results it would seem that at least partial subseries of gradient terms will have to be summed to infinite order. Such a procedure was proposed by Hohenberg and Kohn (1964); we shall discuss a generalization of it for exchange and correlation energy density below. Before doing that, we note that the second-order gradient correction in Eq. (3.9) is also of considerable interest for molecules, though no calculations are available to date with Hartree–Fock–Roothaan quality densities. Mucci and March (1983) have presented arguments (see also Allan *et al.*, 1985; Lee and Ghosh, 1986) that the dissociation energy of homonuclear diatoms should correlate strongly with the density gradient term in Eq. (3.40) (see also Appendix 4.3).

3.6.1. Form of Nonlocal Corrections for Exchange and Correlation Energy

As pointed out by a number of groups, there are difficulties underlying gradient expansions when only exchange is included, that is, within a Hartree–Fock framework (Beattie *et al.*, 1971; Niklasson *et al.*, 1975; Geldart and Rasolt, 1976).

It seems to be the general opinion, though the writer knows of no proof, that provided exchange and correlation are treated together, then a gradient expansion does exist. The lowest-order term (see Herman *et al.*, 1969; Ma and Brueckner, 1968) has the structure

$$\text{constant}\ \frac{e^2(\nabla\rho)^2}{\rho^{4/3}}, \tag{3.40}$$

and the constant has been estimated both numerically and from first-principles calculations (Sham, 1973).

Bearing in mind the comment made above on the need to sum at least partial subseries of gradient terms to all orders, Stoddart *et al.* (1971), considered the quantity E_{xc} to be nonlocal. Their work leads to $E_{xc} = \int \varepsilon_{xc}(\mathbf{r})\, d\mathbf{r}$, where

$$\varepsilon_{xc}^{\text{nonlocal}} = -\tfrac{1}{2}\int d\mathbf{r}' B(\mathbf{r}', \rho(\mathbf{r}))[\rho(\mathbf{r} + \tfrac{1}{2}\mathbf{r}') - \rho(\mathbf{r} - \tfrac{1}{2}\mathbf{r}')]^2, \tag{3.41}$$

the kernel B being a response function which is exactly determined, in principle, by solution of the jellium problem (Stoddart *et al.*, 1971). Of course, in practice only approximate forms are known. The contribution to the one-body potential [Eq. (3.35)] can then be written

$$V_{xc}(\mathbf{r}) = V_{xc}^{0}(\mathbf{r}) + V_{xc}^{\text{nonlocal}}(\mathbf{r}), \tag{3.42}$$

where*

$$V_{xc}^{\text{nonlocal}}(\mathbf{r}) = -\tfrac{1}{2}\int d\mathbf{r}'\, \frac{\delta B(\mathbf{r}', \rho(\mathbf{r}))}{\delta\rho(\mathbf{r})}\,\Delta - 2\int d\mathbf{r}' B\left(\mathbf{r} - \mathbf{r}', \rho\left[\frac{\mathbf{r} + \mathbf{r}'}{2}\right]\right)[\rho(\mathbf{r}) - \rho(\mathbf{r}')], \tag{3.43}$$

*Δ in Eq. (3.43) is simply the square bracket squared in Eq. (3.41).

while the local term $V^0_{xc}(\mathbf{r})$ is completely determined by the solution of the jellium problem already discussed fully in Section 3. As already mentioned, the kernel B is also entirely determined by the jellium problem, so the nonlocal correction in Eqs. (3.41) and (3.43) is completely specified once the electron density $\rho(\mathbf{r})$ in the inhomogeneous system under discussion is given.

Application to Be Metal. This formulation has been applied to Be metal by Stoddart *et al.* (1974) with encouraging results. Thus, for this metal, V^{nonlocal}_{xc} is found to be at most 30–40% of the local density term $V^0_{xc}(\mathbf{r})$ within the unit cell, whereas over much of the volume the correction is very substantially less than this. In this calculation, the measured density of Brown (1972) was used as accurate input data. It would be of obvious interest to study the fermiology of Be with this approach, but to the writer's knowledge this has not, so far, been done.

The most basic of the studies so far carried out on density gradient corrections appears to be that of Ma and Brueckner (1968), already referred to above. But since this has been reviewed earlier (Jones and March, 1986), further details will not be given here. Similarly the later work of Langreth and his co-workers on gradient theory is to be noted (see Section 3.6.2). Since this has been surveyed in some detail by Callaway and March (1984), the related work of Sahni *et al.* (1982) on the density gradient expansion to the exchange energy will be briefly summarized. They note first that the difficulty of expanding in Hartree–Fock theory in gradients is not present in the Kohn–Sham unipotential formalism. However, they go on to study density gradient corrections in atoms, and their conclusion is that even though, as the atomic number increases, the relative errors of the local density and gradient expansion approximations decrease in magnitude (cf. the discussion in Section 3.2 above), the gradient term corrects only a fraction of the error of the local density approximation (see also, Becke, 1992). This is a consequence of the face that the convergence condition $|\nabla\rho|2k_f\rho \ll 1$ is increasingly satisfied as the atomic number increases, but the second convergence condition $|\nabla^2\rho|/2k_f|\nabla\rho$ $24 \ll 1$ is not so well satisfied, k_f being as usual the Fermi wave number.

All this indicates that while density gradient treatments have had some successes, they are not at present sufficiently well developed to provide any final answer to the problem of determining the exchange plus correlation contribution to the ground-state energy functional or to the one-body potential. In this context, it is relevant to mention again the analytical study of Miglio *et al.* (1981) on the Bardeen model of a metal surface. Here, the exchange potential can be calculated exactly, and by comparison with the Dirac–Slater local exchange potential, all the main features, though not fully quantitative values, are found to be given without density gradient corrections. Clearly, since in this model there is a large density gradient near the metal surface, a fuller study of the basic reasons for their findings is needed, in the light of density gradient studies of Langreth, of Perdew, and of others referred to above.

3.6.2. Length Scales and Nonlocal Theory

Although the previous approach enabled, for example, the potential energy in the unit cell of Be metal to be corrected from LDA due to inhomogeneous terms, the underlying physical reasons why one must sum inhomogeneous corrections to all orders need further clarification. Therefore, although connected to the previous arguments, reference will now be made to the work of Langreth, Perdew, and Mehl, which has led to a correction to the local density approximation that seems relatively straightforward to use.

Their discussion can be initiated by a qualitative consideration of length scales. Thus, one writes

$$E_{xc} = E^{0}_{xc} + \gamma \int Z(k_f)(\mathrm{v}k_f)^2 \, d\mathbf{r} + \ldots, \tag{3.44}$$

where $k_f = (3\pi^2\rho)^{1/3}$ is the local Fermi wave number, while Z is a slowly varying function of k_f.

Langreth and Mehl (1981) then focus on a length ξ, say, which measures the scale over which the density varies:

$$\xi^{-1} = \tfrac{1}{2}|\nabla k_f/k_f|, \tag{3.45}$$

the assumption thereby implied being that there is only one important length scale for a given region of the electron charge cloud.

These workers then stress that for the local density approximation to be valid, one must have

$$k_f\xi \gg 1, \qquad q_{\mathrm{TF}}\,\xi \gg 1,$$

where q_{TF}^{-1} is the Thomas–Fermi screening length. For most of the systems of interest in this context, k_f and q_{TF} are not very different. Although the above inequalities are not particularly strongly satisfied, it seems reasonable to proceed.

There is an additional implicit assumption to discuss before tackling the second term on the right-hand side of Eq. (3.44). One must consider E_{xc} as a sum over excitations of varying size, say, λ (see Nozières and Pines, 1958; Hubbard, 1957, for jellium; also Langreth and Perdew (1979) and also (1957) for jellium; see also Langreth and Perdew (1979) and also Peuckert (1976) for inhomogeneous systems). If we put $k = \lambda^{-1}$, then it is necessary to assume in using the inequalities above that the typical k which contributes is not very different from k_f or q_{TF}. The work of Langreth and Perdew (1975, 1977) suggests that this is indeed the case in practice.

Turning to the gradient correction in Eq. (3.44), one evidently needs to ask whether this will be a good approximation to $E_{xc} - E^{0}_{xc} \equiv \Delta E$. If the assumption of a single ξ in a given region holds, then the above ine 5qualities are appropriate in this case also.

Why, then, if the local density approximation is good, does not the addition of the first gradient correlation in Eq. (3.44) improve the theory? The answer lies in the fact that now, in this second term, although a part of the gradient contribution is distributed normally in k, so that k_f is a typical value, a significant fraction is concentrated in a narrow range about $k \sim 0$. Langreth and Perdew (1980) have argued that one must require a stronger inequality, namely $k_f \gg 40$; this results from a detailed numerical calculation.

Langreth and Mehl (1981) then consider the way to circumvent this stronger inequality requirement while still retaining the above inequality. Let the error ΔE in the local density correction be analyzed into Fourier components as

$$\Delta E = \int \Delta E(k)\mathrm{d}k \tag{3.46}$$

$$Z(k_f) = \int_0^\infty z(k, k_f)\mathrm{d}k. \tag{3.47}$$

A relation such as Eq. (3.46) can also be written for the gradient correction in Eq. (3.44). The true $E(k)$ approaches zero rapidly as $k \to 0$ when $k\xi$ becomes less than unity, that is, when λ becomes $> \xi$. This is a consequence of sum rules. However, the Fourier decomposition of the gradient correction in Eq. (3.44) does *not* have this property, i.e., the higher-order gradient terms omitted in Eq. (3.44) are essential to restore the correct behavior of $\Delta E(k)$ when $k\xi < 1$. In contrast, these higher-order terms make little contribution when $k\xi \sim 1$, provided that the inequalities $k_f\xi \gg 1$, $q_{\mathrm{TF}}\xi \gg 1$ are obeyed. Thus one can circumvent $q_{\mathrm{TF}}\xi > 40$ by recognizing that the true $\Delta E(k)$ goes to zero in the region $k\xi < 1$.

Therefore, Langreth and Mehl write,

$$z(k, k_f) = z_{\mathrm{grad}}(k, k_f)\theta(k - \xi^{-1}) \tag{3.48}$$

and they attempt to restore quantitative reliability to the simple form in Eq. (3.48) by using

$$\xi^{-1} = 3f|\nabla k_f/k_f| = f|\nabla\rho/\rho|, \tag{3.49}$$

where f is a fitting parameter. Actually, the value finally adopted by Langreth and Mehl, namely $f = 0.15$, is not very different from the value $f = 1/6$ implied by the simplest considerations.

The quantity $z_{\mathrm{grad}}(k, k_f)$ has been calculated in RPA.* After use of some detailed fitting to simple analytic forms, Langreth and Mehl finish up with the proposed form

$$E_{\mathrm{xc}} = E^0_{\mathrm{xo}} + (4.28 \times 10^{-3})\int d\mathbf{r}(|\nabla\rho|^2/\rho^{4/3})\{2\exp(-F) - \tfrac{7}{9}\}, \tag{3.50}$$

where $F = 0.262|\nabla\rho|/\rho^{7/6}$. That a term like $(\nabla\rho)^2/\rho^{4/3}$ should appear in the energy density was first deduced by Herman *et al.* (1969) on dimensional grounds [see Eq.

*For further discussion of the random phase approximation, see Appendix A3.4.

(3.40) above]. Obviously, since the factor in curly brackets in Eq. (3.50) is dimensionless, a term of that form could not be detected by such an argument. As will be seen in the next section, when one generalizes to spin-dependent potentials, it is true that the local density approximation, generalized to jellium with spin polarization, has been widely useful, even in situations with large density gradients.

3.7. SPIN-DEPENDENT ONE-BODY POTENTIALS

To introduce spin-dependent potentials, it will be useful to summarize below the discussion by Stoddart and March (1971; see also von Barth and Hedin, 1972) of the density functional approach to magnetic systems. Suppose one has spin densities ρ_+ and ρ_- for the electrons of upward and down ward spin, respectively, the total density evidently being the sum $\rho = \rho_+ + \rho_-$. Consider the case of electrons moving in a spin-dependent external potential $V^{\sigma}_{\text{ext}}(\mathbf{r})$, $\sigma = \pm$. For this system, one writes the total energy as a functional of the spin densities ρ_+ and ρ_-,

$$E[\rho_+, \rho_-] = T_s[\rho_+] + T_s[\rho_-] + \tfrac{1}{2}\int\int d\mathbf{r}d\mathbf{r}' \, \frac{\rho(\mathbf{r})\rho(\mathbf{r})}{|\mathbf{r} - \mathbf{r}'|} + \int d\mathbf{r}[V^{+}_{\text{ext}}(\mathbf{r})\rho_+(\mathbf{r}) + V^{-}_{\text{ext}}(\mathbf{r})\rho_-(\mathbf{r})] + E_x[\rho_+] + E_x[\rho_-] + E_c[\rho_+, \rho_-]. \tag{3.51}$$

In Eq. (3.51), $T_s[\rho_\sigma]$ is the kinetic energy of a noninteracting gas of density ρ_σ, and E_x and E_c are the exchange and correlation contributions to the total energy.

Essentially, the spin-dependent external potential V_{ext} is introduced, even though one can switch off the spin dependence of the external potential later, to break the directional symmetry of the problem. For example, the external disturbance could be a magnetic field, applied, say, in the z direction, which would have an energy of interaction with the magnetic moment density $m(\mathbf{r}) = -(1/2c)[\rho_+(\mathbf{r}) - \rho_-(\mathbf{r})]$. Then the total energy may be written

$$E[\rho_+, \rho_-] = E[\rho, m] = G[\rho(\mathbf{r}), m[\rho(\mathbf{r})] + \tfrac{1}{2}\int d\mathbf{r}d\mathbf{r}' \, \frac{\rho(\mathbf{r})\rho(\mathbf{r})}{|\mathbf{r} - \mathbf{r}'|} + \int d\mathbf{r}[V_{\text{ext}}(\mathbf{r})\rho(\mathbf{r})] + E_{\text{field}}, \tag{3.52}$$

where the last term is simply the energy of interaction with the magnetic field while $G[\rho, m]$ has been written for the sum of kinetic, exchange, and correlation energy as a functional of ρ and m.

Let us require next that $E[\rho_+, \rho_-]$ be stationary with respect to arbitrary variations in $\rho_+(\mathbf{r})$ and $\rho_-(\mathbf{r})$, subject to conservation of the total number of electrons. If one defines

$$V_\sigma(\mathbf{r}) = V^\sigma_{\text{ext}}(\mathbf{r}) + \int d\mathbf{r}' \frac{\rho(\mathbf{r}')}{|\mathbf{r} - \mathbf{r}'|} + V^\sigma_{\text{xc}}(\mathbf{r}), \tag{3.53}$$

it can be seen that again one can formally reduce the many-electron problem to solutions of single-particle equations with spin-dependent one-body potentials. Needless to say, the exchange and correlation contributions to the spin-dependent one-body potentials can to date be set up only approximately. Most frequently, one appeals to local spin-density approximations, generalizing the arguments of Section 3.2 to include spin dependence. An example of spin density theory applied to the Stoner criterion for ferromagnetism is set out in Appendix 3.1.

Work Using Monte Carlo Calculations on Jellium. As discussed above, Vosko *et al.* (1980) have fitted the Monte Carlo calculations of Ceperley and Alder (1980), reviewed in Section 5.4 below, to construct local density approximations to spin-dependent potentials. Again, this area has been reviewed by Callaway and March (1984) and therefore only a few applications will be cited here. For example, this data on the correlation energy of the spin-polarized homogeneous electron assembly has been used by Vosko and Wilk (1980) to evaluate cohesive energies of the alkali metals Li, Na, K, and Rb. Good results were obtained for what is a difficult quantity to calculate, and there was, in particular, a significant improvement in the agreement between theory and experiment for Rb over previous calculations. In later work the same local spin-density approximation has been used to calculate successfully the Fermi contact contribution to the Knight shift in solid Mg by Nusair *et al.* (1981). Spin-dependent one-body potentials have also been discussed by Ellis (1982), while a different application of local density theory worthy of note here is that of Delley *et al.* (1983a–c) on the binding energy electronic structure of small Cu particles. Local density theory has also been applied recently by Connolly and Williams (1983) to phase transformations in transition-metal alloys by invoking the cluster expansion of Sanchez and de Fontaine (1981), but it would be too far from the main theme to go into details of this interesting application here.

3.8. DENSITY FUNCTIONAL TREATMENT OF WIGNER CRYSTALLIZATION

After the pioneering work of Wigner (1934, 1938), the problem of electron crystallization in jellium received a considerable amount of attention over the years—resulting, however, in predictions for the coupling strength of r_s^c at freezing spanning almost three orders of magnitude. As will be discussed in some detail in Section 5.5 below, the best estimate for r_s^c comes from Quantum Monte Carlo (QMC) studies which yield $r_s^c = 100 \pm 20$. QMC also establishes that the transition to a *bcc* regular structure takes place from a fully spin-polarized liquid, this phase being lower in energy with respect to the unpolarized liquid for $r_s > 75$. Therefore, in the following, let us be content with a discussion of the coexistence between the spin-polarized liquid and a regular crystalline phase.

A DFT theory of freezing for jellium clearly requires approximation of the exchange-correlation functional $E_{xc}[\rho]$, which is not known exactly. Unfortunately, the energy balance from which the crystallization phenomenon results is very delicate for jellium. A careful investigation of Senatore and Pastore (1990) reveals that the accuracy afforded by LDA does not suffice in this case. However, a different approximation scheme is proposed, which turns out to work quite well. Rather than the full functional $E_{xc}[\rho]$, one can approximate the difference $\Delta = E_{xc}[\rho_s] - E_{xc}[\rho_0]$ between the solid and the liquid phase by appealing to an expansion of Δ in powers of the density difference $\rho_Q(\mathbf{r}) = \rho_s(\mathbf{r}) - \rho_0$ around the liquid. To second order, such an expansion yields

$$E_{xc}[\rho_s] = E_{xc}[\rho_0] + \tfrac{1}{2}\iint d\mathbf{r}d\mathbf{r}'\ [-\chi^{-1}(\mathbf{r}-\mathbf{r}') + \chi_0^{-1}(\mathbf{r}-\mathbf{r}')]\rho_Q(\mathbf{r})\ \rho_Q(\mathbf{r}')], \tag{3.54}$$

with $\chi(\mathbf{r})$ and $\chi_0(\mathbf{r})$ being, respectively, the static response function of the homogeneous liquid and the response function of the noninteracting electrons (i.e., the Lindhard function; see also Appendix 3.4). One of the motivations for using such a *quadratic approximation* is that, for classical liquids, it is known to work surprisingly well (for a review see Baus, 1990).

As already mentioned in Section 3.1, the Euler–Lagrange problem for the ground-state energy functional of interacting particles can be conveniently recast into the so-called Slater–Kohn–Sham problem (i.e., the self-consistent problem of noninteracting particles in a density dependent one-body potential). Once the exchange-correlation functional is known, it is a simple matter to obtain at once the density-dependent Slater–Kohn–Sham potential. From Eqs. (3.24) and (3.54) one gets, in Fourier transform,

$$V(\mathbf{q}) = \rho_Q(\mathbf{q})[-\chi^{-1}(q) + \chi_0^{-1}(q)]. \tag{3.55}$$

It may be worth stressing at this point that, whereas the LDA borrows only a *thermodynamic* property of the homogeneous liquid (i.e., the exchange-correlation energy), the *quadratic approximation* involves—in principle—structural information of the liquid at all the wave vectors that are relevant in the modulated phase. Thus, for a regular solid, one finds that such a region of wave vectors extends from about $2qF$ onward.

To perform calculations, one needs the response function of the homogeneous interacting electron liquid. This quantity, however, is not known exactly. Senatore and Pastore (1990) have therefore employed the so-called STLS decoupling scheme (Singwi *et al.*, 1968; see Appendix A3.5) to construct $\chi^{-1}(q)$ from the structure factor $S(k)$ obtained from the QMC simulations.

The calculation of the ground-state energy of the Wigner crystal requires the self-consistent solution of Slater–Kohn–Sham equations for the Bloch orbitals of a single fully occupied energy band, since there is one electron per unit cell and one is considering the spin-polarized state. This can be accomplished by using standard computational techniques for band-structure calculations. The results obtained can be summarized as follows:

The quadratic approximation predicts freezing into the *bcc* lattice at $r_s = 102$; a value which compares extremely well with the QMC prediction of $r_s^c = 100 \pm 20$. Also a Lindemann ratio γ (rms deviation about the lattice site divided by the nearest-neighbor distance) of 0.34 is obtained, whereas QMC suggests $\gamma = 0.30 \pm 0.02$ for all quantum systems studied to date. The calculated density turns out to be still well localized, even if considerably less so than in classical freezing. This, together with the high symmetry of the periodic structure, suggests the possibility of a *tight binding* approximation in which Bloch orbitals are built from one Gaussian orbital per site with a variational width. By the use of this approximation, calculations simplify considerably, whereas the results for r_s^c and γ change only slightly. In fact, one gets $r_s^c = 107$ and $\gamma = 0.29$.

Senatore and Pastore (1990) have also investigated the stability of the *fcc* electron crystal. They find, within the fuller calculations, that the *fcc* solid is in fact the stable phase between $r_s = 97$ and $r_s = 108$, being in this range lower in energy than the *bcc* crystal (and, of course, also lower in energy than the homogeneous liquid). For higher values of the coupling r_s the *bcc* remains the stable phase in agreement with the findings of harmonic lattice calculations (see, e.g., Foldy, 1971).

Wigner electron solids will be taken up again in the final Chapter 9. There, however, the concern will be with the two-dimensional electron assembly in a GaAs/AlGaAs heterojunction in the presence of a strong applied magnetic field.

3.9. ENERGY GAP IN DENSITY FUNCTIONAL TREATMENT OF CRYSTALS

In this relatively brief introduction to the problem of the energy gap as given by density functional calculations on periodic crystals, it is to be noted that it is characteristic of calculations of bands in the local density approximation that the energies of excited states in the semiconductors and insulators are too low with respect to the valence band. In this context, the work of Wang and Klein (1981) may be cited. These workers made all-electron calculations of energy bands in six semiconductors, including Si, by using an exchange-correlation potential, the correlation being of Wigner form. The occupied bandwidth and X-ray form factors compare favorably with experiment, but the fundamental band gap for Si was only 0.65 eV. A similar problem was discussed by Khan and Callaway (1980) for the band structure of solid neon and argon. There, the calculated band gaps were found to be about one-half of the experimental values, when calculated with an exchange-correlation potential. To obtain the correct band gap it proved necessary to use an $X\alpha$ potential, with $\alpha = 1.2$: ie to vary the coefficient of the Dirac-Slater $\rho^{1/3}$ potential.

Below, the question of the foundations in density functional theory of determining the band gap will be exposed, leaning on the survey by Pickett (1986). Within density functional theory, he notes, the description of single-particle excitation energies can be approached by way of the ground-state energy functional as well as through a convėntional description of excited states.

Excitations from Ground-State Energies. Consider a semiconducting system for which N electrons exactly fill the valence bands. The corresponding total energy will be denoted by E_N. If one electron from the valence-band maximum is removed to infinity, the energy of the system is E_{N-1}, making the minimum energy necessary to excite an electron out of the crystal, namely the ionization potential,

$$I = E_{N-1} - E_N \tag{3.56}$$

If an electron, originally at infinity, is introduced into the lowest conduction-band state, the change in energy is the electron affinity A:

$$A = E_N - E_{N+1}. \tag{3.57}$$

Then I-A, which is given by the difference $E_{N-1} + E_{N+1} - 2E_N$ in ground-state energies is equal (neglecting electron–hole interaction effects) to the energy gap describing the lowest excitation of the semiconductor.

In contrast to the one-body potential $V(\mathbf{r}) = V_{\text{Hartree}} + V_{\text{xc}}(\mathbf{r})$ of density functional theory, in the formally exact theory of single-particle excitations [see March (1972) for a simple example of the jellium model of Section 3.1], the quasiparticle wave function is a solution of a similar Schrödinger equation, as discussed for instance by Pickett and Wang (1984), namely,

$$[-\nabla^2 + V_{\text{ext}}(\mathbf{r}) + V_{\text{Hartree}}(\mathbf{r})]\phi(\mathbf{r}, E) + \int d\mathbf{r}' M(\mathbf{r}, \mathbf{r}', E)\phi(\mathbf{r}', E) = E\phi(\mathbf{r}, E). \tag{3.58}$$

The major difference is the appearance in Eq. (3.58) of a nonlocal energy-dependent potential M, which describes exchange-correlation effects, and the replacement of discrete eigenvalues i with a continuous excitation energy E. Pickett draws attention to the fact that in a semiconductor, however, the low-energy solutions form undamped excitations very reminiscent of those calculated from the Slater–Kohn–Sham (SKS) equations. It is natural then to identify the SKS eigenvalues and eigenfunctions as approximations to actual single-particle excitation energies and wave functions. A similar point of view emerged from Slater's early work (see Callaway and March, 1984), and band theory studies have implicitly assumed this connection.

The relationship between the gap $E_g = I - A$, and the Slater–Kohn–Sham eigenvalues ε_i is derived by expressing the ground-state energy as

$$E = \sum_i \varepsilon_i \theta(\mu - \varepsilon_\iota) - \tfrac{1}{2}\int d\mathbf{r} \int d\mathbf{r}' \, \frac{\rho(\mathbf{r})\rho(\mathbf{r}')}{|\mathbf{r} - \mathbf{r}'|}$$

$$- \int d\mathbf{r}\, \rho(\mathbf{r}) V_{xc}(\mathbf{r}, \rho) + E_{xc}[\rho]. \tag{3.59}$$

In Eq. (3.59), $\theta(x)$ is the Heaviside function, which is 1 for $x < 0$ and zero otherwise. As pointed out by Perdew and Levy (1983) and by Sham and Schlüter (1983)—see also below for details—the gap can now be expressed in the form

$$E_g = \varepsilon_{N+1}(N) - \varepsilon_N(N) + V_{xc}(\mathbf{r}; \rho)\Big|_{N-\delta}^{N+\delta}, \tag{3.60}$$

where $\varepsilon_i(J)$ is the ith eigenvalue in a system containing J electrons and where the limits $N \pm \delta$ indicate that the exchange-correlation potential is to be evaluated for densities $\rho(\mathbf{r})$ containing fractionally more or less than N electrons. In deriving this relation, the notion is used that the densities of the $N + 1$, N, and $N - 1$ electron assemblies differ by order N^{-1} and that $E_{xc}[\rho]$ is continuous for varying electron number. Making the assumption that V_{xc} depends quasi-continuously on N, the last term in Eq. (3.58) vanishes, and this relation equates the Slater–Kohn–Sham gap with the true gap. As reviewed in some detail by Callaway and March (1984), this is not found to be the case from density functional calculations, the available Slater–Kohn–Sham type calculations underestimating the band gap considerably.

A partial resolution of this difficulty has been proposed, which notes that V_{xc} is permitted to have a discontinuity at values of N for which a complete set of bands is occupied. As a result, the Slater–Kohn–Sham gap is increased by the position-independent contribution

$$\Delta V_{xc} = V_{xc}(\mathbf{r}; \rho)\Big|_{N-\delta}^{N+\delta} \tag{3.61}$$

The reasoning behind the discontinuity goes as follows (see Sham and Schlüter (1983); Perdew and Levy (1983)]: The argument rests on the separation of the ground-state energy functional $E[\rho]$ into

i. the electrostatic potential energy,
ii. the kinetic energy $T_s[\rho]$ of a noninteracting system with the same electron density ρ as the true many-body system, and
iii. the exchange-correlation energy E_{xc}.

The variational principle then yields a single-particle Schrödinger equation with the potential given by the sum of the electrostatic potential V_{es} due to external and electronic charges and the exchange-correlation potential V_{xc} given by $\delta E_{xc}/\delta\rho$.

Sham and Schlüter argue that for extensive systems with $N \to \infty$, $\delta T_s/\delta\rho$ for a noninteracting system with an insulating ground state has a discontinuity:

$$\frac{\delta T_s}{\delta \rho +} - \frac{\delta T_s}{\delta \rho -} = \varepsilon_g \tag{3.62}$$

by taking the limits of the Euler equation

$$\left.\frac{\delta T_s}{\delta \rho}\right|_{N \pm \delta} + V = \mu_{N \pm \delta} \tag{3.63}$$

for $N \pm \delta$ particles. While each term $\delta T_s/\delta\rho_\pm$ is position dependent, the difference is a position-independent functional of the density.

Exchange-only: Two Plane Wave Model

They then proceed to calculate the gap correction for a two-plane-wave model including exchange only, the Fermi surface being taken to be a cylinder of equal length and diameter. The gap correction Δ is then plotted against the density functional gap ε_g. This exchange-only model illustrates how the density functional equation underestimates the Hartree–Fock gap, the correction being of the order of the gap itself. Sham and Schlüter (1983) note that correlation contributions to the gap correction will alter the numerical values but will not change the qualitative picture.

Pickett (1986), in his short review, also discusses density functionals for excitations. He refers to an approach that deals directly with the difference between the so-called mass operator M of many-body theory appearing in Eq. (3.58), and V_{xc}. For metals, this difference is rather small; for semiconductors it is an order of magnitude or more smaller than M itself, but the reader must refer to the article by Pickett for further details.

3.10. EXACT EXCHANGE-CORRELATION POTENTIAL IN TERMS OF DENSITY MATRICES

In this section, a brief summary will be presented of two routes to the exchange-correlation potential $V_{xc}(\mathbf{r})$ of density functional theory, set out by Holas and March (1995). These start, respectively, from (i) a differential form of the virial theorem and (ii) an expression for the 'local' energy in terms of density matrices given by Dawson and March (1984) (see also Cohen and Frishberg, 1976).

The differential form of the virial theorem referred to above can be written in the form

$$\nabla v(\mathbf{r}) = -\,\mathbf{f}(\mathbf{r}; [u, \rho, n_1, n_2]). \tag{3.64}$$

Here v(**r**) is the external potential, u is the two-body Coulomb repulsion, ρ is the electron density while n_1 and n_2 are first and second order density matrices respectively. n_1 reduces to the indempotent matrix ρ(**r**, **r**′) in the case of a single Slater determinant.

Holas and March show that the function **f** in Eq. (3.64) can be written explicitly as

$$\mathbf{f}(\mathbf{r}; [u, \rho, n_1, n_2]) = \{-\tfrac{1}{4}\nabla\nabla^2\rho(\mathbf{r}) + \mathbf{z}(\mathbf{r}; [\rho]) + 2\int d\mathbf{r}'[\nabla u(\mathbf{r}, \mathbf{r}')n_2(\mathbf{r}, \mathbf{r}')]\}/\rho(\mathbf{r}) \tag{3.65}$$

Finally, in Eq. (3.65) the vector field **z** is completely specified by the kinetic-energy density tensor.

3.10.1. Path-Integral Form of External Potential v(r)

Eq. (3.64) can be regarded as a differential equation for the external potential v(**r**). Because f(**r**) = − ∇ V(**r**) the force field f(**r**) is conservative. Therefore, it follows that the potential at point $\mathbf{r}_0$, say, is the work done in bringing an electron from infinity to $\mathbf{r}_0$ against the force field **f**(**r**):

$$v(\mathbf{r}_0) = -\int_{\infty}^{r_0} d\mathbf{r}\cdot\mathbf{f}(\mathbf{r}) \tag{3.66}$$

Since f, as noted above, is conservative, the value of the line integral in Eq. (3.66) is independent of the path. It should also be noted that Eq. (3.66) has been written to satisfy the boundary condition v(∞) = 0 which is customary.

Using the Euler equation of density functional theory (Dreizler and Gross, 1990) in terms of the Hohenberg-Kohn functional F[ρ], one can re-express Eq. (3.66) as

$$\frac{\delta F[\rho]}{\delta\rho(\mathbf{r}_0)} = \mu + \int_{\infty}^{r_0} d\mathbf{r}\cdot\mathbf{f}(\mathbf{r}; [u, \rho, n_1, n_2]). \tag{3.67}$$

This Eq. (3.67) can be used, within the framework of Slater–Kohn–Sham (SKS) equations in Appendix 4.5, to write a formally exact expression for the exchange-correlation potential in terms of u, ρ, n_1, n_2 and the idempotent first-order density matrix constructed from SKS orbitals.

The conclusion demonstrated by Holas and March (1995) is that this expression, in the exchange-only approximation, leads to the Harbola-Sahni (1989) work formalism for the exchange-only potential $V_x(\mathbf{r})$. To include correlation, one has to develop a relation between second and first order density matrices transcending that which is obtained for a single Slater determinantal wave function [compare Eq. (3.3)].

3.10.2. Local Energy Relation

Very briefly, the local energy relation given by Dawson and March (1984) can alternatively be used as a route to the exact exchange-correlation potential $V_{xc}(\mathbf{r})$. One then obtains an integral equation to be solved for $V_{xc}(\mathbf{r})$, but now, in addition to the information required in the virial route discussed above, the third-order density matrix of the fully interacting system appears. This is the price one pays to avoid the path integral entering via the virual. Once approximations are made for density matrices in the latter approach, there is no assurance that the integral for $V_{xc}(\mathbf{r})$ will remain path independent. No such uncertainty arises in the local energy approach (Holas and March, 1995).

Finally, dynamic properties of solids, which will be a major component of Chapter 6, are introduced in Appendices 3.4 and 3.5, for the very specific model of jellium which is most appropriate for simple metals such as Na and K.

Chapter 4

Localized versus Molecular Orbital Theories of Electrons

Apart from possible dynamic effects due to electron–electron correlations referred to in Chapter 2, the atomic correlation effects have to be sought by quantitative examination of the problem. In contrast, as already mentioned, one has qualitative consequences in multicenter problems that are worthy of full consideration. Let us start by reviewing the situation in the H_2 molecule, going back to the pioneering work on the theory of the chemical bond by Heitler and London (1927).

This Heitler–London description merely asserted that a useful ground-state symmetric space-wave function could be built up from the atomic orbitals (1*s* functions) centered on nuclei *a* and *b*, namely ϕ_a and ϕ_b. After symmetrization, one is led to the spatial wave function

$$\Psi_{HL}(1, 2) = \phi_a(1)\phi_b(2) + \phi_b(1)\phi_a(2) \tag{4.1}$$

The first term on the right-hand site of Eq. (4.1) evidently corresponds to electron 1 on nucleus *a* and electron 2 on nucleus *b*. The second part is added because of the indistinguishability of electrons.

Turning to the delocalized description, one introduces a molecular orbital ψ_{MO} and in the singlet ground state, one puts into it two electrons with opposed spins. Then the MO total space wave function is written in the form

$$\Psi_{MO}(1, 2) = \psi_{MO}(1)\psi_{MO}(2) \tag{4.2}$$

and in terms of the 1*s* atomic orbitals, in the approximation in which the molecular orbital is built up as a linear combination of atomic orbitals,

$$\begin{aligned}\Psi_{LCAO\text{-}MO}(1, 2) &= [\phi_a(1) + \phi_b(1)]\,[\phi_a(2) + \phi_b(2)]\\ &= \Psi_{HL}(1,2) + \phi_a(1)\phi_a(2) + \phi_b(1)\phi_b(2).\end{aligned} \tag{4.3}$$

In the second part of Eq. (4.3), it has been noted explicitly that the LCAO–MO wave function can be viewed as a linear superposition of the Heitler–

London covalent terms and an equally weighted admixture of ionic terms, $\phi_a(1)\phi_a(2)$ evidently representing both electrons on nucleus a, etc. That Eq. (4.3) is incorrect as the internuclear distance R gets large compared with the size a_0 of the 1s hydrogen orbitals is quite clear; the molecule dissociates into two neutral H atoms, just as described by the original Heitler–London wave function (4.1).

4.1. COULSON–FISCHER WAVE FUNCTION WITH ASYMMETRIC ORBITALS FOR H_2 MOLECULE

An important clarification of the role of electron correlation in molecules came with the work of Coulson and Fischer (1949) on H_2. They asked what was the best admixture of covalent and ionic states at each internuclear distance R by contemplating asymmetric molecular orbitals $\phi_a + \lambda\phi_b$, $\lambda \leq 1$, and $\phi_b + \lambda\phi_a$, the former representing, with $\lambda < 1$, the electron primarily but not wholly belonging to nucleus a, etc. Then they formed the (unsymmetrized) variational wave function

$$\Psi_{\text{Coulson–Fischer}} = [\phi_a(1) + \lambda\phi_b(1)][\phi_b(2) + \lambda\phi_a(2)]. \tag{4.4}$$

Determining λ as a function of R by minimization of $\langle H \rangle$ with respect to the wave function [Eq. (4.4)], H being the total Hamiltonian of the H_2 molecule, they found the following situation:

i. For $R<1.6R_{\text{equilibrium}}$, $\lambda = 1$.
ii. For $R>1.6R_{\text{equilibrium}}$, λ falls quite rapidly to zero as R is increased,

$R_{\text{equilibrium}}$ being the equilibrium internuclear separation. For $\lambda = 1$, Eq. (4.4) becomes identical with Eq. (4.3), whereas for $R>1.6R_{\text{equilibrium}}$ it can be seen that electrons quickly "go back onto their own atoms."

This idea that one should decrease the weight of the ionic configurations in a conventional molecular orbital treatment has been taken up in the work of Gutzwiller (1963, 1964, 1965) for treating strong correlations in narrow energy bands, and the results of his method will be discussed in some detail below.

The important point to be stressed from the above is that electron correlations can have the qualitative effect of driving electrons back onto their own atoms when the internuclear spacing becomes substantially larger than the size of the atomic orbitals involved.

Model for Two-Electron Molecule: Spin and Charge Densities. It is convenient at this point to follow Falicov and Harris (1969) and refer to a model for a two-electron homopolar molecule. These workers discuss the eigenstates of the one-band Hamiltonian for such a two-electron system, and their results are summarized in Appendix 4.1. They then use the exact solution for the ground state as a standard by which to assess the validity of the MO, Heitler–London, and other states having either spin- or charge-density waves.

By definition, of course, the MO approximation is undercorrelated, the Heitler–London states being always overcorrelated. The spin-density and charge-density waves are less easily classified, the under- or overcorrelation depending on the strength of the interaction.

Falicov and Harris construct from spin- and charge-density wave states, which have broken symmetry, symmetrized versions. These symmetrical states were always found in their work to be slightly undercorrelated. This work has been extended by Huang *et al.* (1976).

4.2. GUTZWILLER'S VARIATIONAL METHOD

As mentioned above, a possible way to account for the correlation of antiparallel electrons in a multicenter problem is to partly project out from a given uncorrelated wave function those components corresponding to doubly occupied (ionic) sites. In the case of H_2, this was most simply done by Coulson and Fischer (1949), as set out in Section 4.1. The generalization of such an approach to situations with an arbitrary number of centers is due to Gutzwiller (1963, 1965). Here an outline will be presented of the key points of Gutzwiller's approach, which has also been reviewed by Vollhardt (1984). The next two sections will discuss in some detail the application of the Gutzwiller variational method to treat correlation in molecules, on the one hand, and the Hubbard Hamiltonian, on the other. The latter has received renewed attention in connection with the exciting discovery of high T_c superconductivity (Bednorz and Müller, 1986).

The starting point of Gutzwiller's variational approach is the uncorrelated wave function for the problem under consideration. This is constructed, for a regular lattice with N sites and one Wannier orbital ϕ per site, from the Bloch waves $\Psi_{\mathbf{k}}(\mathbf{r})$,

$$\Psi_{\mathbf{k}}(\mathbf{r}) = \frac{1}{\sqrt{N}} \sum_i \exp(i\mathbf{k}\mathbf{R}_i)\phi(\mathbf{r} - \mathbf{R}_i). \tag{4.5}$$

Using the second-quantization formalism (see Mahan, 1990), the uncorrelated ground-state wave function can be written as

$$|\Phi_0\rangle = \prod_{\mathbf{k}\in K} a^{\dagger}_{\mathbf{k}\uparrow} \prod_{\mathbf{q}\in Q} a^{\dagger}_{\mathbf{q}\downarrow} |0\rangle, \tag{4.6}$$

where $a^{\dagger}_{\mathbf{k}\sigma}$ is the creation operator of an electron in the Bloch wave Ψ_k and with spin projection σ, $|0\rangle$ is the vacuum state, and K and Q are sets of points in reciprocal space, which in general may be delimited by different Fermi surfaces. If one denotes the creation operator of an electron in the Wannier orbital ϕ at the site i by $a^{\dagger}_{i\sigma}$,

$$a^\dagger_{i\sigma} = \frac{1}{\sqrt{N}} \sum_k \exp(-i\mathbf{k}\mathbf{R}_i) a^\dagger_{\mathbf{k}\sigma}, \tag{4.7}$$

and the corresponding number operator by $n_{i\sigma}$, the Gutzwiller wave function is written as (see also Senatore and March, 1994)

$$|\Phi\rangle = \prod_i [1 - (1 - g)n_{i\uparrow}n_{i\downarrow}]|\Phi_0\rangle = g^D|\Phi_0\rangle. \tag{4.8}$$

Above, $D = \Sigma_i n_{i\uparrow}n_{i\downarrow}$ is the operator that counts the number of doubly occupied sites. Clearly, in the wave function Φ, the components containing doubly occupied sites are reduced by a fractional amount 1-g ($0 \le g \le 1$) with respect to their value in the uncorrelated wave function Φ_0, thus reducing the repulsive interaction energy among antiparallel electrons. For a given Hamiltonian H the variational parameter g has to be found by minimizing the ground state energy

$$E_0(g) = \frac{\langle\Phi|H|\Phi\rangle}{\langle\Phi|\Phi\rangle}. \tag{4.9}$$

Hubbard Hamiltonian. To fix ideas, let us briefly consider now the case originally investigated by Gutzwiller, i.e., that of the so-called Hubbard Hamiltonian

$$H = \sum_{ij} \sum_{\sigma} t_{ij} a^\dagger_{i\sigma} a_{j\sigma} + U \sum_i n_{i\uparrow} n_{i\downarrow}, \tag{4.10}$$

which is designed to describe fermions on a lattice in a narrow-band system. The first term on the right-hand side of Eq. (4.10) is the kinetic energy due to the hopping of electrons between sites, and the second term crudely describes the on-site repulsion of electrons of antiparallel spin. It is straightforward to show that for $g = 1$ one simply regains the uncorrelated state, which of course is the exact ground state for zero on-site repulsion, $U = 0$. In this case $D = D_0 \equiv N_\uparrow N_\downarrow/N$. On the other hand, $g = 0$ corresponds to a fully correlated wave function in which the components containing doubly occupied sites are suppressed, while $D = 0$ yields the exact ground state for $U = \infty$. Therefore, for finite repulsion, one will have $0 < g < 1$ and $0 < D < D_0$, since the effect of correlation is precisely to reduce the number of doubly occupied sites present in the uncorrelated wave function.

Thus far, even for the Hubbard Hamiltonian, the evaluation of the averages in Eq. (4.9) has proved too difficult to carry out analytically in the general case, with the notable exception of the one-dimensional lattice (Metzner and Vollhardt, 1987; Gebhard and Vollhardt, 1987). Hence, one has either to employ numerical methods (Horsch and Kaplan, 1983; Yokoyama and Shiba, 1987a,b; Gros *et al.*, 1987a,b) or to resort to approximate treatments, as originally done by Gutzwiller (1963, 1965). It has subsequently been shown that the approximations originally introduced by Gutzwiller to evaluate the averages in Eq. (4.9) are just equivalent to the neglect of the spin-configuration dependence of the various terms appearing in the expansion of both $\langle\phi|H|\phi\rangle$ and $\langle\phi|\phi\rangle$ (Ogawa *et al.*, 1975; see also Vollhardt,

1984). While the interested reader is urged to consult the original papers for the details of Gutzwiller's approximation (GA), it is convenient to record here at least the main results of GA as applied to the Hubbard Hamiltonian.

Within the GA, the average of Eq. (4.9) with the Hamiltonian given in Eq. (4.10) can be written as

$$E_0/N = n_\uparrow q_\uparrow \bar{\varepsilon}_\uparrow + n_\downarrow q_\downarrow \bar{\varepsilon}_\downarrow + Ud. \tag{4.11}$$

Here, $n_\sigma = N_\sigma/N$; q_σ is the discontinuity of the momentum distribution $n_{\mathbf{k}\sigma}$ at the Fermi surface; $d = D/N$, with D the average number of doubly occupied sites; and

$$\bar{\varepsilon}_\sigma = \frac{1}{N_\sigma} \sum_{|\mathbf{k}|<k_{F\sigma}} \varepsilon(\mathbf{k}). \tag{4.12}$$

In Eq. (4.12), $\varepsilon(\mathbf{k})$ is the energy eigenvalue of the Bloch wave Ψ_k,

$$\varepsilon(\mathbf{k}) = \frac{1}{N} \sum_{ij} t_{ij} \exp[-i\mathbf{k}(\mathbf{R}_i - \mathbf{R}_j)], \tag{4.13}$$

and the zero of energy has been chosen so that $\Sigma_\mathbf{k}\varepsilon(\mathbf{k}) = 0$, i.e., $t_{ii} = 0$. The GA simply yields

$$q_\sigma = \frac{\{[(n_\sigma - d)(1 - n_\sigma - n_{-\sigma} + d)]^{1/2} + [(n_{-\sigma} - d)d]^{1/2}\}^2}{n_\sigma(1 - n_{-\sigma})}, \tag{4.14}$$

where g has been eliminated in favor of d, and d has to be determined variationally, by minimizing the ground-state energy $E_0(d)$ as given in Eq. (4.11). In the special case of a paramagnetic half-filled band, for which $n_\uparrow = n_\downarrow = \frac{1}{2}$, $q_\uparrow = q_\downarrow = q$, and $\bar{\varepsilon}_\uparrow = \bar{\varepsilon}_\downarrow = \bar{\varepsilon}_0 < 0$, one obtains, after minimization

$$d = \frac{1}{4}\left(1 - \frac{U}{U_c}\right), \tag{4.15}$$

$$q = 1 - \left[\frac{U}{U_c}\right]^2, \tag{4.16}$$

$$E_0/N = |\bar{\varepsilon}_0| \left[1 - \frac{U}{U_c}\right]^2, \tag{4.17}$$

with $U_c = 8|\bar{\varepsilon}_0|$. This shows that, within the GA, at a finite critical value of the interaction $U = U_c$, d, q, and E_0 all vanish. The vanishing of the discontinuity in the momentum distribution at the Fermi surface, and consequently of the kinetic energy, would signal a metal–insulator transition (Brinkman and Rice, 1970; see also March *et al.*, 1979; March and Parrinello, 1982). In fact, at the critical strength U_c, all sites would be singly occupied and the particles fully localized. This point will be taken up again below. The metal–insulator transition phenomenon, on the other hand, will be discussed at some length in Chapter 7.

Before turning to the discussion of how the Gutzwiller variational method has been generalized to treat interatomic and intra-atomic correlation in molecules, one must mention here the work by Kotliar and Ruckenstein (1986). These authors have proposed a treatment of the Hubbard Hamiltonian that makes use of auxiliary boson fields, in analogy with the so-called slave-boson approach first proposed by Barnes (1976). They have shown that, within the functional integral formalism, a particular choice of the auxiliary fields yields exactly the result of GA as a mean field or, more precisely, a saddle-point approximation to the exact functional integral. The merit of such an approach is that in principle it would allow for systematic improvement on GA, in that one could study the effect of Gaussian fluctuations about the saddle point. However, nothing has been done in this direction, to the author's knowledge, at the time of writing.

4.3. LOCAL APPROACH TO CORRELATION IN MOLECULES

As discussed above, the local approach to correlation was introduced by Coulson and Fischer in the qualitative treatment of the H_2 molecule. Subsequently, Gutzwiller extended the idea to solids to deal with strong correlation in narrow bands originating from very localized orbitals such as *d*-orbitals, within the model description afforded by the Hubbard Hamiltonian. Starting from Gutzwiller's systematic scheme, a further step toward the quantitative treatment of correlation in small molecules was taken by Stollhoff and Fulde (1977, 1978).

Their main point is as follows. One still wants to set up a variational scheme, in which those components of the wave functions that correspond to two electrons being too close are reduced. To this end, the wave function may be described, for instance, by assigning its value on the points of a spatial mesh (the grid) in which the system of interest, say the molecule, is embedded. In practice, one may as well give at such points the amplitudes of basis set functions, in terms of which the uncorrelated wave function is expanded. Within this picture, the wave function of Eq. (4.8) corresponds to a grid whose points coincide with the locations of atoms in the molecule. A quantitative treatment clearly requires that one improves upon the ansatz of Eq. (4.8) in at least two ways. One should (i) choose a finer grid and (ii) correlate the electronic motion not only on site but also between different sites. The latter point can be easily realized by generalizing the variational wave function to read (Stollhoff and Fulde, 1977)

$$|\Phi\rangle = \prod_{i=0}^{I} P_i |\Phi_0\rangle, \tag{4.18}$$

where the P_i are projection operators defined by

$$P_i = \prod_j (1 - \eta_i O_{ij}), \tag{4.19}$$

$$O_{oj} = n_{j\uparrow} n_{j\downarrow}, \tag{4.20}$$

$$O_{ij} = \sum_{\sigma\sigma'} n_{j\sigma} n_{j+i\sigma'}. \tag{4.21}$$

For $I = 0$, Eq. (4.18) yields the original Gutzwiller ansatz. However, for nonzero I, density correlations between different grid points are also taken into account, up to the Ith neighboring points. Notice that in the latter case one is also correlating the motion of parallel-spin electrons on different grid points. The reason for this is that, though the uncorrelated wave function $|\Phi_0\rangle$ is antisymmetrized, the Pauli exclusion principle is completely effective only at very short distances.

The problem of a finer grid can be also dealt with in quite a straightforward way (Stollhoff and Fulde, 1978). Instead of directly making the grid finer, one can alternatively choose to keep as grid points the atomic position in the molecule, while using at each site a number of basis set functions with varying degrees of localization, possibly also off center. Thus, the operator $a^\dagger_{i\sigma}$ now creates an electron in the basis set function i instead of in one of the self-consistent one-particle orbitals in terms of which $|\Phi_0\rangle$ is constructed. Here, the index i labelling the creation operator stands in fact for the pair (i, α), where α indicates either one of the original basis set functions or a suitable new combination; it might be, for instance, an *s–p* hybrid.

4.3.1. Model of H_6 on Ring

The effect of taking into account off-site correlation through the use of the variational wave function given in Eq. (4.18) with $I \neq 0$ has been investigated by Stollhoff and Fulde (1977) by studying the H_6-model of Mattheiss (1961). This model has the advantage that one knows the exact solution, which has been obtained numerically. The model assumes six sites equally spaced on a ring, with one *s*-atomic orbital at each site. From these, one constructs a Wannier orbital, in terms of which the Hamiltonian reads

$$H = \sum_{ij\sigma} t_{ij} a^\dagger_{i\sigma} a^\dagger_{j\sigma} + \sum_{\substack{ijkl \\ \sigma\sigma'}} V_{ijkl} a^\dagger_{i\sigma} a^\dagger_{j\sigma'} a_{l\sigma'} a_{k\sigma}. \tag{4.22}$$

Bloch waves are then built from the Wannier orbital, and from these the uncorrelated Hartree–Fock ground state $|\Phi_0\rangle$ is obtained. The correlation energy of the system was studied by Stollhoff and Fulde for various values of the first-neighbor distance R, with the variational wave function of Eq. (4.18). They found that the simpler Gutzwiller ansatz, i.e., $I = 0$, yields only about 50% of the exact correlation energy at the exact equilibrium distance, R = 1.8 a.u. On the other hand the

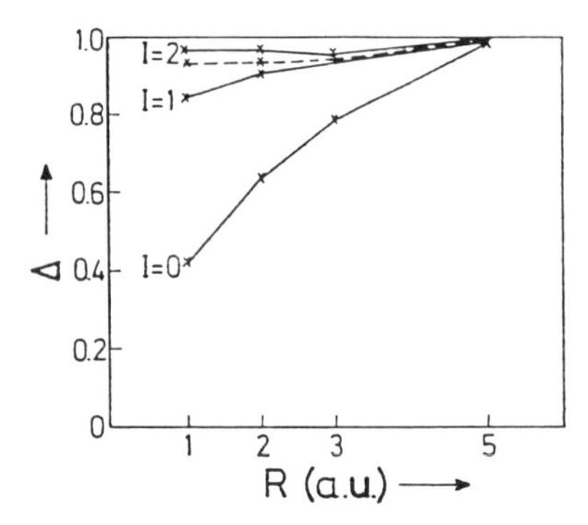

FIGURE 4.1. Gain in correlation energy $\Delta = (E_{H\text{-}F} - E_{Gutz})/(E_{H\text{-}F} - E_{exact})$ for the H_6 model by using the ansatz of Eq. (4.18) to evaluate the ground-state energy E_{Gutz}. Results are shown for $I = 0$ to 2. The dashed line is the approximation given by Eq. (4.23). (Redrawn from Stollhoff and Fulde, 1977.)

situation improves substantially by going to $I = 1$ and even more for $I = 2$. In the latter case, one recovers more than 95% of the correlation energy for any distance. The situation is illustrated in some detail in Fig. 4.1. It should be also noted that a further increase of I from 2 to 3 does not produce anything new, since $I = 3$ corresponds to the largest distance of the atoms in the ring.

Stollhoff and Fulde also investigated the accuracy of a simplified version of Eq. (4.18) obtained by linearization,

$$|\Phi\rangle = (1 - \sum_{i=0}^{I} \eta_i \sum_j O_{ij})|\Phi_0\rangle, \tag{4.23}$$

in order to reduce computational complexity. With this simplified variational ansatz they found that at the equilibrium distance about 90% of the correlation energy was still recovered with $I = 2$. However the agreement with the exact results becomes worse at larger distances, in contrast with the systematic improvement found with the full variational wave function of Eq. (4.18).

It should be mentioned that in general using the linearized wave function of Eq. (4.23) to calculate the correlation energy variationally leads to the so-called size consistency problem. In practice the correlation energy turns out not to be proportional to the electron number, i.e., it is not extensive as it should be in the limit of a large system. A way to restore the correct number dependence is to expand the variational ground-state energy of Eq. (4.9) to a given order in the variational parameters, say the second, starting from the full wave function of Eq. (4.18). In the case of small molecules, the difference that one finds in correlation energy with respect to the use of the linearized wave function in Eq. (4.23) is of only a few percent (Stollhoff and Fulde, 1980). Another possible way to tackle the size consistency problem (see also Appendix 2.3) starts from rewriting the product of projection operators appearing in Eq. (4.18) in an exponential form. This makes it possible to prove a linked cluster theorem (see, for instance, Horsch and Fulde, 1979; also March *et al.*, 1995), which in turn allows a systematic expansion of the correlation energy in terms of connected diagrams. Again, the correct number dependence of the energy is automatically preserved to any order in the expansion in powers of the variational parameters. It should also be mentioned that spin–spin correlation can be dealt with as well within the local approach. To this end one can just enlarge the class of projection operators appearing in Eq. (4.18) by defining further operators $O_{ij} = \mathbf{s}_i \cdot \mathbf{s}_j$, where $\mathbf{s}_i$ is the spin operator at the site i (Stollhoff and Fulde, 1978).

4.3.2. Incorporation of Intraatomic Correlations in Small Atoms and Molecules

Clearly, in the study of the H_6-model only interatomic correlations could be taken into account, and admittedly within an oversimplified model. Therefore, for a more stringent test, Stollhoff and Fulde (1978, 1980) have then studied, within the simplified variational ansatz of Eq. (4.23), some small atoms and molecules, also taking into account intra-atomic correlations. To this end, the radial correlation was treated by considering basis set functions of s type with different degrees of localization. On the other hand, angular correlations were introduced by using appropriate hybrid functions. In particular, s–p mixing was used to construct sets of tetragonal hybrids, and d and f functions were used to obtain hybrids with hexagonal and octagonal symmetry, respectively. Within their simple scheme they were able to recover 93% of the experimental correlation energy for He and 90% for H_2. Thus, starting from the Hartree–Fock uncorrelated ground state, in both cases the results of the local approach based on a Gutzwiller-like ansatz yielded a fraction of the correlation energy very close (a few percent of difference) to the one obtained in configuration interaction (CI) calculations with the same basis sets. Similar results were also obtained for He_2 and Be.

A very detailed study of Ne and CH_4 along these lines (Stollhoff and Fulde, 1980) shows that in more complex situations the local approach is still capable of accounting for the correlation energy obtainable with CI calculations, within a few percent. One should note, though, that the agreement with experiment of the correlation energy yielded by CI calculations with reasonable basis sets tends to worsen with increasing complexity of the system studied.

More recently, Oles *et al.* (1986) have proposed another computational scheme for the treatment of correlations in more complex molecules containing C, N, and H atoms. While the interatomic correlations are still treated within a local approach based on semiempirical self-consistent field calculations, intra-atomic correlations are dealt with by means of a different and simpler scheme. Good agreement is found with experimental results, discrepancies being within a few percent. They have also shown that simple algebraic parametrizations are possible for the various contributions to the correlation energy. In particular, interatomic correlations are found to depend only on bond lengths, whereas intra-atomic correlations are found to be determined for a given atom by its total charge and fraction of p electrons. A further study, on the determination of optimal local functions for the calculation of the correlation energy within the local approach, has also subsequently appeared (Dieterich and Fulde, 1987).

The foregoing has been concerned with the application of a local approach to the calculation of correlation in molecules. However, it is relevant to mention here an important development along these lines in extended systems. The above local approach, in fact, has been suitably modified (Horsch and Fulde, 1979) to treat short-range and long-range correlations in the ground state of solids as well. A first attempt to calculate in this manner the correlation contribution to the ground state energy of diamond (Kiel *et al.*, 1982) suffered the limitation of a poorly converged uncorrelated ground state. When an uncorrelated ground state of good quality is used, however, the local approach turns out to perform remarkably well, reproducing the electronic contribution to the binding energy of diamond with an accuracy of about 2% (Stollhoff and Bohnen, 1988).

4.3.3. Gutzwiller Variational Treatment of the Hubbard Model

As was mentioned above, the Gutzwiller variational method was originally applied to the study of the so-called Hubbard model, characterized by the Hamiltonian of Eq. (4.10). In spite of the apparent simplicity of such a Hamiltonian, the relevant variational calculation is of considerable difficulty. Therefore, Gutzwiller solved the problem in an approximate manner. It was twenty years later that renewed interest in this problem has led to work on the exact solution of the variational problem, by direct numerical evaluation, by analytical means, or by the Monte Carlo method.

Kaplan, Horsch, and Fulde began in 1982 a study to assess the accuracy of the Gutzwiller variational wave function (GVW) in describing the ground state of the single-band Hubbard model with only nearest-neighbor hopping, in the atomic limit (i.e., the limit in which $U \to \infty$). In this case there is no variation to be taken. In fact, $g = 0$, and the Gutzwiller wavefunction reads

$$|\Phi_\infty\rangle = \prod_i [1 - n_{i\uparrow} n_{i\downarrow}] |\Phi_0\rangle. \tag{4.24}$$

They were able to perform, for the half-filled-band situation, the direct numerical evaluation of the spin–spin correlation function

$$q_1 = \langle s_i^z s_{i+1}^z \rangle \tag{4.25}$$

for a number of small regular rings. The actual calculations were done for rings of 6, 10, 14, and 18 sites. By extrapolating from the results for finite rings, they found that in the thermodynamic limit (number of sites N going to infinity), $q_1 = -0.1474$, within 0.2% from the exact result—i.e., within the error bars of their calculation. For q_2, instead, the discrepancy with the exact result was about 7%. They also noticed that the spin–spin correlations they obtained also reproduced qualitative features of the exact ones. Hence, for large U at least in one dimension, Gutzwiller's wave function describes the short-range spin–spin correlations accurately. It should be noted here that, in the atomic limit and for the half-filled-band case, the exact ground state of the one-dimensional Hubbard model is known to be the same as for the antiferromagnetic Heisenberg chain, for which q_1 was exactly evaluated by Bonner and Fisher (1964) and q_2 by Takahashi (1977). In fact, in the large U limit, the Hubbard Hamiltonian goes, to leading order in t/U, into the Heisenberg Hamiltonian with antiferromagnetic coupling,

$$H = J \sum_{\langle ij \rangle} (\mathbf{s}_i \cdot \mathbf{s}_j - \tfrac{1}{4}), \tag{4.26}$$

where J is proportional to t^2/U and t is the hopping energy. This point will be taken up later in Section 4.4.2 in some detail.

Kaplan, Horsch, and Fulde also studied the energy for large but finite U. In this case one has to start from the full variational wave function of Eq. (4.8) and consider an expansion of the energy in powers of the parameter g, to be determined variationally. They found that the leading term of such an expansion was of the form $E = -N\alpha t^2/U$, with α depending on the kinetic energy operator and on the zero- and first-order wave functions obtained from the expansion of the variational wave function of Eq. (4.8) in powers of g. Note that the zero-order wave function, which was given in Eq. (4.24), has no sites with double occupancy, as is clear by inspection. Similarly, the first-order wave function has, on average, only one site doubly occupied. They found that, at variance with the spin–spin correlations, the coefficient α yielded by the GVW was in gross disagreement with the known exact value. It is now known, from the exact solution of the Hubbard model in one-dimension with the GVW (Metzner and Vollhardt, 1987), that α is not a constant but vanishes as $1/\ln(U/t)$ in the limit considered by Kaplan, Horsch, and Fulde. These authors argued that the unsatisfactory result for the energy yielded by the GVW was due to the incorrect description of correlations between doubly occupied sites and empty sites, or holes, as they will be termed in the present context. Therefore, Kaplan *et al.* considered a modification of GVW containing a second variational parameter to improve the treatment of such correlations. In this manner, they were able to reproduce the exact value of α to within 1%.

4.3.4. Open Shell Systems

A further step toward the assessment of the reliability of the GVW was then taken by Horsch and Kaplan (1983). They extended the calculation for finite rings to other values of N, corresponding to open shell systems, and performed calculations for much larger systems with N up to 100, by using the Monte Carlo method to evaluate the relevant averages. This second step was particularly important. First, it fully confirmed the conclusions previously obtained from the exact calculations for small rings. Second, it was the first application of the Monte Carlo method to this problem. Before turning to the presentation of later studies on the Hubbard model, with a variational Monte Carlo technique based on the GVW, the exact results that have been obtained in one dimension for arbitrary band filling and interaction strengths will be summarized.

4.4. EXACT ANALYTIC RESULTS IN ONE DIMENSION

Analytic results for the ground-state energy and momentum distribution function for the Hubbard model with onsite repulsion have been obtained by Metzner and Vollhardt (1987) for the paramagnetic situation. The key point of their attack on the problem lies in a novel approach to the calculation of the expectation values of relevant operators on the GVW.

The calculation of the ground-state energy appears to require the evaluation of the expectation values of both kinetic and potential energy for arbitrary values of the variational parameter g, $0 \leq g \leq 1$. Let us start from the potential energy. It is clear from Eq. (4.10) that, apart from the proportionality constant U, the potential energy operator is the same as the operator D that counts the number of doubly occupied sites. Thus, one has to calculate $\langle D \rangle$ on the wave function of Eq. (4.8). Let us indicate the product $n_{i\uparrow} n_{i\downarrow}$ by D_i. It is then not difficult to show that

$$d = \frac{\langle D \rangle}{N} = g^2 \sum_{m-1}^{\infty} (g^2 - 1)^{m-1} c_m, \tag{4.27}$$

where N is the number of sites and the coefficients $c_m = x_m/N(m-1)!$ are suitable expectation values, which, if one puts $x_m = y_m/\langle \Phi | \Phi \rangle$, are explicitly given by

$$y_m = \sum_{f_1,\ldots,f_m}' \langle \Phi_0 | D_{f_1} \ldots D_{f_m} | \Phi_0 \rangle, \qquad m \leq N, \tag{4.28}$$

with $f_i \neq f_j$. The expectation value appearing above can be transformed into the sum over all the possible pairs of contractions, $P_{ij} \equiv \langle a^\dagger_{i\sigma} a_{j\sigma} \rangle_0$, using Wick's theorem. Such a sum will be denoted by $\{\ldots\}_0$. The important point to realize is that, because of the prime on the sum in Eq. (4.28), one can choose to put $\langle a_{i\sigma} a^\dagger_{j\sigma} \rangle \equiv -P_{ji}$, since in this second kind of contraction one has always $i \neq j$ and therefore there is no δ_{ij} contribution. Equation (4.28) can be rewritten then as

$$y_m = \sum_{f_1,\ldots,f_m}' \{D_{f_1} \ldots D_{f_m}\}_0. \tag{4.29}$$

One can show that the sum $\{\ldots\}_0$ appearing above is in fact the determinant of a matrix having P_{ij} as elements. Hence, for any $f_i = f_j$, the sum $\{\ldots\}_0$ vanishes, since the relative determinant has two identical rows. As a result one has the freedom to remove the prime from the sum in Eq. (4.29). Having done so, one can next observe that in a diagrammatic analysis of x_m with lines corresponding to the P_{ij} the contributions to $\{\ldots\}_0$ arising from disconnected diagrams just cancel the norm $\langle\Phi|\Phi\rangle$. Therefore, one is left with (see also Senatore and March, 1994)

$$x_m = \sum_{f_1,\ldots,f_m}' \{D_{f_1} \ldots D_{f_m}\}_0^c. \tag{4.30}$$

It should be noted that thus far no reference has been made to interaction strength or to dimensionality.

While the above analysis was quite general, to date further progress has proved possible only in one dimension. In particular, in one dimension and for zero total spin ($n_\uparrow = n_\downarrow = n/2$), it is possible to show that c_m is proportional to n^{m+1}. One can restrict attention to $0 \leq n \leq 1$, since because of particle–hole symmetry, for $1 \leq n \leq 2$ the relation $d(n) = d(2 - n) + n - 1$ holds. By using the dependence of d on n and imposing the continuity of its first derivative, the proportionality constant in c_m can be calculated. One obtains

$$c_m = \tfrac{1}{2}(-1)^{m+1} n^{m+1}/(m + 1). \tag{4.31}$$

With the above expression for the coefficients c_m, the series in Eq. (4.27) may be exactly summed to

$$d = \frac{n^2}{2}\left[\frac{g}{1 - G^2}\right]^2 \left[\ln \frac{1}{G^2} + G^2 - 1\right], \tag{4.32}$$

with $G^2 = 1 - n + ng^2$. We stress that, as is clear from Eq. (4.32), for the half-filled band the correlation energy Ud is found to be nonanalytic in g because of the presence of the term $\ln(1/g)$. For strong correlations ($g \to 0$) one finds $d = g^2\ln(1/g)$. Moreover, the double occupancy d is never vanishing, for finite correlations, in contrast with the GA result of Eq. (4.15). This seems to be the case also in two and three dimensions, when the GVW is exactly handled (Yokoyama and Shiba, 1987a).

A similar though more complicated analysis permits the calculation of the momentum distribution function for the same situation considered immediately above, i.e., $n_\uparrow = n_\downarrow = n/2$. It will suffice to simply quote the result for the discontinuity q in the momentum distribution function $n_{k\sigma}$:

$$q = G^{-1}[(G + g)/(1 + g)]^2. \tag{4.33}$$

The overall shape of $n_{k\sigma}$ for half filling and at different values of g is illustrated

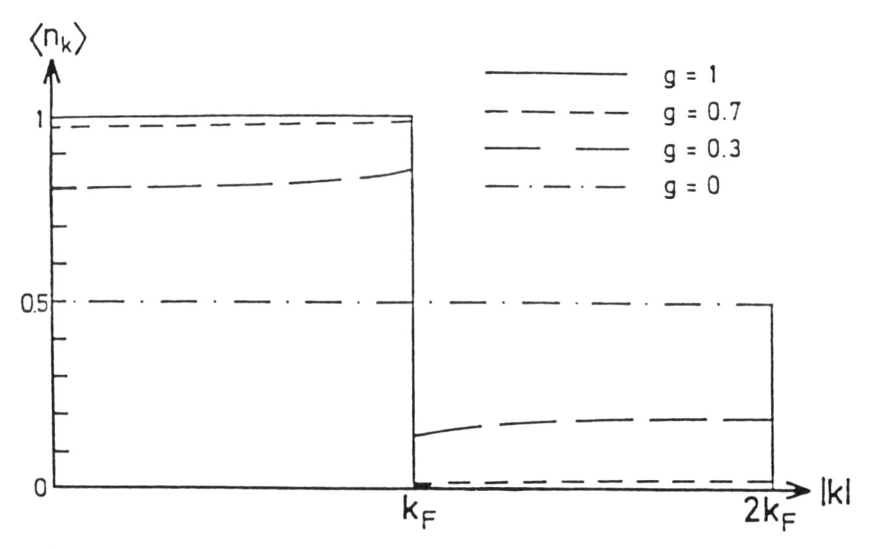

FIGURE 4.2. The momentum distribution $<n_k>$ for the Hubbard model. The results obtained by using the Gutzwiller variational wave function are shown for several values of the correlation parameter g in the case of a half-filled band ($n_\uparrow = n_\downarrow = \frac{1}{2}$). (See Metzner and Vollhardt, 1987.)

in Fig. 4.2. It should be noticed that for $n = 1$ the above expression reduces to $q = 4g/(1 + g)^2$, which is the result of Gutzwiller's approximation (GA). This, however, does not imply that, for this particular value of filling, the kinetic energy coincides with that of GA, since g has to be determined by minimizing the total energy and the expressions for the potential energy are different in the two cases.

By combining Eq. (4.32) with the result for the momentum distribution function one may calculate $E_0(g)$ in one dimension for the paramagnetic ground state. Minimization with respect to g yields the energy for given values of t, U, n. For $n = 1$, the comparison with the result of Gutzwiller's approximate treatment and the exact result of Lieb and Wu (1968; see also Chapter 7) shows that the GVW when exactly handled gives an energy which is intermediate between the other two. This is illustrated in Fig. 4.3. In particular, specializing to the case $n = 1$ and $U/t \to \infty$ with only nearest-neighbor hopping, one finds $\bar{\varepsilon}_0 = -4t/\pi$ and

$$E/N = -(4/\pi)^2)(t^2/U)(\ln \bar{U})^{-1}, \tag{4.34}$$

where $\bar{U} = U/|\bar{\varepsilon}_0|$. It should be noted, with reference to the numerical results of Kaplan *et al.* (1987) discussed above, that the energy is certainly proportional to t^2/U, but the proportionality factor goes logarithmically to zero with U/t. Thus, for the half-filled-band case the GVW gives a result which is qualitatively different from the exact result. It has also to be mentioned that from Eq. (4.33) it is clear

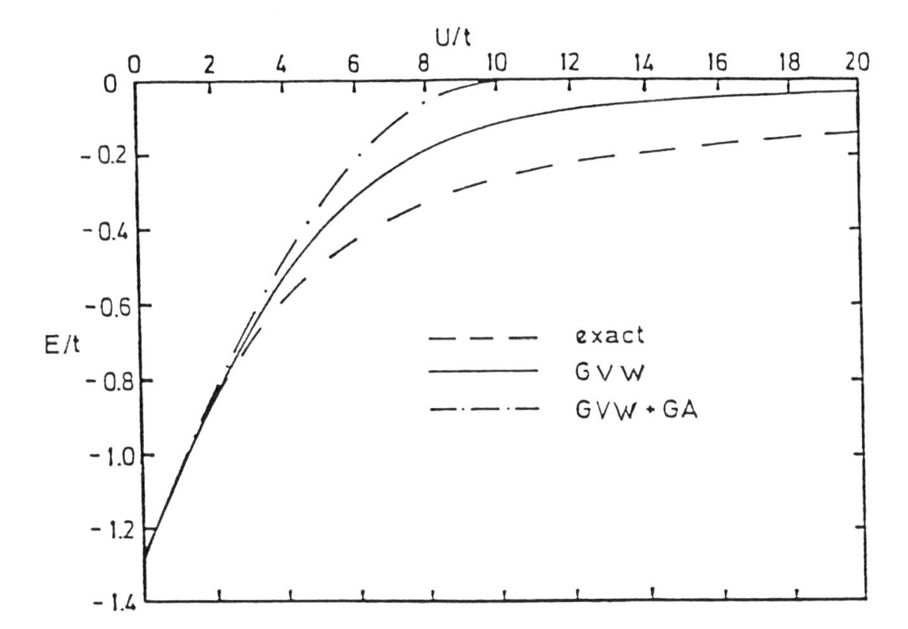

FIGURE 4.3. The ground-state energy E for the one-dimensional Hubbard model with $n_\uparrow = n_\downarrow = \frac{1}{2}$ as a function of U. The results for E, as calculated with the Gutzwiller variational wave function (GVW), are compared with the result of the Gutzwiller approximation (GA). (Metzner and Vollhardt, 1987: reproduced from Senatore and March, 1994)

that a discontinuity in the momentum distribution remains at any finite U, whereas the exact solution of the Hubbard model of Lieb and Wu (1968) gives an insulating system for any nonzero U (see also Ferraz *et al.*, 1978). It would seem from these considerations that the GVW is not a particularly good ansatz for the ground state of the Hubbard model. However, it has already been seen from the results of Kaplan *et al.* that it performs much better in the calculation of spin–spin correlations than for the energy.

4.4.1. Correlation Functions

Using techniques similar to those employed in the energy calculations, Gebhard and Vollhardt (1987) have calculated, for the paramagnetic ground state, a number of correlation functions,

$$C_j^{XY} = N^{-1} \sum_i \langle X_i Y_{i+j} \rangle - \langle X \rangle \langle Y \rangle, \tag{4.35}$$

where X_i, $Y_i = S_i^z$, n_i, D_i, H_i. Here, $n_i = n_{i\uparrow} + n_{i\downarrow}$, and $H_i = (1 - n_{i\uparrow})(1 - n_{i\downarrow})$ is the number operator for the holes. Also $X = N^{-1} \Sigma_i X_i$, and the Fourier transforms of the functions defined in Eq. (4.35) are denoted simply by $C^{XY}(q)$. The results for the spin–spin correlation function $q_1 \equiv C_1^{SS}$ obtained by Gebhard and Vollhardt fully confirm the numerical calculations of q_1 by Kaplan *et al.* However with the analytical solution it is possible to establish that there is in fact a difference of 0.2% between the exact and GVW result in the atomic limit and that such a difference is not due to numerical uncertainties. In addition, the features of the exact q_1 found in numerical investigation (Betsuyaku and Yokota, 1986; Kaplan *et al.*, 1987) on the Heisenberg antiferromagnetic chain are reasonably reproduced, though the agreement worsens at larger distances, as already suggested by Kaplan *et al.* (1982). The comparison between the GVW result and the exact result for the hole–hole correlation function $C^{HH}(q)$ in the atomic limit also shows an overall agreement, which tends to improve as the number of holes tends to zero. Finally, it should be noted that Gebhard and Vollhardt also comment on the correlations between holes and doubly occupied sites. Their conclusion is that this kind of correlation is not described particularly well by the GVW and that this might well be the reason for the logarithmic singularity found in the energy for $g \rightarrow 0$.

4.4.2. Numerical Results including Higher Dimensions

A number of numerical investigations on the Hubbard model with the GVW have been carried out with the help of the Monte Carlo method. One should distinguish between two kinds of investigations. In one case the original Gutzwiller program is implemented; i.e., ground-state properties of the Hubbard Hamiltonian are variationally calculated with the wave function of Eq. (4.8) for arbitrary n, t, U, and dimensionality. In the other case, following the observation of Kaplan *et al.* (1982) that the GVW for $g = 0$ gives an accurate description of spin correlation in the Heisenberg antiferromagnetic chain, one works on the effective Hamiltonian to which the Hubbard Hamiltonian reduces in the limit $U \rightarrow \infty$, thus restricting consideration to the strong-coupling situation.

Yokoyama and Shiba (1987a) have performed a systematic study on the Hubbard model, following the first of the abovementioned approaches. Thus, they have performed calculations in one, two, and three dimensions over the full range of coupling U/t and for both half-filled and non-half-filled bands. However, for technical details of their variational Monte Carlo technique, with which they performed calculations with up to 216 sites (6×6×6 in three dimensions) the original paper should be consulted. Below, a brief review of their main results will be presented:

In one dimension and for $n = 1$, they have calculated E_{kin}/t and $d = E_{\text{pot}}/U$ as functions of g between 0 and 1. For a given value of U one can then construct from such functions the total energy as a function of g and find its minimum by inspection. They have also calculated the momentum distribution. It is merely noted here that in this case the exact solution is available and that, according to

Metzner and Vollhardt (1987), the agreement with their analytic results is excellent. In the limit in which $U \to \infty$, however, they have not been able to isolate the logarithmic correction to the quadratic dependence t^2/U of the energy, but it would have been surprising otherwise. Similar calculations were performed for the test case $n = 0.84$, taken as representative of a less-than-half-filled band. The results for the energy are in fair agreement with the exact result, even if they observe that improvement on the GVW is called for if one wishes to obtain better agreement. The need to take into account intersite correlation is also mentioned, a point that has already been stressed by Stollhoff and Fulde (1977), as discussed at some length above.

In two dimensions, only the case $n = 1$ on a square lattice was considered. They found, for the total energy, that the variational results and those of the Gutzwiller approximation are much closer then they were in one dimension, as can be seen by comparing the two-dimensional case reported in Fig. 4.4 with the one-dimensional case already illustrated in Fig. 4.3. This is in accord with the expectation that GA should become more accurate in higher dimensions. However, comparison with the Hartree–Fock antiferromagnetic energy shows that the

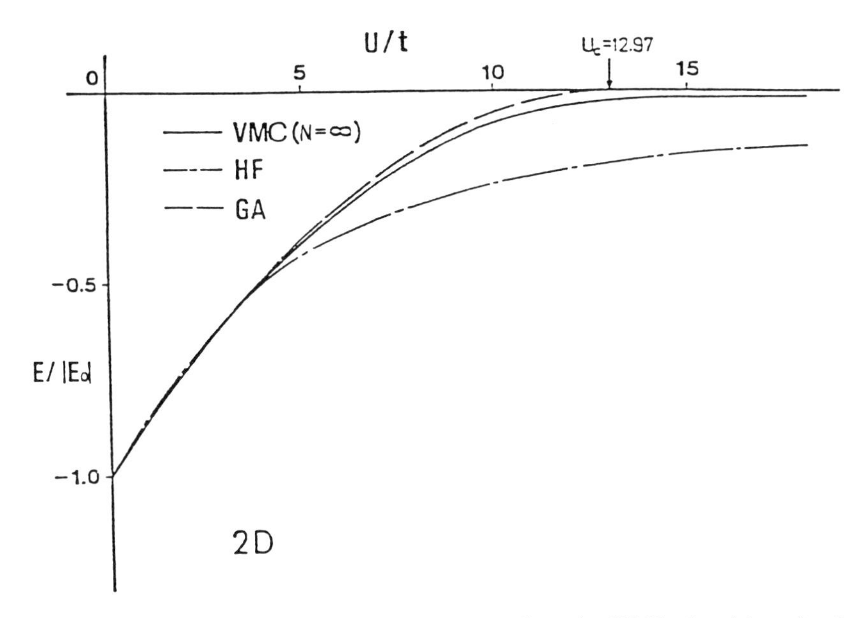

FIGURE 4.4. Normalized ground-state energy for the two-dimensional Hubbard model as a function of U: VMC, Variational Monte Carlo with the Gutzwiller variational wave function (GVW) (extrapolated to an infinite number of sites, $N = \infty$); HF, antiferromagnetic Hartree–Fock; GA, Gutzwiller approximation. The half-filled-band case ($n_\uparrow = n_\downarrow = \frac{1}{2}$) is shown. (Yokoyama and Shiba, 1987a: redrawn from Senatore and March, 1994).

discrepancies with this are sizeable for large U whereas they were much smaller in one dimension.

The calculations in three dimensions were performed for a cubic lattice and again for a half-filled band. The results for the total energy as a function of the interaction strength U/t are similar to those found in two dimensions. Also, in going from two to three dimensions, changes qualitatively similar to those observed in going from one to two dimensions are apparent. Thus the agreement between variational and approximate treatments (i.e., with the GA) improve further, while for large U the discrepancy with the antiferromagnetic Hartree–Fock energy, which is lower, remains sizable. This suggests that, in two and three dimensions, one should perform the variational calculations using a GVW based on the Hartree–Fock ground state of the antiferromagnetic system rather than the paramagnetic one.

To introduce the second of the approaches mentioned above, the derivation of the effective Hamiltonian that can be obtained from the Hubbard Hamiltonian for strong couplings U/t will be briefly summarized. The account follows the very straightforward method given by Gros *et al.* (1987a), but reference should also be made to Castellani *et al.* (1979) and to Hirsch (1985). One can start from the Hubbard Hamiltonian given in Eq. (4.10) and rearrange the kinetic energy T, specializing to the case of nearest-neighbor hopping, to find

$$H = T_h + T_d + T_{\text{mix}} + V, \tag{4.36}$$

with

$$T_h = -t \sum_{\langle ij \rangle,\sigma} (1 - n_{i,-\sigma}) a^\dagger_{i\sigma} a_{j\sigma} (1 - n_{j,-\sigma}), \tag{4.37}$$

$$T_d = -t \sum_{\langle ij \rangle,\sigma} n_{i,-\sigma} a^\dagger_{i\sigma} a_{j\sigma} n_{j,-\sigma}), \tag{4.38}$$

$$T_{\text{mix}} = -t \sum_{\langle ij \rangle,\sigma} n_{i,-\sigma} a^\dagger_{i\sigma} a_{j\sigma} (1 - n_{j,-\sigma}) - t \sum_{\langle ij \rangle,\sigma} (1 - n_{i,-\sigma}) a^\dagger_{i\sigma} a_{j\sigma} n_{j,-\sigma}, \tag{4.39}$$

where T_h and T_d are the kinetic energies describing the propagation of holes and doubly occupied sites, respectively, in their Hubbard bands, and T_{mix} is clearly a mixing term which couples such bands. V is of course the usual onsite repulsion energy, $V = U\Sigma_i n_{i\uparrow} n_{i\downarrow}$. The next step is to apply to the Hamiltonian a suitable unitary transformation

$$H_{\text{eff}} = e^{iS} H e^{-iS} = H + i[S, H] + \ldots \tag{4.40}$$

such that, in lowest order in t/U, T_{mix} vanishes on the right-hand side of Eq. (4.40). This can be accomplished by choosing S such that $i[S, T_h + T_d + V] = -T_{\text{mix}}$, i.e.,

$$S = \sum_{n,m} |n\rangle \frac{\langle n | T_{\text{mix}} | m \rangle}{i(\varepsilon_n - \varepsilon_m)} \langle m|, \tag{4.41}$$

with $|n\rangle$ and $|m\rangle$ being eigenstates of $T_h + T_d + V$. Even if these eigenstates are not known in the general case, for very large U it must be that $\varepsilon_n - \varepsilon_m = \pm U + 0(t)$. Thus one gets

$$S = -\frac{it}{U}\sum_{\langle ij\rangle,\sigma} n_{i,-\sigma}a^{\dagger}_{i\sigma}a^{\dagger}_{j\sigma}(1 - n_{j,-\sigma}) + \frac{it}{U}\sum_{\langle ij\rangle,\sigma}(1 - n_{i,-\sigma})a^{\dagger}_{i\sigma}a_{j\sigma}n_{j,-\sigma}. \tag{4.42}$$

If, as is the case here, one is interested in taking matrix elements of H_{eff} between states with no doubly occupied sites, such as the infinite-U Gutzwiller wave function of Eq. (4.24), it can easily be shown that H_{eff} can be written as

$$H_{\text{eff}} = T_h + i[S, T_{\text{mix}}] + \mathbf{SVS}$$

$$= T_h + \frac{2t^2}{U}\sum_i\sum_{\tau,\tau'}[(a^{\dagger}_{i+\tau}\mathbf{s}\,a_{i+\tau'})\cdot(a^{\dagger}_i\mathbf{s}\,a_i) - \frac{1}{4}(a^{\dagger}_{i+\tau}a_{i+\tau'})\cdot(a^{\dagger}_i a_i)]. \tag{4.43}$$

Above, $a_i\mathbf{s}a_j \equiv \Sigma_{\sigma\sigma'}a^{\dagger}_{i\sigma}(\mathbf{s})_{\sigma\sigma'}a_{j\sigma'}$, and $a_i a_j \equiv \Sigma_\sigma a^{\dagger}_{i\sigma}a_{j\sigma}$, with the vector $\mathbf{s}$ being the spin operator and the indexes τ and τ' running over the first neighbors of i. Two things should be noted: In Eq. (4.43), the term **SVS** has also been considered which would appear to be of higher order in **S**. But the expansion parameter is t/U, and such a term turns out to be of the same order of $i[\mathbf{S},T_{\text{mix}}]$. Also, if three-site terms (i.e., $\tau \neq \tau'$) can be neglected, then Eq. (4.43) simplifies to

$$H_{\text{eff}} = T_h + \frac{2t^2}{U}\sum_{\langle ij\rangle}(\mathbf{s}_i\cdot\mathbf{s}_j - \frac{1}{4}n_i n_j), \tag{4.44}$$

with $n_i = n_{i\uparrow} + n_{i\downarrow}$ being the site number operator. This is exactly true for the half-filled-band case, for which one also has $n_i = 1$. However, for $n \neq 1$, three-site contributions may become important, as has been discussed by Yokoyama and Shiba (1987b). It should be noticed that calculating averages of H_{eff} on $|\Phi_\infty\rangle$ is equivalent to calculating averages of H on a modified wavefunction $|\tilde{\Phi}_\infty\rangle = (1 - i\mathbf{S})|\,\Phi_\infty\rangle$, that is, $\langle|\Phi_\infty|H_{\text{eff}}|\,\Phi_\infty\rangle = \langle|\tilde{\Phi}_\infty|H|\tilde{\Phi}_\infty\rangle$, provided terms of order t^2/U^2 are neglected. In particular, the number of doubly occupied sites is nonzero on this wave function, or more precisely of order t^2/U^2. Notice that, for the half-filled band, if one puts $H_{\text{eff}} = T_h + V_{\text{corr}}$, then

$$D = \langle\tilde{\Phi}_\infty|\frac{1}{U}\,V|\tilde{\Phi}_\infty\rangle = -\frac{1}{2U}\langle\Phi_\infty|V_{\text{corr}}|\Phi_\infty\rangle. \tag{4.45}$$

Before turning to the numerical results that have been obtained by employing the strong-coupling Hamiltonian of Eqs. (4.43) or (4.44), one point should be stressed. As observed first by Kaplan *et al.* (1982), $|\Phi_\infty\rangle$, while giving a good description of the spin–spin correlations, if used to calculate the energy directly from the Hubbard Hamiltonian yields poor results. One should have clearly in mind therefore that $|\Phi_\infty\rangle$ is a good ground state for H_{eff} but not for H. Thus, the combined use of H_{eff} and $|\Phi_\infty\rangle$ is equivalent, as already noted, to working with the Hubbard Hamiltonian, but using an improved Gutzwiller-like wave function.

Using the approach mentioned above, i.e., working in terms of H_{eff} and $|\Phi_\infty\rangle$, Gros *et al.* (1987a) have performed an extensive Monte Carlo study of the Hubbard model for strong coupling. They restrict their investigation to one dimension to obtain good numerical accuracy, even if their interest is (naturally) in the three-dimensional situation. On the other hand, they consider variable band filling, magnetization, and degeneracy N_f of the local state. Their results for the spin–spin correlation function q_1, have established, before the exact solution with the GVW appeared (Gebhard and Vollhardt, 1987), that the small difference between the exact result for this quantity and that obtained from the infinite-U Gutzwiller wave function is beyond the numerical uncertainty of the Monte Carlo calculation. The investigation on the kinetic energy as function of the filling, on the other hand, shows good agreement with the exact result of both Monte Carlo and GA results. Gros *et al.* also discuss the fact that for large U the hole–hole correlation function can be put into correspondence with that of a system of spinless noninteracting fermions. This allows for an assessment of the quality of their results. They find fair agreement, even if some qualitative features of the exact result, related to the presence of a sharp Fermi surface, are missing. Also, an enhancement with respect to the exact result of correlations at short distances is found. Regarding the total energy, for large U and values of the filling very close to 1, the accuracy of the Monte Carlo result is dominated by the kinetic energy contribution, with an accuracy of about 6%. In fact, the potential energy, being determined by q_1, has a much better accuracy, i.e., 0.2%. Finally, they find that when excited-state Gutzwiller wave functions are considered the accuracy of GA is much reduced. It also has to be mentioned that, in the cases in which direct comparison with the analytical results of Metzner and Vollhardt (1987) and Gebhard and Vollhardt (1987) was possible, excellent agreement was found.

One of the reasons for the revived interest in the Hubbard Hamiltonian is its possible relevance in the understanding of the mechanism underlying high-temperature superconductivity, as suggested by Anderson (1987). In particular, investigations in the strong correlation regime are called for. The importance of the scheme briefly outlined above for dealing with the large-coupling situation is then apparent. An investigation in this direction has been made by Yokoyama and Shiba (1987b), within the effective Hamiltonian approach. They have used the effective Hamiltonian of Eq. (4.39) and have also considered both a paramagnetic and an antiferromagnetic ground state. This can be done by simply changing the type of uncorrelated wavefunction $|\Phi_0\rangle$from which the Gutzwiller wavefunction of Eq. (4.24) is built. One and two dimensions were considered. The purpose of the study was to characterize the competition between the two types of ground state. Before briefly summarizing their findings, it should be mentioned that they have also shown that the simple infinite-U Gutzwiller wave function constitutes one of

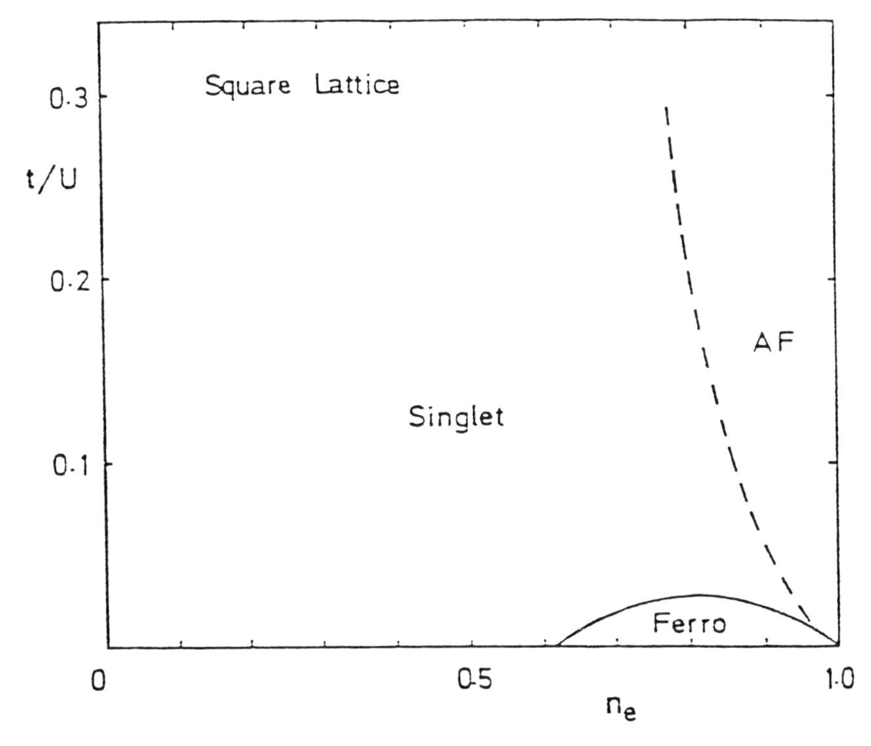

FIGURE 4.5. The phase diagram of the two-dimensional square-lattice Hubbard model, as inferred from VMC using a Gutzwiller-type wave function (Yokoyama and Shiba, 1987b). Here $n_e = n_\uparrow + n_\downarrow$ is the band filling.

the many possible realizations of Anderson's resonating-valence-bond state or singlet state. In one dimension, they find that for both the half-filled and the non-half-filled bands the singlet state is stable against the Néel (antiferromagnetic) state. The situation is quite different in two dimensions. For the half-filled case the singlet state is found to be unstable against the Néel state. Whereas the results for energy and magnetization are in reasonable accord with estimates obtained with different approaches, they question whether the same situation would be found if next-nearest hopping were to be included. Moving from the half-filled band (n=1), at very large U, the ferromagnetic state becomes favorable in energy. However, if n is decreased at some point ($n \cong 0.61$), the singlet state takes over again. On the other hand, if U is decreased the Néel state appears to again decrease in energy. From these results they draw a qualitative phase diagram as shown in Fig. 4.5. This should be considered meaningful only in the small t/U region, where the effective Hamiltonian approach is appropriate.

4.4.3. Cooper Pairing

Finally, the work of Gros *et al.* (1987b) should be mentioned here. By using the effective-Hamiltonian approach, they investigate the stability of generalized Gutzwiller wave functions against Cooper pairing, in the large-U limit and in two dimensions. This is done by evaluating the binding energy of two holes in the variational wave function. As they note, no real attempt was made to optimize their wave function. They find that the paramagnetic or singlet state is stable against s-wave pairing but unstable against d-wave pairing. On the other hand, the antiferromagnetic state is stable against both kinds of pairings. They discuss a possible pairing mechanism and the relevance of their results for high-T_c superconductors.

4.5. RESONATING VALENCE-BOND STATES

In the foregoing it has been seen how the Hubbard Hamiltonian—for large values of the coupling U—can be transformed into an antiferromagnetic Heisenberg Hamiltonian, by means of a suitable unitary transformation. In this connection and also in relation to the high T_c superconductivity, it seems appropriate here to briefly review the mathematical formulation due to Anderson (1973) of the concept of the resonating valence bond (RVB) states first put forward by Pauling (1949: see also Chapter 7, section 7.5 and Appendix A7.1)

In discussing the ground-state properties of the triangular two-dimensional Heisenberg antiferromagnet for $S = \frac{1}{2}$, Anderson (1973, see also Fazekas and Anderson, 1974) proposed that at least for this system, and perhaps also in other cases, the ground state might be the analog of the precise singlet in the Bethe (1931) solution of the linear antiferromagnetic chain. In fact, the zero-order energy of a state consisting purely of nearest-neighbor singlet pairs is more nearly realistic than that of the Néel state.

For electrons on a lattice, one can think of a singlet bond or pair as the state formed when two electrons with opposite spin are localized on two distinct sites. A resonating valence-bond state is a coherent superposition of such singlet bonds; its energy is further lowered as a result of the matrix elements connecting the different valence-bond configurations.

Heuristically, valence bonds can be viewed as real-space Cooper pairs that repel one another, a joint effect of the Pauli principle and the Coulomb interaction. When there is one electron per site, charge fluctuations are suppressed, leading to an insulating state. However, as one moves away from half filling, current can flow; the system becomes superconducting as the valence bonds Bose condense.

Anderson (1987), while stressing the difficulty of making quantitative calculations with RVB states, in fact gives a suggestive representation of them by exploiting the Gutzwiller-type projection technique. Clearly, a delocalized or mobile valence bond can be written as

$$b_\tau^\dagger|\Phi_0\rangle = \frac{1}{\sqrt{N}}\left(\sum_i a_{i\uparrow}^\dagger a_{i+\tau\downarrow}^\dagger\right)|\Phi_0\rangle = \frac{1}{\sqrt{N}}\left(\sum_k a_{k\uparrow}^\dagger a_{-k\downarrow}^\dagger \exp i(k\cdot\tau)\right)|\Phi_0\rangle \quad (4.46)$$

where $b_\tau^\dagger$ is the creation operator for a valence-bond state with lattice vector τ, $a_{i\sigma}^\dagger$ the single-electron creation operator, and N is the total number of sites. A distribution of bond lengths can be obtained by summing $b_\tau^\dagger$ over τ with appropriate weights. One then gets a new creation operator,

$$b^\dagger = \sum_k c_k a_{k\uparrow}^\dagger a_{-k\downarrow}^\dagger, \quad (4.47)$$

with the restriction

$$\sum_k c_k = 0 \quad (4.48)$$

if double occupancy is to be avoided.

Anderson proceeds by (i) Bose condensating such mobile valence bonds

$$|\Phi\rangle = (b^\dagger)^{N/2}|\Phi_0\rangle \quad (4.49)$$

and (ii) projecting out the double occupancy—which would be otherwise present—with an infinite-U Gutzwiller projection operator $P_d = \prod_i(1 - n_{i\uparrow}n_{i\downarrow})$,

$$|\Phi_{\mathrm{RVB}}\rangle = P_d(b^\dagger)^{N/2}|\Phi_0\rangle. \quad (4.50)$$

One can then show that the RVB state written above can be obtained with simple manipulations from a standard BCS state (see March *et al.*, 1995), by projecting on the state with N particles and projecting out, at the same time, the double occupancy (see also Appendix A7.1):

$$|\Phi_{\mathrm{RVB}}\rangle = P_{N/2}P_d\left[\frac{1}{\sqrt{1+c_k}} + \frac{c_k}{\sqrt{1+c_k}}\, a_{k\uparrow}^\dagger a_{-k\downarrow}^\dagger\right]|\Phi_0\rangle. \quad (4.51)$$

Baskaran *et al.* (1987) have subsequently shown that by treating the large-U Hamiltonian of Eq. (4.44) with a mean field (Hartree–Fock) approximation, one obtains precisely a BCS-type Hamiltonian, which—for half filling—yields the RVB state heuristically introduced above. In fact, one also finds that $\Sigma_k c_k = 0$, $|c_k| = 1$, and the c_k change sign across what they call a *pseudofermi* surface. The nature of excitations from such an RVB state will not be dealt with here. The reader is referred instead to the original papers.

4.6. CRITERION FOR LOCAL MOMENT FORMATION AND ITS UTILITY IN STUDYING THE FERROMAGNETISM OF TRANSITION METALS

Stoddart and March (1971) used spin-density functional theory in the context of the Hubbard–Gutzwiller approach to derive a criterion for local moment formation, which has proved useful in the discussions of cooperative magnetism in transition metals given by Hasegawa (1979; 1980a,b), Hubbard (1979a,b), and Roth (1978).

The argument of Stoddart and March is given in some detail in Appendix A3.3; here only the gist of it will be given. Essentially, the Koster–Slater treatment of nonmagnetic impurities is generalized, using spin-density functional theory, to derive a condition for the stability of a local moment.

The idea was to use the above basic formulation in terms of one-body potential theory and to look more closely at the Hartree–Fock problem in order to arrive at a description of correlation effects. In particular, it is useful to think in terms of a probe inserted into a nonmagnetic metal as to scatter electrons of + and − spin selectively.

It is helpful in motivating the model used by Stoddart and March to refer to the treatment of localized moments in metals given by Wolff (1961) following the pioneering work of Friedel (1958) and Anderson (1961). In Wolff's treatment, it turns out that a nontrivial solution remains when one works out self-consistently the response of electrons to an impurity center and then switches off the spin-independent part of the impurity potential at the end of the calculation. The fact that, under some circumstances, the original paramagnetic state is unstable to the formation of a localized imbalance of spin density led Stoddart and March to the derivation of the spin-density functional criterion for local moment formation referred to above, namely,

$$2U\int_{-\infty}^{E_f} dE\, P(E)n(E) \leq -1. \tag{4.52}$$

We recall that U is the repulsive interaction between two electrons of opposite spin in the same atom, while $P(E)$ is the Hilbert transform of the nonmagnetic density of states $n(E)$. Equation (4.52) is different from the Stoner criterion (A3.1.6) for the existence of a ferromagnetic ground state.

For the rectangular density of states

$$\left.\begin{aligned} n(E) &= 1/W \qquad && 0 < E < W \\ &= 0 && \text{otherwise,} \end{aligned}\right\} \tag{4.53}$$

Stoddart and March (1971) find

$$2U\int_{\infty}^{E_f} P(E)n(E)dE$$

$$= \frac{2U}{W}\left[\left(1 - \frac{E_f}{W}\right)\ln\left(1 - \frac{E_f}{W}\right) + \frac{E_f}{W}\ln\frac{E_f}{W}\right]. \tag{4.54}$$

The presence of local moments above the temperature at which ferromagnetic order breaks down is therefore expected, since in many cases the criterion for the formation of localized spins is weaker than that for long-range magnetic order.

Hasegawa (1979) and Hubbard (1979a,b) have used a tight-binding impurity theory similar to that developed by Stoddart and March (1971) to calculate the energy of reversal of a single spin in the lattice. The energy of spin reversal is found to be much less than the energy of local spin formation, in agreement with the small observed values of the Curie temperature T_c. However, these theories neglect fluctuations in the size of the moment. This point is taken up below when reference is made to the work of Sayers (1982), but it is relevant to mention here that the criterion obtained by Stoddart and March was also derived later by Roth (1978) using the coherent potential approximation. Subsequently, Heine *et al.* (1981) and You and Heine (1982) have examined the consequences of the criterion [Eq. (4.52)] using band theory data for a variety of transition metals; the interested reader is referred to their papers for the detailed results.

Referring to the work of Sayers (1982), one notes again that the theories of Hasegawa and Hubbard neglect, in the lowest order, fluctuations in the size of the local moment. However, as Sayers (1982) points out, even below the magnetic ordering temperature there is evidence for rapid spin fluctuations. In α-Mn, for example, the four crystallographic sites have moments of 0.2, 0.6, 1.7, and 1.9 μ_B (Yamada *et al.*, 1970). However X-ray photoelectron spectroscopy reveals a multiplet splitting of the 3*s* and 2*s* electron levels in paramagnetic α-Mn, corresponding to a localized moment of 2.5μ_B on all Mn atoms on a time scale of 10^{-15}sec (McFeely *et al.*, 1974; Nagasawa *et al.*, 1980).

In the work of Sayers, the energy of the ferromagnetic state is compared with that of the nonmagnetic state in which the formation of local moments on such time scales is and is not allowed. He demonstrates that the existence of such moments in the nonmagnetic state leads to a considerable reduction in the energy difference between the nonmagnetic and the magnetically ordered state. The analysis of Sayers is carried out for the case of a single nondegenerate band by means of Gutzwiller's treatment of electron correlation. Although only the difference in energy between the nonmagnetic and ferromagnetically ordered states is considered, the major contribution to this is the intra-atomic spin-formation energy and not the interatomic spin coupling. The conclusions are therefore expected to apply also to the energy difference between the nonmagnetic state and states of more general magnetic order.

4.6.1. Local Moment Formation in the Nonmagnetic State

Consider a single nondegenerate band with n electrons per atom, and as usual let the repulsive interaction between two electrons of opposite spin on the same site be U. In its simplest form, a spin fluctuation may be considered as an attractive interaction between an electron and a hole of opposite spin, which acts to stabilize a local moment on the atom and to reduce the number of doubly occupied sites in the one-band model. The effect is included in the one-band model of Gutzwiller (1965).

Let d_0 denote the average number of doubly occupied sites in the nonmagnetic state and d the optimum number of doubly occupied sites. A value of d smaller than the band value $n^2/4$ corresponds to the formation of local moments. For small U/W (Sayers, 1977), W being the bandwidth, Gutzwiller's method gives

$$d = \frac{n^2}{4} + n^2(1 - n/2)^3 \frac{U}{4\bar{\varepsilon}_0} + O(U^2), \tag{4.55}$$

where $\bar{\varepsilon}_0$ is the average energy per electron in the band.

For the rectangular density of states one has

$$n(E) = \frac{1}{2W}, \qquad -W \leq E \leq W$$

$$= 0 \qquad \text{otherwise}$$

$$\bar{\varepsilon}_0 = -W\left(1 - \frac{n}{2}\right). \tag{4.56}$$

The extent to which this local moment formation lowers the nonmagnetic energy and hence the energy difference between the nonmagnetic and ferromagnetically ordered states is then investigated by Sayers, to whose paper the reader is referred for details. Although his discussion is limited to a comparison between the nonmagnetic and ferromagnetic states, it is expected that the conclusions hold more generally since the most important energy contribution is the intraatomic term.

Following the establishment of the local moment criterion of Eq. (4.52) on the basis of spin density functional theory, there has been a second direction of development in addition to the work of Hubbard. Hasegawa, and Roth already referred to above; this is progress related to the Anderson (1961) model of local moments. This was also considered from the standpoint of spin density functional theory in early work of Stoddart *et al.* (1972). However, they found it necessary to make approximations which did not enable the relation to the Kondo effect to be brought out. Their work has now only some historical interest, for the exact solution of the Anderson model has been given by Wiegmann (1980) and by Kawakami and Okiji (1981).

While the presentation of these detailed solutions is not our purpose here, we note that in subsequent work the contribution of the impurities to the physical properties, at finite temperatures, has been studied numerically on the basis of these exact solutions. In particular, the temperature dependence of the impurity susceptibility and specific heat for the Anderson model have been computed. The results of Kawakami and Okiji (1981) for the magnetic susceptibility are in close agreement with the findings of Krishnamurthy *et al.* (1980).

Finally, returning to spin density fluctuation theory (Hasegawa 1980a,b; see also Hubbard, 1979, 1981), a further interesting application has been made by Evangelou and Edwards (1983) to study the temperature-induced local moments in MnSi and FeSi.

4.6.2. Electron Correlation and Stability of Crystal Structures in the 3*d* Transition Metals

It is highly relevant to the above discussion to conclude this section by referring to work of Sayers and Kajzar (1981), who have used Gutzwiller's method to study the effect of electron correlation on the stability of crystal structures in the 3*d* transition metals. From earlier work, correlation effects are known to be important in determining the bulk modulus and the equilibrium atomic value of the 3*d* metals (Friedel and Sayers, 1977a,b; 1978). This work was closely related to that discussed in Section 4.6.1. in that it was based on the perturbation expansion of Gutzwiller's model (Gutzwiller, 1965; Sayers, 1977). In the later work of Sayers and Kajzar on crystal structure stability, an approximation to the exact second-order correlation correction is used, which, for a small number of electrons or holes in the band, reduces to the weak correlation limit of the result of Kanamori (1963). Their conclusion is that *d–d* correlation acts to stabilize a structure with a high density of states at the Fermi level. This effect reduces the energy difference between the magnetic and nonmagnetic states; one example cited by Sayers and Kajzar in this connection is that this effect will therefore contribute to the body-centered-cubic-hexagonal close-packed transformation of iron at 100 kbar.

It is clear from the work cited above that a number of salient properties of the 3*d* transition metals can be understood when electron correlation is incorporated on the basis of Gutzwiller's variational method. We shall see later that some aspects of the Gutzwiller theory—namely, those directly relevant to metal–insulator transitions—can be built into a rather simple phenomenological theory. Therefore, let us conclude by saying that Gutzwiller's method is a natural consequence of the Coulson–Fischer approach to H_2 discussed in Section 4.1, in that it can be thought of as starting from a single Slater determinant of one-electron wave functions, multiplying this out, and then suppressing somewhat the weight of all those configurations in which two electrons with opposed spins come onto the same atom, this weight being then determined variationally. An alternative, though equivalent, way of thinking about the Gutzwiller method is on the basis of the following two assumptions:

1. Electrons of one spin direction are assumed fixed when one studies the motion of an electron with opposite spin. This assumption is appropriate to narrow bands.
2. Correlations between electrons of opposite spin are assumed to reduce the amplitude of the one-electron wave function on the sites already occupied by an electron of opposite spin, but are otherwise assumed to produce no scattering.

Chapter 5

Quantum Monte Carlo Calculation of Correlation Energy

Over a long period two systematic methods have provided the main routes to the calculation of correlation energies for many-electron systems at zero temperature; namely; configuration interaction (CI) and many-body perturbation theory. The second of these approaches applied to molecules has been dealt with in the review by Wilson (1981). Therefore that ground will not be covered in this chapter. However, there has been much progress using the so-called Quantum Monte Carlo (QMC) method. The goal of this approach [see, e.g., Ceperley and Alder (1984)] is to obtain the exact ground-state wave function of a many-body system by numerically solving the Schrödinger equation in one of its equivalent forms. In practice, this is achieved by means of iterative algorithms, that propagate the wave function from a suitable starting estimate to the exact ground-state result. Thus, in the Diffusion Monte Carlo (DMC) method, one is directly concerned with the evolution in imaginary time of the wave function, which corresponds to a diffusion process in configuration space. In the Green Function Monte Carlo (GFMC) technique, on the other hand, a time-integrated form of the Green function or resolvent is used to propagate the wave function. In fact, in either case the sampling of an appropriate Green function is required, and this is achieved by means of suitable random-walk algorithms.

In another implementation of the QMC method (Blankenbecler *et al.,* 1981; Scalapino and Sugar, 1981; Koonin *et al.,* 1982), the role of the imaginary time propagator $\exp(-\beta H)$ is emphasized more directly. For long times β (low temperatures $T = \beta^{-1}$), the imaginary time propagator is dominated by the ground-state energy E_0. Thus, E_0 can be extracted from knowledge of the partition function $Z = \mathrm{Tr}\exp(-\beta H)$ or, equally well, from the expectation value of the imaginary time propagator on a state that is not orthogonal to the ground state. In either case, the key simplification in the approaches stems from mapping the problem of interacting particles onto that of independent particles; see also the review of Senatore and March (1994), which is followed closely in this chapter. This mapping is accomplished at the expense of introducing auxiliary external fields (Koonin *et al.*, 1982; Hirsch, 1983; Sugiyama and Koonin, 1986) having their own probability density.

In the following, the main features of the QMC method for a many-body system in the DMC and GFMC implementations will first be presented. The complications that arise from antisymmetry when dealing with systems of fermions, the case of interest here, will be discussed. Some applications of the QMC method—the study of the homogenous electron assembly and small molecules—will then be briefly summarized. Finally, the basic ideas underlying the auxiliary external field technique will be pointed out, and some results on the two-dimensional Hubbard model will be presented.

5.1. DIFFUSION MONTE CARLO METHOD

The Schrödinger equation in imaginary time for an assembly of N identical particles of mass m interacting with a potential $V(\mathbf{R})$ is

$$-\frac{\partial\phi(\mathbf{R}, t)}{\partial t} = (H - E_T)\phi(\mathbf{R},t) = (-D\nabla^2 + V(\mathbf{R}) - E_T)\phi(\mathbf{R},t), \tag{5.1}$$

where $D = \hbar/2m$, $\mathbf{R}$ is the $3N$-dimensional vector specifying the coordinates of the particles, t is the imaginary time in units of $\hbar$, and the constant E_T represents a suitable shift of the zero of energy. The (imaginary) time evolution of an arbitrary *trial* wave function is readily obtained from its expansion in terms of the eigenfunctions $\phi_i(\mathbf{R})$ of the Hamiltonian H as

$$\phi(\mathbf{R}, t) = \sum_i N_i \exp[-(E_i - E_T)t]\phi_i(\mathbf{R}). \tag{5.2}$$

Here, E_i is the energy eigenvalue corresponding to $\phi_i(\mathbf{R})$ and the coefficients N_i are fixed by the initial conditions, i.e., by the chosen trial wave function. Clearly, for long times one finds

$$\phi(\mathbf{R}, t) = N_0 \exp[-(E_0 - E_T)t]\phi_0(\mathbf{R}), \tag{5.3}$$

provided $N_0 \neq 0$. Moreover, if E_T is adjusted to be the true ground-state energy E_0, asymptotically one obtains a steady-state solution, corresponding to the ground state wave function ϕ_0. Thus, the problem of determining the ground-state eigenfunction of the Hamiltonian H is equivalent to that of solving Eq. (5.1) with appropriate boundary conditions.

It is not difficult to recognize in Eq. (5.1) two equations that are well known in physics, though combined together. In fact, if only the term with the Laplacian ∇^2 were present on the right-hand side of Eq. (5.1) one would have a diffusion equation (e.g., the equation describing Brownian motion: see, for instance, van

Kampen, 1981). On the other hand, by retaining only the term $[V(\mathbf{R}) - E_T]\phi(\mathbf{R}, t)$, one would obtain a rate equation, that is, an equation describing branching processes such as radioactive decay or birth and death in a population. The discussion below again follows closely that of Senatore and March (1994).

A convenient way of *stimulating* Eq. (5.1) is as follows. Let $\psi(\mathbf{R})$ be the wave function at $t = 0$. One can generate an ensemble of systems (points in the $3N$-dimensional space representing electronic configuration) distributed with density $\psi_T(\mathbf{R})$. Hence, the time evolution of the wave function will correspond to the motion in configuration space of such systems or *walkers,* as determined by Eq. (5.1). In particular, the Laplacian and energy terms in Eq. (5.1) will cause, respectively, random diffusion and branching (deletion or duplication) of the walkers describing the wave function. This schematization of the Schrödinger equation is particularly appealing from a practical point of view. However, a density is nonnegative. Thus, the same should hold for the wave function. This would seem to limit the applicability of such a scheme to the ground state of a bosonic system. For the sake of simplicity, this restriction will be accepted for a while. Later the complications associated with Fermi statistics will be dealt with.

Importance Sampling. Solving the Schrödinger equation in the form [Eq. (5.1)] by random-walk processes with branching is not particularly efficient. In fact, the branching term can become very large whenever the interaction potential $V(\mathbf{R})$ does so, causing large fluctuations in the number of walkers. This slows down the convergence toward the ground state. A more efficient computational scheme is obtained by introducing so-called *importance sampling.* This amounts to considering the evolution equation for the probability distribution $f(\mathbf{R}, t) = \phi(\mathbf{R},t)\psi_T(\mathbf{R})$, rather than directly for the wave function ϕ. By using Eq. (5.1) and the definition of $f(\mathbf{R},t)$, one obtains, after some algebra,

$$-\frac{\partial f(\mathbf{R},t)}{\partial t} = -D\nabla^2 f + [E_L(\mathbf{R}) - E_T]f + D\nabla[fF_Q(\mathbf{R})]. \tag{5.4}$$

Above, $E_L(\mathbf{R}) \equiv H\psi_T/\psi_T$ defines the *local energy* associated with the trial wave function, and

$$F_Q(\mathbf{R}) \equiv \nabla \ln |\psi_T(\mathbf{R})|^2 = 2\nabla\psi_T(\mathbf{R})/\psi_T(\mathbf{R}) \tag{5.5}$$

is the *quantum trial force,* which causes the walkers to drift away from regions where $\psi(\mathbf{R})$ is small. Obviously, when one deals with Eq. (5.4) the walkers are to be drawn from the probability distribution f. With a judicious choice of ψ_T one can make $E_L(\mathbf{R})$ a smooth function close to E_T throughout the configuration space and thus reduce branching. In this respect it is essential that $\psi_T(\mathbf{R})$ reproduce the correct cusp behavior as any two particles approach each other, so as to exactly cancel the infinities originating from $V(\mathbf{R})$. That $F_Q(\mathbf{R})$ is a force acting on the walkers becomes clear by comparing Eq. (5.4) with the Fokker–Planck (FP) equation (see, e.g., van Kampen, 1981). It turns out that, neglecting the rate term, the term containing F_Q has to be identified with the so-called *drift term* of the FP equation and consequently F_Q with the external force acting on the walkers. It is to be stressed again that an important feature of this force is to cause the walkers to drift away from regions of low probability. In fact, from its definition it is apparent that the force becomes highly repulsive in regions where the trial wave function becomes small, diverging where the latter vanishes. Thus, the trial wave function determines the probability with which different regions of configuration space are sampled. Therefore it is important that ψ_T be a good approximation to ϕ_0 in order to keep the walkers mainly in regions of configuration space that are really significant in the statistical averages. It is finally to be noted that the asymptotic solution of Eq. (5.4) is

$$f(\mathbf{R},T) = \psi_T(\mathbf{R})\phi_0(\mathbf{R}) \exp[-(E_0 - E_T)t], \tag{5.6}$$

which becomes a steady-state solution when E_T is adjusted to E_0.

The differential equation Eq. (5.4) for the probability distribution $f(\mathbf{R},t)$ can be recast in integral form as

$$f(\mathbf{R}, t + t') = \int d\mathbf{R}' \, f(\mathbf{R}',t')K(\mathbf{R}', \mathbf{R},t), \tag{5.7}$$

where the Green function $K(\mathbf{R}',\mathbf{R},t)$ is a solution of Eq. (5.4) with the boundary condition $K(\mathbf{R}',\mathbf{R},0) = \delta(\mathbf{R} - \mathbf{R}')$ and is simply related to the Green function $G(\mathbf{R}',\mathbf{R},t)$ for Eq. (5.1), $K(\mathbf{R}',\mathbf{R},t) = \psi_T(\mathbf{R}')G(\mathbf{R}',\mathbf{R},t)\psi_T^{-1}(\mathbf{R})$. The advantage of the above equation is that for short times t it is possible to write approximate simple expressions for $K(\mathbf{R}',\mathbf{R},t)$ (Moskowitz *et al.*, 1982b; Reynolds *et al.*, 1982). The time evolution of $f(\mathbf{R},t)$ for finite t can then be obtained by successive iterations of Eq. (5.7), starting from the initial distribution $f(\mathbf{R},0)$ and using a small time step. In each time iteration, advantage can be taken of the positivity of K for short times so as to interpret it as a transition density. A suitable modified random-walk algorithm can be then constructed, which allows the integration of Eq. (5.7) during a small time interval. The essential steps of such an algorithm are as follows: The walkers representing the initial distribution are allowed to diffuse randomly and drift under the action of the quantum force F_Q. After the new positions have been reached, each walker is deleted or placed in the *new generation* in an appropriate number of copies, depending on the size of the local energy at the old and at the new position relative to the reference energy E_T. Finally, the number of walkers in the new generation is renormalized to the initial population. The interested reader may find a detailed account of such a random-walk algorithm in Reynolds et al. (1982).

Once the long-time probability distribution has been obtained and made stationary by a suitable shift of the constant E_T, one can estimate equilibrium quantities. For the ground state energy E_0, in particular, by using the fact that H is Hermitian one finds that

$$E_0 = \frac{\int \psi_T H \phi_0}{\int \psi_T \phi_0} = \frac{\int E_L f}{\int f} = \frac{\Sigma_i E_L(\mathbf{R}_i)}{\Sigma_i}. \tag{5.8}$$

The sum on the right-hand side of the above equation runs over the positions of all the walkers representing the equilibrium probability distribution.

It should be stressed that the short-time approximation to the Green function introduces a systematic error in the computations. However, this error should decrease with decreases in the time step. Moreover, in the calculation by Reynolds *et al.* 1982) an acceptance/rejection step was used which, within the calculational scheme outlined above, should correct for the approximate nature of G.

5.2. GREEN FUNCTION MONTE CARLO TECHNIQUES

It has been seen that in the Diffusion Monte Carlo method one starts from the Schrödinger equation in imaginary time in order to construct an integral equation that can then be solved iteratively. Such an equation contains a time-dependent Green function. In the Green Function Monte Carlo technique one proceeds in a similar fashion, starting however from the Schrödinger eigenvalue equation. The resulting integral equation, as will be seen below, differs from Eq. (5.7) mainly through the absence of time. In fact, it could be directly obtained from Eq. (5.7) by a straightforward time integration (Ceperley and Alder, 1984). Here, however, it will be convenient to follow another route (Ceperley and Kalos, 1979; Kalos, 1984), which in the present context is perhaps more instructive.

The Schrödinger eigenvalue equation for an N-body system, described by the Hamiltonian H introduced in Eq. (5.1), is

$$H\phi(\mathbf{R}) = E\phi(\mathbf{R}). \tag{5.9}$$

It will be assumed that the potential $V(\mathbf{R})$ appearing in H is such that the spectrum of H is bounded from below, E_0 being the lowest eigenvalue, i.e. the ground-state energy. This is a natural assumption for a physical system, in the nonrelativistic limit. One can choose positive constant V_0 such that $E_0 + V_0 > 0$ and rewrite Eq. (5.9) as

$$(H + V_0)\phi(\mathbf{R}) = (E + V_0)\phi(\mathbf{R}). \tag{5.10}$$

The resolvent $G(\mathbf{R},\mathbf{R}')$—i.e., the Green function for Eq. (5.10)—is then defined by

$$(H + V_0)G(\mathbf{R},\mathbf{R}') = \delta(\mathbf{R} - \mathbf{R}'). \tag{5.11}$$

Due to the manner in which V_0 has been chosen, it is clear that the operator $H + V_0$ is positive definite. This is immediate in the energy representation. The same, of course holds for the inverse operator $G = 1/(H + V_0)$. Actually, a much stricter inequality involving G can be established, namely,

$$G(\mathbf{R},\mathbf{R}') = \langle \mathbf{R}|G|\mathbf{R}'\rangle \geq 0 \qquad \text{for all } \mathbf{R},\mathbf{R}', \tag{5.12}$$

which is of central importance for the calculational scheme that will shortly be described. A brief sketch of how one can prove the inequality in Eq. (5.12) is given in Appendix 5.1. Here, it will merely be noted for future reference that G is integrable. Therefore, because of the above inequality, it can be regarded as a probability, or more precisely as a transition density (see, for instance, McWeeny, 1989).

A space integration of Eq. (5.10), after multiplying by $G(\mathbf{R},\mathbf{R}')$ and utilizing Eq. (5.11), yields the desired integral equation,

$$\phi(\mathbf{R}) = (E + V_0) \int d\mathbf{R}'\, G(\mathbf{R},\mathbf{R}')\phi(\mathbf{R}'). \tag{5.13}$$

In this equation one has to solve simultaneously for E and $\phi(\mathbf{R})$. However, if one has a trial wave function $\psi_T(\mathbf{R})$ and, consequently, a trial energy E_T, it is tempting to try solving Eq. (5.13) by iteration. One would generate a sequence of wave functions, according to

$$\Phi_n(\mathbf{R}) = (E_T + V_0) \int d\mathbf{R}'\, G(\mathbf{R},\mathbf{R}')\Phi_{n-1}(\mathbf{R}') \tag{5.14}$$

and with $\Phi_0(\mathbf{R}) = \psi_T(\mathbf{R})$. In fact, this series converges precisely to the ground-state wave function, and it does so exponentially fast. This can easily be seen by expanding both $G(\mathbf{R},\mathbf{R}')$ and $\Phi_0(\mathbf{R})$ in eigenfunctions of H:

$$G(\mathbf{R},\mathbf{R}') = \sum_i \frac{\phi_i(\mathrm{R})\phi_i(\mathbf{R}')}{E_i + V_0}, \tag{5.15}$$

$$\Phi_0(\mathbf{R}) = \sum_i c_i \phi_i(\mathbf{R}). \tag{5.16}$$

One immediately finds

$$\Phi_n(\mathbf{R}) = \sum_i \left(\frac{E_T + V_0}{E_i + V_0}\right)^n c_i \phi_i(\mathbf{R}). \tag{5.17}$$

Thus, provided $c_0 \neq 0$, one obtains

$$\lim_{n\to\infty} \Phi_0(\mathbf{R}) \propto c_0 \phi_0(\mathbf{R}), \tag{5.18}$$

which is the desired result.

As in the case of the Diffusion Monte Carlo method, one has cast the problem of finding the ground-state wave function and energy of a given Hamiltonian in integral form, suitable for iterative techniques. For the sake of simplicity, let us temporarily restrict the discussion to bosonic systems, for which not only $G(\mathbf{R},\mathbf{R}')$ but also the ground-state wave function is nonnegative. If one assumes for a while that $G(\mathbf{R},\mathbf{R}')$ is known, a random-walk algorithm similar to the one described for the Diffusion Monte Carlo technique is readily constructed. One can proceed as follows: First, at random, an initial population of walkers is drawn from the probability distribution $\Phi_0(\mathbf{R})$, let us say at the positions $\{\mathbf{R}_i\}$. Then, a new set of configurations $\{\mathbf{R}'_j\}$ is generated at random, each one with a conditional density $\sum_i(E + V_0)G(\mathbf{R}_i,\mathbf{R}'_j)$. This density determines the number of walkers corresponding to each new configuration. Finally, a renormalization of the new walkers' population to the initial size yields the new generation. The above series of steps corresponds to the first iteration of Eq. (5.14). The successive iterations can be performed in the same manner, changing only the first step. In the nth iteration, in fact, one has to take as the initial population of walkers the new generation of the previous iteration.

It is clear from Eq. (5.17) that the change in size of the walkers' population is determined by the trial energy E_T. Depending on whether this is larger or smaller than the true ground-state energy E_0, the population will asymptotically grow or decline. This suggests a way of estimating the ground-state energy (Ceperley and Kalos, 1979). In fact, from the space integration of Eq. (5.17), it follows that asymptotically

$$E_0 = E_T + V_0\left(\frac{N_n}{N_{n+1}} - 1\right), \tag{5.19}$$

with N_n the walkers' initial population in the nth iteration and N_{n+1} the walkers' new population before size renormalization. The above energy estimator, known as the *growth* estimator, is unfortunately biased even in the limit in which Φ_n is converging to ϕ_0, as discussed by Ceperley and Kalos (1979). A better energy estimator is the one introduced with Eq. (5.8), when describing the Diffusion Monte Carlo method. However, in order to use that estimator one has to work with the density $f(\mathbf{R}) = \phi(\mathbf{R})\psi_T(\mathbf{R})$ rather than directly with the wave function. This can be readily arranged by multiplying Eq. (5.13) by $\psi_T(\mathbf{R})$ on both sides, so as to obtain the new integral equation

$$f(\mathbf{R}) = (E + V_0)\int d\mathbf{R}'K(\mathbf{R},\mathbf{R}')f(\mathbf{R}'), \tag{5.20}$$

where

$$f(\mathbf{R}) = \phi(\mathbf{R})\psi_T(\mathbf{R}) \tag{5.21}$$

and

$$K(\mathbf{R}, \mathbf{R}') = \psi_T(\mathbf{R})G(\mathbf{R},\mathbf{R}')\psi_T^{-1}(\mathbf{R}'). \tag{5.22}$$

Clearly, Eq. (5.20) can still be simulated by means of the random-walk algorithm already described, with $f(\mathbf{R})$ and $K(\mathbf{R},\mathbf{R}')$ replacing, respectively, $\phi(\mathbf{R})$ and $G(\mathbf{R},\mathbf{R}')$. Within such a calculation scheme, successive iteration of Eq. (5.20) will produce a walkers' population which asymptotically tends to be distributed with density $f = \phi_0\psi_T$. Consequently, the *local* energy estimator of Eq. (5.8) will also tend to the exact ground-state energy.

In the foregoing, knowledge of $G(\mathbf{R},\mathbf{R}')$ has been assumed. In practice, for a system of interacting particles, G is not known and must also be sampled by means of suitable techniques. This can be achieved by relating the exact Green function G to some trial or reference Green function G_T, which is known analytically or is numerically calculable. In the Domain Green Function method of Kalos *et al.* (1974) (see, also, Ceperley and Kalos, 1979; Moskowitz and Schmidt, 1986) the trial Green function is taken to be that appropriate to independent particles in a constant potential, with the motion restricted to a subdomain of the whole space. In another implementation of GFMC due to Ceperley (1983) (see, also, Ceperley and Alder, 1984), a better trial Green function is introduced, which is much closer to the exact one and is defined over the whole of space. In either case, the integral equation relating G_T to G is such that a random-walk algorithm, somehow similar to the one outlined above for the calculation of f, can be used. In fact, a global random-walk algorithm can be devised that combines the iterations of Eq. (5.20) with those needed to correct for the difference between G and G_T. Details as to how this is done in practice are to be found in the references quoted above. Here, it will simply be observed again that, in the approach due to Ceperley and Alder (1984), Eq. (5.20) is obtained as a time average or Laplace transform of Eq. (5.7). In addition, the sampling of the time-integrated Green function [$K(\mathbf{R},\mathbf{R}')$ in the present notation] is also obtained in practical calculations by summing its time-dependent counterpart. This makes such an approach very similar to the Diffusion Monte Carlo method, while removing the truncation error due to the use of a finite time step.

5.3. QUANTUM MONTE CARLO TECHNIQUE WITH FERMIONS: FIXED-NODE APPROXIMATION AND NODAL RELAXATION

Here, the modifications that one can make to the calculational schemes described above, when the restriction of a nonnegative wave function is relaxed will be discussed. Of course, this is necessary if one is to deal with Fermi systems and in general with excited states of the many-body Hamiltonian H introduced in Eq. (5.1).

Let us briefly recall the relation between the eigenfunctions of H and those of a many-body system described by H but also obeying Bose or Fermi statistics. For bosons (with zero spin), only those wave functions of H that are symmetric under exchange of any two particle coordinates are admissible solutions. On the other hand, the ground-state wave function of H is completely symmetric, since the ground state energy is nondegenerate and H commutes with any permutation of particles. Thus the ground state of H which of course is characterized by a non-negative wave function, is also the bosonic ground state. Needless to say, the bosonic excited states will have wave functions with positive and negative regions. For fermions, the symmetry of the wave functions is determined by the chosen spin configuration, which we shall denote here by s. In general, for many fermions it will always be true that a certain number of particles, say M, have the same spin projection. The admissible wave functions of H are then only those which are antisymmetric with respect to the permutation of any two particle coordinates, within the group of particles having the same spin projections. This is equivalent to saying that only excited states of H can be considered, since as has already been observed the ground-state wave function is completely symmetric. It follows at once that the fermionic wave functions must have positive and negative regions, separated by nodal surfaces in the $3N$-dimensional space of the particle coordinates.

The need for dealing with wave functions $\phi(\mathbf{R})$ that change sign would seem to preclude the use of random-walk algorithms such as those discussed above, in that they require a positive wave function. In fact, if one knew the location of the nodal surfaces* and consequently of the connected domains into which they subdivide the whole space, the problem would not be a real one. Since the equations involved in the evolution of the wave function are linear, one could just consider in each domain the evolution of the modulus of ϕ. However, such nodal surfaces are not known in more than one dimension, and the problem remains. A simple remedy is to take as nodal surfaces those of the trial wave function $\psi_T(\mathbf{R})$, assumed to be a good approximation to the exact wave function at least in the location of such surfaces. This approximation corresponds to a solution of the Schrödinger equation within a restricted class of functions and so should have variational character. As such it should yield energies being upper bounds to the exact ones, in the absence of other approximations or sources of error.

A practical way of realizing this so-called fixed-node approximation, within calculations based on random-walk algorithms, is to delete those walkers which, during the calculation, cross a nodal surface of the trial wave function (see, for instance, Reynolds *et al.*, 1982; see also Umrigar *et al.*, 1991). This is a simple way of enforcing the vanishing of the wave function at the nodal surfaces. In fact, this procedure corresponds to solving the Schrödinger equation separately in each domain, the evolution of the walkers in one domain having become independent from that in the other domains. More precisely, if $\{V_\alpha\}$ denote the domains bounded by the nodal surfaces of ψ_T, one is finding the ground state of the Hamiltonian H in each domain V_α, according to

$$H\phi_\alpha(\mathbf{R},s) = \varepsilon_\alpha \phi_\alpha(\mathbf{R},s), \tag{5.23}$$

*Note added in proof. In this context, the early work of D. J. Klein and H. M. Pickett (1976: J. Chem. Phys. *64*: 4811) should be cited.

and

$$\phi_\alpha(\mathbf{R},s)\psi_T(\mathbf{R},s) > 0 \qquad \mathbf{R} \in V_\alpha, \tag{5.24}$$

$$\phi_\alpha(\mathbf{R},s) = 0 \qquad \mathbf{R} \notin V_\alpha. \tag{5.25}$$

Above, the spin configuration s has also been indicated. It should be clear that within the random-walk approach, the asymptotic stationarity of the walkers' population can only be attained if $E_T = \varepsilon_m$, with ε_m being the minimum among all the ε_α corresponding to volumes V_α which were populated at the beginning of the random walk. Therefore, if the trial energy E_T is suitably adjusted to yield stationarity, the asymptotic walkers' population will be distributed with density

$$f_\infty(\mathbf{R}) = f(\mathbf{R},t \to \infty) = c_m\phi_m(\mathbf{R},s)\psi_T(\mathbf{R},s). \tag{5.26}$$

5.3.1. Variational Bound of Fixed-Node Approximation

One wishes next to make the statement about the variational nature of the fixed-node approximation more precise. First of all, let us note, that since H is completely symmetric under exchange of any two particle coordinates, the symmetry of wave functions is preserved during the evolution, i.e., the random walk. Thus from any one of the ϕ_α, which are defined in a given domain, a wave function with the antisymmetry dictated by the chosen spin configuration can be formally constructed by summing the given ϕ_α over all the permutations P of the electrons, according to

$$\hat{\phi}_\alpha(\mathbf{R},s) = \sum_P (-)^P \hat{\phi}_\alpha(P\mathbf{R},s). \tag{5.27}$$

From the variational principle applied to the Hamiltonian H it follows that the variational energy associated with the above wave function satisfies

$$\frac{\int d\mathbf{R}\,\hat{\phi}^*_\alpha H \hat{\phi}_\alpha}{\int d\mathbf{R}\,\hat{\phi}^*_\alpha \hat{\phi}_\alpha} = \varepsilon_\alpha > E_0, \tag{5.28}$$

where E_0 is the minimum eigenvalue among those relative to the eigenfunctions of H with the given symmetry (i.e., is the exact energy of the fermionic ground state) for the given spin configuration. On the other hand, from Eq. (5.26) and the Hermiticity of H, it also follows that the local energy estimator, which was explicitly given in Eq. (5.8), tends asymptotically to ε_m. This completes the proof that the fixed-node approximation yields a variational upper bound to the exact ground-state energy of a fermionic system.

Before discussing to what extent it is possible to improve on the fixed node approximation, it is worth noting a detail which may appear technical at this point but will prove useful in what follows (see also Senatore and March, 1994). In applying the above scheme, one is solving the Schrödinger equation by means of random walks separately in each of the domains fixed by the trial antisymmetric function ψ_T. In practice the sign of ψ_T is needed so as to delete walkers that change domains. It should be clear to the reader that with regard to importance sampling nothing changes if one defines a *guidance* function ψ_G to be positive everywhere, and takes as the distribution $f = \phi\psi_G$, provided that the sign ψ_T is still used to locate the nodal surfaces. In fact the fixed-node approximation, as presented above, corresponds to taking $\psi_G = |\psi_T|$. However, nothing changes in the foregoing discussion if one makes a different choice for ψ_G. It is to be stressed that, to yield an efficient sampling, ψ_G should be a good approximation to the modulus of the exact ground state. This is equivalent to saying that ψ_G should not differ too much from $|\psi_T|$ since ψ_T was assumed from the outset to be a good approximation to the exact ground state.

It was mentioned above that the difficulties in applying a random-walk algorithm to the calculation of the fermionic ground state arise from the fact that the corresponding wave function possesses positive and negative regions and therefore cannot be regarded as a walkers' density. However, this difficulty is easily remedied, at least formally. The manner in which this can be done is as follows: One can simply regard positive and negative regions of the trial wave function as determining the positive densities of different objects, the white and black walkers. In fact, an antisymmetric wave function can be always written as the difference of two positive functions. The partitioning into white and black walkers mentioned above corresponds to a particular choice of such functions. It is clear from the linearity of Eqs. (5.1) and (5.14) that one can consider the evolution of each one of these functions separately. At each step of the iteration process yielding the evolution, the difference between the two functions gives an antisymmetric wave function (Kalos, 1984).

The approach outlined above to the problem of Fermi statistics is the essence of the so-called nodal relaxation method (Ceperley and Alder, 1980, 1984; Ceperley, 1981). However, while it seems relatively simple and straightforward, such a method turns out to be unstable when numerically implemented. The reason for this is now also clear. The two positive functions into which the antisymmetric wave function is divided are no longer antisymmetric. As such they acquire a projection onto the symmetric Bose ground state,. which evidently has a lower energy than the Fermi ground state. As the evolution takes place, the Bose component of each of the functions grows exponentially with respect to the antisymmetric component. In practice, due to the finite numerical precision, after a sufficient number of iterations the latter component is completely lost, and the difference between the two functions is simply noise. To avoid this, one must restrict the evolution to a time region such that the Fermi component can still be isolated. Therefore, it is important that the trial wave function shall already be close to the exact Fermi ground state, since the closer it is the shorter the evolution time can be made. For the reason as given above, the estimates of the Fermi ground state based on the nodal relaxation method have also been termed transient estimates (Kalos, 1984).

5.3.2. Implementation of Nodal Relaxation Method

Below a brief outline of the way the nodal relaxation method has been implemented in practice will be given (see, for instance, Ceperley and Alder, 1984). To fix ideas, let us consider the application of the method within the Green function formalism with importance sampling. Thus, the starting point is Eq. (5.20), which gives the evolution of the distribution $f(\mathbf{R}) = \phi(\mathbf{R})\psi_T(\mathbf{R})$. Let us, however, distinguish between a trial wave function, which provides the initial nodal surfaces, and a positive guidance function, to be used in the importance sampling. In this way one can deal with the evolution of a positive distribution, for which a random-walk algorithm can still be used, and attach to each walker the appropriate sign at the end, in a manner which depends on the number of crossings of the walker through the nodal surfaces of ψ_T. Since the importance sampling is made according to ψ_G, it is useful to introduce a corresponding Green function

$$D(\mathbf{R},\mathbf{R}') = \psi_G(\mathbf{R})G(\mathbf{R},\mathbf{R}')\psi_G^{-1}(\mathbf{R}'), \tag{5.29}$$

and an appropriate weight function

$$W(\mathbf{R}) = \psi_T(\mathbf{R})\psi_G^{-1}(\mathbf{R}). \tag{5.30}$$

The relation between the above Green function and the one defined in Eq. (5.22) is

$$K(\mathbf{R},\mathbf{R}') = W(\mathbf{R})D(\mathbf{R},\mathbf{R}')W^{-1}(\mathbf{R}'). \tag{5.31}$$

If one maintains for f the definition $f = \phi\psi_T$ and introduces correspondingly a new distribution $g = \phi\psi_G$ relative to the guidance wave function one can rewrite Eq. (5.20) as

$$g(\mathbf{R}) = (E + V_0)\int d\mathbf{R}'D(\mathbf{R},\mathbf{R}')g(\mathbf{R}') \tag{5.32}$$

and of course,

$$f(\mathbf{R}) = W(\mathbf{R})g(\mathbf{R}). \tag{5.33}$$

Equation (5.32) must be iterated, starting from a suitable initial distribution. The best that one can do is to start from the fixed-node distribution $g_{FN} = \phi_{FN}\psi_G$, obtained, for instance, from a previous calculation with guidance function ψ_G. To this fixed-node distribution corresponds an initial g,

$$g_0(\mathbf{R}) = \sigma_0(\mathbf{R})|g_{FN}(\mathbf{R})| \equiv \sigma_0(\mathbf{R})|g(\mathbf{R})|, \tag{5.34}$$

where

$$\sigma_0(\mathbf{R}) = \frac{W(\mathbf{R})}{|W(\mathbf{R})|} = \mathrm{sgn}(\psi_T(\mathbf{R})). \tag{5.35}$$

From the computational point of view, one can just begin iterating Eq. (5.32) starting from $g = |g_0|$ and assigning to each new walker a *counter* in which is stored the sign σ_0 of the initial fixed-node parent. In this way only positive distributions enter the random walk. The final distribution $f = \phi\psi_T$ is then readily obtained from the walkers' population corresponding to the final g by assigning to each walker a weight and a sign according to $W(\mathbf{R})\sigma_0$, where $\mathbf{R}$ is the position of the walker and σ_0 is the sign associated with the position of the initial fixed-node parent at the beginning of the random walk. This immediately follows from Eqs. (5.32–5.34).

Once the nodal relaxation has been performed, the fermionic ground-state energy can be evaluated by means of the local energy estimator, introduced with Eq. (5.8), which in the present case can be written as

$$E_0 = \frac{\Sigma_i W(\mathbf{R}_i)\sigma_0 E_L(\mathbf{R}_i)}{\Sigma_i W(\mathbf{R}_i)\sigma_0}. \tag{5.36}$$

It is to be stressed once again that, in applying the nodal relaxation method to practical calculation, it is essential to start from a very good guess for the fermionic ground state, in order to minimize the length of the random walk associated with the relaxation of the nodes. In fact, during the evolution of the modulus of the wave function the fermonic signal decreases, in favour of the bosonic component. This leads, for instance, to an exponential growth of the variance of the energy with respect to its average value, as can be seen from the definition of average implied by the local energy estimator of Eq. (5.36) (Schmidt and Kalos, 1984; Kalos, 1984).

Finally, a comment should be made on the choice of the guidance wave function: ψ_G should be sufficiently close to $|\psi_T|$ to ensure an efficient sampling of configuration space, ψ_T having been chosen so as to be a good approximation to the exact ground state. However, in the case in which nodal relaxation is performed, other considerations come into play. In particular, within the framework of the Diffusion Monte Carlo method it emerges from Eqs. (5.4–5.5) that a zero of the guidance function corresponds to an infinitely repulsive force acting on the walkers. This means that the walkers are pushed away from the nodal surfaces of ψ_T and cannot cross them. In other words the statistical system associated with the walkers is nonergodic, in that the walkers cannot redistribute themselves among the various domains determined by the nodal surfaces. On the other hand, within the Green function Monte Carlo procedure the effect of the vanishing of ψ_G is to yield an excessive branching in the walkers' generation (Ceperley and Alder, 1984). In either case, a simple way to remedy this inconvenience is to take $\psi_G(\mathbf{R})$

$= |\psi_T(\mathbf{R})|\,[1 + s(\mathbf{R})]$, where $s(\mathbf{R})$ is a nonnegative function that is very small away from nodal surfaces and vanishing at infinity. Near a nodal surface, however, $s(\mathbf{R})$ behaves in such a way as to make $\psi_G(\mathbf{R})$ small but finite. A reasonable choice of $s(\mathbf{R})$ is obtained by compromising between reduction of branching and efficiency of the sampling (Ceperley and Alder, 1984).

5.4. MONTE CARLO COMPUTER EXPERIMENTS ON PHASE TRANSITIONS IN UNIFORM INTERACTING ELECTRON ASSEMBLY

The ground-state energy of a uniform interacting electron assembly has been computed by Ceperley and Alder (1980) using the Diffusion Monte Carlo method described in Section 5.1. In particular, these authors have calculated the ground-state energy for four distinct phases of this system of charged particles, at various densities. They have considered

(i) the unpolarized Fermi fluid,
(ii) the fully polarized Fermi fluid,
(iii) the Bose fluid, and
(iv) the Bose crystal on a *bcc* lattice.

With this information, which can be regarded as the most accurate to date, they were able to predict the transitions between the various phases with reasonable accuracy.

For each phase, the calculations were performed by first generating fixed-node wave functions, or more precisely the corresponding populations of random walkers, and then applying to such fixed-node distributions the nodal relaxation scheme discussed in Section 5.3.2. The trial wave functions for the Fermi phases were chosen as a product of Slater determinants, one for each spin projection population, and a Jastrow factor ensuring the cusp condition as any two electrons approach each other. The Slater determinants were constructed from plane waves with the wave vector lying within the Fermi sphere. Of course at given density the Fermi wave vector of the fully polarized electron fluid is $2^{1/3}$ larger than that of the unpolarized system. For the crystalline phase, the one-particle orbitals were chosen to be Gaussians centered around the lattice sites with a width chosen variationally.

The analysis of the convergence of the nodal relaxation shows that in the present case the Hartree–Fock nodes, which were employed in the calculations, constitute a good approximation to the nodes of the exact wave function. In fact, it was found that the convergence of the relaxation process was relatively rapid. The effect of the finite number of particles and finite time step on the results of the calculations was also systematically studied, and extrapolations to an infinite number of particles and zero time step were performed. It is worth mentioning that the systematic error originating from the finiteness of the *sample* was found to be one order of magnitude larger than the statistical error, in spite of the fact that the interaction between the particles and their images in the periodically extended space were used in an Ewald summation procedure (see, e.g., Ceperley, 1978) to eliminate the major surface effect.

The results of Ceperley and Alder (1980) for the charged Fermi and Bose systems are summarized in Table 5.1. A more direct interpretation of their findings is obtained from Fig. 5.1, where the quantity $r_s^2(E - E_{\text{BOSE}})$ is plotted against r_s. Thus, the energy of each phase is referred to the Bose ground state. It can be seen that the two curves corresponding to the paramagnetic ground state and to the Wigner crystal, respectively, intersect when $r_s \approx 80$ Bohr radii.

The interest in the fully polarized or ferromagnetic state was indicated by work of Bloch (1928) within the Hartree–Fock approximation. Bloch's theory represents the simplest example of spin density functional theory. Of course, it is now well known that Hartree–Fock theory predicts too readily the existence of ferromagnetism. This is because it correlates parallel-spin electrons essentially correctly through the Fermi hole, whereas antiparallel-spin electrons are uncorrelated. Thus, the energy of the fully ferromagnetic state is predicted more accurately than that of the paramagnetic state. In particular, Bloch's theory leads to ferromagnetism for $r_s > 6$. This is only a slightly larger r_s than for metallic cesium,

TABLE 5.1. Ground-state Energy of the Charged Fermi and Bose Systems.

r_s[a]	PMF[b]	FMF[c]	BF[d]	*bcc*[e]
1.0	1.174(1)[f]	—	—	—
2.0	0.0041(4)	0.2517(6)	-0.4531(1)	—
5.0	-0.1512(1)	-0.1214(2)	-0.21663(6)	—
10.0	-0.10675(5)	-0.1013(1)	-0.12150(3)	—
20.0	-0.06329(3)	-0.06251(3)	-0.06666(2)	—
50.0	-0.02884(1)	-0.02878(2)	-0.02927(1)	-0.02876(1)
100.0	-0.015321(5)	-0.015340(5)	-0.015427(4)	-0.015339(3)
130.0	—	—	-0.012072(4)	-0.012037(2)
200.0	—	—	-0.008007(3)	-0.008035(1)

[a]Wigner sphere radius in units of Bohr radii.
[b]Paramagnetic (unpolarized) Fermi fluid.
[c]Ferromagnetic (polarized) Fermi fluid.
[d]Bose fluid.
[e]Bose crystal with *bcc* lattice.
[f]Digits in parentheses represent the error bar in the last decimal place.

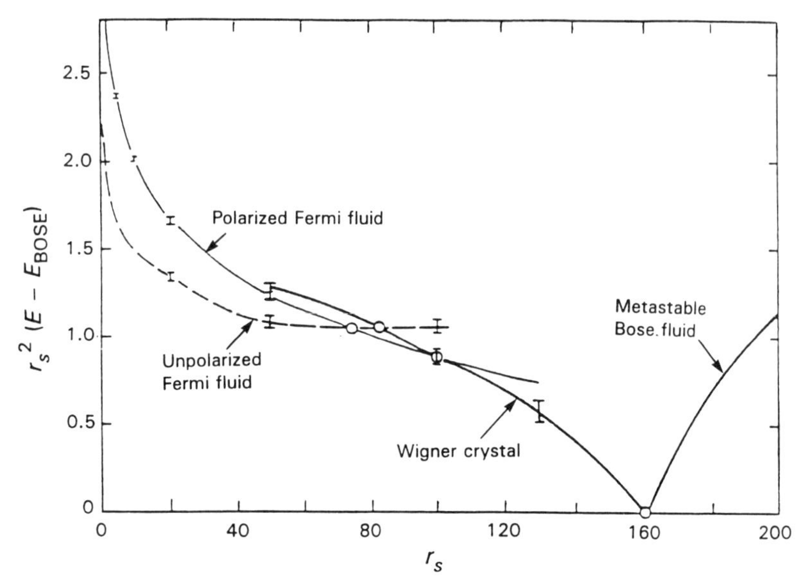

FIGURE 5.1. The energy of the four phases studied relative to that of the lowest boson state times r_s^2 in rydbergs versus r_s in Bohr radii. Below $r_s = 160$ the Bose fluid is the most stable phase, while above, the Wigner crystal is most stable. The energies of the polarized and unpolarized Fermi fluid are seen to intersect at $r_s = 75$. The polarized (ferromagnetic) Fermi fluid is stable between $r_s = 75$ and $r_s = 100$, the Wigner crystal above $r_s = 100$, and the normal paramagnetic Fermi fluid below $r_s = 75$. (After Ceperley and Alder, 1980; redrawn from Senatore and March, 1994).

the simple metal of lowest electron density. There is no sign of a tendency to ferromagnetism in the physical properties of this metal.

In fact, as comparison of the curves from computer experiments displayed in Fig. 5.1 shows, the ferromagnetic state does not become stable with respect to the paramagnetic state until $r_s > 70$, indicating the vital importance of electron correlation in discussing this magnetic transition (see also Herman and March, 1984). This ferromagnetic state intersects the Wigner crystal at $r_s \approx 100$. It is to be noticed that around such values of r_s, the Bose and Fermi crystals differ in energy by an amount which is less than 1.0×10^{-6} Rydbergs.

While Ceperley and Alder point out that their computer studies need refinement, there can be little doubt that the Wigner crystal becomes stable in the range $70 < r_s < 100$ and this, of course, is very valuable information. It is to be noted that one area of obvious importance if the Wigner crystal is to be unambiguously identified in three dimensions is the question of the temperature at which melting of the electron crystal occurs (see Ferraz *et al.*, 1979; March, 1988).

5.5. CALCULATION OF CORRELATION ENERGY FOR SMALL MOLECULES

Configuration interaction (CI) calculations have been able to account typically for about 80% of the correlation energy of molecules such as water (see, for example, Rosenberg and Shavitt, 1975; Meyer, 1971). However, interesting chemistry occurs on an energy scale of only a fraction of the correlation energy. For example, the O—H bond strength in water is about 50% of the correlation energy. Thus, the correlation energy computed using large-CI wave functions differs from the exact (nonrelativistic, Born–Oppenheimer) energy by this same order of magnitude. Furthermore, it can be difficult to improve CI results because convergence to the exact result is slow and can be nonuniform (see, however, very accurate energies by Mårtensson-Pendrill *et al.*, 1991). In addition, the numerical effort involved in CI calculation for a system of N electrons increases with a power of N which is between 4 and 5.

The quantum Monte Carlo method, on the contrary, appears to be free of the limitations inherent to an expansion procedure. However (see Reynolds *et al.* (1982), Ceperley, 1986; Hammond *et al.* (1987) in QMC the computational effort increases strongly with the number of electrons. Though this approach was developed and used primarily in the fields of nuclear and condensed matter theory (cf. Section 5.4), chemical calculations have subsequently been carried out (Anderson, 1975, 1976, 1980; Mentch and Anderson, 1981; Alder *et al.*, 1982; Moskowitz and Kalos, 1981; Moskowitz *et al.*, 1982 a & b; Ceperley and Alder, 1984; Moskowitz and Schmidt 1986). In the following some of these calculations will be briefly reviewed. In particular, the quality of calculations performed

(i) with diffusion Monte Carlo,
(ii) with the Green function Monte Carlo,
(iii) within the fixed node approximation, and
(iv) allowing for nodal relaxation,

will be illustrated

The DMC method has been applied by Reynolds *et al.* (1982) to the calculation of the ground-state energy of some small molecules H_2, LiH, Li_2, H_2O, within the fixed-node approximation. Trial wave functions of different sophistication were considered by these authors to demonstrate the importance of ψ_T on the efficiency of the sampling. All their importance functions were in the form of a product: a Slater determinant for each of the two groups of electrons with given spin projection times a correlation factor of Jastrow type. Of course, the Slater determinants were included to ensure the symmetry associated with the particular choice of the total spin projection, whereas the Jastrow factor was such as to reproduce the correct cusp behavior of the wave function as the electrons approach one another. They have considered three kinds of trial wave function corresponding to Slater determinants constructed, respectively, from

(i) a minimal basis set of Slater-type atomic orbitals (see, for instance, March and Mucci, 1993),
(ii) a somewhat enhanced basis set and/or an optimized version of (i), and
(iii) localized Gaussian orbitals.

In case (iii) the Jastrow factor contained additional terms to reproduce also the cusp behavior associated with the electron-nuclear Coulomb attraction: this should have the effect of making the local energy associated with ψ_T even smoother.

Ground-State Energies of Small Molecules. The fixed node (FN) QMC ground-state energies for some molecules are recorded in Table 5.2 for the three choices of trial function ψ_T listed above. Also reported for comparison are the energies obtained using the Hartree–Fock (HF) approximation, the best CI calculations, and the exact clamped nuclei or Born–Oppenheimer approximation, all in the usual nonrelativistic framework afforded by the many-electron Schrödinger equation. All energies are recorded in hartrees. It is clear that, with the exception of the water molecule, the fixed-node energy accounts for most of the correlation energy. There are improvements in going from simpler to more sophisticated wave functions, so that with the best trial wave function (III), the fixed-node energy accounts for 95% of the correlation energy or more. It should be noted that in the case of the hydrogen molecule the ground-state wave function has no nodes. Therefore, differences between fixed node and exact energies of H_2, which are beyond the statistical error, give a measure of the numerical error associated with the use of a finite time step in the diffusion Monte Carlo calculation. In the case of the water molecule, it seems that none of the trial functions used is of especially

TABLE 5.2. Total Ground-State Energy of Some Small Molecules[a]

	Energy (hartrees)			
	H_2	LiH	Li_2	H_2O
HF[b]	−1.1336	−7.987	−14.872	−76.0675
FN-I	−1.1745(8)[c]	−8.047(5)	−14.985(5)	−76.23(2)
FN-III		−8.059(4)	−14.991(7)	−77.377(7)
FN-III	−1.174(1)	−8.067(2)	−14.990(2)	
Best CI	−1.1737	−8.0647	−14.903	−76.3683
Exact	−1.17447	−8.0699	−14.9967	−76.4376

[a]From Reynolds *et al.* (1982).
[b]Symbols explained in the text.
[c]Figure in parentheses is the statistical error.

good quality. In this case, only about 80% of the correlation energy is accounted for by the fixed-node calculations.

Reynolds *et al.* (1982) have also performed fixed-node calculations for the ground-state energy of the Li diatomic molecule at various internuclear distances around the equilibrium value. The results of such calculations, based on a trial function ψ_T of type II, are recorded in Table 5.3, along with the Hartree–Fock and exact energies. It is found that 90% or more of the correlation energy is recovered in this example as well.

As noted by Reynolds *et al.* (1982), the Born–Oppenheimer approximation, adopted throughout their work, can also be relaxed. This can be achieved by allowing the nuclei, as well as the electrons, to diffuse. The diffusion constant for each nucleus is then $\hbar/2M$, where M is the nuclear mass. Thus the nuclei diffuse considerably more slowly than the electrons; this fact may make the calculations lengthier.

To briefly summarize, it is found that using relatively simple trial functions ψ_T and only modest computational effort, one can obtain with the fixed node QMC at least as much, and often more, of the correlation energy than proves possible by CI calculations to date, for simple molecules.

Fixed-node calculations for small molecules using the Domain Green function (DGF) method have been performed by Moskowitz and Schimdt (1986). As these authors stress, the use of the DGF method is free of the systematic error present in the DMC because of the finite time step. It is worthwhile to comment briefly about their results. Among the systems that they consider are the LiH molecule, the Be atom, and the BeH_2 molecule. With their calculations they can reproduce ground-state correlation energies within chemical accuracy, i.e. to within a few percent, when using the best trial functions. The same degree of accuracy is also found in the prediction of excitation energies for the Be atom and the energy barrier for insertion of Be in H_2.

TABLE 5.3. Ground-State Energies at Selected Nuclear Separations for Li_2

	Energy (hartrees)		
r(Bohr)	HF[a]	FN-II[b]	Exact
3	−14.786	−14.905	−14.915
4	−14.853	−14.968	−14.983
5.05 (equil.)	−14.872	−14.991	−14.997
6	−14.869	−14.985	−14.992
7	−14.859	−14.976	−14.982

[a]Symbols as in Table 5.2.
[b]Typical statistical uncertainty, 0.005 hartree.

It should be stressed once again that the quality of fixed-node energies depends on the capability of the chosen trial wave function to reproduce the nodal surfaces of the exact ground state. Thus, while the fixed-node approximation yields upper bounds to the exact energy, the systematic optimization of such results does not appear to be easy. In this respect the nodal relaxation technique, which has been developed by Ceperley and Alder (1980) for the electron gas in the first instance, may prove valuable. One has to keep clearly in mind, though, that such a technique gives only transient estimates for the energy. In other words, if the relaxation is performed for overly long times, it becomes intrinsically unstable, with the interesting signal that decreases exponentially while the noise remains constant. Thus, the nodal relaxation technique crucially depends on the availability of a good starting guess for the ground-state wave function, and this may be provided by a preliminary (or in actual practice contemporary) fixed-node calculation.

Ceperley and Alder (1984) have performed further calculations on some of the small molecules previously considered with the DMC method, improving on them in two ways. They have switched to Green function Monte Carlo, utilizing a technique, introduced by Ceperley (1983), to sample the exact Green function, thus eliminating any error associated with the use of a finite time step. Furthermore, they have also implemented the nodal relaxation. In these calculations they have also considered the H_3 molecule. A brief summary of their study will be given as an example of how nodal relaxation can work. Since a description in some detail of the main points of the nodal relaxation method has already been given, it will suffice for present purposes to report their results for the ground-state energies of the molecules considered. Then the problem of the convergence of the nodal relaxation procedure will be referred to. It is further to be noted that for these calculations, with the exception of those for the H_2O molecule, trial wave functions of type III as used in the work of Reynolds *et al.* were employed. For H_2O a more sophisticated trial function ψ_T was used. The interested reader must refer to the original paper by Ceperley and Alder (1984) for the details of their trial functions.

In Table 5.4, the ground-state energies for LiH, Li_2, and H_2O are recorded, as well as those for three different configurations of H_3 denoted by I, II, and III. At first glance, the nodal relaxation appears in all cases to reduce the fixed-node energy to agree with the exact energy. However, a few comments are called for. It is true that the relaxed-node energies coincide with the exact values to within the statistical error. Nevertheless, the statistical error increases with increasing total energy. In addition, as has already been mentioned, the nodal relaxation can provide only transient estimates of the fermionic ground-state energy. Therefore, it appears necessary to examine the convergence of the nodal relaxation in detail. To this end one can consider the total energy as a function of the generation

TABLE 5.4. Comparison of Fixed-Node and Relaxed-Node Ground-State Energies with CI and Exact Results

Molecule	Energy (hartree)				
	FN[a]	RN[b]	Δ[c]	Exact	CI
H_3(I)	−1.6581(3)	−1.6591(1)	0.0009(2)	−1.65919	−1.65876
H_3(II)	−1.6239(3)	−1.6244(3)	0.0005(2)	−1.62451	−1.62337
H_3(III)	−1.6606(2)	−1.6617(2)	0.0011(2)	−1.66194	−1.66027
LiH	−8.067(1)	−8.071(1)	0.004(1)	−8.0705	−8.0690
Li_2	−14.990(2)	−14.994(2)	0.004(1)	−14.9967	−14.903
H_2O	−76.39(1)	−76.43(2)	0.04(1)	−76.437	−76.368

[a]Symbols (except RN) as in Table 5.3 and 5.3.
[b]Relaxed-node energy.
[c]$\Delta = E_{FN} - E_{RN}$.

number, starting from the first generation after the beginning of nodal release. Such curves for two typical cases are shown in Figs. 5.2–5.5. In Fig. 5.2 the relaxing of the total energy for LiH is depicted. It is clear that in this case one can confidently speak of convergence. It is to be stressed that in the figure are also reported the results of runs in which the trial wave function was deoptimized to show how this destroys the convergence. The case of H_2O is illustrated in Fig. 5.3. Although the relaxed energy at the end of the run is in agreement with the exact energy, within the statistical uncertainty, there is no indication that convergence was reached, in that the slope of the energy curve does not seem to diminish.

A better way of studying the convergence of the nodal relaxation is to examine the energy difference between successive generations. Of course, a sign of convergence should be the vanishing of such a quantity after a sufficient number of generations. This quantity can also be evaluated more accurately than the total energy itself, as discussed by Ceperley and Alder (1984). In Fig. 5.4 the difference in release-node energy is shown for LiH. It is clear that with a good trial wave function the node relaxation can be considered as converged. However in the case of H_2O, reported in Fig. 5.5, it is not possible to say very much since the error bars increase too fast with the generation number even in the difference calculation.

The conclusion to be drawn from the above discussion is that whereas the nodal relaxation appears to work well with light molecules, there are still problems with the heavier ones. This is related to the fact that in the heavier molecules, because of the larger nuclear charges Z, the total energies are also larger and hence so are the error bars. Moreover, with increasing Z the difference between the Bose

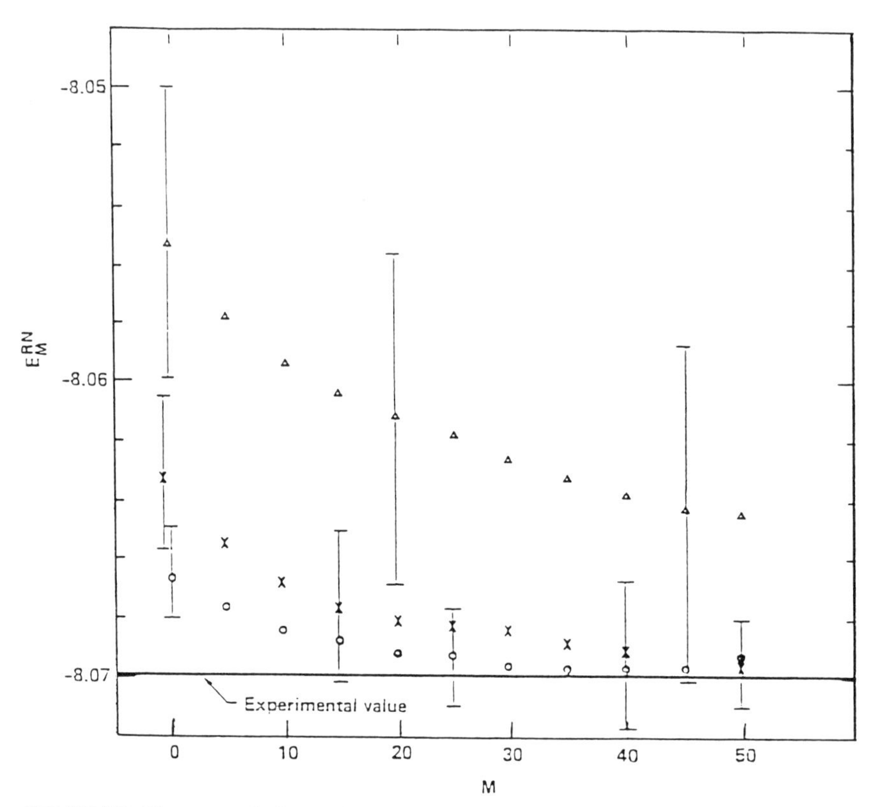

FIGURE 5.2. The energy (in hartrees) versus the number of generations after node release for the molecule LiH. The results for three different trial functions are shown. (O) indicates the results obtained with the best trial function. The parameters in the other two trial functions were deliberately deoptimized to raise the fixed-node energy. (From Ceperley and Alder, 1984: redrawn from Senatore and March, 1994).

and Fermi ground-state energies also increases, and correspondingly the rate (exponential!) at which the Bose component obscures the Fermionic one.

A number of ways to partially improve the QMC computations for heavier molecules are listed by Ceperley and Alder (1984). One way is based on a different treatment of inner electrons, possibly by means of pseudopotentials, so as to deal in practice with an equivalent problem with lower Z. Another route is to consider the possibility of deleting all the random walks that cross the nodes frequently. They also cite the possibility of calculating energy differences directly by means of correlated random walks. For references to practical calculations on heavier atoms and molecules, mainly exploiting the separation of electrons into core and valence, the reader may consult Senatore and March (1994).

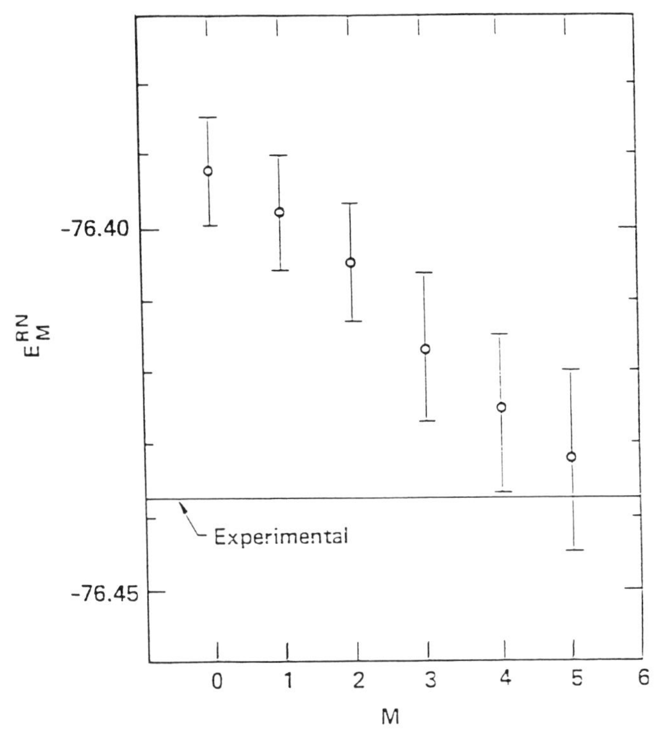

FIGURE 5.3. The energy (in hartrees) versus the number of generations after node release for the molecule H_2O. (From Ceperley and Alder, 1984: redrawn from Senatore and March, 1994).

5.6. AUXILIARY FIELD QUANTUM MONTE CARLO (AFQMC) METHOD AND HUBBARD MODEL

The central idea of the AFQMC method, as has already been anticipated, is to exactly rewrite the propagator of a many-particle system—with two-body interactions—in terms of a propagator for independent particles interacting with auxiliary external fields. While this procedure introduces the need for averaging over the values taken by such extra fields, it allows use to be made of well-established techniques to treat the propagator for independent particles. The strategy is relatively simple and consists of three main steps:

(i) Find the appropriate transformation that substitutes for the coupling between the particles a coupling to suitably chosen classical fields;

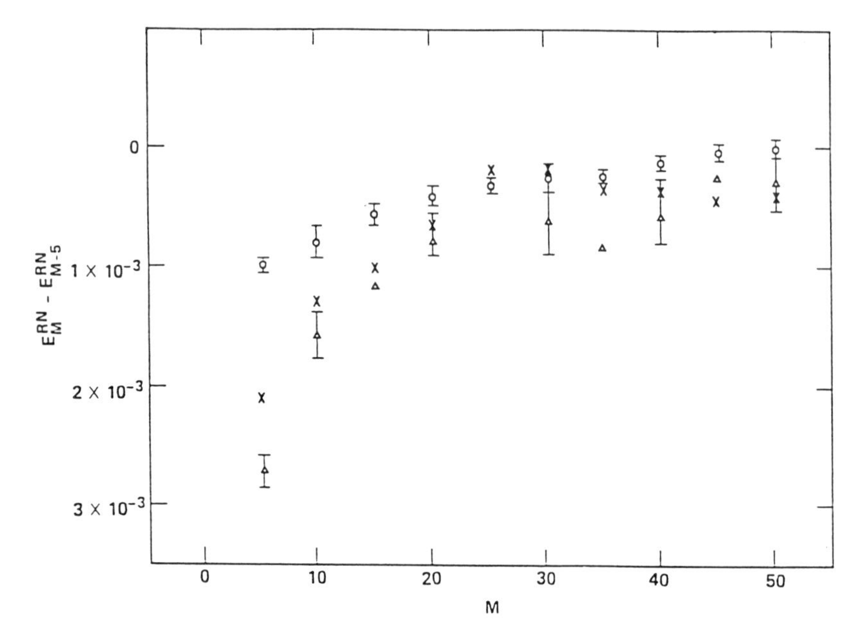

FIGURE 5.4. The change in release-node energy (in hartrees) every five generations after node release for the molecule LiH. (From Ceperley and Alder, 1984: redrawn from Senatore and March, 1994).

(ii) Solve the new problem of independent particles, i.e., calculate the trace (partition function) or a suitable matrix element of the propagator; and

(iii) Perform the average over the auxiliary fields, usually with techniques borrowed from classical statistical mechanics.

For the sake of simplicity, rather than considering the general case of an arbitrary Hamiltonian with quadratic interactions (see, e.g., Sugiyama and Koonin 1986; Negele and Orland, 1988), here the focus will be on the specific case of the Hubbard model, for which some applications will then be discussed.

The basic relation, which allows for the introduction of the auxiliary fields, is the Hubbard–Stratonovich (HS) identity (Hubbard, 1959; Stratonovich, 1958), valid for any Hermitian operator $\hat{O}$

$$e^{1/2\alpha^2\hat{O}^2} = \frac{1}{\sqrt{2\pi}} \int_{-\infty}^{\infty} dx\, e^{-(1/2)x^2} e^{-\alpha x \hat{O}} \tag{5.37}$$

or appropriate extension of it (see below). The Hubbard Hamiltonian of Eq. (4.10) contains a one-body term, $K = \sum_{ij\sigma} t_{ij} a^{+}_{i\sigma} a_{j\sigma}$, and a two-body interaction $V = U\sum_i n_{i\uparrow} n_{i\downarrow}$. Therefore it appears as a good candidate for the HS transformation. However, the propagator $e^{-\beta H}$ contains the sum $H_0 + V$, rather than merely V. Here

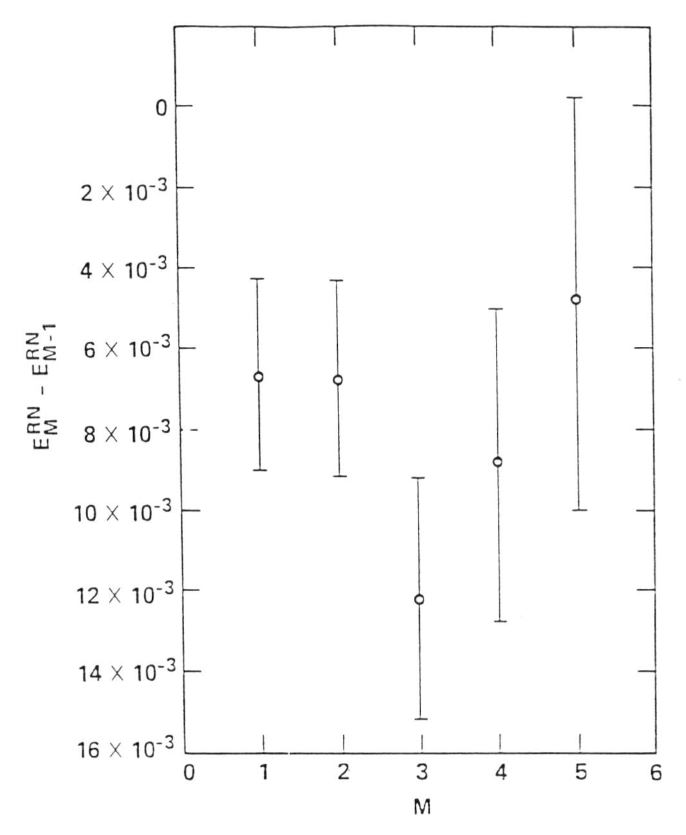

FIGURE 5.5. The change in release-node energy (in hartrees) every five generations after node release for the molecule H_2O. (From Ceperley and Alder, 1984: redrawn from Senatore and March, 1994).

$H_0 = K - \mu \sum_{i\sigma} a^+_{i\sigma} a_{i\sigma}$. The trick to circumvent this difficulty is to resort to the Trotter formula

$$e^{-\beta(H_0+V)} = \prod_{r=1}^{L} e^{-\varepsilon(H_0+V)} = \prod_{r=1}^{L} e^{-\varepsilon H_0} e^{-\varepsilon V} + O(\varepsilon), \tag{5.38}$$

with $\varepsilon = \beta/L$, so that the exponential of V appears explicitly. In fact, V still needs some rearrangement. One possibility is to rewrite $e^{-\varepsilon V}$—in each of the time slices generated by the Trotter formula—using the relation (Hirsch, 1983)

$$n_{i\uparrow} n_{i\downarrow} = -\tfrac{1}{2}(n_{i\uparrow} - n_{i\downarrow})^2 + \tfrac{1}{2}(n_{i\uparrow} + n_{i\downarrow}), \tag{5.39}$$

which is valid for fermions ($n^2_{i\sigma} = n_{i\sigma}$). Then, use of the HS transformation in Eq. (5.38) plus elementary manipulations yield the propagator in a new form:

$$U(\beta) = e^{-\beta H} = \left[\frac{\varepsilon}{2\pi}\right]^{NL/2} \int \prod_{r=1}^{L} \prod_{i=1}^{N} dx_{ri} e^{-(\varepsilon/2)x_{ri}^2} \prod_{r'=1}^{L} e^{-\varepsilon h_{r'}(x)}, \tag{5.40}$$

valid to order ε, with

$$h_r(x) = H_{0r} + \sum_i \left[\sqrt{U} x_{ri}(n_{ri\uparrow} - n_{ri\downarrow}) + \frac{U}{2}(n_{ri\uparrow} + n_{ri\downarrow}) \right] \tag{5.41}$$

and N the number of sites of the finite lattice on which the Hubbard model is considered. It is evident that the Hamiltonian $h_r(x)$ contains only one-body operators: in particular, it describes independent particles moving in an external field $\sqrt{U}x_{ri}$. Thus, through a Gaussian averaging, the imaginary time propagator $U(\beta)$ is related to a new propagator for independent particles

$$U_\varepsilon(\beta, x) = \prod_{r=1}^{L=\beta/\varepsilon} e^{-\varepsilon h_r(x)}. \tag{5.42}$$

A number of comments are now called for. The accuracy of Eq. (5.40) increases with increasing $L = \beta/\varepsilon$, and this alternative representation of the propagator becomes exact only in the limit $\varepsilon \to 0$ ($L \to \infty$). However, one can study such a convergence systematically and accordingly estimate the systematic error introduced by a finite Trotter time ε. The reader will have noted the presence in Eq. (5.40) of a time index denoted by r. This new labelling simply keeps track of the Trotter time slice.

5.6.1. Partition Function

As anticipated, quantities that one typically wants to calculate are the partition function $Z = \mathrm{Tr}\, U(\beta)$ or a *projected* partition function $Z_T = \langle \Phi_T | U(\beta) | \Phi_T \rangle$, where Φ_T is a trial wave function nonorthogonal to the ground state. In either case, for $\beta \to \infty$ one obtains a limiting behavior $Z(Z_T) \propto e^{-\beta E_0}$. Thus, one can obtain E_0 as

$$E_0 = -\lim_{\beta\to\infty} \frac{1}{\beta} Z(\beta) = -\lim_{\beta\to\infty} \frac{1}{\beta} Z_T(\beta). \tag{5.43}$$

From Eqs. (5.40–5.41) it is clear that the evaluation of Z requires that of

$$\tilde{Z}(x) = \mathrm{Tr}\, U_\varepsilon(\beta, x). \tag{5.44}$$

This can be evaluated (Blankenbecler *et al.*, 1981) in terms of a determinant involving the matrices associated with the Hamiltonians $h_r(x)$ of Eq. (5.41). Once $\tilde{Z}(x)$ is found, the problem of calculating Z is reduced to that of evaluating a multidimensional integral over the variables $\{x_{ri}\}$. In fact, one readily obtains

$$Z = \int dx e^{-\beta \tilde{U}(x)} \operatorname{sgn}[\tilde{Z}(x)], \tag{5.45}$$

where

$$\int dx \equiv \left[\frac{\varepsilon}{2\pi}\right]^{NL/2} \int \prod_{r=1}^{L} \prod_{i=1}^{N} dx_{ri}, \tag{5.46}$$

and

$$\tilde{U}(x) \equiv \frac{1}{2L} \sum_{r=1}^{L} x_{ri}^2 - \frac{1}{\beta} \ln |\tilde{Z}(x)|. \tag{5.47}$$

Thus, the problem of calculating the partition function for a quantum system is reduced to the classical problem of calculating a configurational integral. One may merely note that numerical problems can, however, arise from the possible presence of nodal surfaces in the function $\tilde{Z}(x)$. These can give

(i) problems of ergodicity in sampling the integral of Eq. (5.44) as well as
(ii) problems of signal-to-noise ratio if Z results from the cancellation of large contributions of opposite sign.

This is a manifestation of the so-called fermion sign problem.

The considerations above, and in particular Eqs. (5.44)–(5.46), remain valid if one calculates a projected partition function—the sole difference being that Z and $\tilde{Z}(x)$ are replaced by Z_T and $\tilde{Z}_T(x) = \langle \Phi_T | U_\varepsilon(\beta, x) | \Phi_T \rangle$, respectively. Appropriate techniques are available (Sugiyama and Koonin, 1986; see also Sorella, 1989) to evaluate $Z_T(x)$. With manipulations similar to those summarized above, it is also possible to reduce the calculation of relevant correlation functions to that of pseudoclassical averages. It seems appropriate to add here that, in particular for the Hubbard Hamiltonian, it has proved possible (Hirsch, 1983) to write a discrete transformation of the HS type, i.e.,

$$\exp(-\varepsilon U n_{i\uparrow} n_{i\downarrow}) = \frac{1}{2} \sum_{\sigma=\pm 1} \exp(-\varepsilon [J\sigma(n_{i\uparrow} - n_{i\downarrow}) + U(n_{i\uparrow} + n_{i\downarrow})/2] \tag{5.48}$$

with $\cosh(\varepsilon J) = \exp(\varepsilon U/2)$. This transformation then maps the quantal Hubbard problem onto an equivalent classical problem, where one has to sample over Ising variables rather than continuous ones. According to Hirsch (1986), in calculations using the transformation of Eq. (5.48) the sign problem is not severe.

5.6.2. Results for Two-Dimensional Hubbard Model

To give a practical illustration of the AFQMC, work on the two-dimensional Hubbard model, which was stimulated by its possible relevance to the high T_c superconductors will next be discussed (see also Appendix 6.1).

Hirsch and Tang (1989) studied the two-dimensional Hubbard model with

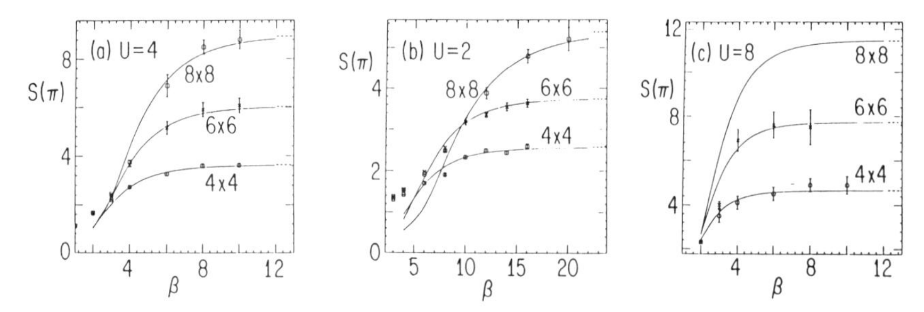

FIGURE 5.6. Magnetic structure factor for $S[\mathbf{q} = (\pi, \pi)]$ vs. imaginary time β for lattices of size 4×4, 6×6, and 8×8, at half-filling. The solid line shows results of spin-wave theory (Hirsch and Tang, 1989). The dashed lines at the right side indicate the $T = 0$ limits of the spin-wave results. (After Hirsch and Tang, 1989: redrawn from Senatore and March, 1994).

nearest-neighbor hopping for lattices as big as 8×8 and for imaginary times up to $\beta = 20t$, with $-t < 0$ being the hopping energy. These workers performed AFQMC simulations based on the sampling of the partition function Z for different values of the coupling U and of the filling ρ, defined as the ratio between the number of electrons N_e and the number of sites in the lattice $N : \rho = N_e/N$. Because of electron–hole symmetry, one can take $\rho \leq 1$. The doping can be conveniently defined as $\delta = 1 - \rho$. It is to be noted that, since the lattice can accomodate $2N$ electrons, $\rho = 1$ corresponds to half-filling, and similarly $\rho = 0.5$ to a quarter filling. Hirsch and Tang were especially interested in the magnetic properties of the system, which can be characterized by the magnetic structure factor

$$S(\mathbf{q}) = \frac{1}{N} \sum_{ij} e^{i\mathbf{q}\cdot(\mathbf{l}_i - \mathbf{l}_j)} \langle (n_{i\uparrow} - n_{i\downarrow})(n_{j\uparrow} - n_{j\downarrow}) \rangle, \tag{5.49}$$

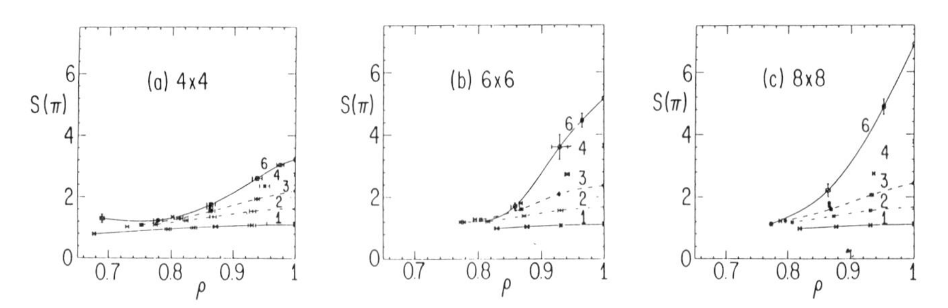

FIGURE 5.7. Magnetic structure factor for $S[\mathbf{q} = (\pi, \pi)]$ vs. band-filling for $U = 4$, with 4×4, 6×6, and 8×8 lattices at various imaginary times. The numbers next to the curves indicate β, the curves being drawn through the points to guide the eye. (After Hirsch and Tang, 1989: redrawn from Senatore and March, 1994).

and in particular by its value at the wave vector $\mathbf{q}_m = (\pi, \pi) \equiv \pi$, $S(\pi)$. They find that at half-filling the magnetic structure factor exhibits a sharp peak at $\mathbf{q}_m$. Such a peak, shown in Fig. 5.6, tends to become more and more pronounced with increasing lattice size or coupling U. Also apparent—from the same figure—is the saturation of $S(\pi)$ with increasing simulation time β. In fact, as the time becomes shorter the drop in $S(\pi)$ becomes more and more pronounced for larger lattices and couplings. The presence of such a peak in $S(\mathbf{q})$ is interpreted as a sign of anti-

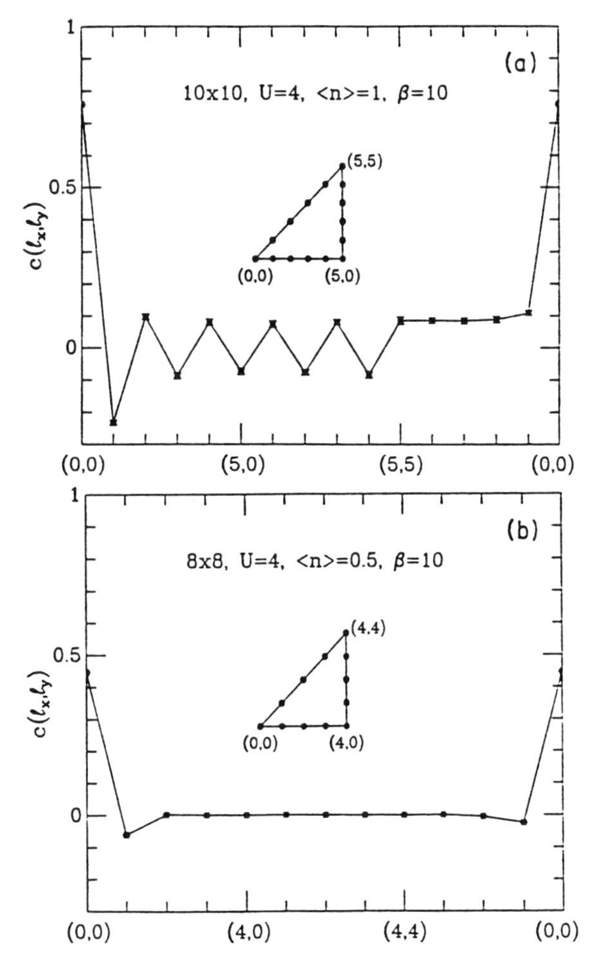

FIGURE 5.8. Spin–spin correlation function $c(l_x, l_y)$. The horizontal axis traces out the triangular path shown in the center of the figure. Strong antiferromagnetic correlations are visible in (*a*), which is for a half-filled band ($<n> \equiv \rho = 1.0$), but nearly absent in (*b*), which is at quarter-filling ($<n> \equiv \rho = 0.5$). (After White *et al.*, 1989: redrawn from Senatore and March, 1994).

ferromagnetic long-range order. As they move from ρ = 1, they find a rapid suppression of such ordering, as can be seen from Fig. 5.7, and they conjecture that the antiferromagnetic order disappears immediately away from half-filling. Hirsch and Tang also find that at low temperatures (long imaginary times) the properties of the two-dimensional Hubbard model at half-filling are well described by spin-wave theory with renormalized local moment and spin-wave velocity.

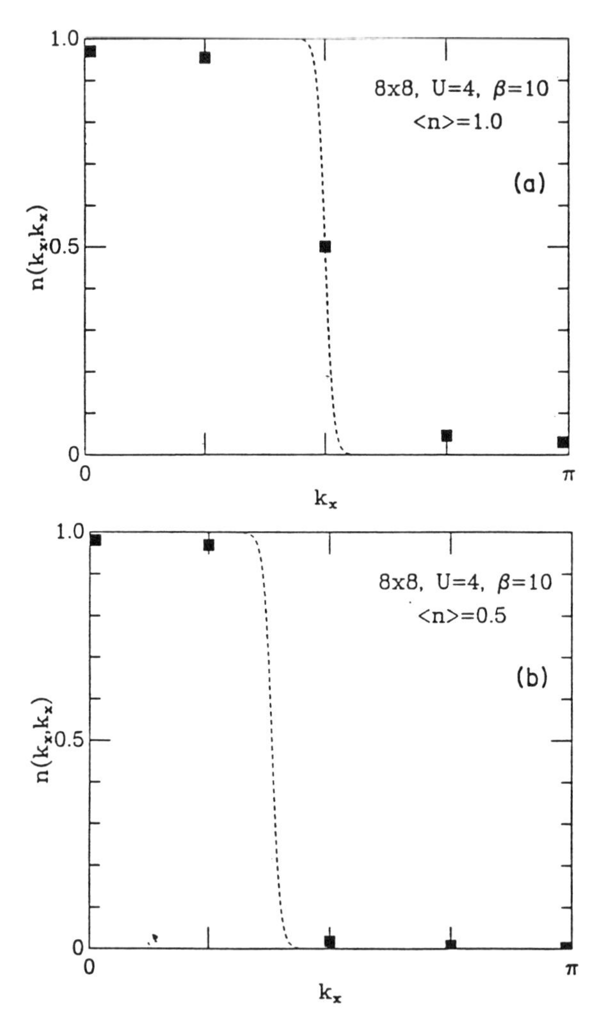

FIGURE 5.9. The momentum distribution $n(\mathbf{k})$ for an 8×8 lattice with $U = 4$ and $\beta = 10$ at (a) half-filling ($\langle n\rangle \equiv \rho = 1.0$) and (b) quarter-filling ($\langle n\rangle \equiv \rho = 0.5$). The dashed curves are the $U = 0$ results. (After White *et al.*, 1989: redrawn from Senatore and March, 1994).

The results above are fully confirmed by independent investigations. In Fig. 5.8 the results of White *et al.* (1989) for the spin–spin correlation function

$$c(l_x, l_y) = \frac{1}{N} \sum_i \langle m(\mathbf{l}_i + \mathbf{l}) m(\mathbf{l}_i) \rangle, \tag{5.50}$$

are shown, where $m(\mathbf{l}_i)$ is the local magnetic operator along the z-axis. Clearly, while staggered spin correlations are clearly visible for half-filling, these are nearly absent for the quarter-filled case. In fact results of Sorella (1989), obtained using the projected partition function technique and Langevin dynamics to sample the pseudoclassical partition function, show that even with a doping δ of only 12% such correlations are practically absent. Thus the conjecture of Hirsch and Tang on the suppression of antiferromagnetic order away from half-filling seems fully confirmed.

White *et al.* have also presented results for the single-particle momentum occupation $n(k) \equiv \langle n_{k\sigma} \rangle = \langle a^+_{k\sigma} a_{k\sigma} \rangle$. As can be seen from Fig. 5.9, the one-electron momentum distribution for the interacting electrons at half-filling is broadened more than at a quarter-filling, with respect to the noninteracting electron case. This is confirmed also by a subsequent study of Moreo *et al.* (1990), which in addition, from investigation of the compressibility and of the electron self-energy, provide evidence for a one-electron gap at half-filling. Of course, this is consistent with the presence of antiferromagnetic long-range order. On the contrary, there is no evidence of a gap at a quarter-filling, where the system appears to be in a Fermi-liquid state—as is also suggested by the sharpness of the one-electron momentum distribution, which more closely resembles that for noninteracting electrons. Thus, one can conclude that in two-dimensions the Hubbard model predicts antiferromagnetic long range order for any finite U at half-filling behaving like an insulator, whereas such order disappears on moving away from $\rho = 1$ and the system behaves rather like a paramagnetic Fermi liquid.

Chapter 6

Quasiparticles and Collective Excitations (Especially Plasmons)

The methods of the preceding chapter are very powerful for calculating the correlation energy of many-electron systems. In the present chapter, as well as the next, attention will be focused primarily on the interpreation of specific physical phenomena that arise solely from electron–electron interactions.

Already, in Chapter 3, references has been made to the organized charge-density oscillations that can occur in a homogeneous electron fluid because of the long-range Coulomb interaction e^2/r_{ij} between electrons ij at separation r_{ij}. In the present chapter, such density oscillations, with their associated quanta, plasmons, will be studied, but now in the inhomogeneous electron assembly existing in a periodic crystal. Then, as with all wave propagation in period structures, one can expect, at least in principal, allowed and forbidden bands, i.e., a plasmon band structure (see, for example, March and Tosi, 1972, 1973).

Before discussing "state of the art" studies on plasmons, however, a brief review will first be given relating to collective excitations and quasiparticles (see also March and Parrinello, 1982; Mahan, 1989; and Engel *et al.,* 1991a, b).

6.1. ELEMENTARY EXCITATIONS

It is useful in a many-electron system to classify elementary excitations in two categories:

i. Particle-like excitations, and
ii. Collective phenomena.

The related quanta of category (i) excitations may be quasielectrons, quasi-holes, polarons, or the like, while those of the latter category are plasmons, phonons, magnons, etc. While in the present chapter, as already mentioned, plasmons will be singled out from the collective excitations, other such quasiparticles as magnons and phonons are discussed fully in the book by March and Parrinello (1982: for a more mathematical account, see also Sewell, 1986). Of course, elsewhere in the present volume, other quasiparticles will come up in specific contexts, e.g., with relevance to theories that exist at the time of writing for high T_c superconductivity (see Appendices 7.3 and 7.4 in particular).

As already implied, not only quasiparticle band structures in crystals (see, for example, the case of solid iodine in Chapter 7), but also knowledge of plasmon band structures, affords valuable information on the many-electron system under discussion. Plasmon energies are now readily measured—e.g., in fast-electron experiments, where such electrons lose energy to the plasmon excitations.

The theory of plasmon excitations and their dispersion relations has been widely studied for the uniform electron fluid (see, for example, the extensive review of Singwi and Tosi, 1981). The results have been applied to simple metals and are touched on in Section 6.2. However, the main focus will be directed here to plasmons in semiconductors. In particular, a two-band model (Engel *et al.*, 1991b) will be treated in some detail.

As is exemplified in Appendix 6.1, following March and Tosi (1972), one can treat plasma waves in periodic crystals either by an $\mathbf{r}$ space approach, in which one focuses on equations of motion for the density $\rho(\mathbf{r}, t)$, or from a $\mathbf{k}$ space viewpoint. Here, some attention will also be directed to the validity and usefulness of the plasmon picture in describing the dielectric properties of crystalline materials. This will necessitate an introduction to a particular many-electron approximation, the so-called GW approach (see especially Section 6.4 below). This will be introduced, somewhat intuitively. First by referring to the work of Inkson and Bennett (1977) on silicon, and in particular to the role of exchange and correlation. Then a more formal approach to the GW technique will be presented and developed with specific reference to plasmons in semiconductors.

6.2. EQUATION OF MOTION FOR ELECTRON DENSITY $\rho(\mathbf{R}, t)$

In early work, Bloch (1933, 1934) generalized the Thomas–Fermi electron density theory referred to in Chapters 2 and 3 to treat the density associated with plasma oscillations (the so-called hydrodynamic theory). An excellent account of this is given by Lundqvist (1983). Following March and Tosi (1972), this approach can be generalized within the framework of density functional theory, thereby removing at least formally all semiclassical Thomas–Fermi approximations from the foundations of the theory. This treatment was designed specifically to deal with plasmons propagating in periodic lattices and has been reviewed by the writer elsewhere (March, 1982). It will suffice therefore here to write down only the elementary basis for this approach, which has been applied to plasmon losses in nearly-free electron metals (March and Tosi, 1973).

Let us summarize the above procedure first for the homogeneous electron fluid of density ρ_0, say. Suppose this density of electrons is displaced relative to the nonresponsive positive neutralizing jellium background and allowed to oscillate such that the time-dependent density $\rho(\mathbf{R}, t)$ can be written as

$$\rho(\mathbf{R}, t) = \rho_0 + \rho_1(\mathbf{R}, t), \tag{6.1}$$

where $\rho_1(\mathbf{R}, t)$ is assumed to be small compared to ρ_0; that is, one is concerned only with the problem of small oscillations around the ground-state density. Then it is straightforward to derive from Maxwell's equations an equation of motion for ρ_1, which has the elementary form

$$\frac{\partial^2 \rho_1(\mathbf{R}, t)}{\partial t^2} = -\omega_p^2 \rho_1(\mathbf{R}, t), \tag{6.2}$$

where, as usual, ω_p denotes plasma frequency, related to the density of the electron fluid ρ_0 by

$$\omega_p = (4\pi e^2 \rho_0/m)^{1/2}. \tag{6.3}$$

It is also known that the next degree of approximation, in which dispersion of the plasma waves is incorporated into the homogenous system, can be written

$$\frac{\partial^2 \rho_1(\mathbf{R}, t)}{\partial t^2} = -\omega_p^2 \rho_1(\mathbf{R}, t) + \alpha \nabla^2 \rho_1(\mathbf{R}, t). \tag{6.4}$$

If one now seeks a solution of Eq. (6.4) by writing

$$\rho_1(\mathbf{R}, t) = n_k \exp(i\mathbf{k} \cdot \mathrm{R}) \sin \omega t, \tag{6.5}$$

then one is led almost immediately to the result

$$\omega^2(k) = \omega_p^2 + \alpha k^2. \tag{6.6}$$

The aim of the work of March and Tosi was to obtain, essentially, a generalization of the equation of motion [Eq. (6.4)] for a periodic crystal lattice. Here their main result will simply be stated, so that the general structure of the **R** space theory is exposed. Denoting by $V_i(\mathbf{R})$ the one-body (Slater–Kohn–Sham) potential energy of density functional theory, and defining the potential $V_1(\mathbf{R}, t)$ created by the density $\rho_1(\mathbf{R}, t)$ through Poisson's equation

$$\nabla^2 V_1(\mathbf{R}, t) = -4\pi e^2 \rho_1(\mathbf{R}, t), \tag{6.7}$$

the equation of motion proposed by March and Tosi [their Eq. (4.6)] reads:

$$\frac{\partial^2 \rho_1(\mathbf{R}, t)}{\partial t^2} + \omega_p^2(\mathbf{R})\rho_1(\mathbf{R}, t) - \frac{1}{m} \nabla \rho_0(\mathbf{R}) \cdot \nabla V_1(\mathbf{R}, t) =$$

$$= \frac{1}{m}\,\rho_1(\mathbf{R}, t)\nabla^2 V_i(\mathbf{R}) + \frac{1}{m}\,\nabla\rho_1(\mathbf{R}, t)\cdot\nabla V_i(\mathbf{R})$$

$$- \frac{1}{m^2}\int dx'\,dt'\,D(\mathbf{R}, \mathbf{x}', t - t') \times V_1(\mathbf{x}', t'). \tag{6.8}$$

This involves the periodic lattice through

1. the local plasma frequency $\omega_p(\mathbf{R})$, obtained from Eq. (6.3) by inserting the periodic density $\rho_0(\mathbf{R})$ for ρ_0;
2. the one-body potential V_i, which of course formally includes exchange and correlation (see Chapter 3); and
3. the kernal D.

In the approximation proposed by March and Tosi (1972), D is determined completely by the Bloch wave functions and the corresponding eigenvalues of the perfect lattice. In Bloch's hydrodynamic theory referenced above, the last term of Eq. (6.8) is further approximated by a local form [see March and Tosi, 1972, Eq. (4.7)].

Crystalline alkali metals: experiment and theory. The experimental study of the excitation spectrum of conduction electrons in the alkali metals has two main basic motivations: (i) the relative weakness of the electron-electron interactions in these metals apart from Li makes them the closest existing realization of the 3D jellium model (see Chapter 3), and (ii) their relatively low electron density ($3.93 \leq r_s \leq 5.62$ in going from Na to Cs) enhances the role of electron–electron correlations. Nevertheless, the role of the ionic cores in the electrons' dynamics can by no means be neglected.

The electron–ion interactions enter to modify the predictions made on jellium in several ways (see also March and Tosi, 1995). Firstly, the ionic cores provide a polarizable background tending to screen the plasma oscillations (Sturm 1983; Sturm, Zaremba and Nuroh 1990). Secondly, interband transitions induce shifts in the energy of plasmons and provide an additional mechanism for plasmon decay (for a review see Sturm 1982). Thirdly, the opening of gaps in the single-pair continuum introduces new collective states associated with zone boundaries (Foo and Hopfield 1968).

Of special relevance for their theoretical implications have been the high-resolution electron-energy-loss experiments carried out on the alkali metals from Na to Cs by vom Felde, Sprösser-Prou and Fink (1989). Their data on plasmon dispersion can be accurately fitted by the expression $\omega_p(k) = \omega_0 + \alpha\,\hbar\,k^2/m + Ck^4$, up to the critical wavenumber for the onset of Landau damping. A relatively weak anisotropy is revealed by measurements on a single crystal of Na for propagation along the [100] and [110] directions. The measured values of ω_0 lie somewhat below the theoretical value $(4\pi n e^2/m)^{1/2}$, the discrepancy being accounted for, to a good extent, by the inclusion of core polarizability from data on alkali metal ions. The dispersion coefficient α in K, Rb, and Cs increasingly departs from theoretical predictions and takes a sizably negative value in Cs.

The theoretical work on the problem posed by these data has examined explanations for the discrepancy requiring (a) a revision of the theory of plasmons in jellium, or (b) a prominent role of the electron–ion interactions. In the work of Lipparini *et al.* (1994) on jellium the inclusion of two-pair excitations suffices to account for the measured values of α in K and Rb, but not in Cs (see also Kalman, Kempa, and Minella 1991). On the other hand Taut and Sturm (1992) have used local and nonlocal density functional approaches within existing theories of exchange and correlation to obtain a variety of theoretical values for α which extends down to the measured values.

Turning to plasmon decay in the alkali metals as revealed in the experiments of vom Felde *et al.*, a finite plasmon half-width is observed at $k = 0$ and is shown to be proportional to the square of the main Fourier coefficient of the electron–ion pseudopotential. This indicates that electron scattering at Bragg planes, rather than scattering against phonons or grain boundaries, provides the main mechanism for the coupling of the long-wavelength plasma excitation in the metal with the single-pair continuum. According to theory (Sturm and Oliveira 1981) this band-structure contribution decreases with increasing k, but its decrease is overbalanced by the appearance of electron–electron scattering. While the observed plasmon damping in Na and K increases like k^2 as predicted, there is a serious discrepancy in its magnitude with the calculations of Sturm and Oliveira, especially for K. The underestimate of the dependence of plasmon damping on wavenumber is confirmed by the calculations of Bachlechner *et al.* (1993) on electron–electron scattering in jellium. Even qualitative agreement between theory and experiment is lacking in Rb and Cs, the data being fitted by a 1.5 ± 0.2 power of k in the former metal and showing a linear increase with k in the latter.

So far, the above treatment has been applied mainly to nearly free electron metals. Below, we shall give a lot of attention to semiconductors, and it is important therefore to introduce this by referring to treatments of exchange and correlation. This will be done, in an introductory manner, by discussing work on silicon.

6.3. EXCHANGE AND CORRELATION IN Si CRYSTALS

The work of Inkson and Bennett (1977) is the source of much of the present section. It is used here, specifically, to introduce a useful approximate technique (so-called GW, to be explained below; see also Appendix 6.8) into the many-electron problem, stemming largely from early work by Quinn and Ferrell (1958) and Hedin (1965). The problem of plasmons will then be taken up again within this GW framework (Engel *et al.*, 1991a,b).

6.3.1. Screened Exchange Part of Self-Energy

The method employed by Bennett and Inkson (1977) to obtain the screened exchange part of the self-energy was to find the Fourier components $\Sigma\,(\mathbf{k}+\mathbf{G}, \mathbf{k}+\mathbf{Q}; E(\mathbf{k}))$ of the self-energy potential $\Sigma(\mathbf{r}, \mathbf{r}'; E(\mathbf{k}))$. Their model for these Fourier components is diagonal and depends only on $|\mathbf{k}+\mathbf{G}|$ and the band index.

The dynamic form of the self-energy, following Hedin (1965), is approximated by

$$\Sigma\,(\mathbf{r}, \mathbf{r}', \omega) = \frac{i}{2\pi}\int_{-\infty}^{\infty} d\omega' G(\mathbf{r}, \mathbf{r}', \omega-\omega') W(\mathbf{r}, \mathbf{r}', \omega') \exp(-i\delta\omega'), \quad (6.9)$$

G being the interacting Green function and W the dynamic screened interaction referred to above, while δ is a positive infinitesimal. The presence of δ ensures that unoccupied states do not contribute to the screened exchange part of the self-energy.

Bennett and Inkson use the standard form (see also Appendix 6.3):

$$G(\mathbf{r}, \mathbf{r}', \omega) = \frac{\Sigma_{\mathbf{k}}[\phi_{\mathbf{k}}(\mathbf{r})\phi_{\mathbf{k}}^{*}(\mathbf{r}')]}{[\omega - E(\mathbf{k}) + i\,\delta(\mu-\omega)]} \quad (6.10)$$

for the Green function, the $\phi_{\mathbf{k}}$ being a complete orthonormal set of eigenfunctions for the system, having eigenvalues $E(\mathbf{k})$. They take these in their calculations as an empirically fitted set of pseudowave functions. (Here their approach is followed by using $\mathbf{k}$ in the extended zone scheme, unless states are labelled $(\mathbf{k}, \nu)$ when $\mathbf{k}$ is in the extended zone scheme, unless states are labelled $(\mathbf{k}, \nu)$ when $\mathbf{k}$ is in the first Brillouin zone and $\nu = 1,2,3,4$ is a band index.) In Eq. (6.10) for G, the quantity μ denotes the chemical potential.

Following Bennett and Inkson, the screened interaction is assumed to be diagonal, and thus one can write

$$W(\mathbf{q}, \mathbf{q}', \omega') = (2\pi)^3\, U(q)\delta(\mathbf{q}-\mathbf{q}')\varepsilon(\mathbf{q}, \omega'), \quad (6.11)$$

where $U(q)$ is simply the Coulomb form $4\pi e^2/q^2$ in Fourier transform.

Using the fact that the $\phi_{\mathbf{k}}$ are Bloch functions, we may write their Fourier components as

$$\langle \mathbf{q}|\phi_{\mathbf{k}}(\mathbf{r})\rangle = (2\pi)^3 \sum_{\mathbf{G}} \delta(\mathbf{k}+\mathbf{G}-\mathbf{q})a_{\mathbf{G}}(\mathbf{k})/V^{1/2}, \quad (6.12)$$

V denoting the volume of the crystal and, as usual, the $\mathbf{G}$ being reciprocal lattice vectors.

Taking the double Fourier transform of Σ, one finds

$$\sum(\mathbf{k}+\mathbf{G}, \mathbf{k}+\mathbf{Q}, \omega) =$$
$$\frac{i}{2\pi}\int_{-\infty}^{\infty} d\omega' \sum_{\mathbf{l},\mathbf{s}} \frac{a_{\mathbf{s}+\mathbf{G}}(\mathbf{l})a_{\mathbf{s}+\mathbf{Q}}(\mathbf{l})}{\omega-\omega'-E(\mathbf{l})+i\delta(\mu-\omega-\omega')} \times \frac{U(|\mathbf{l}+\mathbf{s}-\mathbf{k}|)}{\varepsilon(|\mathbf{l}+\mathbf{s}-\mathbf{k}|, \omega')\exp(-i\delta\omega')}, \tag{6.13}$$

the **s** being reciprocal lattice vectors and the **l** labelling the eigenstates. As usual, translational symmetry of the lattice has ensured that only Fourier components whose arguments differ by reciprocal lattice vectors may be nonzero.

Following Hedin and Lundqvist (1969), it is useful to split the contour integration in Eq. (6.13) into two parts:

$$\begin{aligned}\Sigma &= [\text{Residues of } G \times W(\omega - E(\mathbf{l}))] + [G(\omega_p) \times \text{Residues of } W] \\ &= \text{Screened exchange part} + \text{Coulomb hole part.}\end{aligned} \tag{6.14}$$

Then one obtains for the screened exchange (sx) contribution Σ_{sx} of the self-energy

$$\sum_{sx}(\mathbf{k}+\mathbf{G}, \mathbf{k}+\mathbf{Q}, E(\mathbf{k})) = \frac{\Sigma_{\mathbf{l}}^{occ}\Sigma_{\mathbf{s}}\, a_{\mathbf{s}+\mathbf{G}}(\mathbf{l})a^*_{\mathbf{s}+\mathbf{Q}}(\mathbf{l})U(\mathbf{l}+\mathbf{s}-\mathbf{k})}{\varepsilon(\mathbf{l}+\mathbf{s}-\mathbf{k}), E(\mathbf{k})-E(\mathbf{l})}, \tag{6.15}$$

where only states below the Fermi level are to be included in the sum over **l**. The coefficients $a_G(\mathbf{l})$ are taken to be those of the pseudowave functions; this is a useful approximation since the overlap with the core states is not large.

For the frequency-dependent dielectric function $\varepsilon(\mathbf{q}, \omega)$ Bennett and Inkson employ the Inkson (1972) model, namely,

$$\varepsilon_l(q, \omega) = \frac{1 + (\varepsilon(Q) - 1)}{[(1 + q^2/q_{TF}(\varepsilon(Q)-1) - \omega^2/\omega_p^2)(\varepsilon(Q)-1)]}, \tag{6.16}$$

where q_{TF} is the Thomas–Fermi wavenumber (cf. Appendix 6.2), ω_p as usual denotes the plasma frequency, and $\varepsilon(Q)$ is the static dielectric constant.

The expectation value of the screened exchange part is given by

$$\begin{aligned}\langle\sum_{sx}(\mathbf{k})\rangle &= \int d\mathbf{r}\,d\mathbf{r}'\,\phi^*_{\mathbf{k}}(\mathbf{r})\sum_{sx}(\mathbf{r}, \mathbf{r}', E(\mathbf{k}))\phi_{\mathbf{k}}(\mathbf{r}') \\ &= \sum_{\mathbf{G},\mathbf{Q}} a_{\mathbf{G}}(\mathbf{k})\sum_{sx}(\mathbf{k}+\mathbf{G}, \mathbf{k}+\mathbf{Q}, E(\mathbf{k}))\, a_{\mathbf{G}}(\mathbf{k}).\end{aligned} \tag{6.17}$$

Inserting Σ_{sx} into this equation, one arrives at a first-order estimate of the screened exchange energy.

6.3.2. Coulomb Hole Term

To evaluate the Coulomb hole term in Eq. (6.14), the dielectric function [Eq. (6.16)] is usefully rewritten as

$$\varepsilon_l^{-1}(q, \omega) = \frac{\omega_r(q))^2 - \omega_p^2 - \omega^2}{\omega_r^2(q) - \omega^2,}, \tag{6.18}$$

where

$$(\omega_r(q))^2 = \omega_p^2 \frac{\varepsilon(Q)}{\varepsilon(Q) - 1}\left[1 + \frac{q^2}{q_{\mathrm{TF}}^2}\frac{\varepsilon(Q)-1}{\varepsilon(Q)}\right]. \tag{6.19}$$

Thus, the dynamic screened interaction has poles at $\omega = \pm\, \omega_r(q)$ with residues $\pm\,(\omega_p^2/2\omega_r(q))U(q)$. Bennett and Inkson choose the contour of integration so as to include only the positive frequency pole. Since the energies of interest in the integral are now well above the Fermi level, one can approximately treat the electrons as free and therefore replace a_{G} by $\delta_{\mathrm{G}0}$ and $E(\mathbf{l})$ by l^2. Then one obtains for the Coulomb hole part Σ_{CH} of the self-energy the result

$$\sum_{\mathrm{CH}}(\mathbf{k}, E(\mathbf{k})) = \frac{-\omega_p^2}{4k(2\pi)^2}\int_0^\infty \frac{dq\, q\, U(q)}{\omega_r(q)} \ln\left|\frac{\omega_r(q) + 2kq + q^2}{\omega_r(q) - 2kq + q^2}\right|, \tag{6.20}$$

where units have been employed in which $\hbar^2 = 1$, $2m_e = 1$, and $2\pi/a = 1$, with the lattice constant being 5.43 Å in Si.

In the static approximation, $\omega_r(|\mathbf{l} - \mathbf{k}|)$ is assumed to increase so rapidly with $|\mathbf{l} - \mathbf{k}|$ that in comparison $E(\mathbf{k})-E(\mathbf{l})$ will be negligible. Then the simpler formula, corresponding to the constant Coulomb hole over the zone, results, namely,

$$\sum_{\mathrm{CH}} = -\tfrac{1}{2}\,\omega_p^2 \sum_{\mathbf{q}} U(q)/\omega_r(q)^2. \tag{6.21}$$

For example, the results of Bennett and Inkson for the Fourier components of the screened exchange energy in Si at X_4, $\mathbf{k} = (110)$, are shown in Table 6.1.

Bennett and Inkson found that, at least out to the free-electron Fermi surface, the Coulomb hole term was fairly constant (it changed in magnitude from 6.5 to 7.2 eV).

TABLE 6.1. Screened Exchange Energy Σ_{SX} in Silicon at X_4[a]

Reciprocal lattice vectors		Length	
G	**Q**	$|\mathbf{G} - \mathbf{Q}|$	Σ_{SX} (eV)
000	000	0	−4.62
000	111	3	0.57
000	−111	3	−0.79
000	−1 −1 −1	3	0.80
000	−2 0 0	4	−0.0002
000	2 0 0	4	−0.0001

[a]After Bennett and Inkson (1976).

Their further findings are summarized usefully by recording the exchange and correlation energy (in eV) in various approximations at principal points in the valence band in Table 6.2. Comparison with earlier work of Kane (1971) shows that the agreement with Kane's dynamic calculation is good.

Additional problems arise in extending the work to the conduction bands: the reader is referred to the account of Bennett and Inkson for a discussion of this.

Subsequent work by Fiorentini and Baldereschi (1992) on the GW method is summarized in Appendix 6.4; we move now to a fuller discussion of this area following Engel *et al.* (1991a,b).

6.4. SCREENED INTERACTION FUNCTION *W* AND ONE-PARTICLE GREEN FUNCTION

Most of the following is based on the observation that there is an analogy between the equations governing the behavior of the one-particle Green function on the one hand and the dynamically screened interaction function on the other. To be specific, there is a Dyson-type integral equation for the screened interaction function W similar to the one for the one-particle Green function G in which the role of the unperturbed Green function G^0 is played by the bare Coulomb interaction v and the role of the self-energy operator (function) Σ by the polarization function P. Put otherwise, in the case of the one-particle Green function one has, symbolically $G = G^0 + G^0\Sigma G$, whereas in case of the screened interaction function we have $W = v + vPW$. Interestingly, like the unperturbed Green function G^0, which satisfies a differential equation with the Dirac δ function as the source term, the bare Coulomb interaction satisfies an equation, Poisson's equation, with almost the same structure. This analogy suggests that, just as for the one-particle Green function G, which can be given in terms of a biorthonormal representation, there also exists a biorthonormal-type representation for the dynamically screened interaction function W. More importantly, and this is what one is really aiming at, this similarity suggests that if some particular approximation would work for the

TABLE 6.2. Exchange and Correlation Energy (in eV) at Principal Points in the Valence Band of Silicon[a]

	Γ'_{25}	L'_3	X_4
Bennett–Inkson (1977)	−10.9	−11.2	−11.5
Dynamic (Kane, 1972)	−11.0	−11.1	−10.9
Static (Kane, 1972)	−12.1	−12.1	−12.1
	−12.6	−12.6	−12.6

[a]After Bennett and Inkson (1977).

biorthonormal representation of G, the same might apply to W. As is known, the usual quasiparticle approximation for G is considered sufficiently accurate, at least as far as the behavior of G at low excitation energies is concerned. The question arises whether a similar type of approximation does exist for W, which would then offer substantial computational advantages for applications in real materials.

Let us start by giving the basic relations (see Engel *et al.*, 1991a) concerning the bare and screened Coulomb interaction functions $v(1, 2)$ and $W(1, 2)$, where $j = 1, 2$ stands for a space–time point (r_j, t_j). Poisson's equation reads

$$-\frac{\varepsilon_0}{e^2}\nabla_1^2 v(1, 2) = \delta(1, 2), \tag{6.22}$$

while the relation between W and v can be expressed as

$$W(1, 2) = v(1, 2) + \int d(3)d(4)v(1, 3)P(3,4)W(4, 2). \tag{6.23}$$

This relation can alternatively be written in terms of the inverse dielectric function ε^{-1}:

$$W(1, 2) = \int d(3)\varepsilon^{-1}(1, 3)v(3, 2). \tag{6.24}$$

The solution of Eq. (6.22) is

$$v(1, 2) = \frac{e^2}{4\pi\varepsilon_0}\frac{1}{|\mathbf{r}_1 - \mathbf{r}_1|}\delta(t_1 - t_2), \tag{6.25}$$

while the function P in eq. (6.23) is related to the dielectric function ε by

$$\varepsilon(1, 2) = \delta(1, 2) - \int d(3)v(1, 3)P(3, 2). \tag{6.26}$$

The functions ε and ε^{-1} are connected by

$$\int d(3)\varepsilon(1, 3)\varepsilon^{\,1}(3, 2) = \int d(3)\varepsilon^{-1}(1, 3)\varepsilon(3, 2) = \delta(1, 2). \tag{6.27}$$

The similarity of Eq. (6.27) to the Dyson equation (see, for instance, Mahan, 1989) relating perturbed and unperturbed Green functions and of Eq. (6.23) with the Schrödinger-type equation defining G^0 is apparent [see also Eqs. (10) and (9) of Engel *et al.*, 1991a).

Let us next concentrate on the first relation given in Eq. (6.23). In view of Eqs. (6.22) and (6.25), it will be clear that, basically, Eq. (6.23) is the integral equation to be satisfied by the function W. Starting from Eq. (6.23) is therefore equivalent to starting from Eq. (6.24), in which the formal solution of the integral equation for W has been written down.

Let us here pursue somewhat further the analogies between quasiparticles and collective excitations. These analogies can be made explicit (Engel *et al.*, 1991a) by deriving a representation for the screened interaction function W in terms of particle-like excitations in complete formal analogy with the quasiparticle representation of the one-particle Green function, the particle-like entities now being plasmons. The practical utility of such a description for the purpose of calculating the self-energy operator can be demonstrated. The present section, following Engel *et al.* (1991b), puts this particle-like description of W into the wider context of existing approximate schemes used in many-body calculations by analyzing in more detail the analytic properties of G, W, and Σ. By doing so, it turns out, a practical scheme can be set up which permits the calculation of accurate quasiparticle-like representations of W, Σ, and (to some extent) G itself (Engel *et al.*, 1991b).

Briefly, the concepts of quasi particle and collective excitation are well established in describing the physical properties of systems containing large numbers of interacting particles. Quasiparticles describe the low-lying excited states in, e.g., a solid; they can be interpreted as particle-like entities consiting of electrons "dressed" with a polarization cloud (see, for instance, March and Parrinello, 1975). In contrast to the bare (i.e., unscreened) electrons themselves, these entities are now only weakly interacting, and one can anticipate that their lifetimes will be long, corresponding to a more or less well-defined energy. Similarly, long-lived collective excitations of the system, such as plasmons or phonons, can often be identified and described like independent particles (see Peierls, 1956).

The quasiparticle picture arises naturally enough from many-body theory. Quasiparticles are related to poles in the analytic continuation of the one-particle Green function G (see Appendix 6.4) to the complex energy plane, and Engel *et al.* (1991a) explicitly demonstrate how plasmon excitations are related to poles in the screened interaction function W. Most of their discussion applies equally to quasiparticles and collective excitations, and therefore the terms *quasiparticle* and *collective excitation* are interchangeable throughout most of the present chapter.

Below, some attention will be given to the mathematical foundations involved in the quasiparticle concept. The account below follows that of Engel *et al.* (1991b) rather closely.

Though it is a somewhat formal statement, one can argue that what one observes in an experiment is at least closely related to the energy dependence of the functions G or W along the real energy axis. But it will be seen below that, in particular for the case of W, it is easy to find many different sets of poles in the complex energy plane, corresponding to variable numbers of plasmons with different energies and lifetimes, which describe this energy dependence equally well. The view will be adopted that the most sensible definition of quasiparticle and plasmon energies is in terms of the positions of the peaks in the spectra (which are related to the imaginary parts of G or W, respectively, in the coordinate representation) measured at real energies.

In practice, the concept of collective excitation is also used for numerical purposes in the evaluation of an approximate expression for the electron self-energy within the GW scheme already outlined and utilized above. The self-energy operator relates the true many-body one-particle Green function G to a Green function G^0 of a fictitious system of noninteracting electrons. Calculating it within the GW framework involves knowledge of the screened interaction function W. Still, for such a calculation it is necessary in reality to employ some kind of approximation for the energy dependence of the latter function. The simplest and most widely used approximation at the time of writing is the so-called plasmon-pole approximation (for references, see Engel *et al.*, 1991b) in which the continuous spectrum of charge density fluctuations for positive energies is modelled, within a plane-wave representation of the dielectric matrix, by describing the energy dependence of each of these matrix elements by a simple plasmon pole at some real positive energy. In fact, it proves possible to generalize the ideas underlying the plasmon-pole model, and Section 6.5 will tackle this topic.

As will be demonstrated, the usefulness of the plasmon-pole approximation can be related mathematically to its being a type of Gauss-integration rule for energy integrations. To be somewhat more precise, the plasmon-pole approximation is constructed in such a way as to make the first frequency (or energy) moment of the screened interaction function exact. Such moments also underlie the theory of Gauss integrations. The plasmon-pole approximation thus ensures that energy integrations like the one occurring in the evaluation of the electron self-energy in the GW scheme are adequately approximated for most purposes.

By recognizing this connection with Gauss-integration rules and the closely related mathematics of continued-fraction expansions (CFEs) and orthogonal polynomials, it becomes clear how to find somewhat better approximations to W: one has simply to ensure the correctness of higher-order moments (see Engel *et al.*, 1991b). In Section 6.5 below, therefore, a brief outline of the mathematics of continued fractions will be presented, along with a discussion of its possible merits for representing the screened interaction W. Though, at the time of writing, such a representation has not been turned into actual numerical calculations, it does motivate the approximation set out in Section 6.6 and also makes possible the clarification of the limits of validity of plasmon-pole-type approximations in the description of the screened interaction function W.

In Appendix 6.7, a procedure to describe the energy dependence of W will be summarized, one which uses the above method of moments only to make general statements about the positions of poles. These are then fixed *a priori* along a given path in the complex energy plans and the weights are computed to give the best agreement possible between the exact function and its approximation both along the real and the imaginary energy axes. It turns out that this type of representation is not confined to the screened interaction W but also works for other functions with a similar analytical structure. In particular, it allows a representation of the energy-dependent self-energy operator Σ in terms of simple poles. This is implementable numerically because it requires the evaluation of W (Σ) at a fairly small number of complex energy values only.

In Section 6.8, the effect of using plasmon-pole-type approximations for W in an evaluation of the self-energy will be discussed and compared with the results of Section 6.3.

Though it is true that the emphasis of this discussion is on numerical aspects of the calculation of the self-energy operator, the findings also have some consequences for the definition of quasiparticle energies. Section 6.9 is therefore devoted to this topic. There it will be stressed that the standard procedure employed to obtain a quasiparticle description of the one-particle Green function has deficiencies in relation to its full energy dependence. Historically, it appears that the quasiparticle description of the one-particle Green function goes back to Layzer (1967). Since then, it has been customary to assume that there is a set of quasiparticle energies, which can be derived from the energy-dependent eigenvalues of a Schrödinger-like wave equation, in which the self-energy operator Σ plays the role of an energy-dependent nonlocal potential. However, just as in the case of plasmon energies, direct solution of the equations relating the complex-valued quasiparticle energies to these eigenvalues [see Eq. (6.41)] does not permit a complete description of the full energy dependence of the Green function: in addition to well-defined quasiparticles, there is always a continuous part of the spectrum. This is mathematicaly related to a branch cut in the Green function and cannot be found as a solution to the equations defining quasiparticle energies. The techniques summarized in the present chapter also offer the possibility of an accurate description of this continuous part.

Numerical tests of these proposals have been carried through by Engel *et al.* (1991b) using a simple quasi-one-dimensional two-band model akin to that introduced by Lannoo *et al.* (1985), with some modifications by Toet (1987). The main features of this model are summarized in Appendix 6.3. The advantage of such a model is that a numerically exact calculation of W and of the self-energy operator Σ is relatively straightforward, thereby allowing a direct test of the quality of the various approximations proposed.

6.5. PLASMON ENERGIES AND PLASMON POLES

As already indicated above, this section will present an outline of the way approximations to the energy dependence of the diectric function ε, the screened interaction W, the Green function G, the self-energy function Σ, and other functions having similar analytic structure can be set up systematically, using CFEs and orthogonal polynomials. This, in turn, will clarify the meaning of plasmon energies and plasmon poles. To this end, it is important to first examine the analytic structure of the functions mentioned above (see also Engel *et al.*, 1991b).

The analytic structure of the functions G and W is apparent from their Lehmann-type representations. The one-particle Green function G at zero temperature is defined (see Mahan, 1990) as

$$G(1, 2) = -i_H\langle\Psi_N|\mathcal{T}\{\hat{\psi}(1)\,\hat{\psi}^\dagger(2)\}|\Psi_N\rangle_H. \tag{6.28}$$

Here the arguments 1 and 2 denote the space-time points $(\mathbf{r}, t_1)$ and $(\mathbf{r}_2, t_2)$, respectively; $|\Psi_N\rangle_H$ is the full N-particle state vector in the Heisenberg picture, $\hat{\psi}$ and $\hat{\psi}^\dagger$ are the usual Fermion annihilation and creation field operators (also in the Heisenberg picture) and $\mathcal{T}\{\hat{\psi}(1)\,\hat{\psi}^\dagger(2)\}$ equals either $\hat{\psi}(1)\,\hat{\psi}^\dagger(2)$ or $-\hat{\psi}^\dagger(2)\,\hat{\psi}(1)$, depending on whether $t_1 > t_2$ or $t_1 < t_2$ ($\mathcal{T}$ is the Fermion time-ordering operator). The Fourier transform with respect to $t_1 - t_2$ of this function can be written as

$$G(\mathbf{r}_1, \mathbf{r}_2; \varepsilon) = \hbar \sum_s f_s(\mathbf{r}_1) f_s^*(\mathbf{r}_2) \left\{ \frac{\Theta(\varepsilon_s - \mu)}{\varepsilon - \varepsilon + i\eta} + \frac{\Theta(\mu - \varepsilon_s)}{\varepsilon - \varepsilon_s - i\eta} \right\}, \tag{6.29}$$

where η is an infinitesimally small positive number and $\Theta(x) = 0$ or 1 depending on whether $x < 0$ or $x > 0$. For semiconductors, μ equals some energy in the energy-gap region of the system. The latter are excitation energies of the $(N + 1)$-particle systems with respect to the ground-state energy of the N-particle system. An equivalent representation can be given for the noninteracting Green function G^0. In the case of a finite system, the Lehmann energies, which are related to eigenvalues of the many-body Hamiltonian, are discrete and the Green function has simple poles along the real energy axis. For an infinite system, however, these poles form a dense set (a branch cut), except inside the gap, where there are no eigenstates. The parameter η in Eq. (6.29) ensures that the physically relevant Green function is obtained by approaching the branch cut from below for energies below the chemical potential μ and from above for energies above it (Fig. 6.1a).

A similar analysis can be carried out for the functions related to dielectric properties of the system. All these can be obtained from the knowledge of the density–density correlation function $\chi(\mathbf{r}_1, \mathbf{r}_2; \varepsilon)$, which in the time domain is defined as

$$\chi(\mathbf{r}_1 t_1, \mathbf{r}_2 t_2) = -\frac{i}{\hbar} {}_H\langle\Psi_N|\mathcal{T}\{\hat{\rho}'(\mathbf{r}_1 t_1)'(\mathbf{r}_2 t_2)\}|\Psi_N\rangle_H, \tag{6.30}$$

in which $\hat{\rho}'(\mathbf{r}t)$ represents the density-deviation operator

$$\hat{\rho}'(\mathbf{r}t) = \hat{\psi}^\dagger(\mathbf{r}t)\hat{\psi}(\mathbf{r}t) - {}_H\langle\Psi_N|\hat{\psi}^\dagger(\mathbf{r}t)\hat{\psi}(\mathbf{r}t)|\Psi_N\rangle_H, \tag{6.31}$$

For example, the relation between χ and W is given by

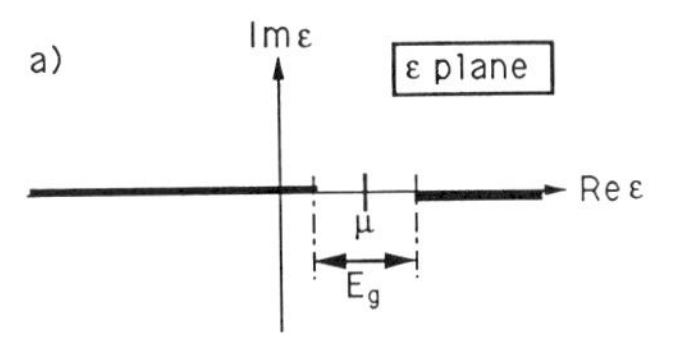

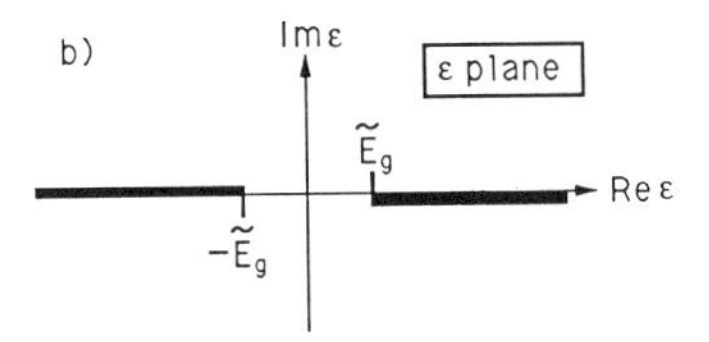

FIGURE 6.1. (a) Shows analytic structure of Green function defined in Eq. (6.29). The parameter there ensures that the physically relevant Green function is obtained by approaching the branch cut from below for energies below and from above for energies above the chemical potential μ. (b) Depicts analytic structure of density–density response function χ and dielectric response functions related to it. Note that χ has a branch cut slightly above the real axis for energies below $-E_g$ and slightly below the real axis for energies above E_g. χ is a symmetric function of energy ε.

$$W(\mathbf{r}_1, \mathbf{r}_2; \varepsilon) = v(\mathbf{r}_1, \mathbf{r}_2) + \int d^3\mathbf{r}'\, d^3\mathbf{r}''\, v(\mathbf{r}_1 - \mathbf{r}')\chi(\mathbf{r}', \mathbf{r}''; \varepsilon)v(\mathbf{r}'' - \mathbf{r}_2), \tag{6.32}$$

where $v(\mathbf{r}_1 - \mathbf{r}_2)$ is the bare electron–electron interaction. The Lehmann-type representation of $\chi(\mathbf{r}_1, \mathbf{r}_2; \varepsilon)$ is given by

$$\chi(\mathbf{r}_1, \mathbf{r}_2, \varepsilon) = 2 \sum_s \rho'_s(\mathbf{r}_1)\rho'^*_s(\mathbf{r}_2)\left\{\frac{1}{\varepsilon - \varepsilon'_s + i\eta} - \frac{1}{\varepsilon + \varepsilon'_s - i\eta}\right\}. \tag{6.33}$$

For a definition of the density-fluctuation amplitudes $\rho'_s(\mathbf{r})$ and the excitation energies ε'_s (which are excitation energies of the N-particle system), the reader may consult Mahan (1990). Defining $\tilde{E}_g = \min_s\{\varepsilon_s'\}$, one again finds that for infinite systems, χ (and consequently all the dielectric response functions related to it) has a branch cut slightly above the real axis for energies below $-\tilde{E}_g$ and slightly below the real axis for energies above $\tilde{E}_g$ (Fig. 6.1b). It is to be noted that χ is a symmetric function of energy: $\chi(\mathbf{r}_1, \mathbf{r}_2, \varepsilon) = \chi(\mathbf{r}_1, \mathbf{r}_2, -\varepsilon)$.

Finally the positions of the branch cuts of the self-energy Σ can be deduced as a superposition of branch cuts of G and G^0. This follows from the relation

$$\Sigma = (G^0)^{-1} - G^{-1}, \tag{6.34}$$

which is readily deduced from the Dyson equation, which in turn can be written symbolically as

$$G = G^0 + G^0 \sum G. \tag{6.35}$$

Computationally, the profound nature of the energy dependence of the screened interaction W and of the Green function G along the real axis constitutes the main obstacle to a model-free calculation of the electron self-energy Σ in the

GW approximation [see Eq. (6.35)]. But Engel *et al.* (1991b) note that none of the functions W, G, Σ, etc. has any (complex) singular points on the physical Riemann sheet, their only analytic features being branch cuts and possibly poles on the real axis. They therefore propose, as already mentioned above, the use of the mathematics of CFEs and Gauss integrations. However, with a view to clarifying the limits, and the merits, of the plasmon-pole approximation, which is the most widely used to date for GW calculations, it will be valuable at this point to review the mathematics of continued fractions.

Let us consider a function $f(z)$ that is analytic everywhere except within a finite interval $[a, b]$ along the real axis, where it has a branch cut, and which decays sufficiently rapidly as $|z|$ tends to infinity. The functions G, W, Σ, etc. can each be viewed as a superposition of two such functions, one having the branch cut along some part of the negative real axis and the other having it along some part of the positive real axis. Use of Cauchy's theorem for the contour indicated in Fig. 6.2 leads, for z not lying on the branch cut itself, to the form

$$f(z) = \frac{1}{\pi} \int_a^b dx \, \frac{g(x)}{x - z}, \tag{6.36}$$

where

$$g(x) = \frac{1}{2i} \lim_{\eta \downarrow 0} [f(x + i\eta) - f(x - i\eta)] \tag{6.37}$$

is the discontinuity across the branch cut. Let us for the moment assume that g is real-valued and does not change sign within $[a, b]$. The general case follows trivially by writing g as the sum of two or more functions, each of which satisifies this requirement separately. Then it can be shown (Engel *et al.*, 1991b) that f can be represented in the form of a CFE, and that this expansion converges everywhere except on the branch cut itself:

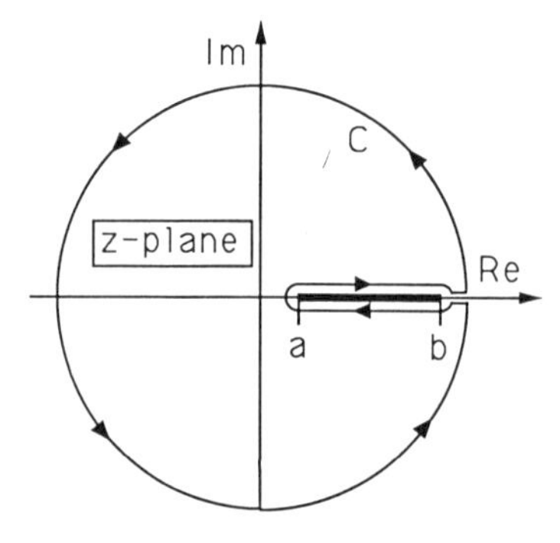

FIGURE 6.2. Function $f(z)$, analytic everywhere except within finite interval $[a, b]$ along the real axis, where it possesses a branch cut. Figure depicts contour used [see Eq. (6.36)].

$$f(z) = -\frac{1}{\pi} \frac{b_0^2}{z - a_0 - \cfrac{b_1^2}{z - a_1 - \cfrac{b_2^2}{\cdots}}}, \tag{6.38}$$

The coefficients a_i and b_i can be evaluated recursively (see Appendix 6.2).

The CFE in Eq. (6.38), if terminated at a finite N, can equivalently be expressed as a sum over residues:

$$f(z) \approx -\frac{1}{\pi} \sum_{i=1}^{N} \frac{w_i}{z - x_i + i\eta}, \tag{6.39}$$

with $a < x_i < b$, $i = 1, 2, \ldots, N$. The x_i are the zeros of the polynomial P_N. Formally, for integrations involving g given in Eq. (6.37) along the branch cut itself, one finds, from Eq. (6.39),

$$\int_a^b dx\, g(x) h(x) \approx \sum_{i=1}^{N} w_i h(x_i), \tag{6.40}$$

where $h(x)$ is an arbitrary real function. It can be shown that this integration formula is exact if h is a polynomial of degree not more than $2N - 1$. Equation (6.40) is, in fact, a Gaussian-quadrature rule for the general weight function $g(x)$.

An obvious shortcoming of the representations presented [see especially Eqs. A6.2.1 and (6.40)] is that they do not converge on the branch cut itself, where the previously continuous function $g(x)$ has been replaced by a sum of δ-functions. This limitation can readily be dealt with (Engel *et al.*, 1991b) by introducing a terminating function $t(z)$ into the expansion of Eq. (6.38) to yield

$$f(z) = -\frac{1}{\pi} \frac{b_0^2}{z - a_0 - \cfrac{b_1^2}{\cdots - \cfrac{\cdots}{z - a_{n-3} - \cfrac{b_{n-2}^2}{z - a_{n-2} - b_{n-1}^2 t(z)}}}}. \tag{6.41}$$

The terminating function $t(z)$ is chosen to incorporate the asymptotic behavior of the continued-fraction coefficients a_i and b_i for $i \to \infty$. Specific forms of terminating functions are discussed by Engel *et al.* (1991a,b). They can be generalized to deal with cases where the spectrum $g(x)$ has discontinuous derivatives at a finite set of points, which is a feature shown by all of the functions G, W, and Σ within the GW scheme due to the presence of van Hove singularities.

One may now attempt to represent the matrix elements of a particular representation of the functions W, Σ, and G by a CFE. For example, the plane-wave matrix elements $\widetilde{W}_{\mathbf{K},\mathbf{K}'}(\mathbf{q};\varepsilon) \equiv W_{\mathbf{K},\mathbf{K}'}(\mathbf{q};\varepsilon) - v_{\mathbf{K},\mathbf{K}'}(\mathbf{q})$ obey the following Kramers–Kronig-type relation:

$$\tilde{W}_{\mathbf{K},\mathbf{K}'}(\mathbf{q};\,\varepsilon) = -\frac{1}{\pi}\int_0^\infty d\varepsilon'\left\{\frac{\mathrm{Im}[W_{\mathbf{K},\mathbf{K}'}(\mathbf{q};\,\varepsilon')]}{\varepsilon-\varepsilon'+i\eta} - \frac{\mathrm{Im}[W_{\mathbf{K},\mathbf{K}'}(\mathbf{q};\,\varepsilon')]}{\varepsilon+\varepsilon'-i\eta}\right\}. \tag{6.42}$$

This can readily be seen using Cauchy's theorem for a contour similar to that in Fig. 6.2 and the fact that W does not have poles except possibly on the real axis (cf. Engel *et al.*, 1991a). Here and in the following, it has been assumed, in the interests of simplicity, that the discontinuity across the branch cut is given by the imaginary part of $W_{\mathbf{K},\mathbf{K}'}(q;\,\varepsilon')$, which can be strictly true only for systems with inversion symmetry. The general case is obtained by considering real and imaginary parts of the discontinuity separately.

Identifying the spectral function Im $W_{\mathbf{K},\mathbf{K}'}$ with the function g in Eq. (6.37), a CEF can be constructed. It is to be noted incidentally that this is possible only in the case where the imaginary part of W falls off at least exponentially, as will always be the case in a computer calculation where a finite basis set is used for the evaulation of the Green function G^0, leading to a finite number of conduction bands and thus to a high energy cutoff in the electron–hole excitation spectrum. On the other hand, the imaginary part of the full dielectric function and hence of its inverse behaves like $O(\varepsilon^{-3})$ as $|\varepsilon|$ tends to infinity and all the moments except the zeroth and first do not exist [see Appendix 6.2: the integrals in Eq. (A6.2.13) diverge]. This should also be borne in mind when discussing sum rules for higher-order moments of the dielectric function. Of course, for practical purposes it is always safe to employ such a high-energy cutoff, as the high-energy spectrum does not affect the dielectric properties at low energies, which represent our main interest here.

In the plasmon-pole approximation, the matrix elements $\tilde{W}_{\mathbf{K},\mathbf{K}'}(\mathbf{q};\,\varepsilon)$ of the screening part of the screened interaction function W are approximated by the expression

$$\tilde{W}^{pp}_{\mathbf{K},\mathbf{K}'}(\mathbf{q};\,\varepsilon) \equiv -\tfrac{1}{2}\,e_{\mathbf{K},\mathbf{K}'}(\mathbf{q})\tilde{W}_{\mathbf{K},\mathbf{K}'}(q;0)\times\left\{\frac{1}{\varepsilon-e_{\mathbf{K},\mathbf{K}'}(\mathbf{q})+i\eta} - \frac{1}{\varepsilon+e_{\mathbf{K},\mathbf{K}'}(\mathbf{q})-i\eta}\right\}\cdot\ \eta\downarrow 0, \tag{6.43}$$

with $e_{\mathbf{K},\mathbf{K}'} > 0$ being the plasmon pole pertaining to the $\mathbf{K}$, $\mathbf{K}'$ element of the inverse of the dielectric matrix. The *static* function $\tilde{W}_{\mathbf{K},\mathbf{K}'}(q;0)$ on the right-hand side of Eq. (6.43) is assumed to be exact. The plasmon poles are found by employing the f-sum rule, which establishes a relationship between the first frequency moment of $\tilde{W}$ and the Fourier components of the ground-state charge density. Both the plasmon-pole approximation and a Gaussian-quadrature rule of order one (one pole and one weight for each branch of W) result in expressions of the general form give in Eq. (6.43) with two parameters $\tilde{W}_{\mathbf{K},\mathbf{K}'}(\mathbf{q};\,0)$ and $e_{\mathbf{K},\mathbf{K}'}(\mathbf{q})$, one of which is fixed in both cases by the requirement that the first frequency moment is to be exact. They differ in that in the former approximation, the value of $\widetilde{W}$ at $\varepsilon = 0$ is fixed, while in the latter the zeroth frequency moment is made exact.

Instead of employing the *f*-sum rule, odd frequency moments can also be obtained from the asymptotic behavior of $\widetilde{W}$. To this end, one writes Eq. (6.43) as

$$\widetilde{W}_{\mathbf{K},\mathbf{K}'}(\mathbf{q};\, \varepsilon) = -\frac{2}{\pi} \int_0^\infty d\varepsilon' \, \frac{\varepsilon'}{\varepsilon^2 - \varepsilon'^2} \operatorname{Im}(W_{\mathbf{K},\mathbf{K}'}(\mathrm{q};\, \varepsilon')). \tag{6.44}$$

Assuming $\operatorname{Im}(W_{\mathbf{K},\mathbf{K}'}(\mathbf{q};\, \varepsilon')$ to fall off exponentially as $|\varepsilon'|$ tends to infinity, this can be expanded in powers of $1/\varepsilon^2$ as

$$W_{\mathbf{K},\mathbf{K}'}(\mathrm{q};\, \varepsilon) = -\frac{2}{\pi} \sum_n \left(\frac{1}{\varepsilon^2}\right)^{n+1} \left(\int_0^\infty d\varepsilon' \, \varepsilon'^{2n+1} \operatorname{Im}(W_{\mathbf{K},\mathbf{K}'}(\mathrm{q};\, \varepsilon')\right). \tag{6.45}$$

Hence the expansion coefficients of $(1/\varepsilon^2)^{n+1}$ can be identified with the $(2n + 1)$th moment. It is to be noted that this expansion is valid only if $|\varepsilon'/\varepsilon| < 1$ within the range of ε' where $\operatorname{Im} W_{\mathbf{K},\mathbf{K}'}(\mathrm{q};\, \varepsilon')$ is nonzero. In principle, Eq. (6.45) and its generalization to apply to other functions with similar analytic structure provide another route to the calculation of inner products [e.g., those of Eqs. (A6.2.10–12)], without having to evaluate integrals over the discontinuity across the branch cut of the function under consideration, simply by analyzing their asymptotic behavior.

It is a simple task to show that the plasmon-pole approximation becomes almost identical to a Gauss-integration formula of order 1 if the imaginary part of W is significant only within a small interval (see Appendix 6.5). Recognizing this analogy to Gauss integrations, one can now understand in more detail why the plasmon-pole approximation for W works fairly well in an evaluation of the GW self-energy. Equation (6.40) shows that the results of energy integrations are reproduced well by such expressions. The self-energy operator is obtained from G and W by means of such an energy integration [Eq. (6.58) below]. A discussion of the consequences of using a plasmon-type approximation with real poles for evaluation of Σ will be set out in Section 6.7 below.

6.6. ILLUSTRATIVE EXAMPLE USING A TWO-BAND MODEL

CFEs with an appropriate terminating function for the matrix elements of the screened interaction W have been calculated in the modified two-band model of Lannoo *et al.* (1985), details of this being summarized in Appendix 6.6. For the head-element $W_{0,0}(q = \pi/2a;\, \varepsilon)$, the result is shown in Fig. 6.3; all other elements can be represented to within the same accuracy. The approximate functions are virtually indistinguishable, graphically, from the exact ones even for a very small order of the CFE (in this case 8). The units used in this and other figures are defined in Appendix 6.5.

Figure 6.3 also illustrates the fact that the plasmon-pole approximation (dash-dotted line) is not an adequate approximation to the function W itself along the real energy axis. This failure is particulary pronounced whenever the imaginary part of the exact W is nonzero. This should occasion no surprise because CFEs, and thus the plasmon-pole approximation as a special case, do not converge to the true functions on the branch cut itself unless they contain a terminating function to describe the behavior of the coefficients for large n.

In Table 6.3, the standard plasmon-pole approximation is compared with a CFE of order 1 (first and third columns). Both the position of the poles and the

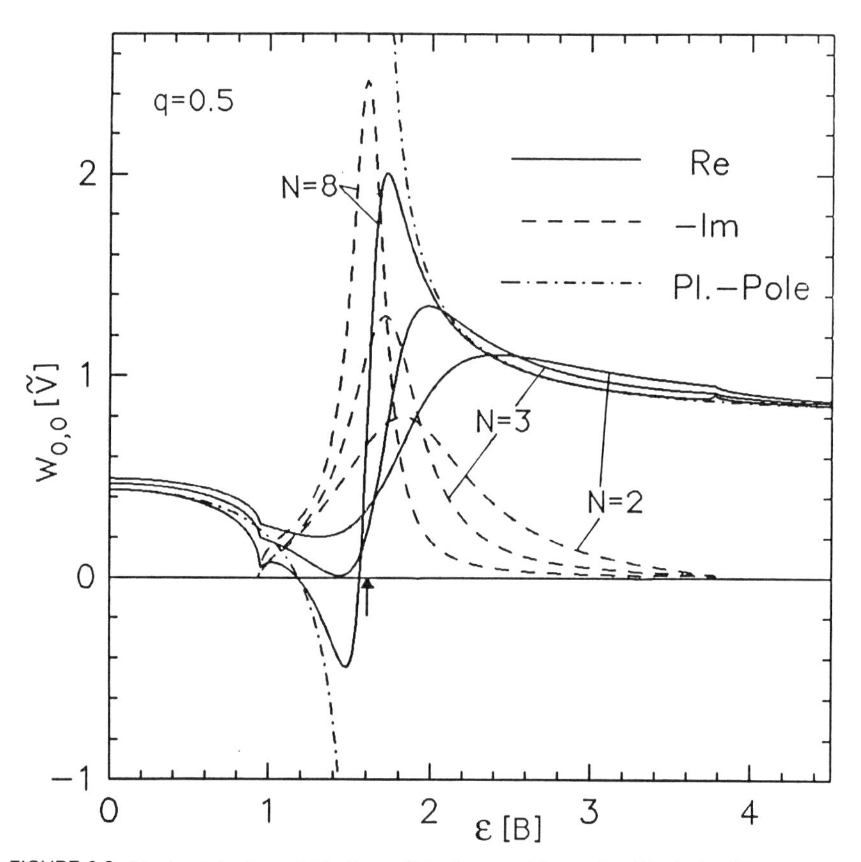

FIGURE 6.3. Head matrix element $W_{0,0}(\mathbf{q} = \pi/2a)$ of screened interaction W calculated in two-band model of Lannoo *et al.* (1980). Plasmon-pole approximation is shown by dash-dotted line (after Engel *et al.*, 1991b).

TABLE 6.3. Comparison of the pole positions $e_{K,K'}(q)$ and the approximation to the static value of $W_{K,K'}(q;\ \varepsilon)$ for $q = \pi/2a$ as given by the three different plasmon-pole models. Pl. pole 1: Standard plasmon-pole approximation, fixing the first frequency moment of the imaginary part of $W_{K,K'}(q;\varepsilon)$ and its static value, at $\varepsilon = 0$. Pl. pole 2: Plasmon pole. CFE: Poles and weights chosen such as to ensure the correctness of the zeroth and first moments of the imaginary part of $W_{K,K'}(q;\ \varepsilon)$ (continued-fraction expansion of order 1).

	$e_{K,K'}(q = \pi/2a)$ (units of B)			$W_{K,K'}(q = \pi/2a;\ \varepsilon = 0)$ (units of $\tilde{V}$)		
(K,K') (units of $2\pi/a$)	Pl. pole 1	Pl. pole 2	CFE	Pl. pole 1	Pl. pole 2	CFE
(−1, −1)	2.6151	2.3442	2.7319	0.2167	0.2167	0.2276
(0,0)	1.6054	1.5510	1.6313	0.4351	0.4351	0.4476
(1,1)	2.6191	2.4522	2.7124	0.1278	0.1278	0.1285

values of the approximations at zero frequency are listed for the diagonal elements of W in the two-band model adopted here. One sees that the position of the plasmon poles depends noticeably on the criterion used: i.e., whether in addition to fixing the first frequency moment, the approximation ensures the correctness of the static value or of the zeroth moment of W. The second column in Table 6.3 shows the results obtained by numerically solving the implicit equation for $e_{\mathbf{K},\mathbf{K}'}(\mathbf{q})$:

$$\tilde{W}_{\mathbf{K},\mathbf{K}'}(\mathbf{q};\ ie_{\mathbf{K},\mathbf{K}'}(\mathbf{k}) = \tfrac{1}{2}\tilde{W}_{\mathbf{K},\mathbf{K}'}(\mathrm{q};\ 0), \tag{6.46}$$

which should be almost identical to the usual plasmon-pole approximation in the case where the latter gives a good description of the behavior of W along the imaginary energy axis [see Eq. (6.43)]. Figure 6.4a compares the goodness of the three single-pole approximations describing the energy dependence of $\tilde{W}_{-2\pi/a,-2\pi/a}$ along the imaginary energy axis. It can be seen that the asymptotic behavior is described correctly by all approximations except the one obtained from Eq. (6.21), which gets the coefficient of $(1/\varepsilon^2)$, describing the aymptotic behavior, slightly wrong. Furthermore, Fig. 6.4b shows that an extension of the plasmon-pole model to three poles and weights (fixing the zeroth and first five moments), as obtained from a CFE, makes the approximation virtually indistinguishable from the exact function along the imaginary energy axis.

To get a reasonable choice for the pole positions $\Im_i$, one can go back to the definition of the function $f(z)$ in Eq. (6.36). Assuming the discontinuity $g(x)$ to be a smooth function along the branch cut, it can itself be approximated by an analytic function $\tilde{g}(z)$. Deforming the contour c into the lower half-plane (Fig. 6.5), one finds a function $\tilde{f}$ defined by

$$\tilde{f}(z) = \frac{1}{\pi}\int_{\mathrm{C}} \frac{\tilde{g}(z')}{z' - z}\, dz' + 2i \sum_j \frac{\mathrm{Res}\ \tilde{g}(\zeta_j)}{(z - \zeta_j)}, \tag{6.47}$$

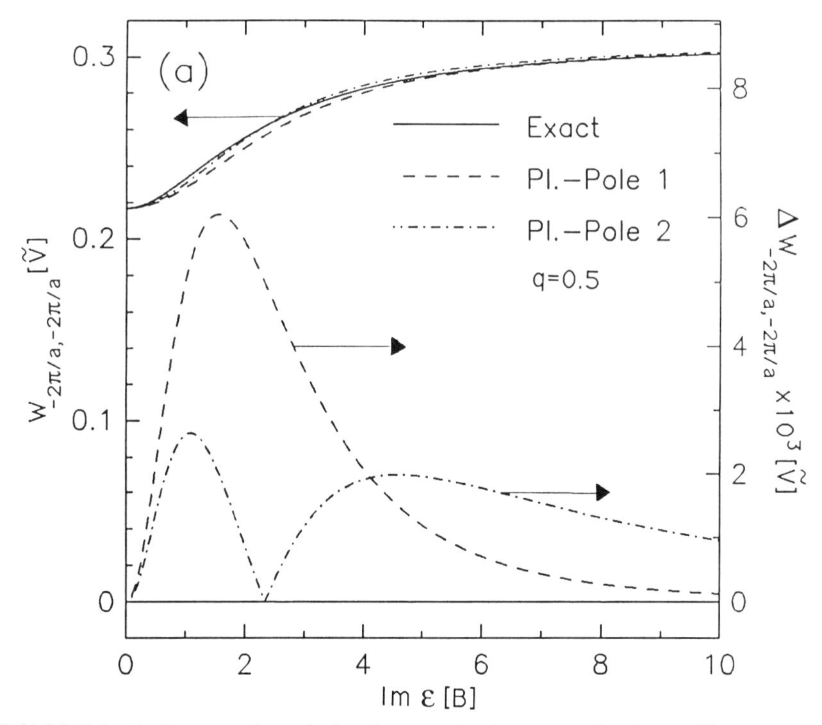

FIGURE 6.4. (a) Compares three single-pole approximations to matrix element $W_{-2\pi/a,-2\pi/a}$ for two-band model.

which agrees with the original function *f* in the upper half-plane and along the real axis. However, the branch cut between *a* and *b* has been replaced by a branch cut along the curve C and residue contributions from possible poles ζ_j in $\tilde{g}$ (see also Appendix 6.7).

6.7. CALCULATON OF PLASMON DISPERSON

Plasmon dispersion in a homogeneous electron fluid has already been considered in Section 6.2; see especially Eq. (6.6). Daling *et al.* (1991) have calculated plasmon energies for silicon by analytically continuing the determinant of the dielectric function ε in the plane-wave representation through the branch cut by means of a Taylor expansion. The zeros of this analytic continuation correspond to plasmon energies and thus to poles in *W*. It was shown by Daling *et al.* that such a method allows the accurate and unambiguous identification of certain well-

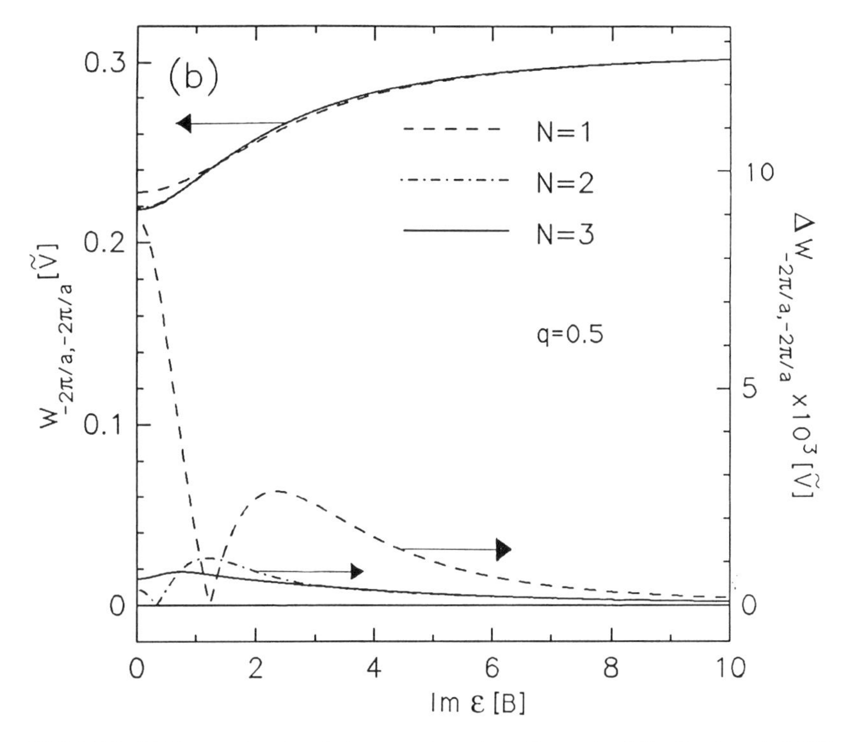

FIGURE 6.4. (b) Shows extension of plasmon-pole model to three poles (after Engel *et al.*, 1991b).

defined plasmon energies and lifetimes. Identifying the analytic continuation of the matrix elements of $\tilde{W}$ with the function $\tilde{f}$ in Eq. (6.54), these well-defined plasmons correspond to the residue part of the analytic continuation of $\tilde{W}$ [the second term in Eq. (6.47)]. However, an exact analytic continuation of W through the branch cut always leads to a function which, in addition to these residue contributions, has itself a branch cut (compare Fig. 6.5). It is not possible to obtain, by means of Taylor expansions, this branch cut as a solution to the equation det $\varepsilon_{\mathbf{K},\mathbf{K}'}(\mathbf{q};\, \varepsilon) = 0$. Above it has been seen that a description of the full energy dependence of W in terms of plasmon-like excitations is, nevertheless, possible; however, the positions of the poles in this representation do not allow a direct physical interpretation and are chosen with a certain arbitrariness. Even a "well-defined" plasmon cut, for practical purposes, always be replaced by a cluster of plasmons with longer lifetimes but different energies without affecting the energy dependence of W along the real axis. In view of all this, it would seem that the best definition of plasmon energies is the positions of the peaks in the spectral part of W; this is presumably what one should bring into contact with experiment.

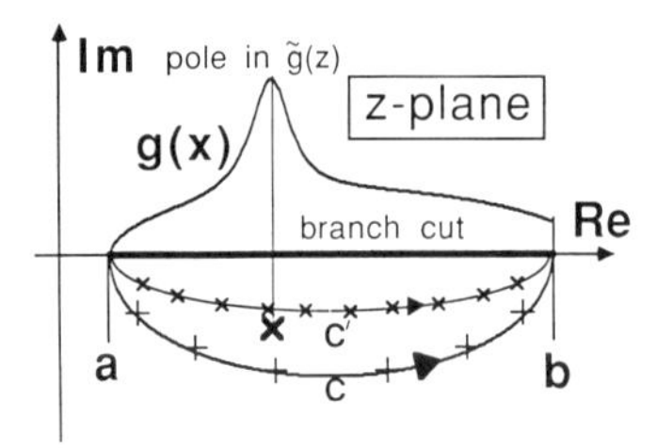

FIGURE 6.5. Shows contour referred to in Eq. (6.47).

The above procedure can readily be used to determine plasmon dispersion according to this "experimental" definition. Figure 6.6 shows the plasmon dispersion curve obtained by taking the position of the peak in the head-element $W_{0,0}(q;\ \varepsilon)$. It is compared to the dispersion of the plasmon pole obtained by employing Eq. (6.18) and that of the pole position of a CFE of order 1 (see Section 6.2). It should be noted that the latter two plasmon definitions have to be qualified by saying that the poles in these are largely mathematical tools. Nevertheless, the agreement among the three curves is quite reasonable for the head element, the discrepancies becoming large for other matrix elements of the screened interaction.

6.8. THE SELF-ENERGY OPERATOR Σ

6.8.1. Quantitative Evaluation

For the purpose of evaluating the self-energy operator Σ, one still must perform a Brillouin-zone integration. Engel *et al.* (1991b) calculated the values of $\widetilde{W}^{m}_{\mathbf{K},\mathbf{K}'}(\mathbf{k})$ for $m = 1, \ldots, 40$ at a number of equidistant **k**-points and used an integration algorithm based on finite differences. Rather than calculating the self-energy directly, they chose the same type of representation as for W, replacing the two branch cuts by poles chosen as in Eq. (6.55), thus having to evaluate it only at a relatively small number of energy values. The result is shown in Fig. 6.7 for one of the matrix elements of Σ at wave number $q = \pi/2a$. It was verified that the self-energy obtained is well-converged with respect to the number of **k**-points used for the Brillouin-zone integration and the number of poles and weights used to represent W and Σ. In this manner, an accurate representation of the full energy dependence of the self-energy operator can be obtained within the simple model adopted.

Due to the one-dimensionality of the system, the plasmon excitations prove to lie much lower in energy than one would expect for a real three-dimensional system. To understand the structure of the GW self-energy, it is best to examine the plane-wave matrix elements of

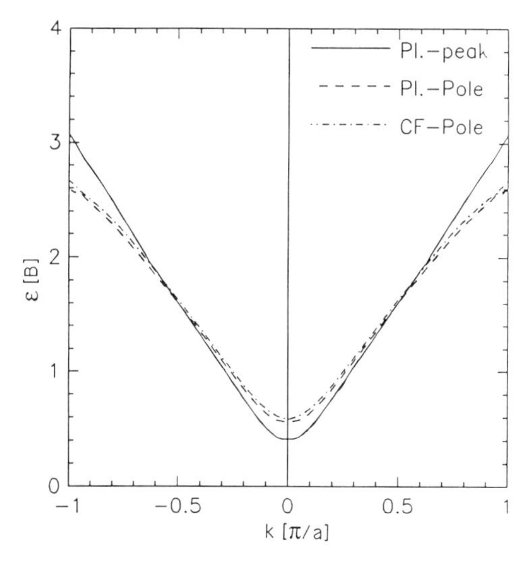

FIGURE 6.6. Plasmon dispersion curve. Continuous curve: from position of peak in head matrix element $W_{0,0}(\mathbf{q};\,\varepsilon)$ of screened interaction W. Dashed curve: plasmon pole result. Dash-dotted curve: continued fraction (order 1) (after Engel *et al.*, 1991b).

$$\tilde{\Sigma} = \Sigma - \Sigma_{\mathrm{HF}}, \tag{6.48}$$

where Σ_{HF} denotes the Hartree–Fock exchange contribution to the self-energy. These can be written in the form

$$\tilde{\Sigma}_{\mathbf{G},\mathbf{G}'}(\mathbf{q};\,\varepsilon) = \frac{i}{2\pi\hbar\Omega}\sum_{\mathbf{k}}\sum_{\mathbf{K},\mathbf{K}'}\sum_{l} d_{l,\mathbf{q}-\mathbf{k}}(\mathbf{K})\,d^{*}_{l,\mathbf{q}-\mathbf{k}}(\mathbf{K}')$$
$$\times\int_{-\infty}^{\infty} d\varepsilon'\,\frac{\widetilde{W}_{\mathbf{G}-\mathbf{K},\mathbf{G}'-\mathbf{K}'}(\mathbf{k};\,\varepsilon')}{\varepsilon - \varepsilon' - \varepsilon_l(\mathbf{q}-\mathbf{k}) - i\eta\,\mathrm{sgn}\,[\mu - \varepsilon_l(\mathbf{q}-\mathbf{k})]}. \tag{6.49}$$

Here the $d_{l,\mathbf{q}}(\mathbf{K})$ and $\varepsilon_l(\mathbf{q})$ are plane-wave coefficients and eigenvalues of the unperturbed system, respectively. The self-energy operator $\{\Sigma_{\mathbf{K},\mathbf{K}'}(\mathrm{q};\,\varepsilon)\}$ becomes non-Hermitian whenever the integrand of the ε' in this equation becomes singular, i.e., for energies ε for which

$$\varepsilon \le \max_{\mathbf{k},l_v}\{-\widetilde{E}_g(\mathbf{k}) - \varepsilon_{l_v}(\mathbf{q}-\mathbf{k})\} \quad \text{or} \tag{6.50a}$$

$$\varepsilon \ge \min_{\mathbf{k},l_c}\{\widetilde{E}_g(\mathbf{k}) + \varepsilon_{l_c}(\mathbf{q}-\mathbf{k})\} \tag{6.51b}$$

holds, where $\widetilde{E}_g(\mathbf{q})$ is the $\mathbf{q}$-dependent lower edge of the electron–hole excitation region, i.e., the excitonic gap (the region where $W(\mathbf{q};\,\varepsilon)$ is non-Hermitian). If the

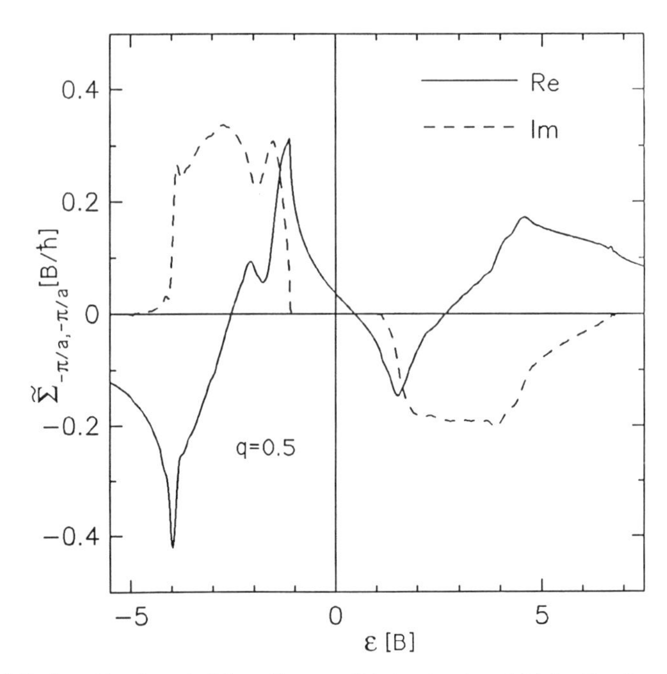

FIGURE 6.7. A matrix element of the self-energy Σ in two-band model (after Engel *et al.*, 1991b).

plasmon energies are small (as is the case in the model system), the non-Hermitian part of the self-energy operator becomes large even for energies close to the quasiparticle gap, where in a real three-dimensional system it would be rather small. It should, however, be noted that surface plasmons can be considerably lower in energy than bulk plasmons, so that the effect of the non-Hermiticity in the self-energy operator may be more pronounced in surface-state calculations and should be taken into account carefully, using the techniques discussed in the present chapter.

6.8.2. Results for Σ using Plasmon-Pole Representation for *W*

Figure 6.7 shows the result obtained for the self-energy Σ using plasmon-pole type approximations for the screened interaction *W*. Figure 6.8a corresponds to the standard approximation [Eq. (6.43)] whereas Fig. 6.8b has been obtained by representing *W* by poles and weights obtained by constructing a CFE of order 1 (see Section 6.2).

The finding is that the plasmon-pole self-energy is of modest accuracy for energies around the mid-gap, but that it fails within the self-energy operator has

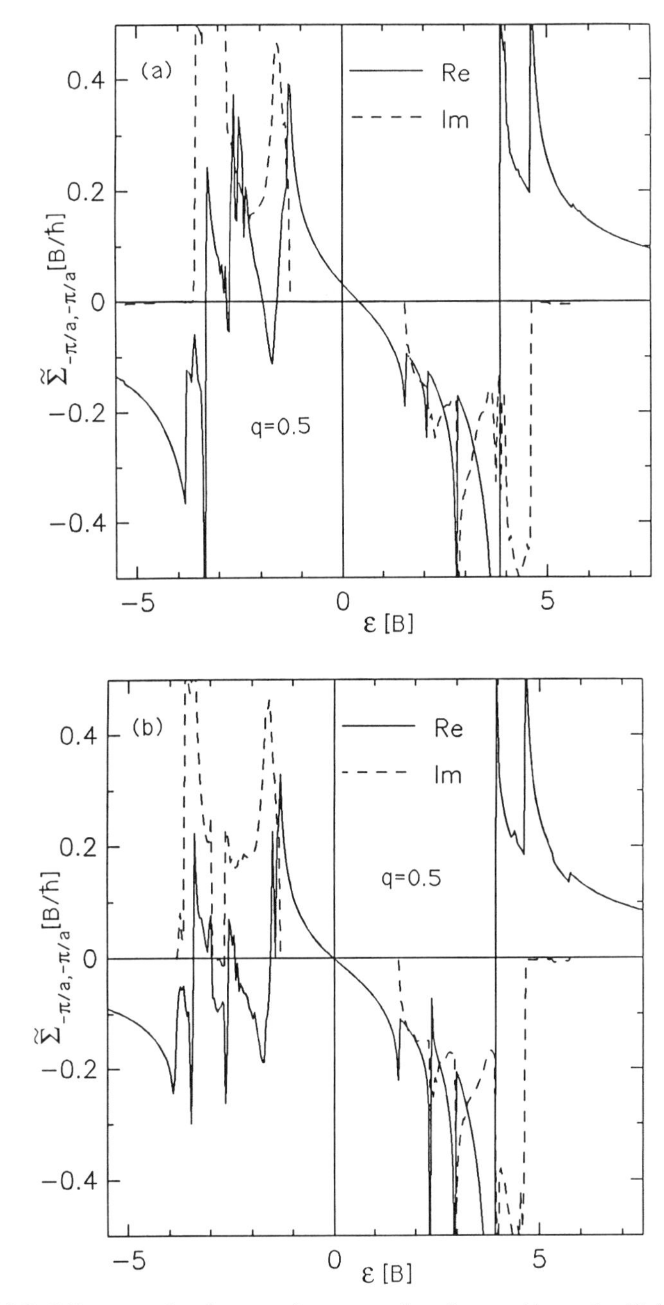

FIGURE 6.8. Self-energy using plasmon-pole representation of screened interaction W. (a) Standard approximation [Eq. (6.43)]. (b) Using continued fraction representation (order 1) (after Engel *et al.*, 1991b).

a large imaginary part. This means that calculations of renormalized band gaps and effective masses using the plasmon-pole approximation should be fairly reliable but that some caution is needed in the evalaution of quasiparticle energies within the region where the true GW self-energy becomes appreciably non-Hermitian (e.g., in calculations of valence-band widths). In obtaining Fig. 6.9, the necessary Brillouin-zone integrations were performed to convergence. In any actual calculation, the necessity of replacing the continuous **k**-integration by a discrete summation over special **k**-points leads, in the context of a plasmon-pole method, to a large number of artificial poles in the self-energy on the real energy axis in energies whose distance from the mid-gap energy is comparable to the plasmon-pole energies [of Eq. (6.38)]. It is therefore not possible within the plasmon-pole approximation to obtain good results for the renormalized band structure at these energies.

It is worth adding here a note on a relationship between the moments of Σ and W (Engel *et al.,* 1991a,b). It can be shown that the nth moment of the self-energy operator Σ is, in general, evaluated correctly only if an approximation for W is employed whose zeroth and first n moments agree with that of the exact function (see Engel *et al.,* 1991b). This condition is not fulfilled by the standard plasmon-pole approximation, which does not fix the zeroth moment of W.

For the purpose of calculating a renormalized band structure near the Fermi level, one is primarily interested in a good approximation to the self-energy operator in the vicinity of the mid-gap energy. In a real solid, most of the weight of the imaginary part of Σ comes at energies far away from the mid-gap, and it is mainly the lowest moments which contribute to the static value of the self-energy. Hence one can expect the standard plasmon-pole self-energy to be inferior to that derived from a modified plasmon-pole description of W. For any purpose, the representation of W proposed in Section 6.3 is to be expected to yield the most accurate results.

6.9. QUASIPARTICLE ENERGIES

At no stage in the above discussion has it been assumed that there exist well-defined quasiparticles or plasmons though naturally, if one wishes, the poles of the above representation [Eq. (6.33)] can be viewed as plasmon energies and the weights can be related to the amplitudes of the charge-density oscillations (see also Section 6.1). The representation in Eq. (6.33) for W accurately describes both the so-called radiative part of the plasmon spectrum (corresponding to short-lived electron–hole excitations) and the peaks related to well-defined charge-density oscillations, i.e., plasmons.

Following Engel *et al.* (1991a,b), we will now address the question of whether a similar conclusion holds for the poles of the one-particle Green function

G, which are generally associated with the particle-like excitations of the electronic system. G can be written in the form

$$G(\mathbf{r}_1, \mathbf{r}_2; \varepsilon) = \hbar \sum_n \frac{\phi_n(\mathbf{r}_1; \varepsilon)\psi_n^*(\mathbf{r}_2; \varepsilon)}{\varepsilon - E_n(\varepsilon)}, \tag{6.52}$$

where $\phi_n(\mathbf{r}; \varepsilon)$ and $\psi_n(\mathbf{r}; \varepsilon)$ are solutions of

$$\left[-\frac{\hbar^2}{2m} \nabla^2 + u(\mathbf{r}) + z_l(\mathbf{r}) \right] \phi_n(\mathbf{r}; \varepsilon) + \int d^3r' \, z_{nl}(\mathbf{r}, \mathbf{r}')\phi_n(\mathbf{r}'; \varepsilon)$$

$$- \hbar \int d^3r' \, \Sigma'(\mathbf{r}, \mathbf{r}'; \varepsilon)\phi_n(\mathbf{r}'; \varepsilon) = E_n(\varepsilon)\phi_n(\mathbf{r}; \varepsilon), \tag{6. 53a}$$

$$\left[-\frac{\hbar^2}{2m} \nabla^2 + u(\mathbf{r}) + z_l(\mathbf{r}) \right] \psi_n(\mathbf{r}; \varepsilon) + \int d^3r' \, z_{nl}(\mathbf{r}, \mathbf{r}')\psi_n(\mathbf{r}'; \varepsilon)$$

$$- \hbar \int d^3r' \, \Sigma'^{\dagger}(\mathbf{r}, \mathbf{r}'; \varepsilon)\psi_n(\mathbf{r}'; \varepsilon) = E_n^*(\varepsilon)\psi_n(\mathbf{r}; \varepsilon) \tag{6.53b}$$

where u is the usual Hartree potential and z_l and z_{nl} are local and nonlocal potentials, which include the external potential plus terms accounting for exchange and correlation effects. The symbol Σ' has been used above for the self-energy to make clear that its definition in Eq. (6.6a) depends on the types of many-body effects which have aready been included in the choice of these local and nonlocal potentials. $\Sigma'^{\dagger}(\mathbf{r}, \mathbf{r}'; \varepsilon)$ is the Hermitian adjoint of the self-energy operator Σ'. Equation (6.57) is the biorthornormal representation of G. The quasiparticle picture arises from this expression by replacing the energy denominator in Eq. (6.57) by simple complex poles ε_n, i.e.,

$$G(\mathbf{r}_1, \mathbf{r}_2; \varepsilon) \simeq \hbar \sum_n g_n \frac{\phi_n(\mathbf{r}_1; \varepsilon_n)\psi_n^*(\mathbf{r}_2; \varepsilon_n)}{\varepsilon - \varepsilon_n}, \tag{6.54}$$

where

$$g_n \equiv \left[\frac{1 - dE_n(\varepsilon)}{d\varepsilon} \bigg|_{\varepsilon = e_n} \right]^{-1}. \tag{6.55}$$

The $\{\varepsilon_n\}_n$ should follow from

$$\varepsilon_n = E_n\,(\varepsilon_n). \tag{6.56}$$

Strictly speaking, this equation in general has no real-valued solutions. Only within the region where Σ is Hermitian can one expect real solutions, though even in that case there is no guarantee that such a solution exists. In semiconductors, if Σ has been obtained from a GW calculation, the position of the branch cuts in

Σ can be deduced from Eq. (6.36); hence Eq. (6.67) can be expected to have real solutions only for energies near the gap. For all other energies, just as in the case of the biorthonormal representation of W, one has to look for possible poles of G in its analytic continuation through the branch cut (see Engel *et al.*, 1991a,b for further details).

6.10. SHARPNESS OF QUASIPARTICLE CONCEPT

The question of whether quasiparticles are well-defined entities within a given energy range depends mainly on the degree of non-Hermiticity of the self-energy operator. Figure 6.9 shows the spectral function $\mathrm{tr}\{\mathrm{Im}G(\mathrm{q}; \varepsilon)\}$ for a number of **q**-values, where the trace Tr denotes a sum over the diagonal elements of the Green function in the plane-wave representation. G in this calculation has been obtained by inverting the Dyson equation

$$G = G^0 + G^0(\Sigma - \Sigma_{\mathrm{HF}})\, G, \tag{6.57}$$

assuming that the unperturbed Green function G^0 already includes the effect of Hartree–Fock exchange interactions. Engel *et al.* (1991a,b) find that for all **q**-

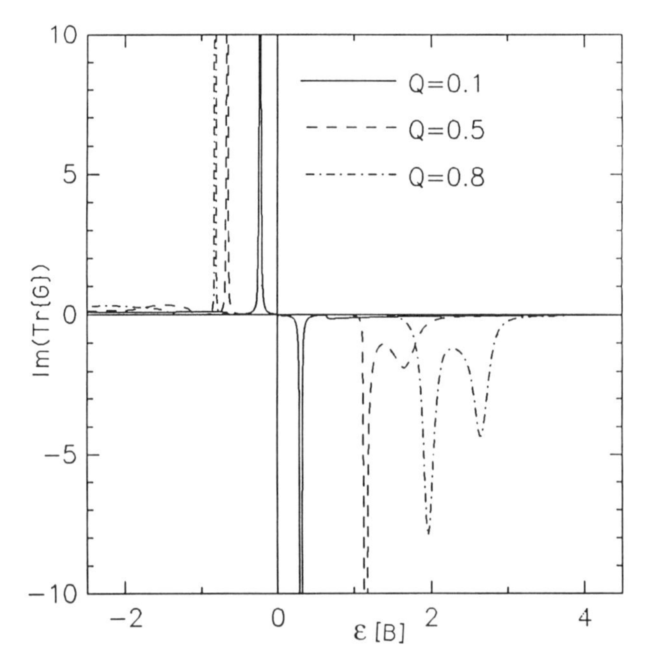

FIGURE 6.9. Spectral function for a number of Q values, as shown (after Engel *et al.*, 1991b).

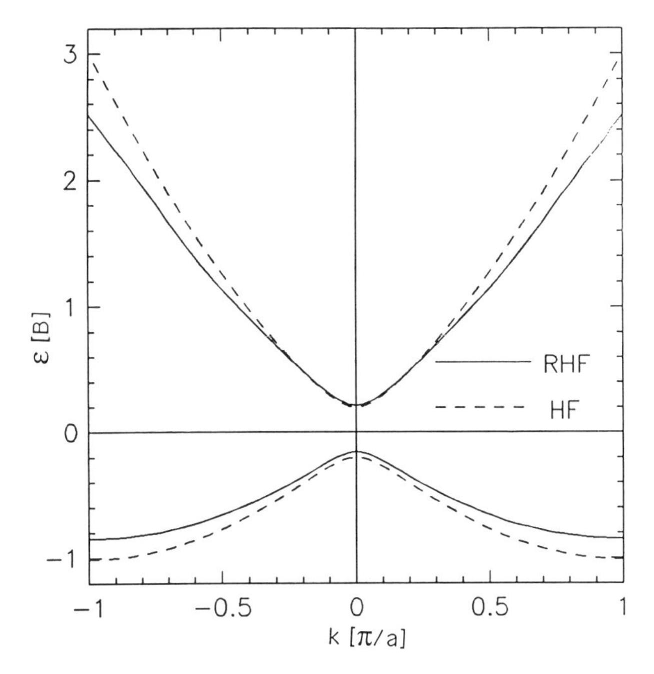

FIGURE 6.10. Compares renormalized Hartree–Fock (RHF) band structure with that of the unperturbed Hartree–Fock (HF) system (after Engel *et al.*, 1991b).

values, the valence-band states lie within the energy region where the self-energy operator is Hermitian [see Eq. (6.62)], and their lifetimes are infinite within the GW approximation. In addition, a small continuous spectrum was found at lower energies. The conduction-band states turned out also to have infinite lifetime for small **q**-values, whereas they were broadened as soon as **q** got sufficiently large. But as an artifact of the model employed, the energies of quasiparticles and plasmons were of comparable size, leading to strong coupling between the two. This was the explanation of a double peak in the conduction-band spectrum at large **q**-values.

Any attempt to evaluate a renormalized band structure using the self-energy operator as a perturbation would, at best, have succeeded in finding the contributions to G arising from poles in its analytic continuation through the branch cut but would have failed to identify the part of the spectrum related to the branch cut in its analytic continuation. This is a mathematical statement of the oft-quoted wisdom that quasiparticles are well-defined entities for low excitation energies only.

Figure 6.10 compares the renormalized Hartree–Fock (RHF) band structure with that of the unperturbed Hartree–Fock (HF) system. The renormalized band structure has been obtained by determining the peak position in the spectral function. The direct band gap has been reduced, though by a smaller amount than has been reported for three-dimensional systems.

For a review of the dimensionality dependence of collective effects in condensed conducting phases, and especially plasmons, the reader may refer to March and Tosi (1995). Mahan (1994) has discussed different forms of GW approximations: a focal point of the present chapter: see Appendix A6.8).

Chapter 7

Metal–Insulator Transitions and the Chemical Bond

Already, in the discussion of the Coulson–Fischer (1949) treatment of H_2 as a function of internuclear distance in Section 4.1 and in the jellium model of Chapter 3 as the background neutralizing density is lowered, the type of transition from delocalized wave functions, belonging to the molecule or solid as a whole, to localized wave functions has been clearly identified.

In the present chapter some further examples of the way electron–electron interactions can affect the qualitative nature of electron states will be studied, the focal point here being chemically bonded solids. The first example, the linear polyenes, arose in the early history of quantum chemistry but remains of considerable interest as it anticipated a phenomenon now referred to by solid-state workers as Peierls's theorem. The second example to be treated is closely linked with the Coulson–Fischer treatment of H_2 as generalized by Gutzwiller to electrons in narrow *d*-bands, as discussed at length in Section 4.2. However, it will suffice here to confine ourselves to a phenomenological treatment given by March *et al.* (1979), which recovers most of the predictions of Gutzwiller's microscopic theory. Because of the rather restricted class of examples focused on in this chapter, the reader is also referred to the older reviews of Mott (1974) and of Brandow (1975) in this general area. A further general reference is Friedman and Tunstall (1978), though their volume concentrates on the transition in disordered systems (see also Chapter 8). Third, pressure-induced transitions in chemically bonded solids will be treated.

7.1. LINEAR POLYENES AND MONTE CARLO STUDIES OF INSTABILITIES IN ONE-DIMENSIONAL SYSTEMS

Turning then to the linear polyenes, there are two points to be made here. The first is that in the early work of Lennard-Jones (1937), who applied molecular orbital theory to a linear chain of C atoms with one π-electron per atom, he tacitly assumed that bond lengths were equal and that the existence of different bond lengths in the known members of the linear polyene series was a consequence of end effects. However, a number of groups later, but almost simultaneously, pointed out that, in accordance with experiment, a lowering of total energy within the molecular orbital framework was brought about by allowing alternating bond lengths. This obviously introduces an energy gap into the excitation spectrum, as is required to understand the absorption spectrum of the finite linear polyenes [see, for example, the account in the book by Murrell (1971)].

The next point to focus on here is that Lieb and Wu (1968), for the admittedly oversimplified case of equal bond lengths, obtained an exact solution of the Hubbard Hamiltonian. Their conclusion is that, with equal bond lengths and with one electron per atom, the case $U = 0$ is a singular point, and for $U \neq 0$ there is always an energy gap due to electron correlation.

The other point to be focused on below for the case of equal bond lengths is that one has here a very simple model system in which the correlation energy can be obtained exactly from the solution of Lieb and Wu. Numerical results for the correlation energy have been given by Johansson and Berggren (1969). Following that, approximate results for correlation energy with alternating bond lengths will be presented.

7.1.1. Correlation Energy in the One-Dimensional Hubbard Model

Because of its interest as a model in which the electron correlation energy can be obtained exactly, a brief summary will be given of the results of the solution of the Hubbard Hamiltonian in one dimension (Lieb and Wu, 1968). First of all, the allowed energies take the form

$$E \propto \sum_j \cos k_j = \int \rho(k, U) \cos k dk, \tag{7.1}$$

where the quasimomenta k_j are determined by a somewhat complicated set of equations given in the paper of Lieb and Wu. As the length of the chain tends to infinity for the case of a half-filled band, the ground-state energy E_g, say, is given by (see also Ferraz, Grout and March, 1978):

$$E_g = -4N \int_0^\infty \frac{J_0(\omega)J_1(\omega)d\omega}{\omega[1 + \exp(\frac{1}{2}\omega U)]}, \tag{7.2}$$

J denoting a Bessel function of the first kind. For small U, one regains from Eq. (7.2) the Hartree–Fock result that the energy per particle is $-4/\pi + \frac{1}{4}U + \ldots$, but at higher order in U it turns out that there are terms which cannot be expanded in powers of U. The complete energy expression has been evaluated numerically as a function of U by Johansson and Berggren (1969), and their results for the total energy are reproduced in Fig. 7.1, which also includes comparison with some approximate treatments.

The quantity $\rho(k)$ appearing in Eq. (7.1) has been calculated by Ferraz *et al.* (1978) as a function of k for various values of U, and their results are shown in Fig. 7.2. This figure shows that $U = 0$ is a singular point. These results imply the absence of a metal–insulator transition for a half-filled band in one dimension. Figure 7.3 shows results for, essentially, the energy gap as a function of interaction strength. It can be seen that the curve is very flat for small U. It is relevant here to mention that the influence of the long-range Coulomb interaction on the energy spectrum of infinite polyenes has also been studied by Tyutyulkov *et al.* (1980).

In summary, in the linear polyenes with one electron per atom a gap can arise from two mechanisms; alternating bond lengths and electron correlation. The system is always insulating, this being the analogue in solid-state language of the theorem of Peierls that one-dimensional metals cannot exist.

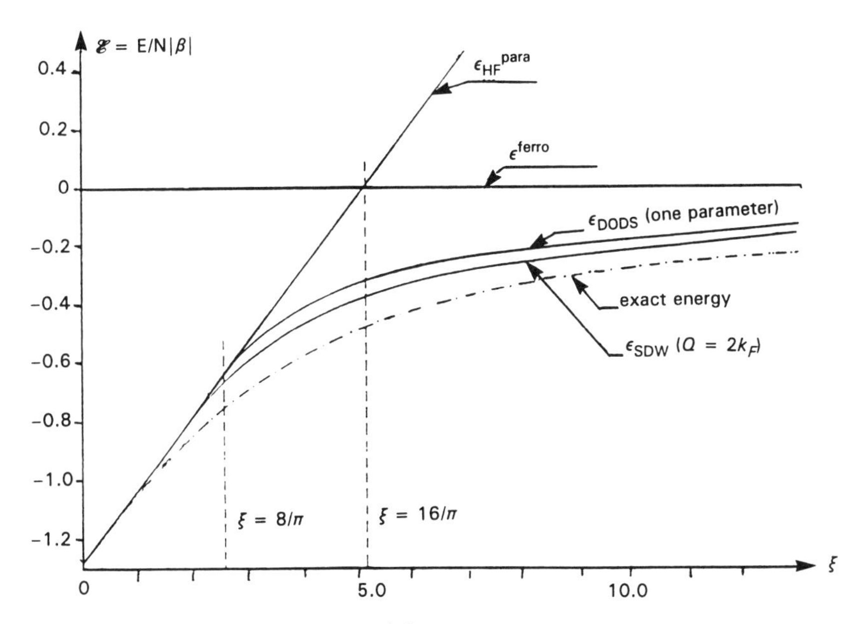

FIGURE 7.1. Total energy per particle vs. $U/|\beta|$ for the Hubbard Hamiltonian in one dimension. Here, 4β is the width of the electronic band in the absence of on-site repulsion, $U = 0$. The labels refer to various approximations and magnetic states (redrawn from Johansson and Berggren, 1969). In particular, the exact energy obtained from the solution of Lieb and Wu is shown, together with the result obtained from a spin density wave treatment (SDW).

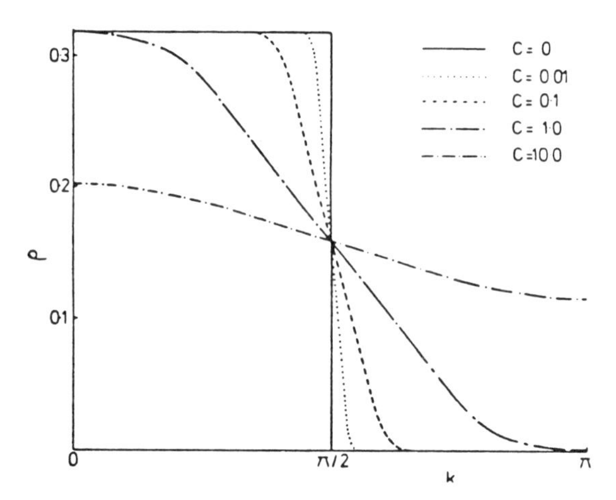

FIGURE 7.2. The function $\rho(k,C)$ entering the exact solution of the Hubbard model in one dimension versus k, at several values of the interaction strength C defined by $U/|\beta|$ and for half filling (after Ferraz, Grout, and March, 1978).

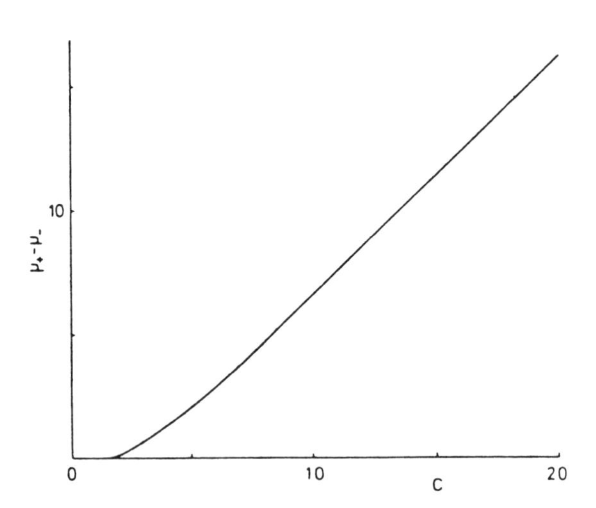

FIGURE 7.3. $\mu_+ - \mu_-$ (i.e., the energy gap) as function of the interaction strength C defined as $U/|\beta|$, and for half filling. Only for $C = 0$ is $\mu_+ = \mu_-(= 0)$. However, note the flatness of the curve for small C showing the absence of term of order C in $\mu_+ - \mu_-$ (after Ferraz, Grout, and March, 1978).

7.1.2. Computer Study of Instabilities

In connection with instabilities in one-dimensional systems, it is noteworthy that Hirsch and Scalapino (1983) have studied charge-and spin-density response functions in the one-dimensional extended Hubbard model by Monte Carlo techniques (see also Chapter 5).

Specifically, they report results obtained from Monte Carlo simulations of a one-dimensional one-quarter-filled Hubbard band with on-site and nearest-neighbor repulsion. Their results show that the $2k_f$ charge-density response is suppressed by the Hubbard interaction in the physically relevant temperature region, while the $2k_f$ spin susceptibility is enhanced. (As usual, k_f denotes the Fermi wave number.)

This is of interest in that in certain quasi-one-dimensional organic conductors like TTF–TCNQ, results of diffuse X-ray experiments (Pouget *et al.*, 1976) reveal scattering at a wave number $4k_f$ in addition to $2k_f$ scattering. It has been suggested by Torrance (1978) and by Emery (1976) that the $4k_f$ response is due to strong Coulomb interactions.

This then was the motivation for the study by Hirsch and Scalapino, though they treat short-range Hubbard interactions (rather than long-range Coulomb forces) and the effect of such short-range forces on the response functions of a one-dimensional electron gas. Specifically, the system they study is defined by the Hamiltonian

$$H = -t \sum_{i\sigma} (C^{\dagger}_{i+1\sigma} C_{i\sigma} + C^{\dagger}_{i\sigma} C_{i+1\sigma}) + U \sum_{i} n_{i\uparrow} n_{i\downarrow} + V \sum_{i} n_{i+1} n_{i}. \tag{7.3}$$

Here $C^{\dagger}_{i\sigma}$ creates a particle of spin σ on site i and the transfer matrix element is denoted by t, while U and V are the on-site and near-neighbor interactions, respectively. For the free system, the bandwidth is $4t$ and units are employed in which $t = 1$. As already mentioned, a one-quarter-filled band was the specific case considered.

For the Hubbard model, with $V = 0$, and for an on-site interaction of strength $U = 4$, both the charge- and spin-density susceptibilities, denoted by $N(q)$ and $\chi(q)$, respectively, have been studied. The sections in Fig. 7.4, which shows these results, would correspond for TTF–TCNQ, where $4t \sim 0.5$ eV, to temperatures of approximately 500, 400, 300, 200, and 100 K. It can be seen in Fig. 7.4a for the spin susceptibility that the peak at $2k_f$ increases as the temperature is lowered. A similar plot for the charge-density response is shown in Fig. 7.4b. Here the peak at $2k_f$ is barely visible and a very weak peak appears at $4k_f$ at intermediate temperatures.

Figure 7.5a and b, taken for ~200 K for TTF–TCNQ, show the dependence on U, and the main result of interest here is the suppression of the charge-density $2k_f$ response with increasing U, but the reverse trend for the spin-density response. Again $V = 0$ so that it is the original Hubbard model which is being simulated. A

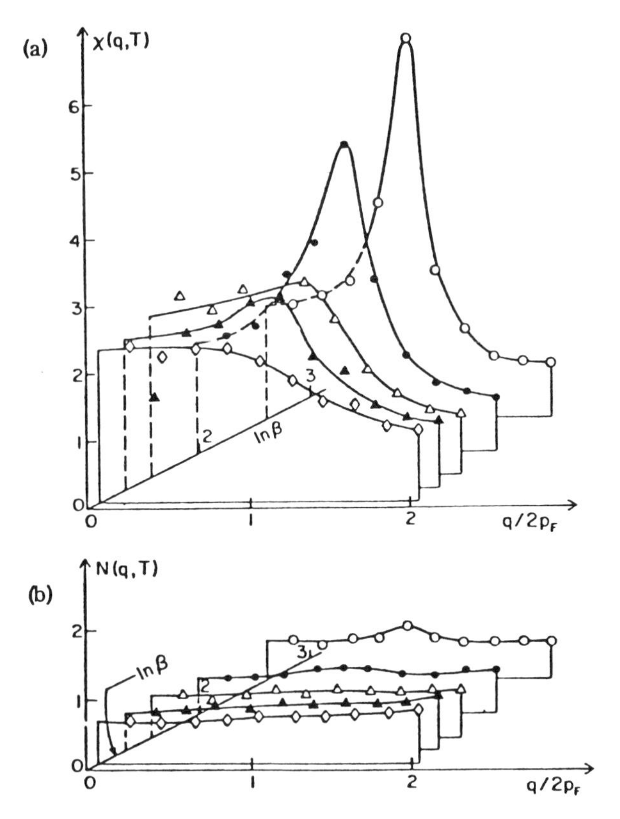

FIGURE 7.4. (a) Spin density and (b) charge density susceptibilities normalized to $2/\pi v_F$ versus q for β = 3, 3.75, 4.75, 7.25, and 14.5 for the Hubbard Model with U = 4, V = 0. Energies are in units of t, the hopping energy. The statistical errors in the Monte Carlo data are about twice the size of the points. The solid lines are drawn through the Monte Carlo points to guide the eye (Hirsch and Scalapino, 1983).

separate, small $4k_f$ peak is also evident for large U in the charge-density susceptibility $N(q)$.

The effect of adding a near-neighbor repulsive interaction V, in addition to the on-site interaction U, has also been investigated: the reader is referred to the original paper for details. To summarize:

1. Increasing Hubbard interaction rapidly suppresses the $2k_f$ charge-density instability, but enhances the $2k_f$ spin density of the one-quarter-filled Hubbard band.

2. With moderate on-site and near-neighbor interactions corresponding to U = 4, V = 2, a strong peak appears at $4k_f$ in the charge-density response. The $4k_f$

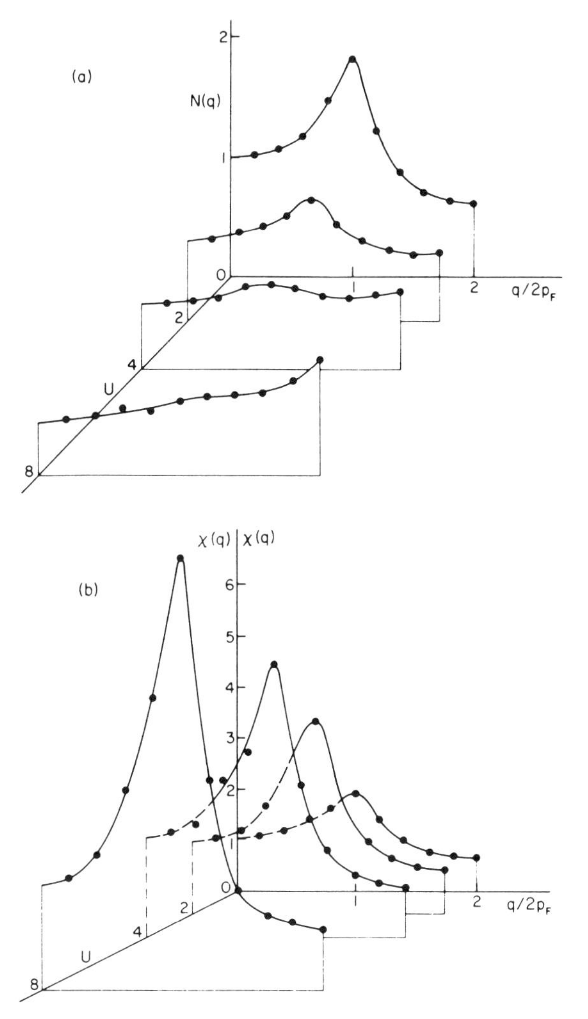

FIGURE 7.5. (a) Charge density and (b) spin susceptibilities at $\beta = 7.25$ for several values of U. Units and uncertainties as in Fig. 7.4 (after Hirsch and Scalapino, 1983).

response shows a weak temperature dependence in the range studied ($T = 500$ to 100K if $4t = 0.5$ eV) while the spin-density response at $2k_f$ shows a strong temperature dependence.

3. The most striking result is the absence of a region of parameters in which structure at both $2k_f$ and $4k_f$ coexist in the charge-density-wave susceptibility.

Reference should finally be made to the earlier work of Chui *et al.* (1974) who also studied the effect of interactions on the Peierls instability.

7.2. ELECTRON–ELECTRON CORRELATION IN ONE-DIMENSIONAL CONJUGATED POLYMERS WITH ALTERNATING BOND LENGTHS

Kajzar and Friedel (1987) have studied the influence of electron correlation on the ground-state energy and the elastic properties of a one-dimensional polymer chain with alternating bond lengths.

As was discussed in Section 4.6, electron–electron interactions are important to the cohesion, magnetism, and stability of crystal structure of d-electron metals. A similar situation can be expected for organic metals and semiconductors. The focal point here will be polymers like polyacetylene or polydiacetylene, which are models of conjugated π-electron systems with one-dimensional delocalization. The polymerization leads to the creation of a long-chain of carbon atoms and it is linked at the same time, with a significant decrease in energy gap (from, for example, 5 to 1.9 eV in the case of polyacetylene). One also observes a substantial increase in the cohesive energy and Young's modulus between monomer and polymer, the latter quantity approaching a value comparable to that of d-electron metals in the polymer chain direction ($E = 5 \times 10^{10}$ N m^{-2}) and is an order of magnitude larger than for nonconjugated polymers (e.g., $E = 2.5 \times 10^{9}$ N m^{-2} for high-density polyethylene).

In their work, Kajzar and Friedel study the contribution of electron correlation to the ground-state energy and elastic constant up to second order in the ratio U/W, where as usual U is the Coulomb correlation energy while W is the bandwidth, in a perturbation expansion as a function of gap energy.

7.2.1. Hamiltonian Incorporating Bond Alternation Property

The Hamiltonian for a linear chain including electron correlations may be written as

$$H = H_0 + V, \tag{7.4}$$

where

$$H_0 = \beta_1 \sum_j (|2j\rangle\langle 2j+1| + |2j+1\rangle\langle 2j|) + \beta_2 \sum_j (|2j\rangle\langle 2j-1| + |2j-1\rangle\langle 2j|). \tag{7.5}$$

describes an alternate linear chain with hopping energies β_1 and β_2 [see Fig. 1a of Kajzar and Friedel (1987)] between orbitals $|j\rangle$ on sites j and

$$V = \sum_{\substack{i,j \\ (i>j)}} \frac{e^2}{r_{ij}} \tag{7.6}$$

evidently represents electron–electron repulsion. For simplicity, Kajzar and Friedel consider the π-electrons of a polyacetylene chain with one electron per atom in the p-band.

The Hamiltonian H_0 incorporates bond alternation through two values of hopping energy, β_1 and β_2, and in the basis of Bloch wave functions the $\varepsilon(k)$ relation is

$$\varepsilon(k) = \pm\, [\beta_1^2 + \beta_2^2 + 2\beta_1\beta_2 \cos\,(2kd)]^{12}, \tag{7.7}$$

where d is the chain repetition unit. Thus for a chain with alternating bond length one always obtains a gap

$$\varepsilon_g = 2|\beta_1 - \beta_2| \tag{7.8}$$

and the bandwidth W is given by

$$W = 2(\beta_1 + \beta_2). \tag{7.9}$$

The perturbation expansion for the ground-state energy in the electron–electron correlation is then written

$$\varepsilon = \varepsilon_0 + \varepsilon_1 + \varepsilon_2 + \ldots, \tag{7.10}$$

where ε_0 is the band energy.

With the usual on-site interaction, ε_1 is the Hartree–Fock solution (cf. Section 7.1.1.)

$$\varepsilon_1 = -\frac{U}{4}, \tag{7.11}$$

where U is the Coulomb interaction: the exchange interaction, being much weaker than U, will be subsequently neglected.

With $\hat{\mathbf{K}}$ the reciprocal-lattice unit vector, Kajzar and Friedel obtain

$$\varepsilon_2 = \frac{-U^2}{N^3} \sum_{\substack{|\mathbf{K}_1|,|\mathbf{K}_2|(\leq \mathbf{K}_4): \\ |\mathbf{K}_3|,|\mathbf{K}_4|(>\mathbf{K}_4)}} \sum_{\hat{\mathbf{K}}} \frac{\delta(\mathbf{K}_1 + \mathbf{K}_2 - \mathbf{K}_3 - \mathbf{K}_4 - \hat{\mathbf{K}})}{\varepsilon_{K_{1\uparrow}} + \varepsilon_{K_{2\uparrow}} - \varepsilon_{K_{3\uparrow}} - \varepsilon_{K_{4\uparrow}}} \tag{7.12}$$

Whereas the first-order correction ε_1 depends only on the number of electrons in the band and the Coulomb correlation energy U, the second-order correction ε_2 also depends on the gap energy according to Eq. (7.12). This dependence is studied by Kajzar and Friedel in their paper.

They compute ε_2 numerically; for the sake of simplicity, they limit themselves to the first Brillouin zone*, an approximation known to be justified at least in these dimensions. Figure 7.6 displays the calculated results of the correlation energy ε_2 as a function of gap energy ε_g for several values of hopping energies β_1.

*This corresponds to $\hat{\mathbf{K}} = 0$ in eqn (7.12).

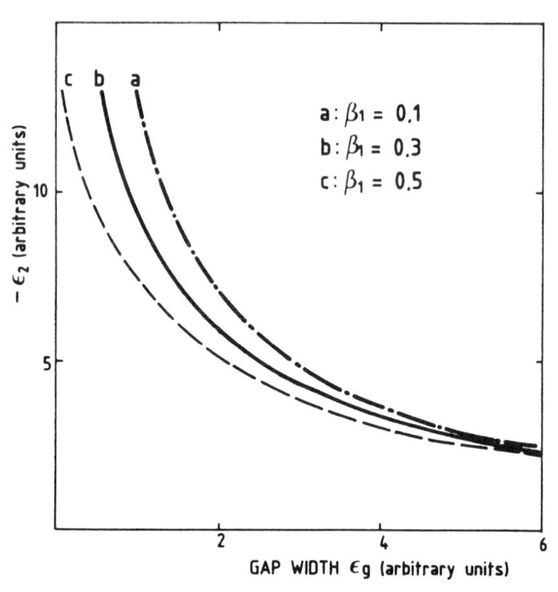

FIGURE 7.6. The absolute value of electron–electron correlation energy ε_2 as function of gap energy for different values of β_1 (after Kajzar and Friedel, 1987).

The correlation energy plainly depends strongly on the gap energy ε_g. The perturbation expansion fails for a vanishing gap, ε_2 then diverging.

For a large gap ($\beta_1/\beta_2 \to 0$) the correlation energy tends to a limiting minimum value given by

$$\varepsilon_2^L/W = -\frac{1}{16}(U/W)^2 = -K(U/W)^2. \tag{7.13}$$

The coefficient K = 0.0625 in Eq. (7.13) is to be compared with the value K = 0.0619 obtained by Baeriswyl and Maki (1985) from the Hubbard model using Gutzwiller's variational approach (described fully in Chapter 4) and K = 0.068 from exact calculations by Economou and Poulopoulos (1979); see also Kivelson and Heim (1982).

Kajzar and Friedel use an approximation to ε_2 discussed by Sayers and Kajzar (1981) which consists of considering only one-electron excitations by putting

$$\varepsilon_{K_1} + \varepsilon_{K_2} - \varepsilon_{K_3} - \varepsilon_{K_4} \sim 2(\varepsilon_{K_1} - \varepsilon_{K_3}) \tag{7.14}$$

in order to get ε_2 in analytical form. Their result is

$$\varepsilon_2 = \frac{-U^2}{16\pi(\beta_1 + \beta_2)} K(\rho), \tag{7.15}$$

where $K(\rho)$ is the complete elliptic integral of the first kind and

$$\rho = \frac{2(\beta_1\beta_2)^{1/2}}{\beta_1 + \beta_2} = (1 - \rho'^2)^{1/2};\ \rho' = \varepsilon_g/2W.$$

Similar variation of ε_2 with ε_g is found as in Fig. 7.6 and the same limiting value ε_2^L/W as in Eq. (7.13) is recovered.

7.2.2. Calculation of Elastic Constant

The elastic constant of the polymer chain is given by the second derivative with respect to $x = (r_1 + r_2)$ (i.e., sum of alternating bond lengths):

$$S = \frac{d^2\varepsilon_{\text{total}}}{dx^2}, \tag{7.16}$$

where

$$\varepsilon_{\text{total}} = \varepsilon_0 + \frac{U}{4} + \varepsilon_2 + \varepsilon_R, \tag{7.17}$$

with ε_R the short-range repulsion energy required for stability of the chain. With some unimportant approximations, Kajzar and Friedel find

$$S = (p/2)^2\,(\varepsilon_0+\varepsilon_2) + \frac{d^2\varepsilon_R}{dx^2}. \tag{7.18}$$

Here p is the exponent appearing in the hopping energies represented by

$$\beta_1 = \beta_0\,(\exp(-pr_1) : \beta_2 = \beta_0 \exp(-pr_2). \tag{7.19}$$

S depends on gap and correlation energy in a way similar to that of the ground-state energy.

The main results of Kajzar and Friedel are supported by nonlinear spectroscopic studies on polydiacetylene and polyacetylene; the reader is referred to the original paper for further discussion.

7.2.3. Phase Diagram for Strongly Correlated Doped *trans*-Polyacetylene Chains

Because the ground state of *trans*-polyacetylene is two-fold degenerate, polyacetylene is a Peierls band-gap insulator at half filling. The degeneracy is a result of the periodic arrangement of alternating double and single bonds, which can be viewed as constituting a commensurate charge-density wave, along the polymer backbone. Su, Schrieffer and Heeger (1980: SSH) have shown that the dimerization of the ground state of this polymer can be accounted for by a one-electron-phonon model with a periodic lattice distortion.

Somewhat surprisingly, subsequent work has shown that in *trans*-polyacetylene, electron–electron correlations enhance dimerization at half-filling (see, e.g., Campbell, DeGrand and Mazumdar, 1984, and other references cited there). This result was first established within the context of the extended Peierls-Hubbard models (see, e.g., Hirsch, 1983). Subsequent work (Campbell, Gammel, and Loh (1988a) has amply substantiated the finding that short-range electron correlations are very significant in the dimerization process in the ground state.

Away from half-filling, there have been relatively few studies of the role of electron correlations in polyacetylene. Below, brief reference will be made to work of Cruz and Phillips (1994). These authors stress that such studies of strong electron–electron correlations are necessary to clarify the mechanism of the insulator–metal transition in this material (see also March and Tosi, 1995).

One relevant observed feature of the transition in polyacetylene is the sudden onset of a Pauli-like susceptibility at a doping level of 5–6% (Moraes, Clen, Chung, and Heeger, 1985). The absence of spin susceptibility below this doping concentration lends support to the viewpoint that charged solitonic rather than electron or hole-like excitations dominate in the lightly doped polymer. Such excitations populate the mid-gap states and do not carry spin.

An early proposal by Kivelson and Heeger (1987) was that the onset of a Pauli susceptibility contribution results from a transition from a soliton to a polaron lattice. However, this model does need some refinement. This is because the subsequent experiments of Kim and Heeger (1989) showed that there are intense infrared active vibrational (IRAV) modes. The intensity of the IRAV mode—a signature of solitonic excitations—increases as the doping level increases.

Nevertheless, it is thought that a transition to some kind of polaron lattice must occur if itinerant spins are to form in the metallic state of *trans*-polyacetylene. It is to be noted here that within the SSH one-electron-phonon model, only soliton excitations are stable, at all doping levels. As a consequence, more work on the metal transition in polyacetylene has focussed on extensions of the SSH model that support polaron formation (Conwell, Mizes, and Choi, 1991; Stafström, 1991; Mizes and Conwell, 1993). For instance, Mizes and Conwell have shown that interchain coupling, as well as chain breaks, stabilize polaron formation. These results were established for short chains containing at most one polaron.

The work of Cruz and Phillips (1994) has addressed the issue of the role of short-range correlations along single strands of *trans*-polyacetylene as a function of doping. These workers incorporate an on-site Hubbard U as well as a nearest-neighbor repulsion V, in order to describe the electron correlations. The continuum model of Takayama, Lin-Liu and Maki (1980: TLM) is used to describe the SSH model. In addition, Cruz and Phillips consider the role of a bond-charge Coulomb repulsion term W, leading to a Hamiltonian in this extended Peierls–Hubbard model of the general form

$$H = H_{TLM} + H_U + H_V + H_W \tag{7.20}$$

where H_{TLM} is the Hamiltonian of the continuum model of TLM referred to above.

Cruz and Phillips then proceed by first solving exactly H_{TLM} for the single-particle states when electrons are added to a single polymer chain. The role of on-site (U), nearest-neighbor (V) and bond repulsion (W) Coulomb correlations is then obtained from a first-order perturbation calculation with the exact single-particle states. By minimizing the total energy, these workers are then able to study the relative stability of polaron and soliton configurations.

Briefly, Cruz and Phillips have shown how a transition from solitons to polarons in *trans*-polyacetylene can be achieved as a function of U alone, or of U and V. The bond-repulsion W seems to break or dissociate polarons into solitons. This is related to other findings at half-filling, in which W increases dimerization (Campbell, Gammel, and Loh, 1988b).

This then leads back to the onset of Pauli-like susceptibility and to the question as to what mechanism drives this onset of spins. It has been proposed by several authors that interchain coupling effects have to be included to describe the transition to the metallic state in polyacetylene. Cruz and Phillips offer the view that it is unlikely that such interchain effects are ultimately responsible for the transition from a soliton to a polaron lattice in *trans*-polyacetylene at the insulator-metal transition, but the issue remains open at the time of writing.

(a) Metal-Insulator and Insulator-Insulator Transitions in Organic Conductors

Phase transitions in the quarter-filled band organic conductors have been discussed by Ung, Mazumdar, and Toussaint (1994). The motivation here is in conducting organic charge-transfer solids. Nearly forty of these materials are superconducting and evidence exists for strong Coulomb interactions among the fermions in these systems (see, e.g., Saito and Kagashima, 1990). Ung *et al.* point out that, for a complete understanding of their normal state, it is essential that the spatial broken symmetries in the quasi-1D superconducting materials be understood. In addition to the $2k_f$ Peierls instability, many of these conductors exhibit a $4k_f$ instability (see Pouget and Comes, 1989).

Ung *et al.*, whose account is followed below, discuss the transitions in the experimentally observable quasi-1D systems. The focus of their work is on the quarter-filled band, where $\rho = 0.5$, ρ being the number of electrons per site. For this commensurate case, the bond order wave (BOW), with periodic modulation of the intersite distances, and the charge density wave (CDW), with modulation of the intrasite charge densities, are distinct. Each of these can have periodicities $2k_f$ (period 4) and $4k_f$ (period 2). Furthermore, each period 4 density wave can occur in two forms, characterized by different phase angles. Table 7.1, taken from

Ung *et al.*, shows the period 4 and period 2 patterns. For zero Coulomb interactions, the ground state comprises coexisting $2k_f$ BOW1 and CDW1 (Ung, Mazumdar, and Campbell, 1992). Table 7.1 also records the observed bond distortion pattern below the $2k_f$ transition in MEM(TCNQ)$_2$, where this pattern is known (Visser, Oostra, Vettier, and Voiron, 1983). The bond distortion in TEA(TCNQ)$_2$ is identical and is accompanied by a modulation of intramolecular charge that resembles the $2k_f$ CDW2 in Table 7.1 (Kobayashi, Ohashi, Marumo, and Saito, 1970).

(b) Instabilities in Real Materials

Ung *et al.* present what seems to be a rather complete picture of the $2k_f$ and $4k_f$ instabilities in the real materials. The free energy of these materials is dominated by high spin states at high temperatures, whose electronic behavior is similar to that of spinless fermions. For nearest-neighbor Coulomb repulsions smaller than a critical value, the lattice dimerizes to the $4k_f$ BOW below the $4k_f$ transition temperature. At still lower temperatures, the free energy is dominated by low spin states. Dimerization of the dimerized lattice now occurs, and the overall

TABLE 7.1. The possible bond distortions and charge modulations in a quarter-filled chain. The single bonds correspond to undistorted bonds, while the double and dotted bonds correspond to short and long bonds, respectively. The last column gives the bond distortion pattern in MEM(TCNQ)$_2$ below the $2k_f$ transition. Here the double dotted bond is a long bond which is shorter than the single dotted bond. In the case of the CDWs, the lengths of the vertical bars on the sites correspond to the charge densities. From Ung *et al.* (1994).

BOW and CDW patterns in $\rho=0.5$			
Period 4	Period 4	Period 2	MEM(TCNQ)$_2$
$2k_F$ BOW1	$2k_F$ BOW2	$4k_F$ BOW	BOW
$2k_F$ CDW1	$2k_F$ CDW2	$4k_F$ CDW	CDW

bond distortion pattern resembles that shown in the last column of Table 7.1. Coexistence of these states with the $2k_f$ CDW2 also explains the observed modulation pattern in $TEA(TCNQ)_2$.

The above account has focussed on the quasi-1D materials. Ung *et al.* discuss some implications for the quasi-2D superconducting TMTSF and BEDT-TTF based materials. The tendency to the phase transitions discussed above is considerably weakened in the quasi-2D superconductors (see the book edited by Saito and Kagashima, 1990). In the quasi-2D regime, the $4k_f$ CDW with alternate occupied sites is still possible. However, Ung *et al.* comment that their demonstration that organic conductors have nearest-neighbor site repulsion below a certain critical value would explain the absence of this CDW transition. On the other hand, the absence of the BOW is a 2D effect. The BOW transition here is analogous to the spin Peierls transition in $\rho = 1$, which is weakened in 2D and gives way to antiferromagnetism [Mazumdar (1987); Tang and Hirsch (1988)]. The CDW2 is a consequence of the BOWs and is thus also expected to be absent in 2D. The only remaining transition that is then possible is to the $2k_f$ SDW, and it has been argued (Mazumdar, Lin, and Campbell, 1991) that for ρ away from 1 the SDW first requires a minimal 2D hopping, but then gradually weakens as the extent of two-dimensionality increases. This would be supported by the occurrence of a spin Peierls transition in the quasi-1D TMTTF materials, the occurrence of SDW in the weakly 2D TMTSF, its vanishing under pressure, and the absence of both spin Peierls and SDW transitions in the even more strongly 2D BEDT–TTF. Thus all possible spatial broken symmetries are weakened in $\rho = 0.5$ in the quasi-2D regime for nearest-neighbor Coulomb repulsions less than a critical value. Whether or not superconductivity is connected with the suppression of spatial broken symmetries remains open at the time of writing.

(c) Phonon-Induced Nonmetal-Metal Transition in Polyaniline

Joo, Prigodin, Min, MacDiarmid, and Epstein (1994) have reported the observation of a phonon-induced nonmetal–metal transition of a doped polyaniline.

Let us first make some general observations on transport near metal-nonmetal transitions. There are two basic mechanisms of single-particle transport in the solid state. One (band transport) is coherent diffusion: electrons move ballistically and are scattered by phonons and by disorder. Such transport takes place mainly in metals with weak disorder or in the more general case for electronic states above a mobility edge. The other (nonmetallic) mechanism is phonon-assisted hopping: electrons jump between localized states by absorbing and emitting phonons. The hopping motion occurs for electrons with energies below the mobility edge.

The temperature dependence of the electrical conductivity $\sigma(T)$ is qualitatively different for these two transport mechanisms. For hopping, σ increases with increasing temperature in contrast to the decrease of σ for band transport. In practical situations it is sometimes difficult to determine the transport mechanism on the basis of $\sigma(T)$ alone. This is the case with conducting polymers. They exhibit relatively high conductivities and at the same time often have a positive $d\sigma/dT$.

(d) Dielectric Measurements and "Catastrophe"

Dielectric constant (ε) measurements can be employed to identify the mechanism of charge transport. For band transport, ε is negative. This corresponds to retardation of the current response due to the inertial mass of the carriers. The value of the negative ε is limited by the rates of relaxation processes or by the measurement frequency. In the hopping regime, electrons remain bound within localized states: at low frequency they follow the variation of the external field. For this situation, ε is positive and proportional to the square of the size of the localized states. In the critical regime for the states near the mobility edge anomalous diffusion takes place and ε has a large positive value—the *dielectric catastrophe.* This cross-over of ε from positive to negative values at the nonmetal–metal transition has been experimentally observed in other systems: e.g., for Si:P and Na_xNH_3 as a function of carrier concentration.

Joo *et al.* report the observation of a T-dependent metal–nonmetal transition, based on such a change of sign of ε from a metallic transport regime to hopping, for samples of doped polyaniline. They have observed that the microwave dielectric constant ε_{mw} is large and negative at room temperature as expected for a normal metal. The absolute value of ε_{mw} decreases with decreasing T and at a boundary temperature of $\approx$ 20 K reaches zero. Below this temperature, ε_{mw} is positive, a fingerprint of an insulator.

Joo *et al.* attribute the ability to observe this behavior of $\varepsilon(T)$ to the key role of 1D chains electronically linking 3D metallic regions in the polymers. It is well known that in a metallic chain the localization of carriers occurs even for weak disorder because of quantum interference of static scattering. At high temperatures, phonon scattering dominates impurity scattering and destroys the elastic scattering interference responsible for electron localization and explains the high σ and very large negative ε_{mw} at room temperature. With T decreasing, the localization effects become strong, and the interplay between the disorder and the phonon scattering leads to the observed behavior of $\sigma(T)$ and $\varepsilon_{mw}(T)$.

Joo *et al.* use a quasi-1D variant of a self-consistent theory of localization to describe the nonmetal–metal behavior of $\varepsilon_{mw}(T)$ (compare Vollhardt and Wölfle, 1980). The frequency-dependent conductivity is written in the form

$$\sigma(\omega) = \sigma_0(\omega)\, \alpha(\omega) \tag{7.21}$$

where $\sigma_0(\omega)$ is a traditional metal form (e.g., the usual Drude–Zener formula, modified for the 1D case) while the factor $\alpha(\omega)$ describes the renormalization of the conductivity due to the quantum interference of impurity scattering. For the 1D system it takes the form (Vollhardt and Wölfle, 1980; Prigodin and Roth, 1993)

$$\alpha(\omega) = 1 - 2\,\{1 + [1 - 4i\omega\tau_{imp}(2k_f)]^{1/2}\}^{-1}. \tag{7.22}$$

Here $\tau_{imp}(2k_f)$ is the impurity backward-scattering time, which is the only effective one for a quasi-1D system. Equation (7.22) reproduces correctly both high and low frequency limits: $\alpha(\omega) \to 1$ for $\omega \to \infty$ while $\alpha(\omega) \approx -i\omega\tau_{imp}(2k_f)$ as $\omega \to 0$ and hence according to Eq. (7.22) the system is an insulator.

The effect of finite temperature is reflected through phonon scattering, which can be expected to be highly anisotropic since (see Conwell, 1980; Kivelson and Heeger, 1988)

$$\tau_{ph}(2k_f, T)/\tau_{ph}(0, T) \approx \exp[\omega(2k_f)/T] \tag{7.23}$$

and $\omega(2k_f)$ for acoustic phonons is taken as ≈ 0.1 eV. The phonon backward scattering contributes to the transport time in the Drude approximation $\sigma_0(\omega) = \sigma_0/(1 - i\omega\tau_{tr})$:

$$\tau_{tr}(T) = \tau_{imp}(2k_f)/[1 + \tau_{imp}(2k_f)/\tau_{ph}(2k_f,T)]. \tag{7.24}$$

The role of phonon forward scattering is to destroy the phase coherence of the impurity scattering. This delocalization effect can be included in Eq. (7.22) by the replacement $\omega \to \omega + i/\tau_{ph}(0,T)$ (Gogolin, 1988; Prigodin and Roth, 1988).

Using the Drude formula and Eq. (7.22) the dc conductivity is obtained in the whole temperature range as

$$\sigma_{dc}(T) = \frac{\omega_p^2}{4\pi}\,\tau_{tr}(T)\,\frac{f(T) - 1}{f(T) + 1} \tag{7.25}$$

where ω_p is the plasma frequency and $f^2(T) = 1 + 4\,\tau_{imp}(2k_f)/\tau_{ph}(0,T)$. The T-dependent dielectric constant at low frequency, $\omega\tau_{ph}(T) \ll 1$ follows as

$$\varepsilon(T) = [\omega_p\tau_{imp}(2k_f)]^2\,\frac{4 - f(T)[f^2(T) - 1]}{f(T)\,[f(T) + 1]^2}. \tag{7.26}$$

The dc conductivity given in Eq. (7.25) exhibits a characteristic maximum at $T = T_m$ say, where $\tau_{ph}(2k_f,T_m)$ becomes comparable with $\tau_{imp}(2k_f)$. At $T = T_m$, $\varepsilon(T)$ has a minimum. Joo *et al.* have also observed finger-prints of the change in the mechanism of electron transport in the T-dependent thermopower and in $\sigma_{dc}(T)$.

To summarize, the experimental results of Joo *et al.* show a transition from a metallic to a nonmetallic phase in doped polyaniline at ≈ 20 K. The dominance of the metallic component of charge transport above this temperature is clear from the negative microwave dielectric response and the linear dependence of thermoelectric power on temperature. Below ≈ 20 K, the carrier mobility is due to phonon-assisted hopping. The metal–nonmetal transition of the system can be interpreted as a cross-over from phonon-controlled metallic transport to phonon-assisted hopping in a weakly localized phase. The precise role of disorder vs. Coulomb localization still needs clarification here (see also Chapter 8 below).

7.3. PLASMON DISPERSION AND STRUCTURE FACTOR OF π-ELECTRONS IN POLYACETYLENE

Reference will be made here to the importance of Coulomb correlations between π-electrons in the organic polymer $(CH)_x$, namely polyacetylene (March, 1985a).

The interest stems from the results of the fast-electron energy-loss experiments of Ritsko *et al.* (1980). In this work, energy-loss spectra for 80-keV electrons transmitted through thin-film samples of polyacetylene were reported, showing that the spectrum of $(CH)_x$ at momentum transfer $q = 0.1$ Å^{-1} is dominated by a broad peak at 4.1 eV. The Kramers–Krönig analysis of Ritsko *et al.* of both optical and energy-loss data demonstrated that this peak occurred after the real part ε_1 of the dielectric function has passed through a node and in a region where the imaginary part ε_2 is relatively small and decreasing rapidly. This work can leave no doubt that the broad peak referred to above at 4.1 eV for $q = 0.1$ Å^{-1} is a plasmon associated with collective electron density oscillations of the π-electron assembly. Therefore it can certainly be concluded that the long-range Coulomb interaction has qualitatively important consequences in the quasi-one-dimensional solid polyacetylene.

But one further striking result was observed in the work of Ritsko *et al.* By following the movement of the peak at 4.1 eV in the loss spectrum at $q = 0.1$ Å^{-1} when q was varied, they could plot the π-plasmon dispersion relation. They found for polyacetylene a linear dispersion over the range $0 < q < 0.7$ Å^{-1}, and this has subsequently been observed in polydiacetylene as well.

Use of Feynman Model of Dynamical Structure Factor

In three-dimensional jellium, the assumption that the dynamical structure factor $S(k,\omega)$ has the form

$$S(k\omega) = S(k)\,\delta[\omega-\omega(k)], \tag{7.27}$$

when combined with the quantum-mechanical first moment, leads to a Feynman-like model of the collective mode dispersion relation $\omega(k)$ in terms of the static structure factor $S(k)$ representing in k space the pair correlation function $g(r)$ of electrons:

$$2m\,\omega(k) = \frac{\hbar k^2}{S(k)}. \tag{7.28}$$

As discussed in detail by Holas and March (1987) for three-dimensional jellium in the high-density metallic phase, the small-angle scattering form of $S(k)$ is given by

$$S(k) = a_2 k^2 + a_4 k^4 + a_5 k^5 + \dots . \tag{7.29}$$

Inserting this into Eq. (7.28) leads, with the first-principles choice of a_2, to

$$\omega(k) = \omega_p + b_2 k^2 + b_3 k^3, \tag{7.30}$$

with ω_p the plasma frequency, the nonanalytic term $b_3 k^3$ in $\omega(k)$ coming from the k^5 term in $S(k)$ in Eq. (7.29).

By use of the same model (March, 1985a) and viewing the π-electrons as a quantal liquid with ground-state structure factor $S(k)$, with k now along the polymer chain, arguments paralleling those in three dimensions lead, with input of the observed dispersion relation by Ritsko *et al.*, namely,

$$\omega(k) = \omega_p + c_1 k + \text{higher order terms} \tag{7.31}$$

to a small-k expansion of $S(k)$ given by

$$S(k) = \frac{\hbar k^2}{2m\omega_p} + a_3 k^3 + \dots . \tag{7.32}$$

This mirrors the three-dimensional jellium result in Eq. (7.29), except that, presumably because of dimensionality dependence, the nonanalytic term now appears at $0(k^3)$. This is why, in the above interpretation, one can see the nonanalyticity in one-dimensional solids much more readily than in the simple nearly-free-electron *sp* metals. Of course, first-principles work on the theory is still required to calculate quantitatively the magnitude of the linear dispersion.

7.4. PHENOMENOLOGICAL THEORY OF METAL–INSULATOR TRANSITION AT T = 0 AND RELATION TO GUTZWILLER'S VARIATIONAL METHOD

Reference has already been made (in Section 4.6) to the use of Gutzwiller's variational method for the calculation of correlation energy in the 3*d* transition metals. Here, a phenomenological theory of March *et al.* (1979) will be discussed, which closely parallels some features of the Gutzwiller theory. Since the phenomenological theory has been reviewed elsewhere (March and Parrinello, 1982), a very brief summary of only the assumptions underlying it, together with the main results, will be given below.

The idea is again to start out from a narrow energy band (or set of such bands). The electron–electron interactions are then introduced through the Hubbard energy U, which, as defined above, is the energy it takes to bring two electrons with opposed spins onto the same site. Of course, the Pauli exclusion principle, or more precisely the Fermi hole discussed in Chapter 3, already keeps two electrons with parallel spin from coming close together.

March *et al.* (1979) then base their phenomenological treatment on the following assumptions:

a. A suitable, generalized order parameter (cf. Chapter 3) is the discontinuity, say q, at the Fermi surface in the metallic phase, of the single-particle occupation number n_k (see Daniel and Vosko, 1960, for high-density jellium).
b. At a critical strength, say U_0, of the Hubbard interaction U, a metal–insulator transition occurs, at which the discontinuity q is reduced to zero.
c. The average number of doubly occupied sites, say d, is a function of q, i.e., the internal energy is $Ud(q)$.

The ground-state energy $E(q)$ is, as is usual in such phenomenological treatments, expanded about the singular point $q = 0$ as

$$E(q) = E_0 + E_1\, q + E_2\, q^2 + \dots . \tag{7.33}$$

With the usual assumptions, namely $E_1(U) = \alpha(U_0 - U)$, i.e., $E_1(U_0) = 0$, $\alpha < 0$ while $E\ (U_0) > 0$, minimization gives the dependence of q and of the difference in energy, say ΔE, between metallic and insulating states on the difference of U from its critical strength U_0: (compare eqns (4.15) to (4.17))

$$q = Q\left[1 - \frac{U}{U_0}\right]; \qquad \Delta E = (\alpha U_0 Q + E_2 Q^2)\left[1 - \frac{U}{U_0}\right]^2 :$$

$$Q = -\frac{\alpha U_0}{2E_2}, \tag{7.34}$$

while from Feynman's theorem

$$d = \frac{d(\Delta E)}{dU}\,\alpha\left[1 - \frac{U}{U_0}\right]. \tag{7.35}$$

These dependencies of q, ΔE, and d on $1 - U/U_0$ are equivalent to those given by Gutzwiller's variational calculation (see Brinkman and Rice, 1970).

Related to this account, the complementary study to the above phenomenology by Spalek *et al.* (1983) should be noted. These workers discuss a simple thermodynamic theory of the metal–insulator transition and local moments in a narrow band, using related ideas which have their origin in Gutzwiller's variational method.

In connection with this discussion of the metal–insulator transition, it is relevant also to mention a renewal of interest in the simple theoretical model introduced by Falicov and Kimball (1969; see also Ramirez *et al.*, 1970) to describe the transition observed in many transition-metal oxides and rare earth compounds. Relevant in this context especially is the work of Wio (1983), where other references are also given to collaborative work with Alascio and Lopez.

Refinement of Gutzwiller Approximation

Weger and Fay (1986) have proposed a relatively simple approach to refine the Gutzwiller approximation by replacing the "step function" for the momentum distribution with a more physical form.

As stressed already, the Gutzwiller approach predicts that electron correlations bring about a metal–insulator transition. As the transition is approached, the effective mass, and hence the electronic specific heat coefficient γ and the spin susceptibility χ diverge, while the compressibility goes to zero. An increase in γ and χ was observed in metallic V_2O_3 under pressure on approaching the metal–insulator transition.

Weger and Fay (1986) note that this behavior is drastically different from that predicted by Hubbard (1964), according to which, as the metal–insulator transition is approached, the density of states at the Fermi energy E_f decreases to zero.

There are, however, some difficulties associated with the Gutzwiller approximation, as Weger and Fay point out. First, for a Hubbard Hamiltonian with interaction parameter U, band width W, and a constant rectangular density of states, the metal–insulator transition occurs at $U/W = 2$. This value is quite large, other instabilities occurring at much lower values of U/W. In particular, the ferromagnetic transition, in the Hartree–Fock approximation, occurs at $U/W = 1$, and the Hubbard metal–insulator transition at $U/W \simeq 1.1$, and an antiferromagnetic transition is predicted by Ogawa *et al.* (1975) to occur at $U/W \simeq 0.8$. At these values of U/W, the mass enhancement predicted by the Gutzwiller treatment is rather small, being around 20%. Second, and as already noted, Gutzwiller took a step function for the momentum distribution. This has merit in leading to the relationship between d, the fraction of doubly occupied sites, and the discontinuity q in the momentum distribution at the Fermi edge. However, more physical distributions have been derived, for example, by Daniel and Vosko (1960). Thirdly, Weger and Fay stress that the enhancement of m^* and χ can be arbitrarily large as one approaches the critical interaction parameter, enhancements of, say, 10^3 and 10^4 being allowed in the Gutzwiller treatment. It is not clear whether such large enhancements are really feasible.

The study of heavy-Fermion systems has certainly played an important role in reviving interest in the Gutzwiller approach, as detailed by Weger and Fay. Therefore, they have re-examined the Gutzwiller approach in the light of the above considerations. Their most important change is using a momentum distribution of the general form derived by Daniel and Vosko; this is compared with the step-function form in Fig. 7.7. However, they find it necessary to make the additional assumption that the Gutzwiller relation still holds between q and d, with q replaced by a suitable average q_{av}. Their conclusion is then that if the metal–insulator transition is assumed to occur when $m^*/m \to \infty$ (i.e., $q \to 0$), the critical condition is $U_c = 0.87W$. The transition in the modified treatment of Weger and Fay is therefore now in agreement with the other estimates mentioned above. Weger and Fay note that d is no longer zero at the transition, ~9% of the sites still being doubly occupied, compared with 25% at $U/W = 0$.

They have tested the sensitivity of their result to the detailed form of the momentum distribution and concluded that the critical values U_c and d_c do not depend strongly on the shape of the momentum distribution for more physically plausible forms than the step function.

Weger and Fay re-examine the spin susceptibility first derived in the Gutzwiller approximation by Brinkman and Rice (1970). They follow the treatment of Vollhardt (1984) by writing χ in terms of the band susceptibility χ_0 and the Landau parameter F_0^a:

$$\frac{\chi}{\chi_0} = q^{-1}/(1 + F_0^a). \tag{7.36}$$

The Landau parameter is given by

$$F_0^a = -\left(1 - \frac{1}{4(1 - 2d)^2}\right) \tag{7.37}$$

In the Gutzwiller approximation, as $U \to U_c$, χ diverges like m^*/m and $F_0^a \to -\frac{3}{4}$. Making the modifications discussed above, Weger and Fay obtain

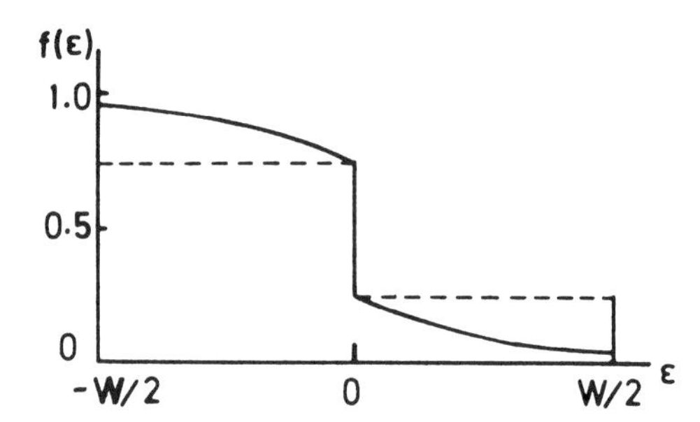

FIGURE 7.7. Single-particle distribution function. Dashed line: Gutzwiller. Solid line: Daniel and Vosko (Weger and Fay, 1986).

$$\frac{\chi}{\chi_0} = \left[\frac{3.5}{2.5 + q}\right]\left[\frac{1}{1 + F_0^a}\right], \tag{7.38}$$

with

$$F_0^a = -\left[1 - \frac{1}{[1 + 0.75(U/W)]^2}\right]. \tag{7.39}$$

This yields $F_0^a = -0.63$ at $U = U_c$. The new point made by Weger and Fay is that at the metal–insulator transition characterized by $q \rightarrow 0$, χ is enhanced by the interaction but tends to the finite value $(\chi/\chi_0)_{U_c} \simeq 4$. A similar modification is found by Weger and Fay for the compressibility.

Weger and Fay are optimistic that although the Hubbard Hamiltonian is certainly an oversimplified model, it is giving a good description of a number of systems. However, they conclude that it will be necessary to find better solutions of this model of electron correlation before one can gain deeper understanding of more complicated models such as the Anderson and Kondo lattices.

7.5. RESONATING VALENCE BOND STATES

In 1973, Anderson proposed that, at least in the triangular two-dimensional antiferromagnet for $S = \frac{1}{2}$, and perhaps in other cases, the ground state might be the analogue of the precise singlet in the Bethe (1931) solution of the linear antiferromagnetic chain. In both cases, the zeroth-order energy of a state consisting purely of nearest-neighbor singlet pairs is more nearly realistic than that of the Néel state. Anderson proposed that higher-order corrections, allowing the singlet pairs to move or "resonate" *à la* Pauling, might make this insulating singlet or resonating valence-bond (RVB) state more stable. Anderson emphasized the distinction between this state and two other locally stable possibilities, the Néel state and the "spin-Peierls" state of a self-trapped, localized array of singlet pairs. Each of these other states has a broken symmetry relative to the high-temperature paramagnetic state and would exhibit a phase transition with temperature. Fazekas and Anderson (1974) improved on Anderson's numerical stability estimates, and Hirakawa *et al.* (1985) and Yamada *et al.* (1985) have proposed application of the RVB state to specific compounds.

Anderson (1987), while stressing the difficulty of representing and calculating with RVB states, gives a representation in terms of Gutzwiller-type projections of mobile-electron states (see also Rice and Joynt, 1987) on the Gutzwiller approximation to the Bethe solution.

A single pair of electrons in a mobile valence bond along a lattice vector τ may be written, according to Anderson (1987: see also section 4.5)

$$b_\tau^+\Psi_0 = \frac{1}{N^{1/2}} [\sum_j c_{j\uparrow}^+ c_{i+\tau\downarrow}] \Psi_0$$

$$= \frac{1}{N^{1/2}} [\sum_k c_{k\uparrow}^+ c_{-k\downarrow}^+ \exp i(k\tau)] \Psi_0, \tag{7.40}$$

where b_τ^+ is the electron-pair creation operator, c_j^+ is the single-electron creation operator, and N is the total number of sites. A linear combination of all nearest-neighbor bonds may be written

$$b_{\mathrm{nn}}^\dagger = \sum_{\tau=(\mathrm{nn})} b_\tau^+, \tag{7.41}$$

where nn indicates summation over nearest neighbors, or in general, if one wants a distribution of bond lengths, the linear combination may be written as

$$b^\dagger = \sum_k a(k) c_{k\uparrow}^\dagger c_{-k\downarrow}^\dagger, \tag{7.42}$$

with the condition on the expansion coefficients

$$\sum_k a(k) = 0, \tag{7.43}$$

if one does not allow double occupancy.

Anderson then seeks a Bose condensation of N electrons in mobile valence-bond states by forming

$$\Psi = (b^+)^{N/2}\Psi_0, \tag{7.44}$$

but, as he points out, this state contains large numbers of empty and doubly occupied sites. The way he attempts to construct a genuine RVB state is sketched in Appendix 7.3. Two roughly equivalent project techniques are summarized there, the simplest being the straightforward Gutzwiller method.

The motivation of this study of the RVB state was the high-temperature superconductivity observed in a number of doped lanthanum copper oxides near a metal–insulator transition (Bednorz and Müller, 1986). The insulating phase is proposed by Anderson to be the RVB state or, as, he puts it alternatively, "quantum spin liquid." The pre-existing magnetic singlet pairs of the insulating state, according to Anderson's ideas, become charged superconducting pairs when the insulator is doped sufficiently strongly. In this picture, the mechanism for superconductivity is hence predominantly electronic and magnetic, though Anderson notes that weak phonon interactions may favor the state.

Extensive developments along these lines have subsequently taken place (see Anderson *et al.*, 1987, and Kivelson *et al.*, 1987). However alternative mechanisms for high $-T_c$ superconductivity in oxides have also been proposed (see, for instance, Emery, 1987, 1989: also Appendix A7.2).

7.6. PRESSURE-INDUCED METAL–INSULATOR TRANSITIONS WITH CHEMICAL BONDS

Let us turn next to consider pressure-induced metal–insulator transitions in chemically bonded solids. It was recognized by Wigner and Huntington (1935) in the early days of the theory of metals that, while the monovalent alkalis Na, K, etc. are stable under normal conditions, hydrogen was quite exceptional in forming a molecular crystal phase at atmospheric pressure. It seemed natural enough, therefore, to inquire what thermodynamic state of hydrogen would be required to dissociate the H_2 molecules and, according to band theory, to lead to a metallic phase quite similar to Na, K, etc. Though the primary conclusion of these workers (namely, that a very high pressure is required to induce the transition) was certainly correct, precise quantal calculation of the pressure turns out to be a difficult many-electron problem. Thus, since this early work, it has been recognized that high-pressure studies, besides being of very considerable value in astrophysics and geophysics, in fact provide a searching test of many-body theories.

At extremely high pressures, when the atoms have been entirely broken down, one expects that the Thomas–Fermi–Dirac theory (see Section 3.3) should be realistic. However, the pressures required to bring about this circumstance for the heavier elements were crudely estimated in the early work of Ramsey to be of the order of 10^{12} atm.

At lower pressures, it is to be expected that the density of all but the heaviest elements will show a series of discontinuous jumps as the pressure increases, corresponding to the breaking down of inner shells. Early work on such a problem was that of ten Seldam (1961), and of Simcox and March (1962), on He (see Siringo *et al.*, 1989).

Another route to estimate the pressure needed to induce a metallic transition in solid He involves the so-called Herzfeld criterion (compare Section 7.3d). This predicts metallization at a volume fraction V_M/V_0 given by

$$\frac{V_M}{V_0} = \frac{n_0^2 - 1}{n_0^2 + 2}, \tag{7.45}$$

where n_0 is the refractive index at zero pressure. This criterion predicts a molar volume V_M at the transition point of ≈ 0.5 cm^3/mole. To estimate the corresponding transition pressure, one must then evidently invoke the equation of state (EOS).

X-ray experiments in the range 15.5 to 23.5 GPa at room temperature have led to an accurate determination of the EOS. On the theoretical side, the work of Loubeyre (1982) has clarified the relevance of three-body exchange contributions and has led to a prediction of the phase diagram.

By using the Herzfeld criterion, complemented by extrapolation of the EOS given by Mills *et al.* (1979), Siringo *et al.* (1989) have estimated the metal–insulator transition pressure P_T in helium as 106 Mbar. This is in good agreement with the value P_T = 112 Mbar obtained by Young *et al.* (1989) with LMTO type calculations for the *hcp* structure.

Following these introductory comments, the continuously bonded semiconductors Si and Ge will next be briefly discussed. The new feature here is that, though initially band gaps reduce with pressure, metallization in fact coincides with a structural phase change.

Then a rather more detailed discussion will be given of solid H_2 and I_2, in both of which the basic building block is the diatomic molecule.

Transition in Covalently Bonded Semiconductors

Experimental Facts. Beginning with the diamond structure monatomic semiconductors, these are characterized by strong tetrahedral coordination, each atom forming two-electron bonds with its four nearest-neighbors. On application of sufficiently high pressure, the monatomic materials, Si, Ge, and Sn go into the more close-packed β-Sn structure. This structural change coincides with the transition to a metallic state. Relevant experimental data have shown through absorption-edge measurements the decreasing band gaps in Si and Ge with pressure and the huge increase in electrical conductivity heralding the structural phase change at the metal–insulator transition.

Further experimental and theoretical studies have exposed a further sequence of structural phase transitions: e.g., in Si

a. diamond to β-Sn at 11 GPa,
b. β-Sn to simple hexagonal phase (*shp*) at GPa,
c. *shp* → hexagonal close-packed (*hcp*) above 40 GPa, and
d. *hcp* → face-centered cubic at ~ 78 GPa.

The highest pressure so far achieved in the study of Si is by Ruoff's group; some 100 GPa.

Change in Electronic Bonding. The electronic properties of Si change through these transitions. In particular, the strong electron pair bonds of the diamond phase are weakened with increasing pressure. One has a coexistence of covalent and metallic bonding in the β-Sn and *shp* phases, followed by the breaking of the covalent bonding in the metallic *fcc* phase. For Ge, some structural differences are found but the general pattern is similar to that of Si (cf. Siringo *et al.*, 1989).

7.7. CAN MOLECULES EXIST IN METALLIC PHASES?

The considerations presented later for normal and expanded alkali metals (see especially the discussion in Chapter 8) can leave little doubt that chemical arguments are very helpful in understanding the strongly correlated electronic behavior in these materials.*

This leads, rather naturally, to the question of whether molecules can exist in ordered metallic phases: i.e., can molecular metallic crystals exist? To be quite specific, one has in mind the behavior of insulating molecular crystals, such as are formed by H_2 and I_2 under normal conditions, as pressure is applied (say, ideally, at $T = 0$), to eventually bring about an insulator–metal transition. Will such high-pressure metallic phases necessarily have no H_2 or I_2 molecules?

To address this question, it will be useful to summarize the results of Ferraz *et al.* (1984), using a model to simulate a hydrogen molecule embedded in jellium.

7.7.1. Heitler–London Theory of H_2 with Screened Coulomb Interaction

To deepen this discussion of whether molecules can exist in metallic phases (e.g., metallic H_2), let us at this point summarize the results of a model of a single H_2 molecule. This model was constructed by Ferraz *et al.* (1984) to study when the screening of the interactions in a Heitler–London theory of a single H_2 molecule (cf. Section 4.1) would result in molecular dissociation. One can view such a procedure as a test of the instability of an electron gas against molecular bonding. Therefore, Ferraz *et al.* (1984) replaced the bare Coulomb interaction $1/r_{ij}$ by $\exp(-k_{TF}r_{ij}) / r_{ij}$ in the usual free-space Heitler–London theory of H_2. The inverse screening length k_{TF} was taken to be that given by the usual Thomas–Fermi result $k_{TF}^2 = 4k_f/\pi a_0$, where k_f is the Fermi wave number of the electron gas (cf. Appendix 5.2), related to the constant density ρ_0 in Eq. (3.36) by $\rho_0 = k_f^3/3\pi^2$. The conclusion was that as the electron density was increased continuously, the binding energy of the molecule was eventually lost. This process is illustrated in Figs. 7.9 and 7.10 for two different values of the screening length k_{TF}^{-1}.

7.7.2. Metal–Insulator Transition in Hydrogen

Ferraz *et al.* (1984) used the critical density at which the H_2 molecule dissociated as an (admittedly rough) measure of the molecular-insulator–metallic hydrogen transition.† However, at this point it must be stressed that, in analogy

Notes added in proof. *See Freeman, G. R. and March, N. H. (1994: J. Phys. Chem. *98*: 9486)
†See Whitman, L. J., Stroscio, J. A., Dragoset, R. A. and Cellota, R. J. (1991: Phys. Rev. Lett. *66*: 1338).

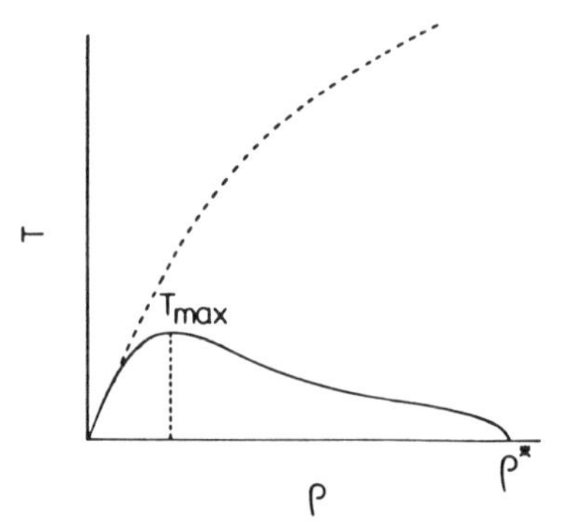

FIGURE 7.8. Reduced melting curve of three-dimensional Wigner electron crystal: see especially Chapter 9 below. The dashed curve is the classical asymptote got from the one-component plasma with T proportional to cube root of the density ρ. ρ* is density of insulator-metal transition at T = 0 (after Ferraz *et al.*, 1979).

with solid I_2 (to be discussed in Section 7.7.3), there may well be metallization in solid hydrogen at a lower pressure than such a model predicts, because of energy-gap closure in the originally insulating molecular H_2 phase.

Such gap closure is a property of the order in the crystal, if not long-range order then at least local coordination. In a limited pressure range, it may well be that molecules of H_2 can remain undissociated even after metallization has occurred (for I_2, this is established; see below, also Siringo *et al.*, 1988a,b; Pucci *et al.*, 1988; Siringo *et al.*, 1989, by combining theory and experiment).

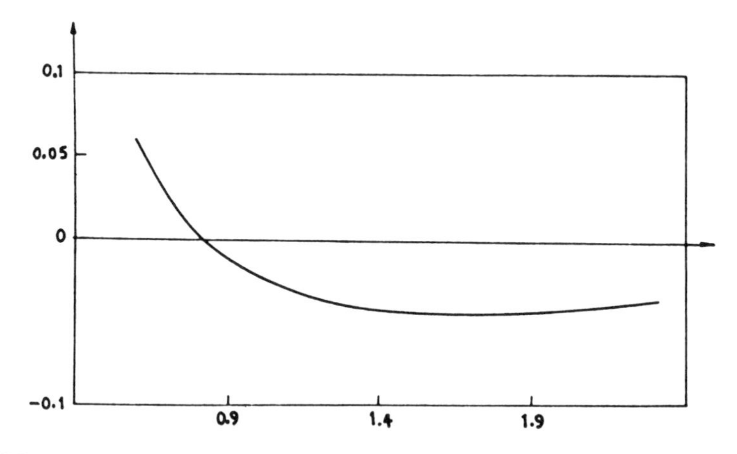

FIGURE 7.9. Potential energy curve for Heitler–London model of H_2 molecule using screened Coulomb interactions with an inverse Thomas–Fermi screening length $k_{TF} = 1$ (after Ferraz *et al.*, 1984).

It is highly relevant, however, to note here that estimates of the critical density r_s made by Ferraz *et al.* (1984) agree quite well with the report of Hawke *et al.* (1978) that the transition to a metallic state of hydrogen with no molecules present has been observed experimentally. However, later authors [see the comments of Ross (1985) and also those of Silvera (see March, 1990)] have raised questions about this reported observation of the metal–insulator transition, and the matter is not settled at the time of writing. On the other hand, there is no dispute that, under sufficiently high pressure, a metallic phase with no molecules present is formed. Ross (1985) estimates the pressure at which this happens to be between 1 and 4 Mbar, and this is consistent with later and sharper estimates. Thus, based on measured equation-of-state extrapolation, Mao *et al.* (1988) estimate 2.3 Mbar, whereas Van Straaten and Silvera (1988) suggest a value of 2.8 Mbar after also invoking the Herzfeld criterion discussed above. Furthermore, the quantum computer simulation (see also Chapter 5) made by Ceperley and Alder (1987) yields a transition to a cubic atomic phase at 3 Mbar.

It is worth supplementing the account of Chapter 5 briefly here by adding, specifically in connection with the study of Ceperley and Alder (1987), that these results rest on three approximations:

1. the restriction to less than a few hundred atoms,
2. incomplete treatment of Fermi statistics, i.e., antisymmetry, and
3. the finite length of the Monte Carlo runs.

The method also has some uncertainty with respect to the crystal structure of the molecular phase. However, these workers carried out calculations for several crystal structures, and their results indicated that the oriented Pa3 phase is preferred at densities higher than that corresponding to $r_s = 1.45a_0$. This orientation-ordering transition has been placed close to 1 Mbar.

7.7.3. Properties of Compressed Solid Iodine

Briefly, solid molecular iodine has a planar structure, with the molecules lying in the planes. The interaction between the planes is weak, as in graphite, and the crystal exhibits two-dimensional behavior.

What is important for what follows is that electrical transport measurements by Drickamer and co-workers (Drickamer, 1965; Drickamer *et al.*, 1966) have demonstrated that iodine becomes metallic under pressure. Specifically, the resistivity ρ at low pressure has the typical behavior

$$\rho \sim \exp(\Delta/2k_BT) \tag{7.46}$$

of a semiconducting material. However, at pressures in excess of 160 kbar, the crystal was found to exhibit metallic conduction in the direction perpendicular to the planes, the temperature variation of ρ now having the form

$$\rho \sim T^{\alpha}; \qquad \alpha > 0. \tag{7.47}$$

In contrast, a small residual energy gap remains parallel to the planes, with semiconducting behavior as a consequence. Finally, as the pressure is increased beyond 220 kbar, the crystal exhibits metallic behavior in all directions and has a smaller resistivity in the direction parallel to the planes.

As to the nature of the second transition at 220 kbar (see, for example, Siringo *et al.*, 1989), from X-ray diffraction studies there can no longer be any doubt that iodine is still a molecular crystal after the first transition at 160 kbar. There is no structural change, but the relevant band gap goes continuously to zero as the pressure approaches 160 kbar (Drickamer, 1965; Drickamer *et al.*, 1966). Below 160 kbar, one has effectively two gaps: $\Delta_{\perp}$ and $\Delta_{\parallel}$, which are related to conduction in directions perpendicular and parallel, respectively, to the planes. The first gap becomes zero at 160 kbar, while the second closes at 220 kbar, giving rise to metallic conduction in the planes as well. In this third pressure range, the resistivity is smaller in the direction parallel to the planes, iodine exhibiting therefore anisotropic behavior.

Siringo *et al.* (1988a,b) have carried out quantum-chemical calculations of the energy bands, including a full account of the variation of the structure of the nuclei with pressure from the experiments of Takemura *et al.* (1982). The chemical treatment of Siringo *et al.* (1988a,b) was based on a two-dimensional model

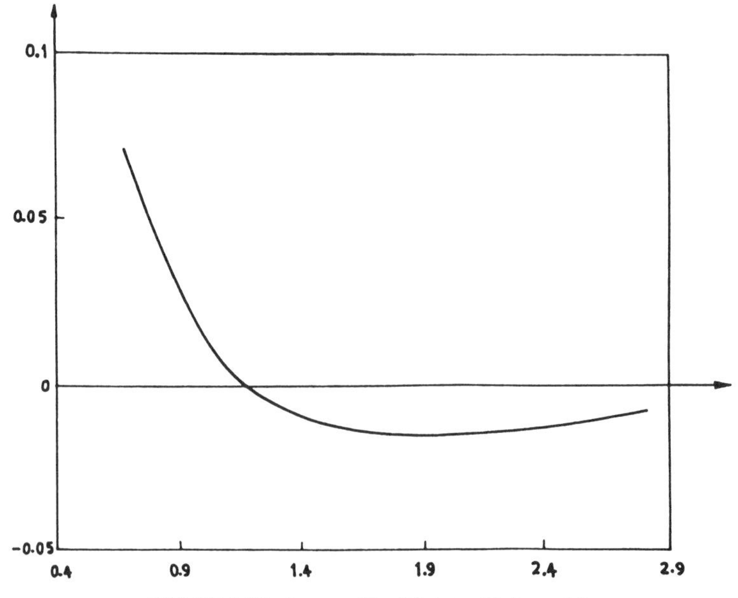

FIGURE 7.10. Same as Fig. 7.9, but with $k_{\mathrm{TF}} = 1.1$.

of $5p$ electron localization going back to Bersohn (1962). Figure 7.11 shows the results of Siringo *et al.* (1988a,b) for the indirect energy band gap between the top of the valence band and the bottom of the conduction band as a function of pressure. It is striking that the band gap reduces from its (fitted) value of 1.35 eV at zero pressure to 0.1 eV at ~200 kbar. Extrapolation suggests that such a two-dimensional structure will become metallic due to band overlap at ~230 kbar, in close accord with the experiments described above.

On the other hand, a complete description of solid iodine under pressure must incorporate the effects of the interaction between planes, which are no longer negligible under high pressure. This problem has subsequently been tackled by Siringo *et al.* (1988a,b), again with good agreement with experiment. Thus, two band gaps are found, and calculating effective masses of electrons and holes from their energy band structure, the main features of the transport measurements by Drickamer and co-workers can be understood in general terms.

Subsequently, Pucci *et al.* (1988) have examined in some detail many-body effects in a model which turns out to be relevant to the above treatments of the molecular solids H_2 and I_2, and it is these latter aspects that will be summarized below.

7.8. ELECTRON–ELECTRON INTERACTIONS AND THE ROLE OF BOND–CHARGE REPULSION IN LOW-DIMENSIONAL SOLIDS

As is clear from the Wigner crystal transition (see melting curve in Fig. 7.8 above), energy band overlap is only one of the possible mechanisms by which metal–insulator transitions can occur. Therefore, Pucci *et al.* (1988) have considered Coulomb interaction (especially bond–charge repulsion) as a function of (high) pressure.

To do so, these workers employed a model Hamiltonian constructed following the pioneering work of Su, Schrieffer, and Heeger (1980; SSH). Adding electron–electron interactions in Hubbard-like fashion results in the Hamiltonian $H = H_{\text{SSH}} + H_{\text{ee}}$:

$$H_{\text{SSH}} = -\sum_{i,\sigma} [t_0 - \alpha(u_{i+1} - u_i)](C^+_{i+1,\sigma}C_{i,\sigma} + \text{h.c.}) + \frac{1}{2} K \sum_i (u_{i+1} - u_i)^2 + \sum_i \frac{P_i^2}{2M}, \tag{7.48}$$

$$H_{\text{ee}} = \frac{U}{2} \sum_{i,\sigma} n_{i,\sigma} n_{i,-\sigma} + V \sum_i n_i n_{i+1}, \tag{7.49}$$

where C_i^+ (C_i) creates (annihilates) an electron at site i, $n_{i,\sigma} = C^+_{i,\sigma}C_{i,\sigma}$, $n_i = \Sigma_\sigma n_{i,\sigma}$ and P_i is the momentum conjugated to the amplitude of the structural distortion u_i,t_0

is the transfer integral, U the on-site repulsion, and V the nearest-neighbor site repulsion.

By varying the relative magnitude of the parameters appearing in Eqs. (7.48) and (7.49), one can have a rich variety of different physical behavior (see, for example, Siringo *et al.*, 1989).

Dimerization in systems such as *trans*-polyacetylene produces a charge-density wave (CDW) with wave vector $k = 2k_f$ (where k_f is the Fermi momentum), i.e., an inhomogeneous charge distribution. One could then expect that Coulomb interaction would oppose the dimerization and screen the $2k_f$ potential in such a way so as to reduce the single-particle gap. However, a number of model calculations have shown that weak interactions tend to enhance the dimerization (Horsch, 1981; Kivelson and Hein, 1982; Campbell *et al.*, 1984).

Kivelson *et al.* (1987) have considered the most general form of H_{ee} and have denoted by $V(n,m,l,p)$ the matrix element of the electron–electron interaction potential $V(r)$:

$$V(n,m,l,p) = \int d\mathbf{r}d\mathbf{r}'\rho_{np}(\mathbf{r})\rho_{ml}(\mathbf{r}')V(\mathbf{r}-\mathbf{r}'), \tag{7.50}$$

where *n,m,l,p are site indices and* ρ_{np} is the matrix element of the electron-density operator $\rho(r)$ in the Wannier representation. With this notation, $U = 2V(0,0,0,0)$, $V = 2V(0,1,1,0)$, the bond charge repulsion is $W = 2V(0,1,0,1)$, and the cross-coupling term between the bond and side-charge densities is $X = 2V(0,1,1,1)$.

By considering the continuum limit of the complete Hamiltonian, Kivelson *et al.* (1987) have suggested that W might tend to suppress dimerization under the condition

$$3W > V. \tag{7.51}$$

While it may well be that the conclusions of Kivelson *et al.* do depend strongly on the use of a δ-function interaction, Pucci *et al.* (1988) have argued that their assumptions do indeed become more appropriate under pressures sufficiently high to markedly increase the screening of the interaction. To demonstrate this, they have calculated W together with V and the on-site repulsion U as functions of interatomic separation by using a Gaussian approximation to the electron–electron interaction potential, with parametrization appropriate to iodine and hydrogen.

The terms relevant to the study of Coulomb dimerization effects are (1) the first-order energy $(3W - V)$ [compare the inequality (7.51), and (2) the second-order contribution $U^2/2t_0$ (see Kivelson and Heim, 1982). In Fig. 7.11 the results of these calculations for $(3W - V)/U$ and $U/2t_0$ as functions of the dimensionless density parameter r_s are displayed. It is seen that $(3W - V)/U$ is positive and greater than $U/2t_0$ at high densities. (The crossover point in Fig. 7.12 is $r_s \sim 2.8$ for I: for H, it is $r_s \sim 0.8$ (Siringo *et al.*, 1988a). One might argue that the disappearance of the dimerization heralds the onset of the monatomic phase (compare Sections 7.1 and 7.2).

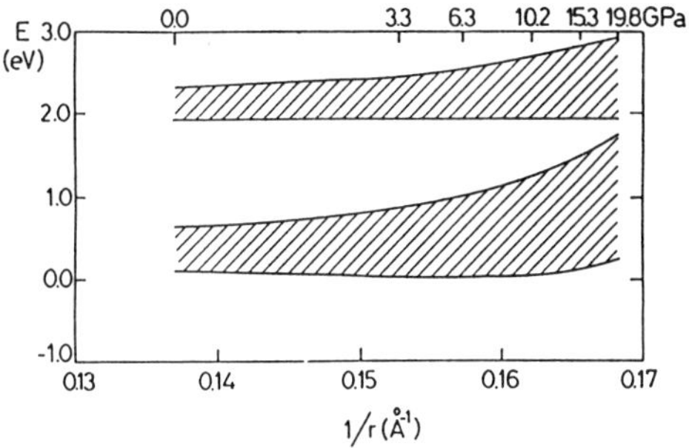

FIGURE 7.11. Results of a two-dimensional band-structure calculation of solid iodine, showing reduction of energy band gap with increasing pressure (top scale, in GPa) and as a function of $1/r$ (lower scale, in $Å^{-1}$), where r is the cube root of the unit-cell volume (after Pucci *et al.*, 1988).

Of course, it must be stressed that one has not only to show that $3W - V > 0$ but also that this positive contribution overcomes the negative contribution derived from $U^2/2t_0$. In the Hartree–Fock approximation, which is valid (Kivelson and Heim, 1982) for small values of U and for $y = (2\alpha/t_0)u_0 < 0.2$, it has been shown (Siringo *et al.*, 1988), for the half-filled band case ($X = 0$), that when

$$\frac{3W - V}{U} > 0.27 \frac{U}{2t_0}, \tag{7.52}$$

Coulomb repulsion opposes dimerization.

One has to be cautious in assessing the quantitative validity of a one-dimensional model for solid H_2 and I_2. One must also note that the model does not include a cross-coupling term between the bond and site-charge densities. This term can be larger than V and W in the extreme screening limit (Kivelson *et al.*,

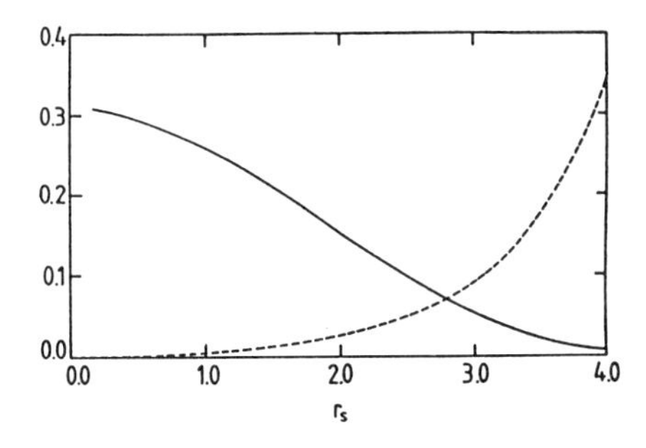

FIGURE 7.12. First-order dimensionless ratio $(3W\text{-}V)/U$ (solid line) and second-order term $U/2t_0$ (dashed line) vs. mean site separation for iodine. The same plots for hydrogen have similar shapes, but curves now cross at $r_s = 0.8$ compared with $r_s = 2.8$ for iodine in this figure (after Pucci *et al.*, 1988; see also March, 1990).

1988), and could increase the pressure at which the onset of the monatomic phase occurs. However, the above mentioned results, where comparison is possible, are in qualitative agreement with the work of Wu *et al.* (1987), who start from the full Coulomb interaction, with general strength and range.

Finally, let us return to H and alkali metals. Both analytic theory and Monte Carlo computer simulation agree that as the density of the jellium model is lowered, the electronic momentum distribution $n(p)$ exhibits in the metallic phase a reduction in the discontinuity at the Fermi momentum p_f, and that this discontinuity q eventually tends (most probably discontinuously) to zero as one passes through the metal–insulator (Wigner) transition to a localized electron crystalline phase.

To attempt to understand the alkali metals as the density is lowered and the electronic correlation becomes increasingly important, magnetic susceptibility data on expanded fluid Cs has been analyzed by Chapman and March (1988), using heavy fermion theory. Their conclusions are related to (a) the predictions from the jellium model of the discontinuity q in the electron momentum distribution and (b) the possibility of directional bonding. Again, it seems that the combination of strong electron–electron correlation with significant electron–ion interaction is usefully represented by a chemical type of picture, even though one is still in the metallic phase.

While the above systems are truly three-dimensional, some discussion has also been given of solid iodine near the metal–insulator transition. This system is quasi two-dimensional, and evidence is presented, both experimental and theoretical, for the existence of a metallic phase under pressure in which, at least over a limited range of pressure, there is a coexistence of I_2 molecules and the metallic state. The possibility that solid hydrogen may have a region of similar behavior has been considered.

To conclude this Chapter, and to lead rather naturally into the next, let us return to the point that at pressures exceeding a few Mbar hydrogen can be expected to become metallic, due to pressure ionization (Wigner and Huntington, 1935; Mao and Hemley, 1989, 1994). Ab initio molecular dynamics simulations of metallic hydrogen at high densities have been carried out by Kohanoff and Hansen (1995). These used the Car-Parrinello (1985) method (for a review, see Galli and Parrinello, 1991). The body-centered cubic structure was investigated at low temperatures and finite r_s and by an admittedly constrained displacements test they found this lattice to be dynamically stable against such displacements at least up to $r_s = 1$ (in units of $a_0 = \hbar^2/me^2$).

Next, Kohanoff and Hansen (1995) investigated the melting of this 'ionic' crystal at $r_s = 0.5$. At this value, diffusion was found to set in at $\Gamma = e^2/ak_BT \sim 230$. They identify this with the limit of mechanical stability of the (overheated) metastable crystal. The thermodynamic transition occurs at a lower temperature, or equivalently a higher value of Γ, which they estimate to be ~290 at $r_s = 0.5$

compared with the one-component plasma (OCP) value $\Gamma = 180$ corresponding to $r_s = 0$. This confirms predictions based on free energy comparisons, made by using an approximate density functional theory (Xu *et al.*, 1994).

The ground-state ($T = 0$) had earlier been explored by diffusion Monte Carlo techniques (Ceperley and Alder, 1987) described in Chapter 5, which then predict a transition between molecular and atomic (metallic) phases at a pressure of about 3 Mbar. Subsequent investigations of the crystal of the atomic phase suggest that the diamond structure is energetically favored (Natoli *et al.*, 1993) near the transition.

Returning to the study of Kohanoff and Hansen (1995), their simulations reveal the remarkable persistence of a weakly damped high-frequency ion-acoustic mode. This observed collective behavior they conclude to be characteristic of the metallic phase of hydrogen. They anticipate that it will change dramatically at the transition towards the molecular phase, thus offering promise of being an efficient diagnostic. This transition begins at a density parameter corresponding to $r_s \simeq 1.3$ by forming clusters of more than 2 atoms, to end up with H_2 molecules at $r_s \simeq 1.8$ (Ceperley and Alder, 1987; Hohl *et al.*, 1993).

Chapter 8

Electronic Correlation in Disordered Systems (Especially Liquid Metals)

In this chapter the focus is on the effects of electronic correlation in disordered systems, with particular reference to liquid metals. As in crystalline solids, it is useful to consider theoretically two limiting cases:

1. The nearly free electron limit, which is most appropriate to the simple *sp* metals Na and K, and
2. The tight-binding limit, which is relevant to the discussion of *d*-electrons—for example, in liquid transition metals.

Before turning directly to these limits, it will be helpful to establish a basic structural description of a pure liquid metal such as Na or Cu, because this immediately establishes a direct link with electronic correlation.

8.1. STRUCTURAL DESCRIPTION OF TWO-COMPONENT LIQUID METAL

It is quite clear that a pure liquid metal such as Na, near its freezing point, is a two-component system, built from equal numbers of Na^+ ions (i) and electrons (e). Clearly, the most basic structural description requires knowledge of three pair correlation functions: $g_{ii}(r)$, $g_{ie}(r)$, and $g_{ee}(r)$. The latter function reveals directly the correlations between the electrons and, in that sense, is closely linked with the discussion of the homogeneous electron fluid in Chapter 3. However, it is quite plain that in a liquid metal this electron–electron correlation function between the valence electrons is reflecting the electron–ion interaction also—a very basic difference from the jellium model, solved by quantum Monte Carlo methods, as set out in Chapter 5.

It is relevant to stress immediately that, at least in principle, these three pair correlation functions can be extracted from diffraction experiments, as set out by Egelstaff *et al.* (1974). These workers pointed out that while the usual ion–ion pair correlation $g_{ii}(r)$ is directly accessible from neutron scattering experiments, it is necessary to combine such measurements with X-ray and electron diffraction studies on the same liquid metal in order to extract $g_{ie}(r)$ and $g_{ee}(r)$ also. From this proposal an area has built up and has been reviewed by Tamaki (1987). Already, experiments reveal the presence of short- and medium-range electron–electron correlations in the simple liquid metals Na and K, and one can expect further important progress in this area in the foreseeable future (see also Johnson *et al.*, 1994).*

8.2. ELECTRON–ION HAMILTONIAN AND DENSITY FLUCTUATION OPERATORS

As a starting point for the theory, let us set up a description of the liquid metal in terms of the local number densities $\rho_e(\mathbf{r}, t)$ and $\rho_i(\mathbf{r}, t)$ of electrons and ions, respectively, at position $\mathbf{r}$ and time t. Then the density fluctuation operators $\rho_\mathbf{k}(t)$ for electrons and ions are introduced through

$$\rho_e(\mathbf{r}, t) = \sum_i \delta(\mathbf{r} - \mathbf{r}_i(t)) = \sum_\mathbf{k} \rho_{e\mathbf{k}}(t)\exp(i\mathbf{k}\cdot\mathbf{r}) \tag{8.1}$$

and

$$\rho_i(\mathbf{r}, t) = \sum_j \delta(\mathbf{r}-\mathbf{R}_j(t)) = \sum_\mathbf{k} \rho_{i\mathbf{k}}(t)\exp(i\mathbf{k}\cdot\mathbf{r}). \tag{8.2}$$

Inverting these relations yields

$$\rho_{e\mathbf{k}}(t) = \mathcal{V}^{-1} \sum_i \exp[-i\mathbf{k}\cdot\mathbf{r}_i(t)] \tag{8.3}$$

and

$$\rho_{i\mathbf{k}}(t) = \mathcal{V}^{-1} \sum_j \exp[-i\mathbf{k}\cdot\mathbf{R}_j(t)], \tag{8.4}$$

$\mathbf{r}_i$ denoting the position of the ith electron, $\mathbf{R}_j$ denoting that of the jth ion, and $\mathcal{V}$ being the total volume of the liquid metal.

Denoting electron–electron, ion–ion, and electron–ion interactions by v, V, and V_{ie}, respectively, and with ions of mass M, the total Hamiltonian reads

$$H = \sum_i \frac{p_i^2}{2m} + \sum_j \frac{p_j^2}{2M} + \frac{1}{2}\sum_{i\neq i'} v(|\mathbf{r}_i - \mathbf{r}_{i'}|) + \frac{1}{2}\sum_{j\neq j'} V(|\mathbf{R}_j - \mathbf{R}_{j'}|) + \sum_{ij} V_{ie}(\mathbf{r}_i - \mathbf{R}_j)$$

*Important computer results have now been reported by de Wijs et al. (Phys. Rev. Lett. 75, 4480, 1995) or liquid metals Mg and Bi and by Magro et al. (Phys. Rev. Lett. 76, 1240, 1996) on dense hydrogen.

$$= \sum_i \frac{p_i^2}{2m} + \sum_j \frac{p_j^2}{2M} + \frac{1}{2} \sum_{\mathbf{k}} v_{\mathbf{k}}(\rho^+_{\mathrm{e}\mathbf{k}}\rho_{\mathrm{e}\mathbf{k}} - n_{\mathrm{e}})$$

$$+ \frac{1}{2} \sum_{\mathbf{k}} V_{\mathbf{k}}(\rho^+_{\mathrm{i}\mathbf{k}}\rho_{\mathrm{i}\mathbf{k}} - n_{\mathrm{i}}) + \sum_{\mathbf{k}} V_{\mathrm{ie}}(\mathbf{k})\rho^+_{\mathrm{e}\mathbf{k}}\rho_{\mathrm{i}\mathbf{k}} \tag{8.5}$$

the density of electrons n_{e} and the ionic density n_{i} being related through the valence Z by $n_{\mathrm{e}} = Zn_{\mathrm{i}}$. It will often be convenient in what follows to work with unit volume; i.e., $\mathcal{V} = 1$.

8.2.1. Response Functions

Let us define the response function

$$\chi_{\mathrm{ee}}(\mathbf{q}, \omega) = \int_0^\infty dt \exp(-i\omega t)\left(\frac{1}{i\hbar}\right)\langle[\rho^+_{\mathrm{e}\mathbf{q}}(0), \rho_{\mathrm{e}\mathbf{q}}(t)]\rangle$$

$$= \beta \int_0^\infty dt \exp(-i\omega t)\langle\dot{\rho}^+_{\mathrm{e}\mathbf{q}}(0); \rho_{\mathrm{e}\mathbf{q}}(t)\rangle \tag{8.6}$$

with appropriate generalizations for χ_{ie} and χ_{ii} (March and Tosi, 1973). Here, as usual, $\beta = (k_B T)^{-1}$, whereas $\langle X; Y\rangle$ is the canonical correlation function of Kubo, namely,

$$\langle X; Y\rangle = \beta^{-1} \int_0^\beta d\lambda\langle\exp(\lambda H)X \exp(-\lambda H)Y\rangle \tag{8.7}$$

with

$$\langle[X(0), Y(t)]\rangle = i\hbar\, \beta\langle\dot{X}(0); Y(t)\rangle. \tag{8.8}$$

It is sometimes useful to define response functions for mass densities. This is then related to the discussion of correlation functions in a classical binary alloy given by Bhatia, Thornton, and March (1974).

8.2.2. Equations of Motion for Density Matrices

Next it will be useful to introduce field operators $\psi(\mathbf{r}, t)$ and $\Psi(\mathbf{r}, t)$ for electrons and ions, respectively, and then to form the equations of motion for the density matrices, defined as $\langle\psi^+(\mathbf{x}, t)\psi(\mathbf{x}', t)\rangle$ and $\langle\Psi^+(\mathbf{x}, t)\Psi(\mathbf{x}', t)\rangle$. These equations are

$$\{i\hbar(\partial/\partial t) - (\hbar^2/2m)(\nabla^2_{\mathbf{x}} - \nabla^2_{\mathbf{x}'})\}\langle\psi^+(\mathbf{x}, t)\psi(\mathbf{x}', t)\rangle$$

$$= -\int d\mathbf{x}''[v(\mathbf{x} - \mathbf{x}'') - v(\mathbf{x}' - \mathbf{x}'')]\langle\psi^+(\mathbf{x}, t)\psi(\mathbf{x}', t)\rho_{\mathrm{e}}(\mathbf{x}'', t)\rangle$$

$$-\int d\mathbf{x}''[V_{\mathrm{ie}}(\mathbf{x} - \mathbf{x}'') - V_{\mathrm{ie}}(\mathbf{x}' - \mathbf{x}'')]\langle\psi^+(\mathbf{x}, t)\psi(\mathbf{x}', t)\rho_{\mathrm{i}}(\mathbf{x}'', t)\rangle \tag{8.9}$$

and

$$\{i\hbar(\partial/\partial t) - (\hbar^2/2m)(\nabla_{\mathbf{x}}^2 - \nabla_{\mathbf{x}'}^2)\}\langle\Psi^+(\mathbf{x}, t)\Psi(\mathbf{x}', t')\rangle$$
$$= -\int d\mathbf{x}''[V(\mathbf{x} - \mathbf{x}'') - V(\mathbf{x}' - \mathbf{x}'')]\langle\Psi^+(\mathbf{x}, t)\Psi(\mathbf{x}', t)\rho_i(\mathbf{x}'', t)\rangle$$
$$-\int d\mathbf{x}''[V_{ie}(\mathbf{x} - \mathbf{x}'') - V_{ie}(\mathbf{x}' - \mathbf{x}'')]\langle\Psi^+(\mathbf{x}, t)\Psi(\mathbf{x}', t)\rho_e(\mathbf{x}'', t)\rangle \quad (8.10)$$

Since these two equations have basically the same form, the calculation need be carried out only for the electronic equation of motion (8.9), and the final result will then be written for (8.10).

8.2.3. Wigner Distribution Functions

With the aim of setting up the theory in terms of current and energy densities, let us next consider the mixed density matrix, or Wigner distribution function, $f(\mathbf{p}, \mathbf{R}, t)$, defined by

$$f(\mathbf{p}, \mathbf{R}, t) = \int d\mathbf{r} \exp\left(\frac{i\mathbf{p}\cdot\mathbf{r}}{\hbar}\right)\left\langle\psi^+\left(\mathbf{R} + \frac{1}{2}\mathbf{r}, t\right)\psi\left(\mathbf{R} - \frac{1}{2}\mathbf{r}, t\right)\right\rangle, \quad (8.11)$$

with a similar function $F(\mathbf{p}, \mathbf{R}, t)$ for the ions.

From this Wigner distribution function, one can now derive the density $n(\mathbf{R}, t)$, the current density $j(\mathbf{R}, t)$, and the "kinetic energy tensor" $\pi(\mathbf{R}, t)$ as

$$n(\mathbf{R}, t) = \sum_{\mathbf{p}} f(\mathbf{p}, \mathbf{R}, t), \quad (8.12)$$

$$\mathbf{j}(\mathbf{R}, t) = \left(\frac{1}{m}\right)\sum_{\mathbf{p}} \mathbf{p} f(\mathbf{p}, \mathbf{R}, t), \quad (8.13)$$

and

$$\pi_{\alpha\beta}(\mathbf{R}, t) = \left(\frac{1}{m}\right)\sum_{\mathbf{p}} p_\alpha p_\beta f(\mathbf{p}, \mathbf{R}, t), \quad (8.14)$$

with analogous definitions of N, $\mathbf{J}$, and Π for the ions in terms of F and the ion mass M.

The equation of motion (8.9) can now be expanded about the diagonal to yield

$$\left\{ i\hbar\left(\frac{\partial}{\partial t}\right) - \left(\frac{\hbar^2}{m}\right)\nabla_{\mathbf{R}}\cdot\nabla_{\mathbf{r}}\right\}\sum_{\mathbf{p}} \exp\left(\frac{-i\mathbf{p}\cdot\mathbf{r}}{\hbar}\right)f(\mathbf{p}, \mathbf{R}, t)$$

$$= -\int d\mathbf{x}''[\mathbf{r}\cdot\nabla_{\mathbf{R}} v(\mathbf{R}-\mathbf{x}'') + O(r^3)]$$

$$\times\left[\langle\rho_e(\mathbf{R},t)\rho_e(\mathbf{x}'',t)\rangle + \left(\frac{m}{i\hbar}\right)\mathbf{r}\cdot\langle\mathbf{j}(\mathbf{R},t)\rho_e(\mathbf{x}'',t)\rangle + O(r^2)\right]$$

$$-\int d\mathbf{x}''[\mathbf{r}\cdot\nabla_{\mathbf{R}} V_{ie}(\mathbf{R}-\mathbf{x}'') + O(r^3)]$$

$$\times\left[\langle\rho_e(\mathbf{R},t)\rho_i(\mathbf{x}'',t)\rangle + \left(\frac{m}{i\hbar}\right)\mathbf{r}\cdot\langle\mathbf{j}(\mathbf{R},t)\rho_i(\mathbf{x}'',t)\rangle + O(r^2)\right], \tag{8.15}$$

the current density operator $\mathbf{j}(\mathbf{R}, t)$ being

$$\mathbf{j}(\mathbf{R},t) = \left(\frac{i\hbar}{2m}\right)\{[\nabla_{\mathbf{R}}\psi^+(\mathbf{R},t)]\psi(\mathbf{R},t) - \psi^+(\mathbf{R},t)[\nabla_{\mathbf{R}}\psi(\mathbf{R},t)]\}. \tag{8.16}$$

Equation (8.15) is the basic equation from which continuity and conservation equations can now be obtained, following March and Tosi (1973).

8.2.4. Continuity Equation and Current

Equating terms independent of $\mathbf{r}$ on both sides of (8.15), one readily finds that

$$\frac{\partial n(\mathbf{R},t)}{\partial t} + \nabla_{\mathbf{R}}\cdot\mathbf{j}(\mathbf{R},t) = 0, \tag{8.17}$$

which is the usual continuity equation.

Terms of $O(r)$ similarly yield

$$\frac{m\partial\mathbf{j}(\mathbf{R},t)}{\partial t} + \nabla_{\mathbf{R}}\cdot\pi(\mathbf{R},t) = -\int d\mathbf{x}''\,\nabla_{\mathbf{R}} v(\mathbf{R}-\mathbf{x}'')\langle\rho_e(\mathbf{R},t)\rho_e(\mathbf{x}'',t)\rangle$$
$$-\int d\mathbf{x}''\,\nabla_{\mathbf{R}} V_{ie}(\mathbf{R}-\mathbf{x}'')\langle\rho_e(\mathbf{R},t)\rho_i(\mathbf{x}'',t)\rangle \tag{8.18}$$

When one considers the jellium model, which corresponds to the granular ions in a real liquid metal being smeared out into a uniform neutralizing background in which interacting electrons move (see Chapter 3), then (8.17) and (8.18) provide the basis for the electron gas theory of Singwi *et al.* (1968, see also Singwi and Tosi, 1981).

If one introduces a velocity field $\mathbf{v}(\mathbf{R}, t)$ through

$$\mathbf{j}(\mathbf{R},t) = n(\mathbf{R},t)\mathbf{v}(\mathbf{R},t) \tag{8.19}$$

and separates the momentum flux tensor into two parts,

$$\pi_{\alpha\beta}(\mathbf{R},t) = mj_\alpha(\mathbf{R},t)v_\beta(\mathbf{R},t) + \pi^0_{\alpha\beta}(\mathbf{R},t), \tag{8.20}$$

then Eq. (8.18) can be written as

$$mn(\mathbf{R}, t)\left(\frac{D\mathbf{v}(\mathbf{R},t)}{Dt}\right) = -\nabla_{\mathbf{R}}\cdot\pi^{0}(\mathbf{R}, t) - \int d\mathbf{x}''\,\nabla_{\mathbf{R}}v(\mathbf{R}-\mathbf{x}'')\langle\rho_e(\mathbf{R}, t)\rho_e(\mathbf{x}'', t)\rangle$$
$$-\int d\mathbf{x}''\,\nabla_{\mathbf{R}}V_{ie}(\mathbf{R}-\mathbf{x}'')\langle\rho_e(\mathbf{R}, t)\rho_i(\mathbf{x}'', t)\rangle \tag{8.21}$$

where

$$\frac{D\mathbf{v}}{Dt} = \frac{\partial\mathbf{v}}{\partial t} + (\mathbf{v}\cdot\nabla_{\mathbf{R}})\mathbf{v}. \tag{8.22}$$

The Navier–Stokes equation for the liquid metal derives from a combination of Eq. (8.21) and the corresponding equation for the ions.

8.2.5. Energy Transport Equation

The final equation can be derived from Eq. (8.15) by working to $O(r^2)$, which yields

$$\frac{i\hbar}{2}\frac{\partial}{\partial t}\left(-\frac{i}{\hbar}\right)^{2}\sum_{\mathbf{p}} p_{\alpha}^{2} f(\mathbf{p}, \mathbf{R}, t) - \frac{\hbar^{2}}{2m}\left(-\frac{i}{\hbar}\right)^{3}\sum_{\mathbf{p}} \mathbf{p}\cdot\nabla_{\mathbf{R}} p_{\alpha}^{2} f(\mathbf{p}, \mathbf{R}, t)$$
$$= -\frac{m}{i\hbar}\int d\mathbf{x}''\,\nabla_{\mathbf{R}_{\alpha}}v(\mathbf{R}-\mathbf{x}'')\langle j_{\alpha}(\mathbf{R}, t)\rho_e(\mathbf{x}'', t)\rangle$$
$$-\frac{m}{i\hbar}\int d\mathbf{x}''\,\nabla_{\mathbf{R}_{\alpha}}V_{ie}(\mathbf{R}-\mathbf{x}'')\langle j_{\alpha}(\mathbf{R}, t)\rho_i(\mathbf{x}'', t)\rangle \tag{8.23}$$

Taking the trace then leads to

$$\left(\frac{\partial}{\partial t}\right)k(\mathbf{R}, t) + \nabla_{\mathbf{R}}\cdot\mathbf{j}_{k}(\mathbf{R}, t) = -\int d\mathbf{x}''\,\nabla_{\mathbf{R}}v(\mathbf{R}-\mathbf{x}'')\cdot\langle\mathbf{j}(\mathbf{R}, t)\rho_e(\mathbf{x}'', t)\rangle$$
$$-\int d\mathbf{x}''\,\nabla_{\mathbf{R}}V_{ie}(\mathbf{R}-\mathbf{x}'')\cdot\langle\mathbf{j}(\mathbf{R}, t)\rho_i(\mathbf{x}'', t)\rangle \tag{8.24}$$

where the kinetic energy density $k(\mathbf{R}, t)$ is

$$k(\mathbf{R}, t) = \sum_{\mathbf{p}}\left(\frac{p^2}{2m}\right)f(\mathbf{p}, \mathbf{R}, t) = \frac{1}{2}\mathrm{Tr}\,\pi(\mathbf{R}, t) \tag{8.25}$$

and the kinetic energy flux $\mathbf{j}_k$ is simply

$$\mathbf{j}_{k}(\mathbf{R}, t) = \sum_{\mathbf{p}}\left(\frac{\mathbf{p}}{m}\right)\left(\frac{p^2}{2m}\right)f(\mathbf{p}, \mathbf{R}, t). \tag{8.26}$$

To see that Eq. (8.24) is, in fact, an energy-transport equation, consider the following identity for the first term on the right-hand side of this equation:

$$-\int d\mathbf{x}''\,\nabla_{\mathbf{R}} v(\mathbf{R}-\mathbf{x}'')\cdot\langle \mathbf{j}(\mathbf{R},t)\rho_e(\mathbf{x}'',t)\rangle$$

$$= -\nabla_{\mathbf{R}}\cdot\int d\mathbf{x}''\,v(\mathbf{R}-\mathbf{x}'')\langle \mathbf{j}(\mathbf{R},t)\rho_e(\mathbf{x}'',t)\rangle$$

$$+\int d\mathbf{x}''\,v(\mathbf{R}-\mathbf{x}'')\langle \nabla_{\mathbf{R}}\cdot\mathbf{j}(\mathbf{R},t)\rho_e(\mathbf{x}'',t)\rangle \tag{8.27}$$

The continuity equation (8.27), expressed in operator form, gives

$$\langle \nabla_{\mathbf{R}}\cdot\mathbf{j}(\mathbf{R},t)\rho_e(\mathbf{x}'',t)\rangle = -\left[\left(\frac{\partial}{\partial t}\right)\langle \rho_e(\mathbf{R},t)\rho_e(\mathbf{x}'',t')\rangle\right]_{t'=t}, \tag{8.28}$$

and hence the last term on the right-hand side of (8.27) can be evaluated as

$$-\int d\mathbf{x}''\,v(\mathbf{R}-\mathbf{x}'')\langle \nabla_{\mathbf{R}}\cdot\mathbf{j}(\mathbf{R},t)\rho_e(\mathbf{x}'',t)\rangle$$

$$= -\frac{1}{2}\int d\mathbf{x}''\,v(\mathbf{R}-\mathbf{x}'')\left(\frac{\partial}{\partial t}\right)\langle \rho_e(\mathbf{R},t)\rho_e(\mathbf{x}'',t)\rangle, \tag{8.29}$$

this following from Eq. (8.28) because of the symmetry of the correlation function $\langle \rho_e(\mathbf{R},t)\rho_e(\mathbf{x}'',t)\rangle$. But the electron–electron potential energy density $v_{ee}(\mathbf{R},t)$ is simply

$$v_{ee}(\mathbf{R},t) = \frac{1}{2}\int d\mathbf{x}''\,v(\mathbf{R}-\mathbf{x}'')\langle \rho_e(\mathbf{R},t)\rho_e(\mathbf{x}'',t)\rangle, \tag{8.30}$$

so the right-hand side of (8.29) becomes $-(\partial/\partial t)v_{ee}(\mathbf{R},t)$.

In the same way, the second term on the right-hand side of Eq. (8.24) can be calculated, yielding

$$-\int d\mathbf{x}''\,\nabla_{\mathbf{R}} V_{ie}(\mathbf{R}-\mathbf{x}'')\cdot\langle \mathbf{j}(\mathbf{R},t)\rho_i(\mathbf{x}'',t)\rangle$$

$$= -\nabla_{\mathbf{R}}\cdot\int d\mathbf{x}''\,V_{ie}(\mathbf{R}-\mathbf{x}'')\langle \mathbf{j}(\mathbf{R},t)\rho_i(\mathbf{x}'',t)\rangle - \left(\frac{\partial}{\partial t}\right)V_{ie}(\mathbf{R},t) \tag{8.31}$$

where

$$V_{ie}(\mathbf{R},t) = \int d\mathbf{x}''\,V_{ie}(\mathbf{R}-\mathbf{x}'')\langle \rho_e(\mathbf{R},t)\rho_i(\mathbf{x}'',t)\rangle. \tag{8.32}$$

Thus, one can rewrite Eq. (8.24) as an energy-transport equation for the electrons

$$\frac{\partial}{\partial t}\,\varepsilon(\mathbf{R},t) + \nabla_{\mathbf{R}}\cdot\mathbf{j}_\varepsilon(\mathbf{R},t) = 0, \tag{8.33}$$

whereas for the ions one has

$$\frac{\partial}{\partial t}\mathcal{E}(\mathbf{R}, t) + \nabla_{\mathbf{R}}\cdot\mathbf{J}_{\mathcal{E}}(\mathbf{R}, t) = 0. \tag{8.34}$$

Here the explicit energy densities for electrons and ions are

$$\varepsilon(\mathbf{R}, t) = \sum_{\mathbf{p}}\left(\frac{p^2}{2m}\right)f(\mathbf{p}, \mathbf{R}, t) + \int d\mathbf{x}\, v(\mathbf{R} - \mathbf{x})\cdot\langle\rho_e(\mathbf{R}, t)\rho_e(\mathbf{x}, t^-)\rangle.$$

$$+ \int d\mathbf{x}\, V_{ie}(\mathbf{R} - \mathbf{x})\cdot\langle\rho_e(\mathbf{R}, t)\rho_i(\mathbf{x}, t^-)\rangle \tag{8.35}$$

and

$$\mathcal{E}(\mathbf{R}, t) = \sum_{\mathbf{p}}\left(\frac{p^2}{2M}\right)F(\mathbf{p}, \mathbf{R}, t) + \int d\mathbf{x}\, V(\mathbf{R} - \mathbf{x})\cdot\langle\rho_i(\mathbf{R}, t)\rho_i(\mathbf{x}, t^-)\rangle.$$

$$+ \int d\mathbf{x}\, V_{ie}(\mathbf{R} - \mathbf{x})\langle\rho_i(\mathbf{R}, t)\rho_e(\mathbf{x}, t^-)\rangle \tag{8.36}$$

Similarly, for the energy fluxes one can write

$$j_\varepsilon(\mathbf{R}, t) = \sum_{\mathbf{p}}\left(\frac{\mathbf{p}}{m}\right)\left(\frac{p^2}{2m}\right)f(\mathbf{p}, \mathbf{R}, t) + \int d\mathbf{x}\, v(\mathbf{R} - \mathbf{x})\langle\mathbf{j}(\mathbf{R}, t)\rho_e(\mathbf{x}, t^-)\rangle.$$

$$+ \int d\mathbf{x}\, V_{ie}(\mathbf{R} - \mathbf{x})\langle\mathbf{j}(\mathbf{R}, t)\rho_i(\mathbf{x}, t)\rangle \tag{8.37}$$

and

$$J_\varepsilon(\mathbf{R}, t) = \sum_{\mathbf{p}}\left(\frac{\mathbf{p}}{M}\right)\left(\frac{p^2}{2M}\right)F(\mathbf{p}, \mathbf{R}, t) + \int d\mathbf{x}\, V(\mathbf{R} - \mathbf{x})\langle\mathbf{J}(\mathbf{R}, t)\rho_i(\mathbf{x}, t)\rangle$$

$$+ \int d\mathbf{x}\, V_{ie}(\mathbf{R} - \mathbf{x})\langle\mathbf{J}(\mathbf{R}, t)\rho_e(\mathbf{x}, t)\rangle \tag{8.38}$$

This completes the derivation of the continuity and conservation equations that was the objective of the above account. These equations are fully microscopic, and the hydrodynamic equations can be obtained from them by gradient expansions (see, for example, March and Tosi, 1973). In March (1990) the theory just set out is applied to a variety of properties of liquid metals. Though the two-component theory set out here is the most basic approach for treating a pure liquid metal such as Na or Be, it remains useful in practice to consider effective (i.e., mediated by the electrons) ion–ion interactions.

8.3. EFFECTIVE ION–ION PAIR POTENTIALS

The outline of this section is as follows. In Section 8.3.1, the one-center displaced electronic charges around a metallic ion are utilized within the framework of density functional theory, as set out in Chapter 3, to construct the pair potential $\phi(R)$; see especially Eq. (8.43). This result was examined by Perrot and March (1990a,b) in two special limits: (a) the linear response regime and (b) the local density approximation. However, these workers also obtained the numerical solution of the Slater–Kohn–Sham (SKS) equations, for it turns out that even for the "nearly–free" electron metal Na the linear response formula is not quantitative. Of course, in solving the SKS equations for the one-center displaced charge, one must adopt a prescription for the exchange and correlation potential discussed in Chapter 3. In Section 8.3.4, the electron theory pair potential $\phi(R)$ for Na, obtained in the way just outlined, will be compared with that extracted from X-ray diffraction measurements on liquid Na just above the freezing point, following the proposal of Johnson and March (1963; see also Reatto, 1988).

8.3.1. Superposition of Single-Ion Densities and Pair Potentials from Density Functional Theory

Having anticipated the inadequacy of the linear response formula for $\phi(R)$ for Na metal, let us now embark on a more fundamental discussion, based on density functional theory (DFT).

Following Perrot and March (1990a,b), let us assume the pair density as a superposition of single-ion displaced charges, but focus directly on the pair problem, utilizing the symmetry about the midpoint of the metallic bond and perpendicular to that bond. By doing so, a fundamental route to the pair potential $\phi(R)$ will be afforded.

Let us therefore start from the DFT equation for a single ion in the conduction electron liquid of density $\bar{\rho}$. This can be written formally as (see, for example, Lundqvist and March, 1983)

$$G'[\Delta + \bar{\rho}] - G'[\bar{\rho}] = -V(r), \tag{8.39}$$

where G' is merely a shorthand notation for the functional derivative $\delta G/\delta\rho(r)$, which in the (completely local) Thomas–Fermi theory (see Chapter 3) is simply proportional to $\{\rho(r)\}^{2/3}$. Here the functional $G = T + E_{xc}$ includes exchange and correlation effects in the local density approximation. V is the usual electrostatic potential created by the displaced charge $\Delta(r)$, namely

$$V = -\frac{Z}{r} + \frac{1}{r} * \Delta, \tag{8.40}$$

where the asterisk is shorthand for the convolution product.

With the two ions at positions R_i and R_j, let us also write the displaced charge $\Delta_i = \Delta(r-R_i)$ and $\Delta G(\Delta+\bar{\rho}) = G[\Delta + \bar{\rho}] - G[\bar{\rho}]$. Then for the pair interaction ϕ between ions i and j

$$\phi = \Delta G[\Delta_i + \Delta_j + \bar{\rho}] - \Delta G[\Delta_i + \bar{\rho}] - \Delta G[\Delta_j + \bar{\rho}]$$
$$+ \left(-\frac{Z}{r_i}\right)\cdot\Delta_j + \left(-\frac{Z}{r_j}\right)\cdot\Delta_i + \frac{1}{2}(\Delta_i + \Delta_j)\cdot\frac{1}{r}*(\Delta_i + \Delta_j)$$
$$-\frac{1}{2}\Delta_i\cdot\frac{1}{r}*\Delta_i - \frac{1}{2}\Delta_j\cdot\frac{1}{r}*\Delta_j + \frac{Z^2}{R}. \tag{8.41}$$

Next note from Eq. (8.40) that

$$ZV(R) = -\frac{Z^2}{R} = \int\frac{Z}{|r'-R|}\Delta(r')\,dr'$$
$$= -\frac{Z^2}{R} + \int\frac{Z}{r''}\Delta(r''+R)\,dr'' = -\frac{Z^2}{R} + \left(\frac{Z}{r_i}\right)*\Delta_j. \tag{8.42}$$

Utilizing Eq. (8.42) and Eq. (8.41), one finds

$$\phi = \Delta G[\Delta_i + \Delta_j + \bar{\rho}] - \Delta G[\Delta_i + \bar{\rho}] - \Delta G[\Delta_j + \bar{\rho}]$$
$$+ \left(-\frac{Z}{r_j}\cdot\Delta_i\right) + \frac{1}{2}\Delta_i\cdot\frac{1}{r}*\Delta_j + \frac{1}{2}\Delta_j\cdot\frac{1}{r}*\Delta_i$$
$$-ZV(R), \tag{8.43}$$

where the dot means the integral of the product of functions through the whole of space—technically the scalar product.

Note that if one works with the exact density of the "molecule," written as $\Delta_i + \Delta_j + \delta$, instead of with $\Delta_i + \Delta_j$ in the superposition approximation, then Eq. (8.43) is correct to $O(\delta^2)$ because of the stationary properties of the energy functional. Let us now turn to examine several particular cases, following Perrot and March (1990).

(a) Linear Response Regime

A linear response regime means $\Delta_i \ll \bar{\rho}$ and $\Delta_j \ll \bar{\rho}$ everywhere. Then one can expand the functionals in the form

$$\Delta G[\Delta_i + \Delta_j + \bar{\rho}] - \Delta G[\Delta_i + \bar{\rho}] - \Delta G[\Delta_j + \bar{\rho}]$$

$$= (\Delta_i + \Delta_j)\cdot G'[\bar{\rho}] - \Delta_i \cdot G'[\bar{\rho}] - \Delta_j \cdot G'[\bar{\rho}] + O(\Delta_i^2, \Delta_i\Delta_j, \Delta_j^2)$$
$$= 0 \quad \text{in first order} \tag{8.44}$$

Thus Eq. (8.43) yields

$$\phi = -ZV(R) + \text{second-order terms.} \tag{8.45}$$

In fact the second-order terms are

$$\tfrac{1}{2}G''[\bar{\rho}]\cdot[(\Delta_i + \Delta_j)^2 - \Delta_i^2 - \Delta_j^2] + \tfrac{1}{2}\Delta_i \cdot V_j + \tfrac{1}{2}\Delta_j \cdot V_i,$$

but from Eq. (8.39)

$$\Delta_i \cdot G''[\bar{\rho}] = -V_i,$$

so that the second-order terms vanish to leave the result

$$\phi_{lr} = -ZV(R) + O(\Delta^3). \tag{8.46}$$

(b) Local Density Approximation

Inside the ion core i, Δ_i is much larger than $\bar{\rho}$ so that expansion in Δ_i is not permissible. But in this same region, Δ_j is very small compared with $\Delta_i + \bar{\rho}$, so one can expand in Δ_j. Next, let us exploit the symmetry about the midpoint of the metallic bond in Eq. (8.43) and replace the integration over the whole of space by twice the integration in the half-space Ω_i containing i.

Then one has

$$\phi + ZV(R) = 2\{\Delta G[\Delta_i + \Delta_j + \bar{\rho}] - \Delta G[\Delta_i + \bar{\rho}] - \Delta G[\Delta_j + \bar{\rho}]\}$$
$$+ \Delta_i \circ V_j + V_j \circ V_i. \tag{8.47}$$

In Eq. (8.47), the volume integrals appearing in ΔG are now restricted to the half-space Ω_i. Now given the assumption that G is local, one can expand for small Δ_j in the half-space Ω_i to obtain

$$\phi + ZV(R) = 2\{\Delta_j \circ G'[\Delta_i + \bar{\rho}] - \Delta_i \circ G'[\bar{\rho}] + O(\Delta_j^2)\}$$
$$+ \Delta_i \circ V_j + \Delta_j \circ V_i$$
$$= -\Delta_j \circ V_i + \Delta_i \circ V_j + O(\Delta_j^2) \tag{8.48}$$

with the help of Eq. (8.40). Equation (8.48) retains the full nonlocality of G around the center i and thus transcends linear response. The integrals in the scalar products of Eq. (8.47) are now restricted to the half-space Ω_i and are denoted by an open circle o. In the second step of Eq. (8.47), use has again been made of the basic Euler Equation (8.40). Using $\nabla^2 V = -4\pi(\Delta - Z\delta)$, one finds

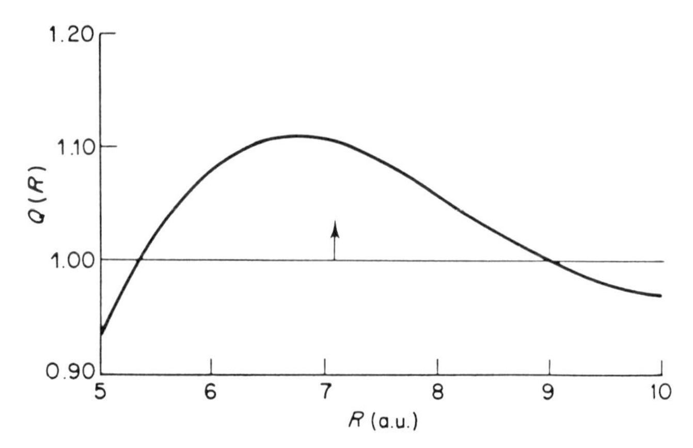

FIGURE 8.1. Screening charge $Q(R)$ round a K^+ ion in liquid K just above the freezing point. Calculated by solving Slater–Kohn–Sham equations with inclusion of an appropriate exchange-correlation one-body potential term. $Q(R)$ tends to unity as R tends to infinity, ensuring perfect screening of the monovalent K ion (after Johnson et al, 1994).

$$\phi + ZV(R) = V_j \circ \left(-\frac{\nabla^2 V_i}{4\pi} + Z\delta(r - R_i)\right) - V_i \circ \left(-\frac{\nabla^2 V_j}{4\pi} + Z\delta(r-R_j)\right). \quad (8.49)$$

Because R_j is outside Ω_i, the term in $V_i \circ Z\delta(r - R_j)$ does not contribute. Therefore

$$\phi + ZV(R) = V(r - R_j) \circ Z\delta(r - R_i) + \frac{1}{4\pi}(V_i \circ \nabla^2 V_j - V_j \circ \nabla^2 V_i), \quad (8.50)$$

and one is left with an integral over the plane Σ through the mid-point:

$$\phi = \frac{1}{4\pi}\int_\Sigma (V_i \nabla_n V_j - V_j \nabla_n V_i)\, ds. \quad (8.51)$$

On this plane one evidently has

$$V_i = V_j, \qquad \nabla_n V_i = -\nabla_n V_j. \quad (8.52)$$

Using the coordinate system in Perrot and March (1990b)

$$\phi(R) = \frac{1}{4\pi}\int_0^\infty d\rho\; 2\pi\rho\; 2V(r)\left(-\frac{dV}{dr}\cos\theta\right). \quad (8.53)$$

But $\cos\theta = R/2r$, $r^2 = R^2/4 + \rho^2$ or $r\,dr = \rho\,d\rho$; hence

$$\phi(R) = \int_{R/2}^\infty dr\; V(r)\frac{dV}{dr}\frac{R}{2} = \frac{R}{4}\left[V\left(\frac{R}{2}\right)\right]^2 + O(\Delta_j^2). \quad (8.54)$$

Thus, in a rather general framework, one finds the local form (8.54). Having discussed fully the local contribution to $\phi(R)$ in Eq. (8.54), it is natural to examine further the status of the simplest nonlocal form in the functional G: this is done in the work of Perrot and March (1990a,b), to which the reader is referred: a brief summary is added here.

8.3.2. Numerical Results for Pair Potentials in Na and Be

Let us first collect the local Thomas-Fermi plus gradient correction (TFG) contributions to the potential. The result is

$$\phi^{(2)}_{\mathrm{TFG}}(R) = \frac{1}{4}R\left\{\left[V\left(\frac{R}{2}\right)\right]^2 - \gamma^2\left[\Delta\left(\frac{R}{2}\right)\right]^2 + \eta^2[\Delta\nabla^2\Delta]_{R/2}\right], \tag{8.55}$$

where η^2 has been calculated by Perrot and March (1990b). These authors have performed the calculation of this potential energy curve for liquid Na and Be near freezing, but they used in Eq. (8.55) the V and Δ provided by the DFT calculation, and never the potential and electron density that would result from the self-consistent solution of the TFG problem.

Figure 8.2 has been constructed for Na. The upper curve at large r is the full

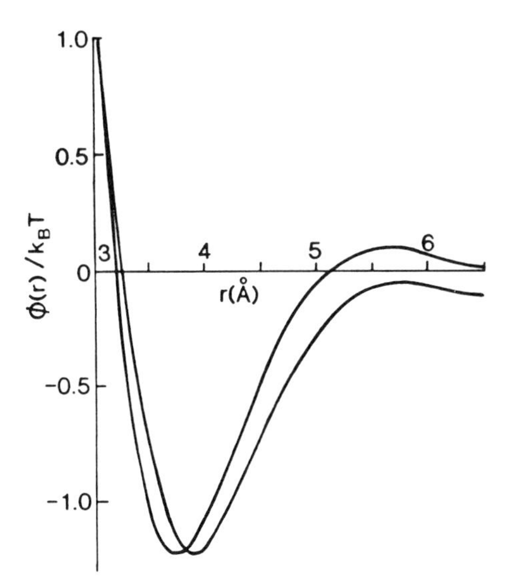

FIGURE 8.2. Pair potential $\phi(r)$ for liquid Na at 100°C and density 0.929 g cm^{-3}. Electron theory, based on $Q(R)$ shown in Fig. 8.1, is used to obtain the upper curve at large r. "Diffraction" potential [inversion of measured liquid structure factor $S(K)$] is shown in lower curve at large r (after Perrot and March, 1990a: Fig. 8.1 is for the other simple alkali K: see Fig. 8.3 for Na).

DFT electron theory potential for liquid Na near freezing. Curves in Fig. 8.4: $\phi^{(1)}_{\mathrm{TFG}}$ and $\phi^{(2)}_{\mathrm{TFG}}$ add T_2, with $\lambda = 1/9$ given by Kirznits, and $T_2 + T_4$ respectively to the curve labeled $\phi_{\mathrm{TF}}(R)$. These curves were calculated as follows: (i) $\phi_{\mathrm{TF}}(R)$ from (8.55) with $\gamma = 0$, $\eta = 0$; (ii) $\phi^{(1)}_{\mathrm{TFG}}(R)$ from Eq. ( ) with $\eta = 0$ and $\phi^{(2)}_{\mathrm{TFG}}(R)$ from Eq. (8.55) with η taken from Perrot and March (1990b).

Similar calculations were carried out for the divalent metal Be. First, the total valence screening charge $Q(R)$ inside a sphere of radius R was calculated from the DFT method. There are no adjustable parameters in this calculation. The input into the full density functional calculation of $Q(R)$ is (a) the atomic number, (b) the mass density at freezing, and (c) an exchange-correlation potential.

All the pair potentials plotted in Fig. 8.4 for liquid Na (also in Fig. 8.5 below for Be) can be regarded as derivable from a single equation:

$$\phi(R) = \Delta G(R) - ZV(R) + \Delta U_1(R). \tag{8.56}$$

Then the common element, for Na say, is the total potential energy change

$$\Delta U(R) = -ZV(R) + \Delta U_1(R) \tag{8.57}$$

which is determined solely by the total valence screening charge $Q(R)$, plotted in Fig. 8.3 for Na and Fig. 8.1 for K.

Thus, the kinetic plus exchange and correlation energy change $\Delta G(R)$ is solely responsible for the differences of the various curves for $\phi(R)$ in Figs. 8.4 and 8.5. Perrot and March (1990b) analyze this in some detail and, in particular, display numerically the remaining nonlocality in the electronic kinetic energy change beyond the fourth-order correction T_4. The remaining question, requiring full investigation, is whether such remaining nonlocality beyond T_4 is determined also by the size and shape of the total screening charge $Q(R)$ around the midpoint

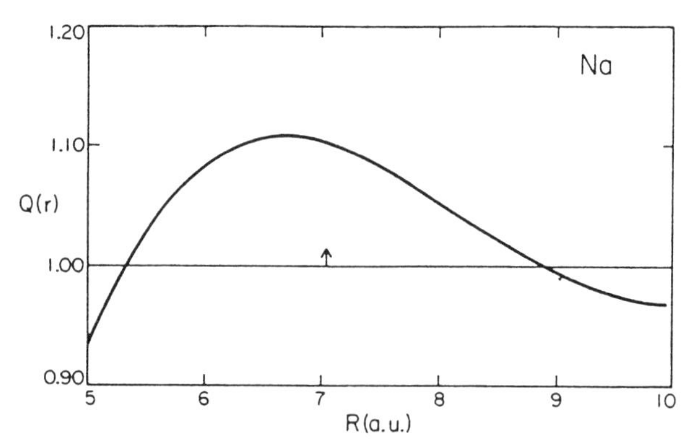

FIGURE 8.3. Total valence screening charge $Q(R)$ defined in Eq. (3.1), for liquid Na near freezing. $Q(R)$ determines the pair interaction $\phi(R)$. Arrow denotes position of principal minimum in "exact" pair potential ϕ.

$R/2$ of the metallic bond. Encouraging this point of view is the following analysis of the long-range oscillatory tail of simple metal pair potentials.

8.3.3. Long-Range Oscillatory Tail of Pair Potentials in Metals

The gradient expansion adopted in Section 8.3.2 deals in wave number k-space with the small k components of the pair interaction, but not with $k \sim 2k_f$, the Fermi sphere diameter. These components are responsible for the asymptotic behavior (Corless and March, 1961) of the ion–ion interaction:

$$\Delta(R) = \frac{A}{R^3}\cos(2k_f R) + O\left(\frac{1}{R^4}\right) \tag{8.58}$$

in linear response. In this asymptotic regime, the potential is

$$V(R) = \frac{\pi}{k_f^2}\,\Delta(R) + O\left(\frac{1}{R^4}\right), \tag{8.59}$$

so it may be written as

$$V(R) = BR^3\left\{\left[\Delta\left(\frac{R}{2}\right)\right]^2 - \frac{1}{4k_f^2}\left[\left.\frac{d\Delta}{dR}\right|_{R/2}\right]^2\right\}, \tag{8.60}$$

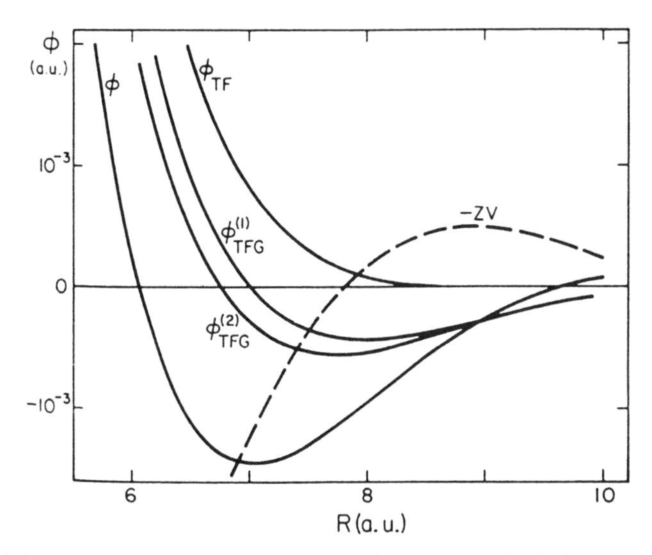

FIGURE 8.4. This shows pair potentials calculated from local density (TF) theory plus density gradient corrections T_2 and T_4 to kinetic energy, for liquid Na near freezing. Various curves were obtained using $Q(R)$ in Fig. 8.3 from different degrees of approximation as follows. ϕ_{TF}, calculated from Eq. (8.54); $\phi^{(1)}_{TFG}$ from Eq. (8.55) with $\lambda =$ in which T_2 only is included; $\phi^{(2)}_{TFG}$ from Eq. (8.55) containing both T_2 and T_4; $\phi_{LR} = -ZV$; $\phi(R)$ is the pair potential obtained.

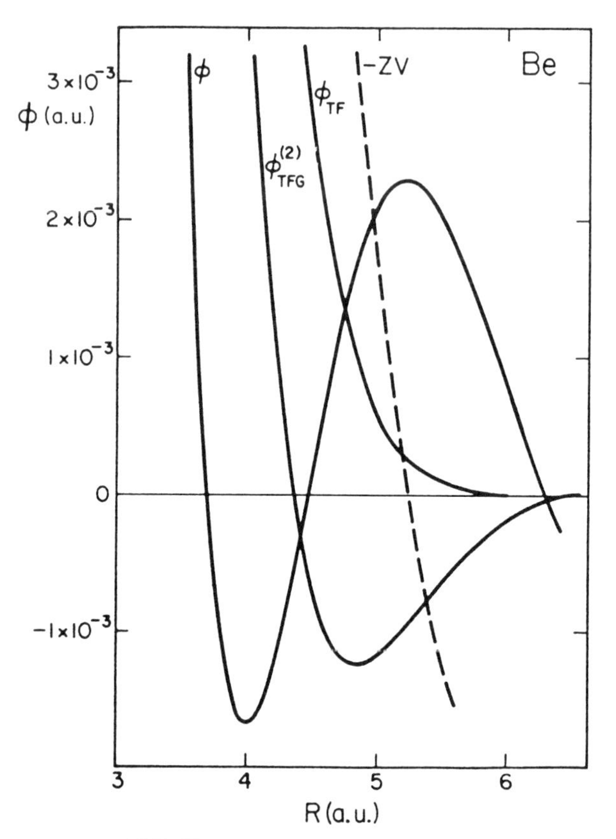

FIGURE 8.5. Same as Fig. 8.4 but for Be.

suggesting that, here again, it may be expressed in terms of the displaced charge Δ at $R/2$. It is not obvious whether this argument holds beyond the linear approximation, when a phase shift has to be introduced into Eq. (8.58). Clearly, further work is called for before a definitive conclusion can be reached on this point. A convenient approach could be the use of the formulation of ϕ in terms of the pressure tensor in the symmetry plane of the "molecule."

8.3.4. Comparison of Electron Theory and "Diffraction" Pair Potential for Liquid Na near Freezing

To confirm the conclusion drawn that the linear response formula (8.45) must be transcended even for the nearly–free electron metal Na, the pair potential $\phi(R)$, from electron theory, is shown in Fig. 8.2 with that extracted by inversion (Johnson and March, 1963) of the measured liquid structure factor $S(k)$ just above the freezing point: this latter potential will be termed the "diffraction" pair potential. This potential is known for Na from the work of Levesque, Weis, and Reatto (1985, see also Reatto, 1988, and Reatto, Levesque, and Weis, 1986). Their work started from the so-called force equation (see, for instance, March 1987):

$$-\frac{\partial U(r_{12})}{\partial r_1} = -\frac{\partial \phi(r_{12})}{\partial r_1} + \rho^2 \int \frac{g_3(r_1 r_2 r_3)}{g(r_{12})} \frac{\partial \phi(r_{13})}{\partial r_1}\, dr_3 \tag{8.61}$$

where the potential of mean force $U(r_{12})$ is related to the pair correlation function $g(r_{12})$ of the ions in the liquid metal by the Boltzmann form

$$g(r_{12}) = \exp\left(\frac{-U(r_{12})}{k_B T}\right). \tag{8.62}$$

Taking $g(r)$ from the X-ray diffraction study by Greenfield, Wellendorf, and Wiser (1971) on liquid Na just above its melting point, Reatto *et al.* have used the modified hypernetted chain approximation as the starting point and then refined the results of $\phi(r)$ using computer simulation. Thereby the need to (approximately) decouple the three-atom correlation function g_3 in Eq. (8.61) in terms of $g(r)$ is bypassed.

Their "diffraction" potential has been compared in detail with the DFT theory $\phi(R)$ plotted in Fig. 8.2 at the same mean density as in the experiment (see Perrot and March, 1990a). It is remarkable that all the main features of the electron theory results are directly reflected in the diffraction $\phi(R)$. This agreement is the more remarkable when it is recalled that the input to the electron theory calculation is merely (a) the atomic number $Z = 11$ for Na, (b) the mean density $\bar{\rho}$ just above the melting point of metallic Na, and (c) the Hedin–Lundquist exchange-correlation potential. On point (c), Perrot and March (1990a) considered also the potential due to Ichimaru (1981). They found that this change was not discernible in the pair potential $\phi(R)$ to graphical accuracy.

Equation (8.56) can be viewed as the basic equation underlying all the pair potentials calculated here. The common element, say for Na, of all the curves shown in Fig. 8.4 is the total potential energy change on bringing two ions from infinite separation in the background conduction electron liquid to separation R. The large differences evident in Fig. 8.4 are plainly a result of the variety of ways in which ΔG is handled.

As noted, linear response fails, somewhat surprisingly for the "nearly–free-electron" metal Na and more severely for Be. Low-order density gradient theory directs attention to the characterization of the pair potential $\phi(R)$ solely by bond midpoint properties through Eq. (8.55). This equation is attractive intuitively, but it is plain from Fig. 8.5 for Be that the large repulsive hump in the DFT $\phi(R)$ calculated by Perrot and March (1990b) is not present in the low-order gradient expansion, the reasons for this being set out in the original work.

8.4. GREEN FUNCTION TREATMENT WITH DIELECTRIC SCREENING AND PSEUDOPOTENTIALS

As seen, one way to obtain a pair potential for liquid metals is to invert measured structure factors, as proposed by Johnson and March (1963), and brought to full fruition for liquid Na by Reatto and co-workers (see Reatto, 1988). As an alternative to **r** space density functional theory, to derive theoretically an effective ion–ion potential consisting of ions and nearly free electrons, one often uses second-order pseudopotential theory. The result is

$$\phi_{ii}^{\text{eff}}(r) = \frac{(Ze)^2}{r} + \int \frac{d\mathbf{q}}{(2\pi)^3} \frac{V_{ei}^2(q)}{V_{ee}(q)} \left[\frac{1}{\varepsilon_{ee}(q, 0)} - 1 \right] \exp(i\mathbf{q}\cdot\mathbf{r}). \quad (8.63)$$

In Eq. (8.63), $V_{ei}(q)$ is a suitable pseudopotential representing the electron–ion interaction, $V_{ee}(q)$ describes the pure Coulomb interaction between the electrons, and $\varepsilon_{ee}(q\ ,0)$ is the static dielectric function of the electronic subsystem. Nevertheless, some workers (e.g., Abe, 1989) have commented that the derivation of the effective pair potential in Eq. (8.63) seems rather obscure. It is therefore noteworthy that Belyayev *et al.* (1989) have discussed the quasiparticle spectra (cf. Chapter 6) and maxima of dynamic ionic structure factors in liquid metals. They showed from two-component theory, set up earlier in this chapter, that Eq. (8.63) can be recovered in the framework of the thermal Green functions developed by Abrikosov *et al.* (1962).

In the foregoing context it is relevant to mention that although a liquid metal is well described as a collection of ions with well-defined valency Z (e.g., Na, $Z = 1$; Be, $Z = 2$; etc.) and electrons for such simple metals when near their freezing points, it has been demonstrated experimentally that by taking such liquids up toward the critical point drastic changes can occur in their electronic properties. Model calculations indicate that these changes can result from the presence of atoms or neutral and charged cluster formation, i.e., to drastic changes in the composition of the fluid. To construct an effective pair potential for expanded liquid metals, one may eventually have to take such compositional changes into account.

However, here, following Schmidt *et al.* (1991), only ions of well-defined valency, plus neutralizing itinerant electrons, will be assumed. Then it will be shown that a somewhat general expression for a screened (effective) ion–ion potential for a liquid metal can be derived. One approximation to this expression then leads back to the result (8.63). Within the same framework, it proves feasible to derive suitable dielectric functions (see also Chapter 6), including the Hubbard (1969) or Geldart–Vosko (1966) local-field corrections. Schmidt *et al.* (1991) present some approximate numerical results for the screened potential derived below, with respect to varying the local-field corrections.

For the viewpoint of Section 8.1 that the liquid metal is a dense two-component plasma of ions and electrons, the thermodynamic Green function technique provides a systematic route to a dynamically screened ion–ion potential. In this framework the derivation of a self-consistent system of equations for the one-particle Green function G, the self-energy (cf. Chapter 6), the polarization function $\cap$, and the screened potential $\mathcal{V}^s$ yields the following equation (Schmidt *et al.*, 1991):

$$V^s_{ab}(12) = V_{ab}(12) + i\hbar \sum_{c,d} \int_0^{-i\hbar\beta} d3\, d4\, V^s_{ac}(13) \cap_{cd}(4343) V_{db}(42), \quad (8.64)$$

where $a, \ldots, d$ denote the different particle species (ions i, electrons e), 1 stands for the space-time point $(\mathbf{r}_1, t_1)$, 2 for $(\mathbf{r}_2, t_2)$, etc., $\beta = (k_B T)^{-1}$, and V_{ab} is the corresponding unscreened potential. The sum in Eq. (8.64) runs over all species (e, i) and spins (σ_e, σ_i). The polarization function $\cap_{cd}$ has to be determined from the following equation:

$$\cap_{cd}(4343) = -\delta_{cd} G_c(43) G_c(34) - \int_0^{-i\hbar\beta} d1'\, d2'\, G_c(41') \frac{\delta \bar{\Sigma}_c(1'2')}{\delta U^{\text{eff}}_d(33)} G_c(2'4), \quad (8.65)$$

with

$$U^{\text{eff}}_d(33) = U_d(33) + \Sigma^H_d(33), \quad (8.66)$$

$$\begin{aligned}\bar{\Sigma}_c(1'2') &= \Sigma_c(1'2') - \Sigma^H_c(1'2') \\ &= i\hbar V^s_{cc}(1'2') G_c(1'2') + i\hbar \sum_b \int_0^{-i\hbar\beta} d\bar{1}\, d\bar{2}\, \mathcal{V}^s_{cb}(1'\bar{2}) G_c(1'\bar{1}) \frac{\delta \bar{\Sigma}_c(\bar{1}2')}{\delta U^{\text{eff}}_b(\overline{22})}, \quad (8.67)\end{aligned}$$

where Σ_c stands for the full and Σ^H_c for the Hartree self-energy, respectively. The effective external field U^{eff} (mean field of interaction) is the sum of the external field U, which produces an inhomogeneity in the system, and of the Hartree self-energy. The one-particle Green functions G_c or G_d are solutions of the Dyson equation (compare Chapter 6)

$$G_c(11') = G^0_c(11') - \int_0^{-i\hbar\beta} d\bar{1}\, d\bar{2}\, G^0_c(1\bar{1})\{U^{\text{eff}}_c(\bar{1}\bar{2}) + \bar{\Sigma}_c(\bar{1}\bar{2})\} G_c(\bar{2}1'). \quad (8.68)$$

In Eq. (8.68), G^0_c describes free particles in the absence of interactions. Equation (8.64) may be expressed by Feynman diagrams:

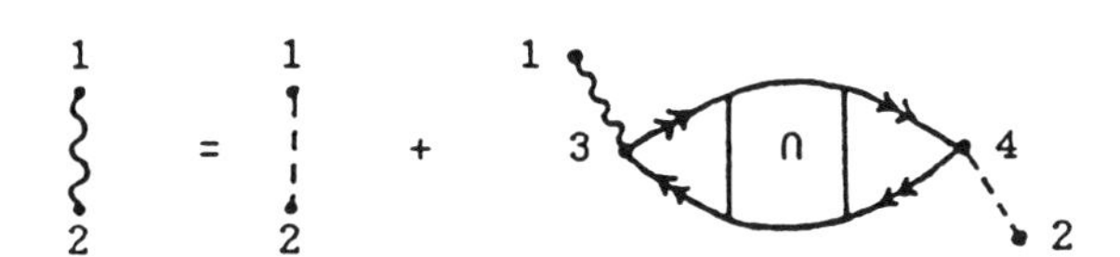

In thermodynamic equilibrium, the quantities $\mathcal{V}^s$, $\cap$, and G are quasiperiodic along the imaginary time axis (Kubo–Martin–Schwinger condition). Therefore, after transforming to the momentum Matsubara frequency representation, analytic continuation, and consideration of a potential V_{ab} that is local in time, i.e., $V_{ab}(12) = V_{ab}(\mathbf{r}_1 - \mathbf{r}_2)\delta(t_1 - t_2)$, one obtains

$$V^s_{\text{ii}}(q, \omega) = V_{\text{ii}}(q) + \hbar \sum_{c,d} V^s_{ic}(q, \omega) \cap_{cd}(q, \omega) V_{di}(q). \tag{8.69}$$

Here ω denotes a complex frequency that arises from the analytic continuation of the functions $\mathcal{V}^s$ and $\cap$ into the complex frequency plane (Matsubara frequencies $\Omega_\nu = i\pi\nu/\hbar\beta$; $\nu = 0, \pm 2, \ldots$; $\Omega_\nu \to \omega$). The particle momentum is just $\hbar q$.

It is evident that an additional equation for the screened ion–electron potential V^s_{ie} is needed to determine V^s_{ii}. In close analogy to Eq. (8.69), one can show that

$$V^s_{\text{ie}}(q, \omega) = V_{\text{ie}}(q) + \hbar \sum_{c,d} V^s_{ic}(q, \omega) \cap_{cd}(q, \omega) V_{de}(q). \tag{8.70}$$

Inserting Eq. (8.70) into Eq. (8.69) and rearranging with respect to $V^s_{ii}(q, \omega)$ yields a general expression for the dynamically screened ion–ion potential for liquid metals that is modified to account for many-particle effects and that contains powers of V_{ie} to arbitrarily high order because of the nonvanishing polarization functions. This can readily be seen from the resulting equation:

$$\begin{aligned} V^s_{\text{ii}}(q, \omega) = \frac{1}{F} \{ & V_{\text{ii}}(q) * [1 - \hbar V_{ee}(q) \cap_{ee}(q, \omega) + \hbar V_{\text{ie}}(q) \cap_{ee}(q, \omega)] \\ & + V_{ie}(q) * [\hbar V_{\text{ei}}(q) \cap_{ee}(q, \omega) + \hbar V_{\text{ii}}(q) \cap_{ei}(q, \omega)] \} \end{aligned} \tag{8.71}$$

where*

$$\begin{aligned} F = {} & [1 - \hbar V_{\text{ii}}(q) \cap_{ii}(q, \omega) - \hbar V_{\text{ei}}(q) \cap_{ie}(q, \omega)] \\ & * [1 - \hbar V_{\text{ee}}(q) \cap_{ee}(q, \omega) - \hbar V_{\text{ie}}(q) \cap_{ei}(q, \omega)] \\ & - [\hbar V_{\text{ie}}(q) \cap_{ii}(q, \omega) + \hbar V_{\text{ee}}(q) \cap_{ie}(q, \omega)] \\ & * [\hbar V_{\text{ei}}(q) \cap_{ee}(q, \omega) + \hbar V_{\text{ii}}(q) \cap_{ei}(q, \omega)] \end{aligned} \tag{8.72}$$

and after expanding the denominator F in powers of $V_{\text{ei}}(q)$ or $V_{\text{ie}}(q)$, respectively. For a system in which ions and electrons interact via pure Coulomb potentials, i.e.,

$$V_{\text{ii}}(q) = -Z V_{\text{ei}}(q) = Z^2 V_{\text{ee}}(q) = \frac{4\pi(Ze)^2}{q^2}, \tag{8.73}$$

with Ze denoting the ionic charge, one can write

$$V^s_{\text{ii}}(q, \omega) = \frac{V_{\text{ii}}(q)}{\varepsilon(q, \omega)}, \tag{8.74}$$

*A frequently useful alternative form of Eqs. (8.71) and (8.72) is that given in Eq. (9) of Schmidt et al (1991).

the liquid metal dielectric function being given by

$$\varepsilon(q, \omega) = 1 - \hbar \sum_{a,b} V_{ab}(q) \sqcap_{ab}(q)\omega) \qquad (a, b = \text{e, i}). \tag{8.75}$$

The first approximation in this development was proposed by Belyayev *et al.* (1989). According to pseudopotential theory, appropriate for weak electron–ion interaction, the contributions of the polarization functions $\sqcap_{ei}(q, \omega)$ and $\sqcap_{ie}(q, \omega)$ are neglected. In this way one obtains the approximate ion–ion interaction (cf. Schmidt *et al.*, 1991):

$$V_i^s(q, \omega) = \frac{V_{ii}(q)\varepsilon_{ee}(q, \omega) + \dfrac{V_{ei}^2(q)}{V_{ee}(q)}[1 - \varepsilon_{ee}(q, \omega)]}{\varepsilon_{ii}(q, \omega)\varepsilon_{ee}(q, \omega) - \dfrac{V_{ei}^2(q)}{V_{ii}(q)V_{ee}(q)}[1 - \varepsilon_{ii}(q, \omega)][1 - \varepsilon_{ee}(q, \omega)]}. \tag{8.76}$$

Here use has been made of the following abbreviation for the ion and electron polarizabilities, corresponding to the usual dielectric function for subsystems of electrons or ions, respectively (a = e, i):

$$\varepsilon_{aa}(q, \omega) - 1 = -\hbar(2\sigma_a + 1)V_{aa}(q)\sqcap_{aa}(q, \omega). \tag{8.77}$$

Introducing the approximation $\sqcap_{ii}(q, \omega) = 0$ into Eq. (8.75), we obtain a screened potential in Fourier space in the form of Eq. (8.63):

$$\mathrm{V}_{ii}^{s}(q, \omega) = V_{ii}(q) + \frac{V_{ei}^2(q)}{V_{ee}(q)}\left[\frac{1}{\mathrm{E}_{ee}(q, \omega)} - 1\right], \tag{8.78}$$

where dynamical effects are still retained through the ω dependence of the dielectric function.

Schmidt *et al.* (1991) have brought the Green function technique for calculating effective interionic potentials into contact with measured structure factors (Hensel *et al.*, 1990) for expanded fluid Cs. The reader is referred to the original paper for details.

8.5. MAGNETIC SUSCEPTIBILITY OF STRONGLY CORRELATED EXPANDED LIQUID Cs RELATED TO ELECTRICAL RESISTIVITY NEAR CRITICALITY

Chapman and March (1988) have given an interpretation of the observed maximum in the susceptibility of expanded fluid Cs in terms of a strongly correlated electron assembly. In their work, the reduced discontinuity, q say, at the Fermi surface in the electronic momentum distribution $n(p)$, due to switching on strong

Coulombic correlations in a Fermi gas where initially $q = 1$, played a major role in the interpretation.

In parallel with this work, calculations of electrical resistivity along the liquid–vapor coexistence curve have been carried out on Cs and Rb by Ascough and March (1990). In both cases, essential input into the self-consistent inclusion of a mean-free path for electron–ion scattering was the measured liquid structure as made available by Hensel and co-workers (1989). Weak scattering theory remains useful for both Cs and Rb over an extended portion of the liquid–vapor coexistence curve. But, naturally enough, as both electron–ion and electron–electron interactions eventually become strong on nearing the critical point, any such approach, whether self-consistent (Ferraz and March, 1979) or in the original perturbative Bhatia–Krishnan–Ziman form (see Ziman, 1961) must fail.

Therefore, in this section attention is shifted to the regime of strong electron–electron correlations. It is already clear from the susceptibility treatment of Chapman and March that the discontinuity q is subsuming both electron–electron and electron–ion effects, at least partially, when both are strong. Therefore the purpose of this section is to motivate a way in which, near the critical point, the electrical resistivity can be linked closely with the magnetic susceptibility, via the discontinuity q.

The argument rests on the so-called force–force correlation formula for electrical resistivity due to Rousseau *et al.* (1971; RSM). In its original form, this is applicable only to independent electrons, though it is, at least in principle, able to handle strong electron–ion correlations. The way this approach might be generalized to incorporate strong electron–electron repulsions has already been set out by the writer (March, 1990) considering the Hall effect in liquid metals (see also section 8.6 below).

Briefly, the argument rests on replacing the Dirac idempotent density matrix ρ in the RSM formula for independent electrons by the fully correlated first-order density matrix γ satisfying $\gamma^2 < \gamma$. Then it is possible to remove all one-body traces from the RSM formula and to obtain the resistivity solely in terms of γ. What becomes clear is that no translationally invariant assumption about the many-body quantity γ can work in the resulting transport theory. A plausible factorization of $\gamma(\mathbf{r}_1,\mathbf{r}_2)$ is to write

$$\gamma(\mathbf{r}_1,\mathbf{r}_2) \approx g(\xi)h(\eta), \tag{8.79}$$

where $\xi = \frac{1}{2}(\mathbf{r}_1 + \mathbf{r}_2)$ and $\eta = \frac{1}{2}(\mathbf{r}_1 - \mathbf{r}_2)$. Some motivation for development can be provided by the form of the one-electron density matrix discussed by March and Sampanthar (1964) in an attempt to describe Wigner-like correlations in jellium (see Chapter 3 and 9). Inserting the resulting γ into the generalized force–force correlation formula then suggests an expansion of the resistivity R in terms of the now small discontinuity q as

$$R = R_0 + qR_1 + \cdots. \tag{8.80}$$

So far, it has not proved possible to make first principles calculations of R_0 and R_1 from the force–force correlation function with strong electron–electron interactions. Therefore, what has been done is to return to the theory of Chapman and March (1988) for magnetic susceptibility. Here it is useful to relate the magnetic susceptibility χ in the strong correlation limit to the Curie limit, χ_c say. Then the quantity $1/\chi - 1/\chi_c$ provides an empirical measure of the discontinuity q and Eq. (8.80) evidently motivates a plot of the measured values of electrical resistivity R versus $1/\chi - 1/\chi_c$. The results are shown in Fig. 8.6, and that there is an intimate correlation between a transport property R and a static property, namely the magnetic susceptibility, is evidently established by this plot. An admittedly long extrapolation in this figure, shown by the dashed curve, suggests that R_0 in Eq. (8.80) is around 1300 μΩcm.

8.6. ELECTRON CORRELATION INTRODUCED VIA FORCE–FORCE CORRELATION FUNCTION THEORY OF LIQUID METAL TRANSPORT

Huang (1948) derived an exact expression in terms of phase shifts $\eta_l = \eta_l(k_f)$ for the excess resistivity of a dilute metallic alloy in which independent free electrons are scattered by a spherical potential $V(r)$. Apart from constants,

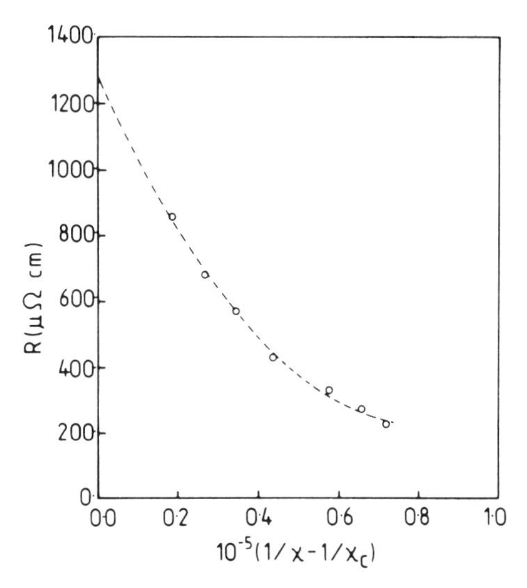

FIGURE 8.6. Shows marked correlation between experimental values of electrical resistivity R and $1/\chi$ measured relative to the Curie limit $1/\chi_c$ (see also March, 1991).

which are omitted here, the nub of the Huang formula is readily rewritten as the sum

$$S = \sum_{l=1}^{\infty} l \sin^2 (\eta_{l-1} - \eta_l). \tag{8.81}$$

Following March (1975b), a result of Gerjuoy (1965) that was rediscovered by Gaspari and Gyorffy (1972) can now be inserted into Eq. (8.81). These workers show that the radial wave functions R_l for such scattering from a potential $V(r)$ of finite range but arbitrary strength satisfy the exact relation

$$\int_0^\infty dr\, r^2 R_{l-1}(r) \frac{\partial V(r)}{\partial r} R_l(r) = \sin (\eta_l - \eta_{l-1}), \tag{8.82a}$$

where, outside the range of V, R_l takes the form

$$R_l(r) = j_l \cos \eta_l - n_l \sin \eta_l, \tag{8.82b}$$

j_l and n_l being spherical Bessel and Neumann functions, respectively. Thus, from Eqs. (8.81) and (8.82a),

$$S = \sum_{l=1}^{\infty} l \int_0^\infty dr_1\, r_1^2 R_{l-1}(r_1) \frac{\partial V(r_1)}{\partial r_1} R_1(r_1)$$
$$\times \int_0^\infty dr_2\, r_2^2 R_{l-1}(r_2) \frac{\partial V(r_2)}{\partial r_2} R_1(r_2) \tag{8.83}$$

Finally, introducing the Dirac idempotent first-order density matrix

$$\rho(\mathbf{r}_1\mathbf{r}_2\mu) = \sum_i^{\mu} \psi_i^*(\mathbf{r}_1)\psi_i(\mathbf{r}_2), \tag{8.84}$$

where the ψ's are merely products of R_l and the spherical harmonics, one can show that S in Eq. (8.81) is simply related to the force–force correlation function $F(\mu)$, which in terms of V and ρ is given by (Rousseau *et al.*, 1972)

$$F(\mu) = \int \frac{\partial V}{\partial \mathbf{r}_1} \cdot \frac{\partial V}{\partial \mathbf{r}_2} \left| \frac{\partial \rho(\mathbf{r}_1\mathbf{r}_2\mu)}{\partial \mu} \right|^2 d\mathbf{r}_1\, d\mathbf{r}_2. \tag{8.85}$$

The burden of the remaining argument is to eliminate the one-body elements in Eq. (8.85) in favor of the density matrix which has a ready many-electron generalization. This can be done (March, 1990) by utilizing the equation of motion of the Dirac density matrix ρ, which in coordinate representation reads

$$\nabla_{\mathbf{r}_1}^2 \rho - \nabla_{\mathbf{r}_2}^2 \rho = \frac{2m}{\hbar^2} [V(\mathbf{r}_1) - V(\mathbf{r}_2)]\rho. \tag{8.86}$$

Dividing both sides by ρ and writing the difference in Laplacian operators as $\Delta L \equiv \nabla_{\mathbf{r}_1}^2 - \nabla_{\mathbf{r}_2}^2$ one finds, with $h^2/2m$ conveniently incorporated in the definition of L:

$$\mathrm{grad}_1 V(\mathbf{r}_1) = \mathrm{grad}_1\left(\frac{\Delta L\rho}{\rho}\right). \tag{8.87}$$

Hence Eq. (8.85) becomes

$$F(\mu) = -\int \mathrm{grad}_1\left(\frac{\Delta L\rho}{\rho}\right)\cdot\mathrm{grad}_2\left(\frac{\Delta L\rho}{\rho}\right)\left|\frac{\partial\rho(\mathbf{r}_1\mathbf{r}_2\mu)}{\partial\mu}\right|^2 d\mathbf{r}_1\, d\mathbf{r}_2. \tag{8.88}$$

The final postulate in this approach is that one can replace the one-body idempotent density matrix ρ in Eq. (8.88) by the many-electron first-order density matrix γ satisfying now the matrix inequality $\gamma^2 < \gamma$ to obtain $F_{e\text{–}e}$ in the presence of electron–electron interactions as

$$F_{e\text{–}e}(\mu) = -\int \mathrm{grad}_1\left(\frac{\Delta L\gamma}{\gamma}\right)\cdot\mathrm{grad}_2\left(\frac{\Delta L\gamma}{\gamma}\right)\left|\frac{\partial\gamma(\mathbf{r}_1\mathbf{r}_2\mu)}{\partial\mu}\right|^2 d\mathbf{r}_1\, d\mathbf{r}_2. \tag{8.89}$$

Since the sum in Eq. (8.81) is proportional to $F(\mu)$, it can be shown, for example, that the deviation ΔR_H of the Hall coefficient from 1/nec is characterized by the first-order density matrix γ (March, 1989; see also Szabo, 1972).

In summary, a formula for ΔR_H can be set up for liquid metals with strong electron–electron interactions, e.g., the expanded alkali metals discussed in Section 8.5 in terms of the first-order density matrix. Given this formula, some consequences follow:

(i) Insertion of any translationally invariant $\gamma = \gamma(|\mathbf{r}_1 - \mathbf{r}_2|)$ whatsoever leads to $F_{e\text{–}e}(\mu) = 0$ and to the simple result $R_H = 1/\mathrm{nec}$.

(ii) Assuming, as in March (1990; see also Eq. (8.79)) for R, that $\gamma(\mathbf{r}_1\mathbf{r}_2) \doteqdot f(\mathbf{r}_1 - \mathbf{r}_2)g(\mathbf{r}_1 + \mathbf{r}_2)$, one must expect ΔR_H to depend on the discontinuity, q say, of the electronic momentum distribution of the liquid metal at the Fermi surface.

(iii) Though further work is clearly required on the many-electron derivation of the formula of ΔR_H, this approach suggests that, even in the presence of strong electron–electron interactions, ΔR_H may be closely correlated with measured conductivity and thermopower. However quantitative numerical studies are still required on this last point at the time of writing (see also, March and Paranjape, 1995).

Chemical Cluster Model*

As this Chapter was nearing completion, the important work of R. Redmer and W. W. Warren (*Phys. Rev.* **B48**, 14892, 1993; referred to below as RW) appeared. These workers considered both electrical transport and magnetic prop-

*Section added in proof

erties of the heavy alkalis at low densities. Applying the model of a partially ionized gas (see also the work by J. P. Hernandez cited in RW), they studied the magnetic susceptibility of Cs and Rb from the vapor to the liquid phase. Their conclusion is that the deviations from the Curie Law are due to the formation of spin paired dimers. The enhancement of the electronic susceptibility at around 2-3 times the critical density in the regime of the expanded liquid can be explained by the formation of Cs_2^+ and Rb_2^+. RW propose that the metal–nonmetal transition occurs in that domain where neutral and charged clusters, and eventually also higher fluctuating clusters, are formed. They note that such a viewpoint is quite compatible with the reverse Monte Carlo calculations of R. L. McGreevy and coworkers (see Nield et al., 1991 cited by RW) who treated the metal–nonmetal transition as a bond percolation problem. Nield et al. found that some finite atomic clusters are present close to the critical point as well as weak links within infinite clusters.

Work of M. Ross and colleagues, cited by RW also on the metal–nonmetal transition in Cs, lends support to a chemical cluster picture. RW contend that the formation of excited neutral clusters at elevated temperatures in the dense vapor is likely to occur, as one may conclude from the total energy calculations of Ross et al. for the Cs_2–Cs_2 cluster. As concluding comments, RW emphasize* their finding that the consideration of electron correlation beyond RPA is essential. They also note that the aspects of disorder which they were not able to consider have however to be treated alongside electronic correlation (see also section 8.8 below).

Below the presentation of mean field theory in disordered systems by Siringo (1995) will be followed. One must note, by way of introduction, the fact that probabilistic one-body approximations were pioneered by Cyrot (1972, see also Economou and coworkers, 1972, 1978, 1983). Relevant work on disorder is reviewed by Stratt (1990: see also Logan and Winn, 1988; Winn and Logan, 1989; Stratt and Xu, 1989; Xu and Stratt, 1989; Bush et al., 1989). Specific works on the use of the Hubbard Hamiltonian for spatially disordered systems are those of Aoki and Kamimura (1976) who employ also the Matsubara–Toyozawa (1961) diagrammatic summation, and of Yonezawa et al. (1974). A physically motivated mean field theory for localization in spatially disordered systems was developed by Logan and Wolynes (1986, 1987; see also Logan 1991). Before turning to Siringo's presentation of mean field theory, within a multicomponent context, some notable work involving lattice modelling of different types includes the studies of Warren and Matthiess (1984), Kelly and Glotzel (1986) and Franz (1986). Rose (1980) employed spin-split self-consistent density functional theory, while the work of Franz (1984) is also relevant in the present context.

*Also, detailed NMR studies by Warren and coworkers (1984, 1987, 1989, these are also reviewed in March Liquid Metals, 1990) fit in generally with the RW scenario.

8.7. DISORDER AND CORRELATION: INCLUDING MULTICOMPONENT MEAN FIELD THEORY

In an early discussion of doped semiconductors* as a medium in which to search for Wigner electron solids, the topic of Chapter 9, Care and March (1971) emphasized the importance of studying the effect on Anderson localization of electronic correlation. The presentation here is that of Siringo (1995).

The minimal Hamiltonian that contains some degree of correlations among the electrons consists of the usual tight-binding form plus the interaction term

$$\hat{H}_1 = \sum_i U\hat{n}_{i\uparrow}\,\hat{n}_{i\downarrow} \tag{8.90}$$

with $\hat{n}_i = C^+_{i\sigma}C_{i\sigma}$, and U is, as usual, the on-site Hubbard interaction between two electrons with opposed spins.

In the atomic limit, when $V_{ij} \to 0$, such an interaction term is essential to recover the correct ground state: in the case of one electron per atom one regains the correct paramagnetic ground state of a monatomic gas. In other words, defining $n_i = \langle\hat{n}_{i\uparrow}\rangle + \langle\hat{n}_{i\downarrow}\rangle$ and $\mu_i = \langle n_{i\uparrow}\rangle - \langle n_{i\downarrow}\rangle$, where the brackets denote a quantum-mechanical average, in the limit $V_{ij} \to 0$ if $n_i = 1$, then one finds $\mu_i = \pm 1$. This situation is guaranteed provided U is greater than zero. The Hamiltonian $H_U = H + H_1$ is the Hubbard Hamiltonian; a useful starting point for the description of some magnetic properties (Stoddart and March, 1970). The case most commonly studied is $n_i \simeq 1$, since in general $\mu_i \leq \min\,[n_i,(2n_i)]$, so correlation effects are unimportant if $n_i \to 0$ or $n_i \to 2$. For $n_i = 1$, the behavior of the model is determined by the ratio between the essential energy parameters U and W. When the band-width W is large compared with U, the system behaves like a simple metal with $\mu_i = 0$. A perturbative mean-field analysis carried out to first order in U/W leads to a semielliptical density of states (DoS), which is merely moved in energy with respect to the unperturbed case.

In the strong U regime, however, in contrast, the DoS is strongly altered and a gap opens at the Fermi level. The system then behaves like a Hubbard–Mott insulator with local moments $\mu_i \simeq n_i = 1$ (Siringo, 1995). Any elementary excitation costs an energy U since, in the ground state, each site is occupied by an unpaired electron. In a liquid the unperturbed bandwidth W is an increasing function of the density ρ; this is clear in the Hubbard approximation since (cf. Siringo, 1995)

$$W = 4J^{1/2} = \text{constant}\ \rho^{1/2}. \tag{8.91}$$

So one is led to expect, in an expansion situation, a metal–insulator transition to occur when ρ is reduced sufficiently. This may be the case, for instance, for liquid

*See also the important early work of Edwards and Sienko (1978, 1981) and the book by Mott (1974). Alloul and Dellouve (1987) use ^{31}P NMR to investigate spin localization in Si: P, while Paglanin et al. (1988) discuss thermodynamic behavior of the same doped system.

Cs taken up the liquid–vapor coexistence curve toward the critical point, a situation discussed already in section 8.5. Warren (1993) has emphasized that in the low-density regime, the magnetic susceptibility exhibits Curie-like behavior, which is what one might expect from a Hubbard–Mott insulator.

In order to fully understand the Hubbard model it is quite instructive to consider the single-impurity problem (the so-called Anderson (1961) model). Suppose that the Hubbard interaction energy U is zero everywhere except for a single site $i = 0$ whose energy level ε_0 is fixed. The corresponding occupation number $n_0 = \Sigma_\sigma \langle \hat{n}_{0\sigma} \rangle$ will be dictated by the relative position between the energy ε_0 and the full band. At the Hartree–Fock level the Hamiltonian reads

$$\hat{H} = \sum_{i\sigma} \varepsilon_i \hat{n}_{i\sigma} + \sum_{i \neq j, \sigma} V_{ij} C^{+}_{i\sigma} C_{j\sigma} + \sum_{\sigma} \varepsilon_0^\sigma \hat{n}_{0\sigma}, \tag{8.92}$$

where the self-consistent σ-spin level $\varepsilon_0^\sigma = \varepsilon_0 + U\langle \hat{n}_{0,-\sigma} \rangle$. If $\varepsilon_0 + U$ is under the bottom of the band, then $n_0 \approx 2$, while if ε_0 is over the top of the band $n_0 \approx 0$. For intermediate values of ε_0 the σ-spin levels ε_0^σ split: for instance, we can have $\langle \hat{n}_{0\uparrow} \rangle \approx 1$, $\langle \hat{n}_{0\downarrow} \rangle \approx 0$ and $\varepsilon_0^\uparrow \approx \varepsilon_0$, $\varepsilon_0^\downarrow \approx \varepsilon_0 + U$. These values are self-consistent provided that the band is roughly contained between ε_0 and $\varepsilon_0 + U$. In this case on the site $i = 0$ (single impurity) one finds a magnetic moment $\mu_0 \approx n_0 \approx 1$. Two important messages emerge: (i) In the presence of a strong U a local moment is present if $n_0 \approx 1$ (correlations are suppressed if $n_0 \approx 0$ or 2); (ii) the occupation number n_0 is basically determined by the position of the energy level ε_0.

These conclusions are relevant for a full understanding of the disordered Hubbard model in the less-studied case $y \ll 1$. For a filling fraction $y \approx 1$ the Hubbard model, even in presence of strong disorder, does not show any relevant difference from the ordered case characterized by a probability distribution $f(\varepsilon) = \delta(\varepsilon)$. A strong U stabilizes the occupation numbers $n_i \approx y$, thus reducing the fluctuations of the charge. On the contrary, in the domain $y \ll 1$ the existence of strong charge fluctuations does not allow one to write $n_i \approx y \to 0$ (y is the ensemble average of n_i). In the light of the previous discussion on the single-impurity problem, it is quite evident that in the presence of strong diagonal disorder a fluctuating level ε_i will imply a fluctuating charge $n_i \neq y$. Even if most of the sites have a negligible charge, a few sites with the lowest levels ε_i will possess a charge $n_i \approx 1$ and, thus, a local moment $\mu_i \approx 1$. Important magnetic properties show up in the disordered Hubbard model just in that domain $y \to 0$ where the usual ordered model has negligible correlation effects. From this point of view the disordered Hubbard model can be regarded as a multicomponent impurity analogue of the single-impurity model (see, for instance, Siringo and Logan, 1992).

8.7.1. Disordered Hubbard Model

As mentioned, the disordered Hubbard model is one obvious candidate for a discussion of the interplay between disorder and correlation. However, it presents some disadvantages because one is, at the time of writing, far from an exact treatment of the model, and the approximations adopted can hide physical content. An interesting point of view has been considered by Kamimura (1980), who chooses, as a set of basis functions, the eigenstates of a one-electron tight-binding Hamiltonian that are Anderson-localized states. However, the nature of these states is merely assumed from the beginning, and the results strongly depend on this initial assumption. Moreover this approach has some validity only in the localization regime and fails to provide a linkage between the metallic and insulator phases.

A more flexible treatment of the Hubbard model can be achieved by use of mean-field theories. In the presence of disorder the formulation of a mean-field theory requires a double averaging process: for instance, one can approximate the operator $\hat{n}_i$ with its quantum expectation value $\langle \hat{n}_i \rangle$; then one is tempted to put $\langle \hat{n}_i \rangle = y$ for any site in the system. The filling fraction y is exactly equal to the average value of $\langle \hat{n}_i \rangle$, but one cannot neglect charge fluctuation at all without losing information about the disordered nature of the environment. When these problems arise, it is better to introduce a proper distribution function, i.e., to work with distributions instead of averages (see also Logan, 1991).

In the light of considerations on the single-impurity problem, it is quite reasonable to assume that the fluctuations of the level ε are the main cause of fluctuation for the charge $\langle \hat{n}_i \rangle$. A mean-field theory could be formulated by requiring the same amount of charge $n(\varepsilon)$ to be found on the sites that have the same level ε. This position allows for some degree of fluctuation of the charge among the local sites. The single-impurity problem expects a strong local moment if $n(\varepsilon) \approx 1$, and a negligible role of on-site correlations if $n(\varepsilon) \approx 0$ or $n(\varepsilon) \approx 2$. In general, the local moment should be at least a function of the level ε, and one introduces the function $\mu(\varepsilon)$ (Siringo, 1995). The definition of these functions is

$$n(\varepsilon) = \langle [\langle \hat{n}_{i\uparrow} \rangle + \langle \hat{n}_{i\downarrow} \rangle] \rangle_{\varepsilon_i = \varepsilon}, \tag{8.93a}$$

$$\mu(\varepsilon) = \langle |\langle \hat{n}_{i\uparrow} \rangle - \langle \hat{n}_{i\downarrow} \rangle| \rangle_{\varepsilon_i = \varepsilon}, \tag{8.93b}$$

where $\langle \cdots \rangle_{\varepsilon_i = \varepsilon}$ is an ensemble average over all sites with $\varepsilon_i = \varepsilon$. The full average over ε must give

$$y = \int f(\varepsilon) n(\varepsilon)\, d\varepsilon, \tag{8.94a}$$

$$\bar{\mu} = \int f(\varepsilon) \mu(\varepsilon)\, d\varepsilon, \tag{8.94b}$$

with $\bar{\mu} \neq 0$ in general since $\mu(\varepsilon)$ is positive. The single-impurity self-consistent σ-spin level of Eq. (8.92) generalizes to

$$\varepsilon_i^\sigma = \varepsilon_i + U\langle \hat{n}_{i,-\sigma} \rangle = \varepsilon_i + \frac{1}{2} n_i U - \frac{\sigma}{2} \mu_i U, \tag{8.95}$$

where $\sigma = \pm 1$. Making use of the functions (8.93), one approximates the σ-spin level with its ensemble-averaged value

$$\varepsilon_i^\sigma = \varepsilon_i + \frac{1}{2} n(\varepsilon_i)U \mp \frac{\sigma}{2} \mu(\varepsilon_i)U, \tag{8.96}$$

where the $\mp$ sign arises from the uncertainty about the orientation of the local moment on the local site. In order to describe a disordered paramagnetic system, one regards both the signs as having the same probability.

The formalism of single-site theories (see, e.g., Blackman and Tagüeña, 1991) can be easily applied to the case of the random σ-spin level: the ensemble-averaged Green function then reads (Siringo, 1995)

$$\bar{G}_\pm^\sigma(\varepsilon,z) = \left[z - \varepsilon - \frac{1}{2} Un(\varepsilon) \pm \frac{\sigma}{2} U\mu(\varepsilon) - S^\sigma(z) \right]^{-1}, \tag{8.97}$$

where $S^\sigma(z)$ is a function of the fully averaged Green function*

$$\bar{G}^\sigma(z) = \int d\varepsilon\, f(\varepsilon)[\tfrac{1}{2}\bar{G}_+^\sigma(\varepsilon, z) + \tfrac{1}{2}\bar{G}_-^\sigma(\varepsilon,z)]. \tag{8.98}$$

The DoS can be extracted from the function $\bar{G}^\sigma(z)$. Moreover, even the partial density of states

$$d_\pm^\sigma(\varepsilon,E) = -\frac{1}{\pi} \operatorname{Im} \bar{G}_\pm^\sigma(\varepsilon,E + i0^+) \tag{8.99}$$

derives from the knowledge of $\bar{G}^\sigma(z)$ and $S^\sigma(z)$. The functions $d_\pm^\sigma(\varepsilon,E)$ give the DoS projected on the local site with level ε and moment $\mu = \pm\mu(\varepsilon)$. The theory is fully self-contained if

$$n(\varepsilon) = \int_{-\infty}^{E_F} [d_\pm^\uparrow(\varepsilon,E) + d_\pm^\downarrow(\varepsilon,E)]\, dE, \tag{8.100a}$$

$$\mu(\varepsilon) = \int_{-\infty}^{E_F} |d_\pm^\uparrow(\varepsilon,E) - d_\pm^\downarrow(\varepsilon,E)|\, dE \tag{8.100b}$$

in accordance with the definitions in Eq. (8.93). The Fermi level E_F is evidently defined by

$$y = \int_{-\infty}^{E_F} [D^\uparrow(E) + D^\downarrow(E)]\, dE. \tag{8.101}$$

This multicomponent mean-field theory only requires a knowledge of the distribution $f(\varepsilon)$, and of the closure relation for the self-energy $S^\sigma(z) = S^\sigma[\bar{G}^\sigma(z)]$. This can be taken to be the Hubbard approximation or some more sophisticated condition (Logan and Winn, 1988; Winn and Logan, 1989). While a full discussion of the model is given by Siringo and Logan (1992), the main results emerging from such a mean-field approximation will be summarized here.

*$f(\varepsilon)$ is the probability distribution of the levels ε. Fig. 8.7 is an example of two delta functions at ε_A and ε_B.

In the following discussion one approximates the diagonal disorder distribution $f(\varepsilon)$ with a cut Lorentzian (see, e.g., Siringo, 1995)

$$f(\varepsilon) = \frac{\text{const.}}{\lambda^2 + \varepsilon^2} \tag{8.102}$$

defined and normalized in the range $-W \leq \varepsilon \leq W$. In the simple Hubbard single-site approximation the problem is characterized by the filling fraction y and by three energies: (i) the bandwidth W of the unperturbed semielliptical DoS; (ii) the half-width λ of the diagonal disorder distribution $f(\varepsilon)$; (iii) the Hubbard on-site Coulomb interaction U. Alternatively one can make use of the three dimensionless variables y, $\tilde{\lambda} = \lambda/W$, and $\tilde{U} = U/W$. For a narrow-band disordered system $\tilde{\lambda} = 0.25$ seems a reasonable value, and U can be comparable with the bandwidth W.

First, let us consider the case $y = 1$ (half-filling). Figure 8.8 shows the DoS and the respective self-consistent charge–moment distributions $n(\varepsilon)$, $\mu(\varepsilon)$ for $\tilde{U} = 0.5$ and $\tilde{U} = 1.0$. In this case $\tilde{\lambda}$ does not affect the result too much, which resembles the usual $\tilde{\lambda} \rightarrow 0$ limit of no disorder. The charge is spread over all sites, and, as was expected, $\tilde{U}$ is the relevant parameter in this limit. When $\tilde{U}$ is small the moments are unstable and the normal restricted Hartree–Fock solution is achieved.

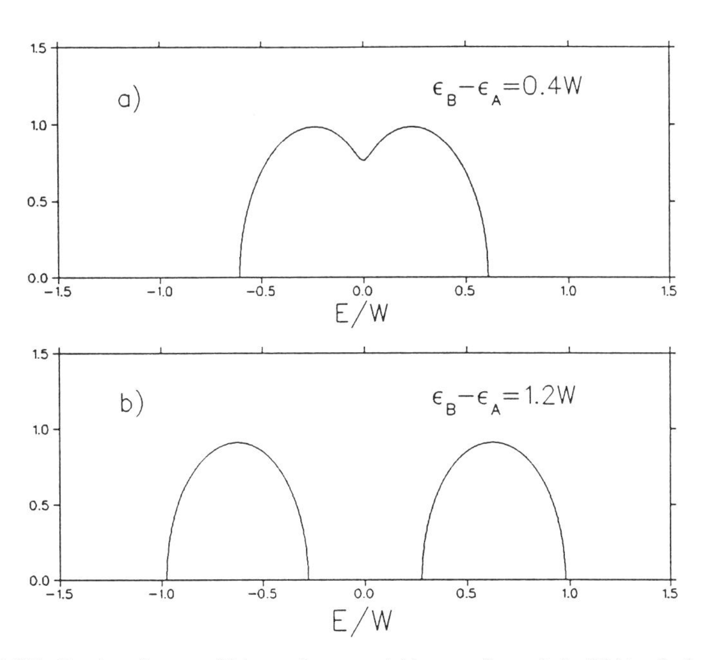

FIGURE 8.7. Density of states of binary alloy at stoichiometry ($n_A = n_B$), in (Hubbard) single-site approximation, for $\varepsilon_B - \varepsilon_A = 0.4W$ (curve a) and for $\varepsilon_B - \varepsilon_A = 1.2W$ (curve b). Note that in curve b the Fermi level falls inside gap only if $y = 1$ (half-filling) (after Siringo, 1995).

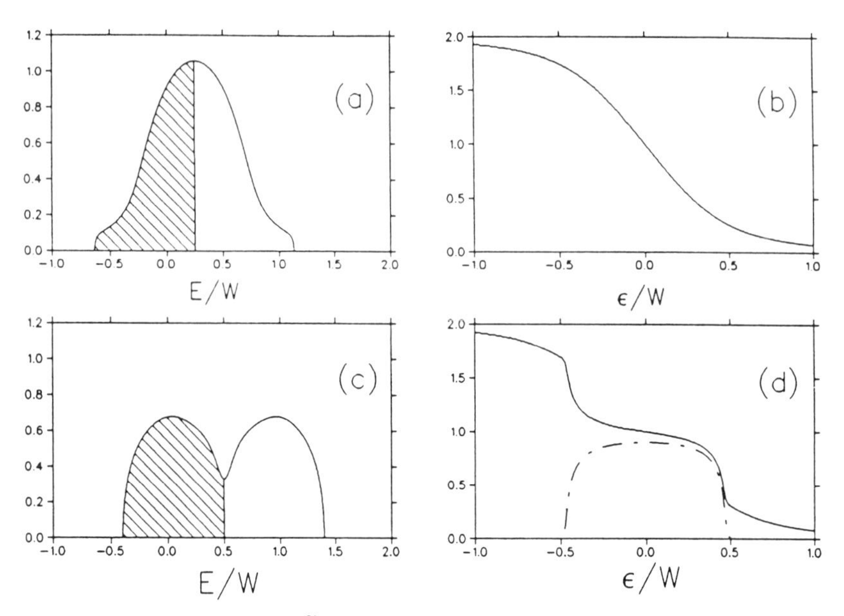

FIGURE 8.8. Density of states for $\tilde{U} = 1.0$ [case (c)]. Charge $n(\varepsilon)$ shown by solid line and moment $\mu(\varepsilon)$ in broken line are also depicted for $\tilde{U} = 0.5$ [case (b)] and $\tilde{U} = 1.0$ [case (d)]. λ is fixed at the value of 0.25 (after Siringo, 1995).

$\tilde{U} = 1.0$ is enough for stabilizing the presence of local moments, and in this case one predicts a paramagnetic ground state with an average local moment $\bar{\mu} \approx 0.7$. A pseudogap opens in the middle of the band, and when $\tilde{U}$ is large the system becomes a Hubbard–Mott insulator.

Once one has made sure that the known phenomenology is recovered in the well established case $y = 1$, one can proceed to examine in detail the limit $y \to 0$. In this limit one always predicts the existence of local moments and a paramagnetic ground state provided that $\tilde{U}$ is nonzero. Let us consider again the case $\tilde{U} = 0.5$, $\tilde{\lambda} = 0.25$ in which correlations do not play a major role at half-filling (Fig. 8.8a,b). As shown in Fig. (8.9), at low filling local moments are stable and they approach the charge value when $y < 0.08$. In the very low filling limit $y \approx 0.02$, one finds $n(\varepsilon) \approx \mu(\varepsilon)$ for most of the filled states, thus indicating a strongly correlated ground state. Charge fluctuations play here a major role since the charge is not evenly spread over all local sites, but only those sites with a very low energy level ε are significantly filled. On the other hand, these sites are few in the system, and it is unlikely that the corresponding local states could overlap each other. In other words, the small amount of charge present in this low-filling limit is trapped

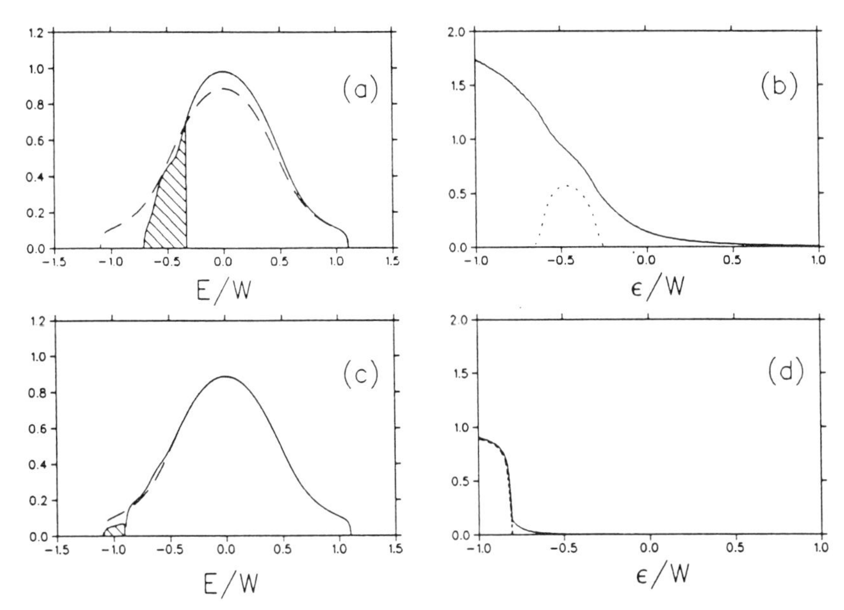

FIGURE 8.9. Density of states for $\tilde{U} = 0.5$, $-\lambda = 0.25$; $y = 0.3$ [case (a)] and $y = 0.02$ [case (c)]. Dashed lines show corresponding density of states for $U = 0$, while shaded areas depict filled portion of bands. Charge $n(\varepsilon)$ shown in solid line and moment $\mu(\varepsilon)$ in broken line for $\tilde{U} = 0.5$ [case (b)] and $\tilde{U} = 1.0$ [case (d)] (after Siringo, 1995).

in a few deep localized states that are singly occupied if the correlation energy U is nonzero (Siringo, 1995).

As $y \rightarrow 0$ the $n(\varepsilon)$ distribution approaches a step function, thus indicating that the atomic limit is correctly achieved. In this limit the $n(\varepsilon)$ distribution can be regarded as a quasiparticle distribution that becomes a step function when the local states are exact eigenstates of the Hamiltonian, or in other words when the filled eigenstates of the Hamiltonian are exactly the local atomic states.

Figure 8.10 shows a turnover for the average value $\langle \mu^2 \rangle$ versus y: when y is small enough the model is not sensitive to the choice of the parameter $\tilde{U}$, but provided that $U \neq 0$ local moments stabilize in the system. It is the analogue of an atomic limit since some local deep states are filled with one electron that cannot jump over the surrounding sites because these have some higher energies. Even a small value for U prevents any double occupancy, thus giving rise to a Curie-law paramagnetic ground state. For large values of the filling fraction y, Fig. 8.10 shows a different behavior for different values of $\tilde{U}$: if $\tilde{U}$ is large the intensity of local moments increases with y and persists even at half-filling where the usual Hubbard–Mott transition is recovered; if $\tilde{U}$ is small any local moments disappear

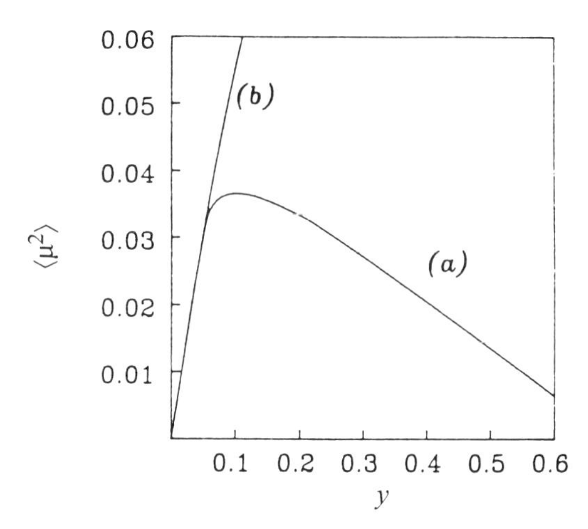

FIGURE 8.10. Averaged squared local moment $\langle\mu^2\rangle$ versus filling fraction y for $-\lambda = 0.25$, $\tilde{U} = 0.5$ [case (a)] and $\tilde{U} = 1.0$ [case (b)] (after Siringo, 1995).

when the band is filled, and the usual metallic restricted Hartree–Fock solution is reached around half-filling.

One would like to map the model on the known experimental results for liquids and amorphous solids by a proper choice of the parameters. For instance, the presence of strong diagonal disorder dominates the intermediate regime occurring at the metal–insulator transition of the liquid alloy Au–Cs. In the Cs-rich phase $(\text{Au–Cs})_{1-x}\text{Cs}_x$ with $0 < x < 0.07$, some experimental evidence has been reported (Dupree *et al.*, 1980) for the existence of localized unpaired electrons. The absence of localization in the solid (Dupree *et al.*, 1985) suggests that one can regard all this as "disorder-induced." Attention will be focused on the upper subband of the alloy, mainly generated by Cs atomic states, and filled with a filling fraction $y = x$ varying with composition. If one ignores the presence of the lower subband (for $x > 0$), the relevant physical aspects can be described by the one-band model discussed by Siringo and Logan (1991). Moreover, the presence of a random distribution of Au^-, Cs^+ ions give rise to disordered Madelung potentials (Logan and Siringo, 1992) that contribute to enlarge greatly the half-width λ of the probability distribution $f(\varepsilon)$: one expects to deal with a strong diagonal disorder induced by the random ionic field. The choice $\tilde{\lambda} = 0.25$, $\tilde{U} = 0.5$ of Fig(s). 8.8–8.10 is quite reasonable for a binary liquid alloy like Au–Cs at ordinary density. The case $y = 1$ (Fig. 8.8) thus reproduces the correct absence of local moments in pure Cs; conversely in the range $0 < y < 0.07$ the observed localizations of unpaired electrons is a consequence of the diagonal disorder in the Cs-rich ionic alloy. The presence of disorder greatly enhances the correlative effect of the Coulomb repulsion, and the observed Curie-law paramagnetic susceptibility is an expected consequence. As is evident from Fig. 8.10 a larger filling fraction y (i.e., a larger amount x of excess Cs) produces a smooth transition toward a different regime, which is metallic, and poorly correlated, in agreement with the experimental findings.

Given the general nature of the approximations adopted for developing such multicomponent mean-field theory, the model by itself could be a useful starting point for understanding the physics of other liquid metals or even very different materials like amorphous semiconductors—for instance, Si:P;B, which behaves like a quenched solid-state gas of phosphorus and boron impurities embedded in a silicon matrix. Even larger magnetic effects may be expected (Siringo, 1995) in this material given the narrow impurity band W, which yields larger values for $\tilde{U}$ and $\tilde{\lambda}$.

The multicomponent mean-field theory clearly shows the importance of correlations in the limit of an almost empty band for a disordered system. When charge fluctuations are negligible, as is often the case in crystals for symmetry reasons, one may assume $n_i = y$ and the largest effect of correlations is expected to appear near the half-filling of the band ($y \approx 1$). It is then commonly believed that in a one-band problem correlation effects are really important only when y is close to 1 and U is large compared with the bandwidth W. It is emphasized here that this is not a general statement, and that correlation effects can become really important in the limit $y \to 0$ in a disordered system. First of all when the band is almost empty a disordered system behaves like an insulator or a dirty metal, so any screening is much reduced and the effective value of U is enhanced with respect to the half-filled band case. But even assuming U constant, in a disordered system one cannot neglect charge fluctuations any longer when y is very small. While in a perfect crystal the charge is spread over Bloch delocalized states at any small value of the filling y, for a disordered system the charge localizes (Siringo, 1995) in atomic-like states. In this atomic limit n_i is zero almost everywhere, but it is $n_i \approx 1$ on a few atomic sites where correlation effects become important.

8.7.2. Liquid Transition Metals and Their Alloys: Experiment and Theory

The measured magnetic susceptibility χ of pure liquid transition metals increases from Mn to Fe, reaching a maximum at Co and decreasing from Ni to Cu (see, for example, Fig. 141 of the book *Electrons in Metals and Alloys* by Alonso and March; 1989). A remarkable result is that only a small change in χ is observed on melting (see, for example, Table 31 of the above-mentioned book).

Jank et al.* have presented self-consistent calculations of the electronic density of states of the molten 3d and 4d transition metals. Their treatment

*W. Jank, Ch. Hausleiter and J. Hafner, *J. Phys. Condens. Matter* **3**, 4477 (1991): see other references to liquid transition metals, both experiment and theory, given there.

employed realistic models of the liquid structure and a linearized-muffin-tin-orbital supercell technique. These workers compared their theoretical results with the magnetic susceptibility experimental data (see also below), as well as with existing experiments on X-ray, photoelectron and Auger-electron spectra.

In agreement with the magnetic susceptibility data, Jank et al. (1991) find that the electronic structure of transition metals undergoes only relatively small modifications on melting. In particular, Jank et al. present the electronic density of states (DOS) of liquid Cr, Mn, Fe, Co and Ni, together with the corresponding DOS of these crystalline transition metals. They also plot the DOS in liquid 4d transition metals; namely Pd, Rh, Mo, Nb, Zr and Y.

For the 3d metals, Jank et al. conclude that the widths of the d bands are the same in the molten state as in the corresponding crystal. They note that some degree of band narrowing to be anticipated from the volume expansion on melting is compensated for by a broadening of the energy bands due to the fluctuations in the interatomic overlap integrals. Numerically, as one follows the series from Cr to Ni, Jank et al. find the band width to be reduced from 6.2eV to 4.4eV. This corresponds, as these authors emphasize, to the experimental findings on the basis of the L_β-X-ray emission spectra.

A further point stressed by Jank et al. (1991) is that the characteristic bonding–antibonding splitting of the d states is still observable in the liquid phase. The splitting found by these authors is larger for the metals with a partially filled d band which crystallize in the bcc structure (Cr, Mn and Fe) and show a strong splitting in the crystal DOS as well. The larger bonding–antibonding splitting in these elements Cr, Mn, and Fe produces the strong covalent bonding effects in elements with a nearly half-filled band.

The general trend of the electronic structure results of Jank et al. for the 4d transition metals is the same as that in the 3d elements, except that the width of the d band is increased due to the more extended character of the 4d orbitals.

8.7.3. Magnetic Susceptibilities

We now return briefly to the magnetic susceptibilities of the 3d transition metals. To allow for electron–electron interactions, the Pauli formula relating the spin susceptibility to the DOS at the Fermi level, $n(E_F)$ say, has to be modified by an interaction J_{eff} to read:

$$\chi = 2\mu_B^2\left[\frac{n(E_F)}{1 - J_{eff}n(E_F)}\right]. \tag{8.103}$$

For reasons set out by Jank et al. (1991) following earlier studies, J_{eff} is taken as 0.05. The procedure then adopted by these authors is to extract the DOS at the

Fermi level from the experimental susceptibilities already referred to. The differences between "experiment" and theory for $n(E_F)$ are but a few per cent for Ni, Co, Fe, and Mn.

There is very extensive measured data on both liquid and disordered metal alloys. The former area has been reviewed by March and Sayers (1979) and the latter by March et al. (1984).* In particular, liquid 3d–3d transition metal alloys have been widely studied experimentally. Very briefly (see, e.g., Alonso and March, 1989), the susceptibility of alloys of transition metals next to one another in the 3d series appears to follow the trend exhibited by the pure metals, which was set out above. However, alloys of 3d elements in the liquid state differing in atomic number by more than unity exhibit strong departures from this behavior. Fe–Ni alloys show only a weak maximum in the region of Fe_3Ni compared with that observed on passing from Fe to Co to Ni, and the susceptibility of Cr alloys decreases in the order $CrFe_3$, $CrCo_3$ and $CrNi_3$. Again a remarkable result is that only a small change in magnetic susceptibility is observed on melting, though Fe is somewhat exceptional in this respect. Much work evidently remains to be done in exploring the way the quantitative details of the magnetism of liquid transition metal alloys depend on electron–electron correlation (see also some useful models and classifications reviewed by March and Sayers, 1979).

8.7.4. Mooji Correlation

The considerations of Mooji (1973), who made notable observations on the temperature coefficient of electrical conductivity have been considered theoretically at a variety of levels, both phenomenological and from first principles. Below we give a brief survey of the work of Girvin and Jonson (1980). The phenomenological approach of Logan and Wolynes (1987) should also be noted here.

Mooji (1973) exposed a correlation between the electrical resistivity of metallic systems and its temperature coefficient (TCR). Girvin and Jonson therefore proposed a theory of transport in strongly disordered metal alloys which focussed on certain model independent features of the electron–phonon dynamics. Their finding was that in the low-resistivity limit the adiabatic phonon approximation has validity and the disorder associated with phonons increases the resistivity. In the opposite limit of high resistivity, where any weak-scattering assumption must break down due to incipient Anderson localization, the adiabatic-phonon assumption no longer is true. Phonons then assist electron mobility, the result being an anomalous TCR. Model calculations presented by Girvin and Johnson suggest that such a mechanism could underly the Mooji correlation. The effect of electron–electron as well as electron–phonon interactions is handled in the sub-

*N. H. March, Ph. Lambin and F. Herman, *J. Magn. Magn. Mater.* **44**, 1 (1984).

sequent work of Logan and Wolynes (1987), see also Logan (1991) by adding inverse relaxation times. At the time of writing, the importance of interplay between these two processes remains to be clarified.

8.8. Interplay between Electronic Correlation and Disorder Near Metal–Insulator Transition*

To conclude this Chapter, mention must be made of the important review by D. Belitz and T. R. Kirkpatrick (*Rev. Mod. Phys.* **66**, 261, 1994, referred to below as BK). These workers start out from a phenomenological discussion of slow modes, based on diffusive electron dynamics which results from the conservation laws for particle number, spin and energy. BK then treat in considerable detail the progress that has resulted from the application of field-theoretic techniques and scaling ideas. Their focus throughout is the metal–insulator (MI) transition proper and its immediate vicinity, excluding effects of weak disorder or deep in the insulator. The pioneering work of F. Wegner (*Z. Phys.* **B25**, 327, 1976) is stressed, in which real-space renormalization group (RG) methods were used to argue that the dynamic conductivity had the scaling form (see also BK)

$$\sigma(t, \Omega) = b^{-(d-2)} f(tb^{1/\nu}, \Omega b^{d}) \tag{8.104}$$

In Eq. (8.104), t is a dimensionless measure of the departure from the critical point, Ω is the frequency, b is an arbitrary scale parameter, f is an unknown scaling function and ν an unknown correlation length exponent. Eq. (8.104) predicts that the static conductivity vanishes at the MI transition with exponent $\nu(d-2)$ and that the dynamical conductivity at the critical point $t = 0$ behaves as $\Omega^{(d-2)/d}$. As BK note, the former behavior results from the choice of scale parmeter $b = t^{-\nu}$, while the latter is found from $b = \Omega^{-1/d}$. An important further step was the mapping by Wegner (*Z. Phys.* **B35**, 207, 1979) of the Anderson localization problem (MacKinnon and Kramer, 1993, see BK; Thouless, 1976) onto an effective field theory: indeed the methods employed by Wegner and his colleagues are the central theme of the review by BK. A. M. Finkelstein (1983, see BK) extended the field-theoretic discussion of the Anderson transition to allow for electron–electron interactions. This application not only permitted the use of RG methods to deal with strong disorder but also proved suitable to treat electron–electron correlations of arbitrary strength.

After extensive discussion of field-theoretic techniques, BK focus on possible scaling scenarios of the MI transition of interacting electrons and in particular how Eq. (8.104) might be generalized to the interacting case. They then review explicit calculations that demonstrate how these scaling scenarios are realized in various universality classes.

*Section added in proof

Universality Classes and Range of Interaction

As BK stress, Wegner (1976, see also Abrahams et al., 1979; see BK) recognized that the MI transition can be viewed as a continuous quantum phase transition that is characterized by three independent critical exponents. The singular behavior of physical quantities near the MI transition are then determined by exponents which in turn are related to the three independent ones by scaling laws.

The identification of distinct universality classes was a further important step towards describing MI transitions. Using methods from either field theory or from many-body perturbation theory, it was established that for noninteracting electrons there are four main universality classes for MI transitions (Efetov et al., 1980, see BK). After embodying electron–electron correlation, it was appreciated by Castellani et al. (1984, see BK) that it was essential to distinguish between the cases of short- and long-range electron–electron correlation. It then turns out that there are at least eight universality classes for the localization transition in disordered Fermion systems with electron–electron interaction (see Table I of BK: the respective three independent critical exponents are also recorded in their Table III). As BK note, one of the main experimental problems is the determination of the relevant universality class for a given system.

Chapter 9

Magnetically Induced Wigner Solid

The phase equilibrium between a two-dimensional electron liquid (Laughlin, 1989; see also Appendix 9.1) and a Wigner electron solid induced by a magnetic field will be the main focal point of this final chapter, following the discussion of electron crystals in zero magnetic field in Chapters 3 and 5. It will be useful, by way of introduction, to extend the zero-field discussion by giving a brief summary of the main physical properties of the electron crystal in zero field. So far, in the quantal regime, at the time of writing, this Wigner crystal has not been unambiguously observed in the laboratory. However, in a GaAs/AlGaAs heterojunction in a high magnetic field, observations have verified the original proposal of Durkan *et al.* (1968) (for a specimen of highly compensated *n*-type InSb) that magnetic fields would aid Wigner crystallization (now referred to as magnetically induced Wigner solid, MIWS). The pioneering experiment of Andrei *et al.* (1988) in this latter area has been subjsequently verified and extended by numerous workers (see, for instance, Buhmann *et al.*, 1991; Andrei, 1992).

9.1. SUMMARY OF PHYSICAL PROPERTIES OF WIGNER ELECTRON CRYSTAL (ZERO MAGNETIC FIELD)

Before introducing the basic physics of MIWS, it will be helpful background to summarize the physical properties that one must expect the Wigner electron crystal to exhibit in its ground and low-lying excited states (Lea and March, 1989; Care and March, 1975):

(i) The ground state is insulating, electrons vibrating about the sites of the Wigner bcc lattice.
(ii) Closely related to (i), the Fermi surface discontinuity in the electronic momentum distribution disappears in the insulating phase.
(iii) Long-range magnetic order is to be expected in the ground state. Herman and March (1984), following earlier work of Carr (1961), used the

Ceperley–Alder (1980) numerical results to argue that, most probably, Néel-type antiferromagnetism exists, upward-spin electrons being located on one of the two interpenetrating simple cubic lattices forming the bcc electron crystal, downward spins being on the other.

(iv) Phononlike, noncurrent carrying, low-lying excited states exist.

(v) Defects (e.g., interstitials and vacancies) can occur at elevated temperatures and, via these, small nonzero electronic conductivity can be anticipated at $T \neq 0$.

(vi) Wigner solidification is strongly influenced by dimensionality (see, for example, Fig. 9.1).

So far, only (i) has been tested by "experiment," and then, as discussed, on the computer rather than in the physics laboratory!

Because of the low critical density, n_c, reflected in the Ceperley–Alder limits $r_c \sim 80$–$100a_0$, it has been clear for a long time that the low-density electrons in a suitable semiconductor provide the most promising medium in which to search for Wigner crystallization. This prompted Durkan and March (1967) and Durkan *et al.* (1968) to study magnetic-field-dependent effects in semiconductors.

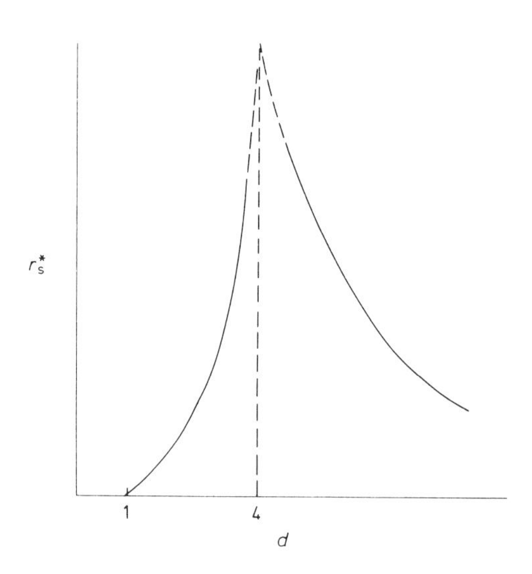

FIGURE 9.1. Critical interelectronic spacing, r_s^* for Wigner crystallization as a function of dimensionality d (schematic only). Note that the electron–electron interaction here satisfies the d-dimensional Laplace equation. r_s^* is equal to zero for one dimension and is infinite for $d = 4$ with this particular interaction (see also Parrinello and March, 1976).

9.2. MAGNETIC-FIELD-INDUCED WIGNER CRYSTALLIZATION: EXPERIMENTAL FINDINGS

It is of some interest to review briefly the arguments presented in the work of Durkan *et al.* (1968). Their work was motivated by the experiments of Putley on *n*-type InSb in a magnetic field, in which a transition was observed which could be interpreted as "conduction in a metallic-like impurity band" formed by overlap of hydrogenic donor impurity wave functions, which, under sufficiently large applied magnetic field *B*, became so narrow because of reduced overlap in directions perpendicular to the magnetic field that metallic conduction ceased. Arguments by Durkan *et al.*, (1968), which were order-of-magnitude in nature, strongly suggested that one could expect magnetic fields to aid Wigner electron crystallization.

Some attention was subsequently directed to ways of observing such Wigner electron crystallization experimentally. The work of Egelstaff *et al.*, (1974), cited in Chapter 8, focused on comparison of X-ray and neutron scattering in (ideally) the lowest-electron-density metals (alkalis). Studies of the theory of neutron scattering specifically from materials with favorable *g* values were carried out by Elliott and Kleppmann (1975).

However, from different directions, evidence began to accumulate much more recently for quantum-mechanical Wigner crystallization in two-dimensional systems (e.g., abnormal behavior observed in direct current (d.c.) transport for a (Landau level) filling factor $\nu < \frac{1}{5}$ (Störmer *et al.*, 1980) and therefore here all attention will be focused on this area.

Andrei *et al.* (1988) tested directly for a defining property of a solid, namely its rigidity to shear. An unbounded charged two-dimensional liquid has only plasmon excitations in zero field, related to the number of electrons per unit area, n_s say, by (see Lea and March, 1989)

$$\omega_p^2 = 2\pi e^2 n_s q/\varepsilon m. \tag{9.1}$$

The application of a perpendicular magnetic field *H* introduces a gap

$$\omega_+^2 = \omega_p^2 + \omega_c^2. \tag{9.2}$$

There are no gapless bulk excitations, as discussed, for example, by Girvin *et al.* (1982).

As stressed by Andrei *et al.* (1988), a system capable of resisting static shear, characterized by modulus μ does, in sharp contrast, exhibit a gapless magnetophonon branch of frequency

$$\omega_- = c(2\pi\mu)^{1/2} q^{3/2}/H, \qquad \omega_c \gg \omega_p. \tag{9.3}$$

Andrei *et al.* (1988; see also Andrei, 1992) take the presence of such a mode as positive evidence of a solid phase. In practice, their experiment consists of causing a longitudinal electric wave of defined wave number q to interact with the electrons and sweeping its frequency. By this technique, they establish beyond reasonable doubt the existence of an electron solid phase.

9.3. MELTING OF WIGNER ELECTRON CRYSTALS

In light of the foregoing experiment, it is of obvious interest to consider in more detail the melting curve of the Wigner electron crystal.

In early work, in zero magnetic field, Parrinello and March (1976) discussed the thermodynamics of Wigner crystallization in D dimensions and drew a (schematic) picture of the critical r_s value for the phase transition versus dimensionality D. In their work the electron–electron interaction was taken to satisfy Laplace's equation in D dimensions (see Fig. 9.1).

Though one is interested here in the quantum-mechanical Wigner transition [classical Wigner crystallization is simulated, according to Ferraz and March (1972); see also Stishov *et al.*, 1960) by the freezing of liquid Na, with relatively uniform conduction electron density, into a body-centered -cubic (bcc) metal crystal], Parrinello and March (1976) pointed out that the melting curve of the Wigner crystal in three dimensions approached the classical (one-component plasma) melting criterion ($\Gamma = e^2/r_s k_B T \sim 170$; see e.g., March and Tosi, 1984) in the extreme-low-density limit, because of the absence of any quantum-mechanical tunneling as r_s tends to infinity. Ferraz *et al.*, (1978, 1979) interpolated between this limit and the $T = 0$ transition, using the Clausius–Clapeyron equation for a first-order electron liquid–solid transition in three dimensions. Their main conclusions have been subsequently supported by the more quantitative work of Nagara *et al.* (1988; see also March, 1989) based on a partial summation to all orders of the Wigner–Kirkwood expansion. Subsequent to the work of Parrinello and March (1976) and of Ferraz *et al.* (1978, 1979), Fukuyama *et al.* (1985) have treated theoretically the melting of the two-dimensional Wigner crystal.

9.3.1. Quantal Three-Dimensional Melting of Wigner Crystal

The early experiments of Somerford (1971) on n-type InSb were interpreted by Care and March (1971; 1975) in terms of a Wigner transition in a magnetic field, and this interpretation was subsequently pressed by Kleppmann and Elliott (1975). In particular, these latter workers could interpret the anisotropy of the conductivity observed by Somerford, even though it is true that the experiments exhibited somewhat large scatter. They concluded that, although the interpretation of Somerford's results is consistent with the Wigner transition, this does not prove its existence (see also, Mansfield, 1971). Therefore, here, most attention will be focused on the interpretation of the experiments of Andrei *et al.* (1988) and of Buhmann *et al.* (1991). Andrei *et al.* studied transitions in a magnetic field in a two-dimensional GaAs heterojunction, as already explained, while Buhmann *et al.* subsequently made a photoluminescence study on the same system. However, before turning to these experiments, it is of interest in the context of the work of Somerford (1971) to consider first the theoretical treatment of melting a three-dimensional Wigner electron crystal in a magnetic field.

9.3.2. Melting in a Magnetic Field

To illustrate the form of the phase diagram to be expected for the Wigner transition in a magnetic field, Fig. 9.2 shows the thermal energy $k_B T_m$ at melting, in units of e^2/r_s. This diagram has been constructed (Lea and March, 1989) from data presented by Kleppmann and Elliott (1975). Adopting their proposal, the melting temperature has been estimated from the difference in ground-state energies of localized and delocalized electrons in an applied magnetic field.

While the indications from Fig. 9.2 are that there could be some (approximate) convergence to a common limit in the high-field regime, the main merit of the plot shown is that the somewhat complicated crossover of curves in the original presentation in the intermediate range (Kleppmann and Elliott, 1975) is avoided by using the convenient variable $n^{2/3}/H$. At zero temperature, melting leads back to values of $n^{2/3}/H$ that only weakly depend on the specimen electron density n, the value of which characterizes curves A–D in Fig. 9.2.

In spite of some variation in shape associated with the change in electron density from curve A ($n = 10^{22}$ cm^{-3}) to curve D ($n = 5 \times 10^{23}$ cm^{-3}), Fig. 9.2 demonstrates the general trend of a decrease from the limit as H tends to infinity to a "critical" value of $n^{2/3}/H$ at $T_m = 0$ in all four cases. It is interesting, in view of these apparently somewhat simple results for three-dimensional melting, to develop next the thermodynamics of Wigner solids in a magnetic field, following Lea *et al.* (1992).

9.4. THERMODYNAMICS OF TWO-DIMENSIONAL WIGNER CRYSTAL MELTING IN A MAGNETIC FIELD

As mentioned, the thermodynamics of Wigner crystallization was considered by Parrinello and March (1976) in zero magnetic field. The starting point of the study of the thermodynamics of an electron crystal to electron liquid first-order melting transition (Lea *et al.*, 1992) is the result (see, for example, Pippard, 1966) for the melting temperature, T_m, as a function of magnetic field H. At constant area Ω, this can be written

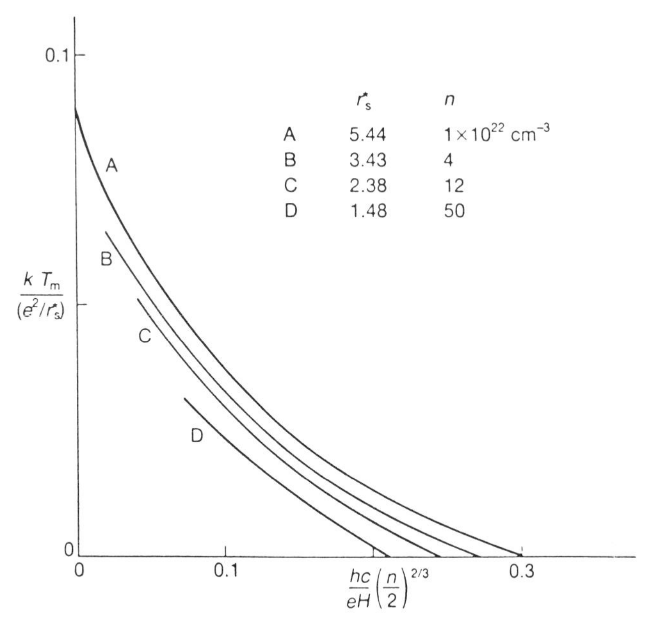

FIGURE 9.2. Melting curves of three-dimensional Wigner electron crystals in a magnetic field H. These were constructed, for different electron densities, from the numerical estimates of Kleppmann and Elliott (1975) of the difference in ground-state energy of localized and delocalized electronic configurations. Here the parameter $r_s^* = r_s a_B$, where r_s is the Wigner–Seitz radius and a_B is the Bohr radius (after Lea and March, 1989).

$$\left(\frac{\partial T_m}{\partial H}\right)_\Omega = \frac{-\Delta M}{\Delta S}. \tag{9.4}$$

If the subscript C denotes the crystal phase and L the liquid phase, then $\Delta M = M_L - M_C$ is the change of magnetization on melting, while $\Delta S = S_L - S_C$ is the corresponding entropy change. Equation (9.4) is readily cast into a form directly useful in analyzing the measurements referred to above, by using the relation between Landau level filling factor ν and H, namely

$$\nu = nhc/eH. \tag{9.5}$$

Then Eq. (9.4) becomes

$$\left(\frac{\partial T_m}{\partial \nu}\right)_\Omega = \left(\frac{H}{\nu}\right) \frac{\Delta M}{\Delta S}. \tag{9.6}$$

Some properties that follow from this equation are now listed:

(i) Turning points in the melting curve in the (ν,T) plane correspond to $\Delta M = 0$, provided the entropy change, ΔS, does not simultaneously go to zero. This result means that maxima (see Fig. 9.3) in the melting curve immediately locate points where the magnetization in the strongly correlated liquid equals that in the Wigner solid.

(ii) Infinite slopes on the melting curve, at specific values of ν away from $\nu = 0$, imply $\Delta S = 0$ for nonzero ΔM.

(iii) Equation (9.6) can also be used to locate points where $\Delta E = 0$, where E is the internal energy. If we write the magnetic Helmholtz free energy, F, relevant to phase equilibrium at constant T, field H, and area Ω, as (Pippard, 1966)

$$F = E - TS - HM, \tag{9.7}$$

the equilibrium condition $F_L = F_C$ evidently then yields

$$\Delta E - T\Delta S - H\Delta M = 0. \tag{9.8}$$

As envisaged in (iii), the existence of a point, or points, on the melting curve where $\Delta E = 0$ yields at such points

$$\Delta M/\Delta S = -T_m/H \tag{9.9}$$

the slope of the melting curve is then given by [from Eq. (9.6)]

$$\left(\frac{\partial T_m}{\partial \nu}\right)_\Omega = -\frac{T_m}{\nu}. \tag{9.10}$$

This equation can only be satisfied for $\partial T_m/\partial\nu$ negative and then whenever the melting curve lies at a tangent to a hyperbola of the form $T_m\nu$ = constant. Hence, one now has a prescription for locating points where ΔM, ΔS, and ΔE are zero, and these have been marked on the phase diagram in Fig. 9.2. At each point the relative signs of the nonzero differentials are then fixed by the relations in Eq. (9.3). Also, Eq. (9.8) requires that ΔS and ΔM cannot simultaneously have opposite signs to ΔE.

Let us next consider these results in the light of the schematic phase diagram (Fig. 9.3) from Buhmann *et al.* (1991), into which they have subsumed the main features compatible with microscopic theory and experiment. It is clear (see Andrei *et al.*, 1988) that, as ν tends to zero, the melting temperature T_m should tend to that of a classical one-component plasma. This has a melting temperature, T_{mc} say, given by (see also Parrinello and March, 1976)

$$T_{mc} = e^2(\pi n)^{1/2}/\varepsilon k_B\Gamma_m, \tag{9.11}$$

where ε is the dielectric constant of the host material while $\Gamma_m = 127 \pm 3$ (Deville, 1988). Hence, Fig. 9.3 displays $t_m = T_m/T_{mc}$ versus ν with $t_m = 1$ at $\nu = 0$. As this point is approached, one anticipates, on physical grounds, that the entropy S_L of the liquid will be greater than that of the solid, S_C. (Should this expectation not be

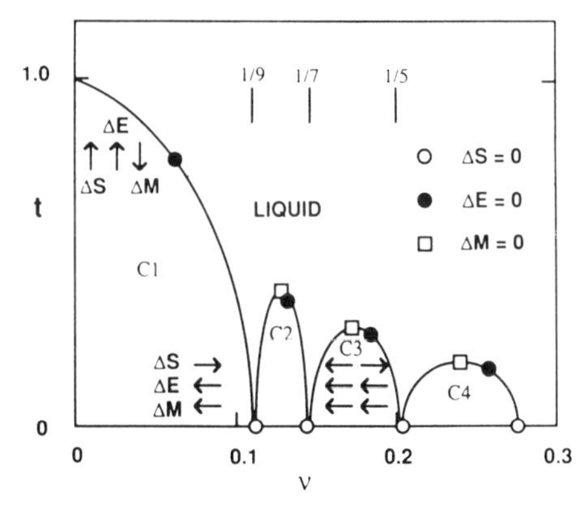

FIGURE 9.3. Schematic phase diagram proposed by Buhmann *et al.* (1991) for two-dimensional Wigner electron crystallization in a magnetic field perpendicular to the electron sheet. The diagram shows four crystalline phases, C1 to C4, and the re-entrant liquid phases at Landau filling factors ν of $\frac{1}{9}$, $\frac{1}{7}$, and $\frac{1}{5}$. The symbols mark the points ΔM, ΔS, and ΔE=0, as deduced from thermodynamics (see text). The arrows show the direction of ΔM, ΔS, and ΔE across the phase boundary nearest to the arrows. The arrows in phases C2 and C4 match those shown in phase C3. Note that, as alternatives, the possibilities of a finite liquid range at $T = 0$ or a solid always stable at the lowest temperatures are not excluded (after Lea *et al.*, 1991).

borne out by experiment, the changes to be made in Fig. 9.3 are straightforward.) Since there is no physical reason for an infinite slope of the melting curve at $\nu = 0$, it is clear from Eqs. (9.5) and (9.6) that, with ΔS not equal to zero, ΔM must approach zero at least as rapidly as ν^2 along the melting curve, at constant density.

Figure 9.3 has been labeled with the points corresponding to $\Delta M = 0$, $\Delta E = 0$, and $\Delta S = 0$. The arrows show the directions in which S, M, and E increase across the melting curve. For phase C1, the arrows are completely determined by the condition $\Delta S > 0$ for the classical crystal at $\nu = 0$, which also implies $\Delta M < 0$ and $\Delta E > 0$ in the same limit. In this picture $\Delta S = 0$ occurs only at $T = 0$ as required by the third law of thermodynamics, which implies that the melting curve approaches the axis with a vertical slope. The arrows in the other regions, C2, C3, and C4 follow from either (i), taking $\Delta S > 0$ at the highest temperatures in each solid phase, or (ii), assuming that the thermodynamic functions behave in a similar way each time the phase diagram approaches $\nu = 0$.

The simplest conceptual picture that emerges from this phase diagram is that the four separate solid-phase regions are essentially the same Wigner solid (presumably a triangular crystal with one electron per lattice point on a two-dimensional hexagonal Bravais lattice) with regions near $\nu = \frac{1}{5}$, $\frac{1}{7}$, and $\frac{1}{9}$ where the liquid phase has a lower free energy. It is conceivable that the four solid phases are different, but this seems somewhat unlikely, particularly since the ν values indicate that the change in ordering is occurring in the liquid. There have been many theoretical calculations of the ground-state energies of the Wigner solid and the fractional quantum Hall effect liquid states (see the review by Isihara, 1989). There is a consensus that the solid has lower energy at very small values of ν but that the FQHE* states with $\nu = 1/q$, as introduced by Laughlin (1983), have lower energies above some critical $\nu_c \sim 1/6.5$ (Lam and Girvin, 1984). The energy of the FQHE states also has cusps at $\nu = 1/q$ with the limiting slopes being related to the energy gap of the quasiparticle excitations in the liquid (Halperin *et al.*, 1984).

Figure 9.3 shows that the entropy change on melting, ΔS, is greater than zero on either side of the liquid state at ν_q. Let us assume that, at low temperatures

$$\Delta S = A(\nu)T^{\alpha}, \tag{9.12}$$

where $A(\nu)$ and α are positive constants which may vary strongly around ν_q. (The entropy from the thermal excitations from the ground state at ν_q of the form $\exp(-\Delta/kT)$, where Δ is the quasiparticle energy gap, will be negligible near $T = 0$.) This entropy could come from low-lying excitations in the liquid or from quasiparticle excitations related to $|\nu - \nu_q|$ (which might give $\alpha = 0$). Then, on the melting curve

$$T_m\,\Delta S = \Delta E - H\,\Delta M = \Delta\mathcal{E} = A(\nu)T_m^{\alpha+1}. \tag{9.13}$$

In Eq. (9.13), $\Delta\mathcal{E}$, the free-energy difference at $T = 0$, has the re-entrant cusp minima following the melting curve (Fig. 9.3). Since $\mathcal{E}_S$ is a monotonically increasing function of ν, this feature is entirely due to the electron liquid, consistent with the theoretical arguments. The magnetization of the liquid, M_L, will be discontinuous at a cusp in the energy at $\nu = 1/q$ and could change sign (see Isihara, 1989). If the change in magnetization on melting is approximately M_L, then the relative signs of ΔM agree with the conclusions of Fig. 9.3, as sketched in Fig. 9.4, assuming the transitions occur each side of ν_q. This field dependence of ΔM is reminiscent of the de Haas–van Alphen effect at integral ν values, suggesting that the magnetism in the electron liquid phase is intimately connected with the exotic variation of ΔM shown here. The conclusion here (see Lea *et al.*, 1992) is that, for a finite width of re-entrant liquid phase at elevated temperature, the magnetization and entropy are properties of the liquid, connected by Eq. (9.13) on the melting line. However, as T decreases toward zero, the entropy must go to zero. If this occurs by some further ordering among the excitations, then ΔM may also go to zero at the same time. If $\Delta M/\Delta S$ remained constant in this process, then $\partial t_m/\partial\nu$ would approach the axis with finite slope, from Eq. (9.6). That the orbital magnetism of the strongly correlated electron liquid could be consistent with the anyon model (see Wilczek, 1990) is demonstrated later, following Lea *et al.* (1992). The presence of minima in the free energy could also lead to the possibility of other phenomena, such as phase separation. However, the relative energies of

*Fractional quantum Hall effect.

the different phases are small compared with the total Coulomb energy of the electron sheet, so the free-energy versus density plot will probably exhibit only a small cusp with no minima at ν_q.

9.5. PHENOMENOLOGICAL AND MICROSCOPIC MODELS OF PHASE DIAGRAM OF TWO-DIMENSIONAL WIGNER SOLID: ANYONS

The appropriate generalization (9.6) of the Clausius–Clapeyron equation to describe the first-order melting transition of two-dimensional Wigner electron solids will be used (a) within a phenomenological framework and (b) in conjunction with microscopic theories relating to anyons and composite fermions, to represent the main features of the melting curve as a function of Landau level filling factor ν, following the work of Lea *et al.* (1992).

9.5.1. Phenomenological Model

It seems natural enough, in view of the zeros in ΔM already referred to, at temperature T_3 and $\nu = \nu_3$ in region C_3, say, of Fig. 9.3, to express ΔM as an expansion in $\nu^{-1} - \nu_3^{-1}$. Assuming that the entropy difference along the melting curve is a maximum near ν_3 and falls to zero as T_m goes to zero, one can write Eq. (9.6) as

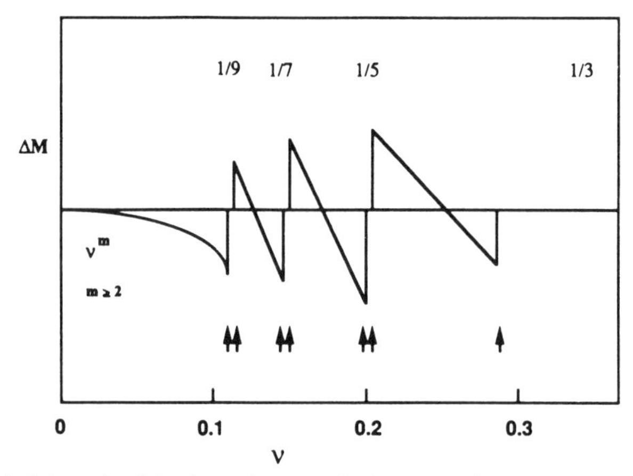

FIGURE 9.4. Schematic of the change in magnetization on melting along the melting curve shown in Fig. 9.3. The arrows mark the liquid–solid transitions (after Lea *et al.*, 1991).

$$\frac{\partial T_m}{\partial \nu} \propto \frac{H}{\nu} \frac{\nu^{-1} - \nu_3^{-1}}{1 - \alpha(\nu^{-1} - \nu_3^{-1})^2}, \tag{9.14}$$

where α is a constant close to unity. Equation (9.14) can now be integrated to give

$$T_m(\nu) = T_3(1 + C \ln(1 - \alpha(\nu^{-1} - \nu_3^{-1})^2)), \tag{9.15}$$

where the parameters C and α must be determined experimentally. The form of $T_m(\nu)$ given by Eq. (9.15) is shown in Fig. 9.5b for the C_3 phase. Further terms in the expansion in $\nu^{-1} - \nu_3^{-1}$ could be included, which would change somewhat the detailed shape of the phase boundaries. This analysis should apply between any two re-entrant phases.

9.5.2. Anyon Model: Classical Limit

Having established a primitive phenomenological form appropriate to represent regions C_2–C_4 in Fig. 9.3, let us now turn to a model that can give further insight into the microscopic origin of the phase diagram. This is the anyon model in which one has fractional statistics, as discussed fully by Wilczek (1990). The

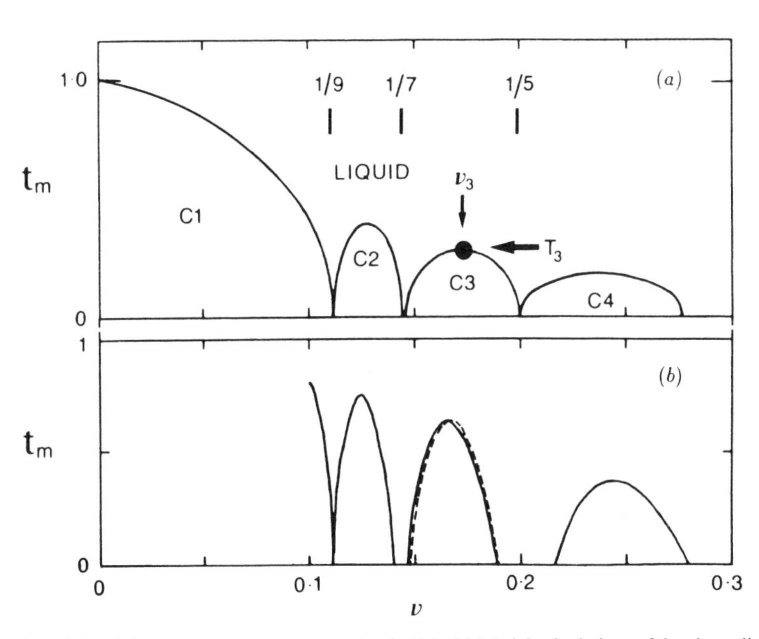

FIGURE 9.5. (a) Schematic phase diagram as in Fig. 9.3. (b) Model calculations of the phase diagram. The solid lines show a model based on Eqs. (9.21), (9.22), and (9.23), as described in the text. The dashed line in phase C3 only shows the schematic form given by the phenomenological equation (9.15) (after Lea *et al.*, 1992).

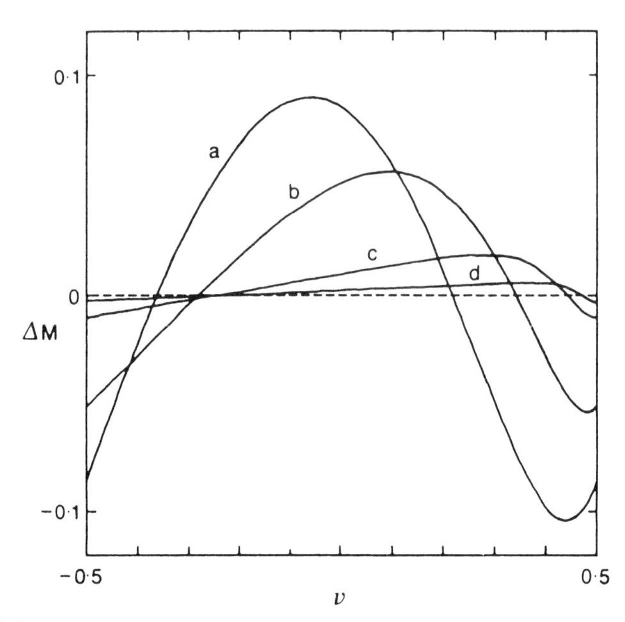

FIGURE 9.6. Magnetism ΔM of an anyon gas versus fractional statistics parameter γ. Different curves (a–d) correspond to values of parameter x $(= h\omega_c/2kT) = 1, 2, 5$, and 10, respectively (after Lea *et al.*, 1992; cf. March, 1993).

anyons are characterized by a parameter γ that is 0 for fermions and $\pm\frac{1}{2}$ for bosons.

A dilute (nondegenerate) gas of noninteracting two-dimensional anyons, having mass m, in a magnetic field has been studied by Johnson and Canright (1990) and by Dowker and Chang (1990). The second virial coefficient B_2 in the equation of state

$$P/nk_BT = 1 + nB_2(T) + O(n^2) \tag{9.16}$$

is

$$B_2(T) = \frac{\lambda^2}{x}\left(\gamma - \frac{\exp(4\gamma x)}{2\sinh(2x)} + \frac{1}{4\tanh(x)}\right), \tag{9.17}$$

where λ is the de Broglie thermal wavelength given by $\lambda^2 = h^2/2\pi mkT$, $x = \hbar\omega_c 2kT$, and ω_c is the cyclotron frequency. Hence, the differential magnetization per unit area of the anyons relative to a classical gas is

$$\Delta M_a(\gamma) = -\frac{n^2}{2m}\frac{\partial B_2}{\partial x}. \tag{9.18}$$

This is shown in Fig. 9.6 as a function of γ for $x = 1, 2, 5$, and 10. Note that the magnetization is not symmetric about $\gamma = 0$, but does have the same value for $\gamma = \pm \frac{1}{2}$.

To relate Eq. (9.18) to a field-dependent magnetization, one must relate the statistics parameter γ to ν. To gain orientation, let us refer to Fig. 9.3. Taking the admittedly schematic form there literally, one notes that $T_m = 0$ at values $\nu = 1/q$, where $q = 9$, 7, and 5. At these very points, microscopic theory predicts a Bose condensate (Lee, 1990), though this is immediately unstable against raising the temperature. However, this motivates the assumption that at $\nu = 1/q$ the value of γ corresponds to the boson value. Hence, as the number of flux quanta per electron, $1/\nu$, increases, the particles alternate between fermions and bosons. Turning to the maxima in the melting curve in the (ν, T_m) plane, one has $\Delta M = 0$, and since the Wigner electron crystal is built from fermions, one expects that the first-order melting transition described by the analogue, Eq. (9.6), of the Clausius–Clapeyron equation will degenerate at these points to a second-order transition, since $\Delta M = 0$ there. Hence, one assumes that the anyons are fermions at the maxima in the melting curve. The assumption that γ is a continuous variable between these limits is related to a mean-field approximation (Johnson and Canright, 1990). If one assumes that for nonintegral values of q the particles are anyons with fractional statistics, then the magnetization will be field dependent from Eq. (9.4).

These ideas can be subsumed in thc rclation

$$\gamma = \frac{1}{2\nu} - j, \tag{9.19}$$

where the integer j is chosen so that γ varies from $-\frac{1}{2}$ to $+\frac{1}{2}$ as ν decreases. This assumption allows plots of the field-dependent ΔM_a versus ν of an anyon gas for various parameter values, from Eqs. (9.16) and (9.17), as shown in Fig. 9.7. The plot is calculated for $x = A/\nu$, which corresponds to a variable field at a fixed

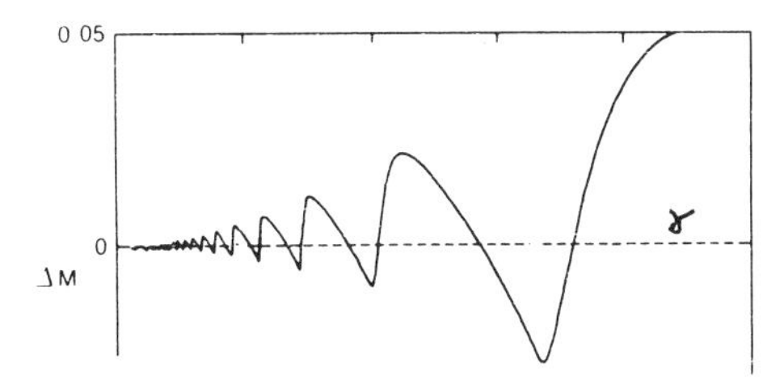

FIGURE 9.7. Magnetization ΔM of the electron liquid as a function of ν: anyon model (after Lea *et al.*, 1992).

temperature, for the arbitrary value $A = 1$. It can be seen that the magnetization of this noninteracting anyon gas has many of the features required to explain qualitatively the phase diagram of the two-dimensional Wigner crystal. In particular, ΔM_a changes sign at $\nu = 1/q$ as required to explain the re-entrant nature of the solid regions C_2, C_3, and C_4. It also predicts that there may be further structure in the phase diagram of the region C_1, at $\nu = 1/q$, where q is an odd integer. It appears likely that the purely statistical effects described here will persist in the presence of interactions and that similar magnetization changes will also occur at the FQHE states, at integral fractions.

While, in view of the simplified nature of the assumption, Eq. (9.19), it does not seem appropriate to press the detail quantitatively, it seems remarkable that the anyon model does indeed extend the de Haas–van Alphen–type singularities referred to by Lea, March, and Sung (1992) into the region $\nu \ll 1$.

9.5.3. Low-Temperature Treatment: Composite Fermion Model

A closely related approach at low temperatures, in contrast to the classical limit, is to use the idea of "composite" fermions (Jain, 1989, 1991; Trivedi and Jain, 1991) in which a fermion plus an even number of flux quanta is also a fermion. These composite fermions are then regarded as separate entities in the magnetic field of the remaining flux lines, with a new filling factor ν_i given by

$$\nu_i = \frac{\nu}{1 - 2i\nu} = \frac{p}{q - 2ip}, \tag{9.20}$$

where $2i$ is number of flux quanta associated with each composite fermion. The second part of Eq. (9.20) gives ν_i for the fractional quantum Hall states with $\nu = p/q$. For each of these states a value of i can be chosen so that ν_i is an integer. The sign of ν depends on whether the flux in the composite fermion is parallel or antiparallel to the applied field. For the FQHE states with $p = 1$, then $\nu_i = 1$ with $i = 1, 2, 3, 4, 5$, for $q = 3, 5, 7, 9, 11$, and so on. This idea has been successfully used by Jain (1991) to map the fractional quantum Hall effect states to equivalent integer quantum Hall effect states (see also Appendix 9.8).

9.5.3.1. Entropy of the Composite Fermions

If the entropy $S(\nu)$ of the liquid goes to zero at the fractional filling factors, then one can relate the entropy to the equivalent integer state based on ν_i. Using the results for the de Haas–van Alphen effect, one has the entropy per electron $S(\nu)$ for a partially filled level:

$$S(\nu) = \frac{-k}{\nu_i} \{\nu_i \ln \nu_i + (1 - \nu_i)\ln(1 - \nu_i)\} \tag{9.21}$$

for $\nu_i \le 1$ and equivalent forms for $\nu_i > 1$, where ν_i is given by Eq. (9.20). The entropy of these fermions for $i = 1$, 2, and 3 is shown in Fig. 9.8 for $0.1 < \nu < 0.35$ and goes to zero at each ordered state of the fractional quantum Hall effect. This entropy expression will only be valid insofar as the system behaves as an assembly of weakly interacting composite fermions. Note, however, that the entropy close to the $\nu = \frac{1}{5}$ state, for instance, has the same form for $i = 2$ and $i = 3$. The functional form of $S(\nu)$ between two neighboring states is close to the phenomenological expression used in Section 9.5.1. According to the third law of thermodynamics, the entropy $S(\nu)$ away from the ordered states must also go to zero in the very low temperature limit. This could occur by further ordering among the excitations or, in real systems, by some form of localization due to random potentials, as in the integer quantum Hall effect, which could also lead to the observed Hall plateaus. The effect of these localizing potentials would presumably also lead to a pinned Wigner crystal as the solid ground state, as also observed experimentally (Glattli *et al.*, 1990).

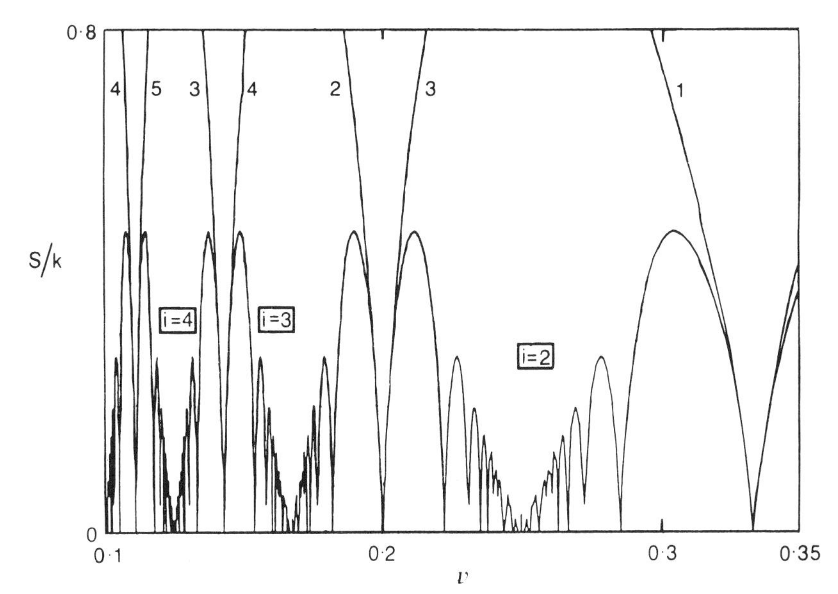

FIGURE 9.8. The entropy $S(\nu)/k$ per electron for the composite fermion model, calculated using Eq. (9.). The various lines are marked by the value of i, the number of flux quanta pairs in each composite fermion. Note that $S(\nu)$ goes to zero at each ordered state $\nu = p/q$ of the fractional quantum Hall effect (see also Appendix 9.1) (after Lea *et al.*, 1992).

9.5.3.2. Energy of the Electron Liquid

The internal energies (including the magnetic potential energy term $-HM$, which is different from the definition employed by Lea, Sung, and March (1992) of the electron liquid and solid, have been calculated by several authors (see Isihara, 1989, for a review). For $\nu \leq 0.1$ the solid is thought to have the lower free energy at $T = 0$ and is then the ground state. The Laughlin liquid states (see also Appendix 9.8) have the lower ground-state energies for $\nu = 1/q$, where $q = 3, 5, 7$, and probably 9. If the temperature T of one of these states is increased, then quasiparticle excitations are thermally excited across an energy gap Δ. Tao (1990) has shown that this energy gap is field dependent and must have the form

$$\Delta = \Delta_\nu \frac{e^2}{l}; \qquad \frac{\Delta}{kT_{mc}} = \Gamma_m \Delta_\nu \frac{\sqrt{2}}{\nu},$$

where $l = (\hbar c/eH)^{1/2}$ is the magnetic length, and Δ_ν is a dimensionless parameter to be calculated or found from experiment. For $\nu = 1/q = \frac{1}{3}$, $\Delta_\nu = 0.018$ (Isihara, 1989). The second part of the equation gives the energy gap in units of the transition temperature of the classical electron gas: $T_{mc} = e^2(\pi n)^{1/2}/k\Gamma_m$, where $\Gamma_m = 127$ (Deville, 1988). If the magnetic field is increased or decreased, then quasiparticle or quasihole excitations will be generated with energies ε_+ and ε_-, respectively, where $\varepsilon_+ + \varepsilon_- = \Delta$ since an increase in temperature creates neutral pairs of excitations. If an excitation is equivalent to an extra or missing flux line, then one can write the energy per electron near $\nu = \nu_q = 1/q$ as

$$E(\nu) = E_q + \frac{|(\nu - \nu_q)|}{\nu_q} q\varepsilon_\pm, \tag{9.22}$$

where E_q is the ground-state energy at $\nu = 1/q$. Hence, the energy $E(\nu)$ will have cusps at $\nu = 1/q$, as shown in Fig. 9.9.

9.5.3.3. Magnetization of the Electron Liquid

The magnetization $M = -(\partial F/\partial H)_T$, where F is the free energy $F = E - TS$. At zero temperature, a cusp in the energy near ν_q, for instance, will give a jump in the magnetization per electron of magnitude $q(\varepsilon_+ + \varepsilon_-)/H = q\Delta/H$ as ν passes through the ordered state. Hence, the magnetization may change sign as ν passes through each of the ordered states indicated by $S = 0$ in Fig. 9.8. The schematic variation of M with ν will therefore be very similar in structure to the diagram for the anyon gas in Fig. 9.7, though many more states may be revealed.

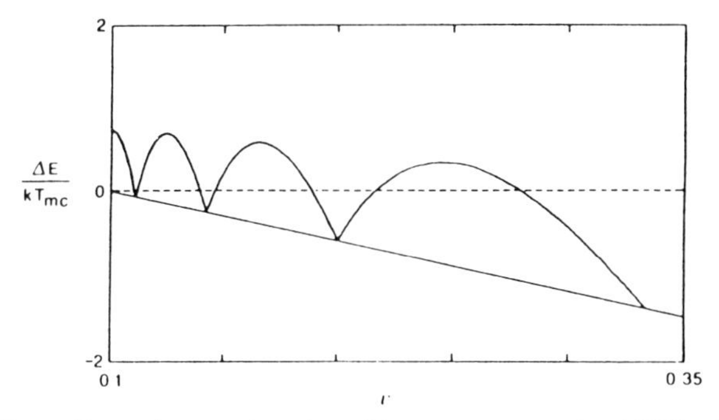

FIGURE 9.9. Form of internal energy along the melting curve, expressed as $\Delta E/kT_{mc}$. Gradients at the displayed cusps in ΔE are known in terms of excitation energies and the energy gaps Δ_ν. This information is used asymptotically in this figure at $\nu = 1/q$. The turning points shown in ΔE versus ν are as required by the thermodynamic arguments (see main text and Lea *et al.*, 1991) (after Lea *et al.*, 1992).

9.5.3.4. The Liquid–Solid Phase Diagram

Let us now return to the discussion of the field dependence of the electron–solid phase boundary. If the liquid phases at $\nu = \frac{1}{3}, \frac{1}{5}, \frac{1}{7}$, and $\frac{1}{9}$ do have lower internal energy than the solid, they will form the ground state at zero temperature. For $\nu < \nu_q$ the energy will increase as excitations are created until $E_S = E_L$ (the subscripts refer to the solid and liquid phases). At this point there will be a phase transition

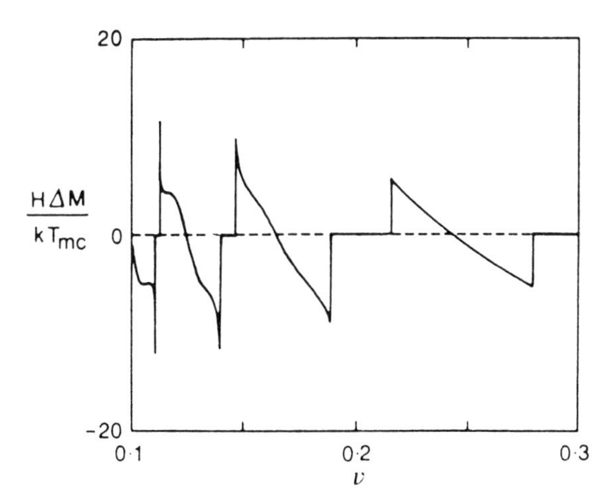

FIGURE 9.10. The magnetization change ΔM along the melting curve, expressed as $H\,\Delta M/kT_{mc}$ (after Lea *et al.*, 1992).

to the solid phase. Hence, the phase diagram will indeed contain re-entrant liquid phases as proposed by Buhmann *et al.* (1991). The experimental evidence indicates that these occur at $\nu = \frac{1}{5}, \frac{1}{7}$, and $\frac{1}{9}$, while Plaut *et al.* (1990) found that the highest value of ν for which the solid was present was $\nu = 0.28 \pm 0.02$. However, it must be noted that in the original results of Andrei *et al.* (1988) a possible solid phase was found for $\nu \approx 0.34$. Glattli *et al.* (1990) also reported a liquid phase close to $\nu =$, which would correspond to $i = 2$, $\nu_i = 1$ in Eq. (9.20). One can use the experimental value of Plaut *et al.* (1990) to determine $\Delta E = E_L - E_S$ at $\nu = \frac{1}{3}$ by assuming $\varepsilon_+ = \varepsilon_- = \Delta/2$ with $\Delta_\nu = 0.009$ (Isihara, 1989) in Eq. (9.22) so that $\Delta E = 0$ at $\nu = 0.28$. If one also assumes that ΔE at the ground states $\nu = \nu_q$ decreases in magnitude until the solid phase becomes the only stable phase for $\nu > 0.1$ (see Isihara, 1989), one can then sketch $\Delta E(\nu)$ along the melting curve, expressed in units of $k_B T_{mc}$ as shown in Fig. 9.9. The gradients at the cusps in ΔE are determined by the excitation energies and the energy gaps Δ_ν. Ordered states at other fractional values of ν may also produce cusps in ΔE, but the energy gaps may well be smaller. Only if $\Delta E = E_L - E_S < 0$ will a re-entrant phase occur.

The phase boundary for $T > 0$ can, in principle, be found by equating the free energies $F_S = F_L$ and hence

$$T_m = \frac{E_L - E_S}{S_L - S_S} = \frac{\Delta E}{\Delta S}, \tag{9.23}$$

where ΔE and ΔS are evaluated along the melting curve. For the purpose of the ensuing discussion (see Lea *et al.*, 1992), let us assume that the entropy of the solid is small compared with the liquid and hence that $\Delta S \approx S_L$, the entropy of the liquid. If $\Delta E > 0$, but S_L decreases near a fractional state, then T_m would increase rapidly and the neglect of the solid entropy would no longer be a good approximation. Indeed if the solid entropy becomes greater than S_L while $\Delta E > 0$, an increase in temperature will stabilize the solid phase. However, along the melting curve, which is the region of interest here, thermal excitations will probably smooth out the variations in S_L shown in Fig. 9.8, and any small cusps in E. To demonstrate the origin of the observed melting curve, the internal energy shown in Fig. 9.9 has therefore been used and an entropy variation that asymptotically approaches the calculated entropy, Eq. (9.21), for values of ν close to the liquid ground states. The energy gaps have been taken to increase with ν as $\Delta_\nu = 0.054\nu$. The resultant phase diagram, as calculated from Eq. (9.23), is shown in Fig. 9.5b for $\nu > 0.1$. This does indeed have some of the features of the melting curve postulated from the experimental results, though further structure may also occur in the phase diagram at other ordered states.

Let us also assume that the magnetization difference between the two phases is dominated by the field-dependent magnetization discussed and that $\Delta M \approx M_L$, the magnetization of the liquid phase. Introducing $t_m = T_m/T_{mc}$, one can then write Eq. (9.6) as

$$\frac{\partial t_m}{\partial \nu} = \pm \frac{2^{1/2}\Gamma_m \Delta_\nu}{\nu^{3/2} S(\nu)} \tag{9.24}$$

close to the liquid ground state. The important property that $\partial t_m/\partial \nu$ is a function only of ν is to be stressed, as required by scaling arguments (Tao, 1990). All the thermodynamic functions should depend only on ν and $t = T/T_{mc}$ in this region of the (t,ν) plane. The magnetization change along the melting curve $\Delta M = -(\partial \Delta F/\partial H)_T$ as calculated from this simple model is shown in Fig. 9.10, and shows the features that were derived from thermodynamic arguments by Lea *et al.* (1991).

To recap:

(i) A phenomenological model can be set up (Lea *et al.*, 1992) for which the analogue, Eq. (9.6), of the Clausius–Clapeyron equation for the melting curve of the Wigner crystal can be integrated. The simplest model of the phase boundaries of the phases C_2–C_4 in Fig. 9.3 gives a re-entrant phase diagram.

(ii) The magnetization of the two closely related microscopic models can be calculated: (a) an anyon gas model in the classical limit and (b) a composite Fermion model, which complements (a) in that the magnetization and entropy are determined at low temperatures.

While the models to date do not allow a fully quantitative prediction of the melting curve of the two-dimensional Wigner electron crystal, the main features of the magnetization of the electron liquid are reproduced by the models (a) and (b). The work of Lea *et al.* (1992) supports the usefulness of the anyon concept in magnetic fields, though further work needs to be done to relate this model to the composite fermion picture. The two models appear, at the time of writing, to be complementary: one being readily calculable in the classical limit, while the other is most tractable at low temperatures. Both models, in the end, relate to an electron associated with an integral number of flux lines, and the decomposition into "fermions plus additional flux lines" as opposed to "anyons" may be a matter of semantics, though as yet one can reach no decisive conclusion on this point. The final point is that, while the aforementioned work supports the model for anyons in applied magnetic fields, no conclusions are to be drawn from all this as to the usefulness of the model of field-free anyons in the context of high-temperature superconductivity. However, the anyon model in the context of this chapter is clearly of sufficient interest to motivate discussion of the statistical distribution function for these fractional spin particles.

9.6. STATISTICAL DISTRIBUTION FUNCTION FOR ANYONS

Particles with fractional statistics, termed anyons by Wilczek (1990), are of considerable current interest in connection with (a) the fractional quantum Hall effect, (b) the phase diagram of two-dimensional Wigner crystals in strong magnetic fields as discussed earlier, and (c) high-temperature superconductivity (Laughlin, 1989).

Though treatments of the thermodynamics of an anyon gas have been given (see Wilczek, 1990) our concern below is with the statistical distribution function $f(\varepsilon)$ for states of energy ε in such a gas. The purpose of this section is twofold. In Section 9.6.1 a form of $f(\varepsilon)$ is proposed for fractional statistics. The argument is based on collision theory, invoking detailed balance and a modification of time-reversal symmetry, which is necessary for anyons. Then in Section 9.6.2 a direct derivation of the same result is given, starting from the Hamiltonian of an anyon gas (Lea *et al.*, 1993).

9.6.1. Collision Theory: Motivation

Let us start from the treatment of collisions in a gas of fermions following, for example, Ma (1985). Suppose we have a collision $1 + 2 \leftrightarrow 3 + 4$; i.e., states 1 and 2 interact to change to states 3 and 4. We note that collisions are necessary to allow thermodynamic equilibrium to be attained, even in an ideal gas. Then, following Ma, this reaction has the rate (compare March, 1993)

$$f_1 f_2 (1 - f_3)(1 - f_4) R,$$

where $1 - f_3$ and $1 - f_4$ are the probabilities that there are no particles in states 3 and 4. These factors must be present for fermions because if states 3, 4 are occupied then the reaction cannot occur. Now one invokes the fact that this rate must equal that in the reverse direction, i.e.,

$$f_1 f_2 (1 - f_3)(1 - f_4)\, R = = f_3 f_4 (1 - f_1)(1 - f_2) R'. \tag{9.25}$$

Ma next uses time-reversal symmetry of quantum mechanics to show that $R = R'$ and this leads to the form [see Ma, Eq. (3.21)]

$$\frac{f(\varepsilon)}{1 - f(\varepsilon)} = \exp(-\alpha - \beta\varepsilon). \tag{9.26}$$

We next note that, for bosons, the factors $1 - f_3$ and $1 - f_4$ in Eq. (9.25) become $1 + f_3$ and $1 + f_4$: it is as though, if states f_3 and f_4 are already occupied, other states would be more inclined to go to them.

Turning to fractional statistics, the interchange of two particles introduces a phase factor of the form $\exp(2\pi i\gamma)$, where we choose the sign such that $\gamma = 0$ for fermions and $\gamma = \frac{1}{2}$ for bosons. The main assumption on which the generalized collision theory presented here is based is that this reduces the tendency $1 + f(\varepsilon)$ for particles to like the same state for bosons to $1 - a(\gamma) f(\varepsilon)$, where the anyon factor $a(\gamma)$ will be taken to have the properties

$$a(0) = 1, \qquad a(\tfrac{1}{2}) = -1, \tag{9.27}$$

$$-1 \le a(\gamma) \le 1. \tag{9.28}$$

Obviously the "boundary conditions" specified in Eq. (9.27) for $\gamma = 0$ and $\frac{1}{2}$ ensure the correct Fermi–Dirac and Bose–Einstein limits.

Then, whereas Ma rewrites Eq. (9.26) for bosons and fermions as

$$\frac{f(\varepsilon)}{1 \pm f(\varepsilon)} = \exp(-\alpha - \beta\varepsilon), \tag{9.29}$$

one now has for anyons

$$\frac{f(\varepsilon)}{1 - a(\gamma)f(\varepsilon)} = \exp(-\alpha - \beta\varepsilon) \tag{9.30}$$

or, with these simple assumptions*

$$\frac{1}{f(\varepsilon)} = \exp(\alpha + \beta\varepsilon) + a(\gamma). \tag{9.31}$$

Equation (9.31) represents the shape of the statistical distribution function for anyons. While the argument of this section (see, however, Section 9.6.2) does not contain within itself a unique procedure for determining the anyon shape factor, the fractional statistics phase factor $\exp(2\pi i\gamma)$ prompts us to assume that $a(\gamma)$ will be expressible as an appropriate Fourier series compatible with condition (9.28). Then in the absence of further information at this stage, use of Occam's razor leads us to propose

$$a(\gamma) = \cos(2\pi\gamma). \tag{9.32}$$

Thus, by combining Eqs. (9.31) and (9.32), one has the result

$$f(\varepsilon) = \frac{1}{\exp(\alpha+\beta\varepsilon) + \cos(2\pi\gamma)}. \tag{9.33}$$

One notes immediately that with the special form (9.33) arising from the shape of $f(\varepsilon)$ in Eq. (9.31), the particular case $\gamma = \frac{1}{4}$ leads back precisely to Boltzmann statistics. Figure 9.11 shows the distribution function $f(\varepsilon)$ for $\gamma = 0$, $\frac{1}{4}$, and $\frac{1}{3}$, giving $a = 1$, 0.5, 0, and -0.5 respectively. In each case the distribution has been normalized to unity, for $\beta = 5$, by choosing the value of α.

9.6.2. Derivation of Occupation Number from Hamiltonian of Free Anyon Gas

From the physical argument, we now proceed to the derivation of exact results by using the Hamiltonian of the anyon gas with statistical parameter n =

*Note added in proof. These assumptions have now been transcended: see e.g. D. Sen and R. K. Bhaduri (Phys. Rev. Lett. *74*, 3912, 1995) and other references there.

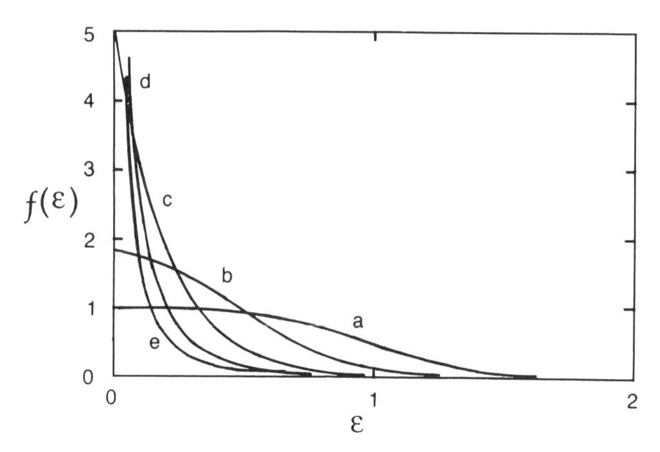

FIGURE 9.11. Statistical distribution function $f(\varepsilon)$ for anyon gas. Different curves correspond to different values of fractional statistics parameter. $f(\varepsilon)$ contains Bose and Fermi–Dirac distribution functions as limiting cases (after March *et al.*, 1993).

$\frac{1}{2}\gamma$ and transformations similar to those employed by Mori (1990). Explicitly the Hamiltonian H takes the form

$$H = \tfrac{1}{2}m \int d^2\mathbf{r}\ \psi^+(r)\ [\mathbf{p} - \mathbf{A}(\mathbf{r})]^2\psi(\mathbf{r}). \tag{9.34}$$

Here $\psi(\mathbf{r})$, $\psi^+(\mathbf{r})$ are the fermion field operators, and $\mathbf{A}(\mathbf{r})$ represents the vector potential

$$\mathbf{A}(\mathbf{r}) = \int d^2\mathbf{r}'\ \mathbf{K}(\mathbf{r}-\mathbf{r}')\rho(\mathbf{r}'), \tag{9.35}$$

where $\mathbf{K}(\mathbf{r}) = (1/n)\text{grad}\,[\theta(\mathbf{r})]$ and $\theta(\mathbf{r})$ is the azimuthal angle of the two-dimensional vector $\mathbf{r}$ when $r > a$ and zero otherwise. After integration the limit a tends to zero is to be taken.

The next step is to employ the operator

$$U(\mathbf{r}) = \exp\left([i/n]\int d^2\mathbf{r}'\ \theta(\mathbf{r} - \mathbf{r}')\rho(\mathbf{r}')\right). \tag{9.36}$$

Here note that if $U(\mathbf{r}')$ is obtained from $U(\mathbf{r})$ after rotation by 2π about the z axis, then $\theta(\mathbf{r}') = \theta(\mathbf{r}) + 2\pi$ and $U(\mathbf{r}') = \exp(2\pi i N/n)U(\mathbf{r})$, where $N = \int d^2\mathbf{r}\ \psi^+(\mathbf{r})\psi(\mathbf{r})$. Following Mori, one then defines the anyon field operators by

$$\tilde{\psi}^+(\mathbf{r}) = \psi^+(\mathbf{r})U(\mathbf{r}), \qquad \tilde{\psi}(\mathbf{r}) = U^+(\mathbf{r})\psi(\mathbf{r}). \tag{9.37}$$

Using standard quantum-mechanical methods, the following equalities can then be derived:

$$U^+(r')\psi^+(r)U(r') = \exp\{[-i/n]\theta(r'-r)\}\psi^+(r), \tag{9.38a}$$

$$U^{+}(r')\psi(r)U(r') = \exp\{(i/n)\theta(r' - r)\}\psi(r), \tag{9.38b}$$

and by explicit calculation the relations concerning the interchange of the field operators follow:

$$\tilde{\psi}^{+}(r)\tilde{\psi}^{+}(r') = -\exp\{[i/n][\theta(r - r') - \theta(r' - r)]\}\tilde{\psi}^{+}(r')\tilde{\psi}^{+}(r), \tag{9.39a}$$

$$\tilde{\psi}(r)\tilde{\psi}(r') = -\exp\{[i/n][\theta(r - r') - \theta(r' - r]\}\tilde{\psi}(r')\tilde{\psi}(r), \tag{9.39b}$$

$$\tilde{\psi}(r)\tilde{\psi}^{+}(r') = \delta(r - r') - \exp\{[-i/n][\theta(r - r') - \theta(r' - r)]\}\tilde{\psi}^{+}(r')\tilde{\psi}(r). \tag{9.39c}$$

These relations coincide with those of Mori when the position variables **r**, **r**′ correspond to z, z' on the same Riemann sheet. However, this is not always the case, as these operators by definitions (9.37) are uniquely defined on the n-fold Riemann sheet, whereas they are multivalued functions in the two-dimensional plane. Thus, inconsistencies arise when this is not taken into account.

The Hamiltonian expressed in terms of these operators is reduced to a free-anyon Hamiltonian.

We define the anyon annihilation and creation operators:

$$a_k = \frac{1}{\sqrt{V}} \int d^2r \exp(i\mathbf{k}\cdot\mathbf{r})\psi(r), \tag{9.40a}$$

$$a_k^{+} = [a_k]^{+}, \tag{9.40b}$$

where V is the volume in two dimensions and integration is only in the first Riemann sheet. Definition on the mth Riemann sheet differs by the phase $\exp\{i(m - 1)2\pi\mathcal{N}/n\}$.

By using the preceding relations, one obtains the thermal average of the anyon occupation number $n_k = a_k^{+}a_k$ by following the standard procedure used for fermion operators (see, e.g., Kittel, 1963). One also needs the relation

$$a_k a_{k'}^{+} = \delta_{k,k'} - \cos(\pi/n)a_{k'}^{+}a_k + \sum Q(q)a_{k'+q}^{+}a_{k+q}, \tag{9.41}$$

where

$$Q(q) = 4/V\int d^2r \sin(2q\cdot r)\sin\{[\theta(r) - \theta(-r)]/n\}. \tag{9.42}$$

Note that the phase difference $\theta(r) - \theta(r')$ is invariant under rotations, although its value is $+\pi$ or $-\pi$. Then one can prove that $Q(q)$ vanishes, and the final result is

$$\langle n_k \rangle_T = \{\cos(\pi/n) + \exp[\beta(k^2/2m - \mu)]\}^{-1}. \tag{9.43}$$

This result is compatible with the axial symmetry of the Hamiltonian; i.e., $\langle n_k \rangle_T$ is independent of the orientation of k.

In conclusion, we emphasize for plane waves, corresponding to ε in Eq. (9.9) equal to $k^2/2m$, that Eq. (9.19) leads back to the form (9.9) with $a(\gamma) = \cos(2\pi\gamma)$, obtained earlier by physical arguments. In connection with the argument of Section 9.6.1, note that, although time-reversal symmetry is not valid for anyons, when combined with parity it leaves the Hamiltonian of anyons invariant. Returning to Eq. (9.25), it then follows that $R = R'$.

The approximations of this last section give Eq. (9.31) for the statistical distribution function for anyons. Of course, this contains the Fermi–Dirac and Bose–Einstein distribution functions in the appropriate limits. Applications of the anyon distribution function have barely commenced as yet; progress is to be anticipated in this area in the foreseeable future (see footnote on p. 271).

Appendix to Chapter 2

A2.1. DENSITY MATRICES FOR MANY-PARTICLE OSCILLATOR MODEL

The purpose of this Appendix is to collect the results of the calculation of the first- and second-order density matrices by Cohen and Lee (1985) for the N-boson model based on an oscillation Hamiltonian. With their exact ground-state wave function, the reduced density matrices have been obtained as follows (ω^2 is a spring constant: δN is for normalization)

$$\rho_s = \binom{N}{s}\left[\frac{\delta N}{\pi}\right]^{3/2}\left\{\frac{N\omega}{(N-s)\omega + s\delta N}\right\}^{3/2}$$

$$\exp\left\{-\frac{1}{2N}[\omega + (N-1)\delta N]\sum_{i=1}^{s}(r_i^2 + r_i'^2) - \frac{1}{N}\sum_{i<j}^{s}(\mathbf{r}_i\cdot\mathbf{r}_j + r_i'\cdot r_j') - C_x R_s^2\right\} \quad \text{(A2.1.1)}$$

where

$$\mathbf{R}_s = \sum_{i=1}^{s}(\mathbf{r}_1 + \mathbf{r}_1') \quad \text{(A2.1.2)}$$

and

$$C_s = -\frac{1}{4N}\frac{(N-s)(\omega-\delta N)^2}{(N-s)\,\omega + s\delta N}. \quad \text{(A2.1.3)}$$

In particular, the first- and second-order density matrices are (Cohen and Lee, 1985)

$$\rho_1(\mathbf{r}_1,\mathbf{r}_1') = N\left[\frac{\delta N\ N\omega/\pi}{(N-1)\,\omega\ + \delta N}\right]^{3/2} e^{-}\ a_1(r_1^2 + r_1'^2) + a_2\mathbf{r}_1\ \mathbf{r}'_1 \quad \text{(A2.1.4)}$$

and

$$\rho_2(r_1, r_2; r_1', r_2') = \frac{N(N-1)}{2}\left[\frac{N\omega\delta^2 N/\pi^2}{(N-2)\omega + 2\delta N}\right]^{3/2} \exp\{-b_1(r_1^2 + r_1'^2 + r_2^2 + r_2'^2)$$

$$- b_2(\mathbf{r}_1\cdot\mathbf{r}_2 + \mathbf{r}_1'\cdot\mathbf{r}_2') + \mathbf{b}_3(\mathbf{r}_1\cdot\mathbf{r}_2' + \mathbf{r}_1'\cdot\mathbf{r}_2' + \mathbf{r}_1\cdot\mathbf{r}_1' + \mathbf{r}_2\cdot\mathbf{r}_2')\}, \quad \text{(A2.1.5)}$$

where

$$a_1 = \frac{1}{4N}\,\frac{(N-1)(\omega^2 + \delta N^2) + 2(N^2 - N + 1)\omega\delta N}{(N-1)\omega + \delta N} \quad \text{(A2.1.6)}$$

$$a_2 = \frac{1}{2N}\,\frac{(N-1)(\omega - \delta N)^2}{(N-1)\omega + \delta N} \quad \text{(A2.1.7)}$$

$$b_1 = \frac{1}{4N}\,\frac{(N-2)\omega^2 + (3N-2)\,\delta N^2 + 2(N^2 - 2N + 2)\omega\delta N}{(N-2)\omega + 2\,\delta N} \quad \text{(A2.1.8)}$$

$$b_2 = \frac{1}{2N}\,\frac{(N-2)\omega^2 - (N+2)\,\delta N^2 + 4\omega\delta N}{(N-2)\omega + 2\,\delta N} \quad \text{(A2.1.9)}$$

$$b_3 = \frac{1}{2N}\,\frac{(N-2)(\omega - \delta N)^2}{(N-2)\omega + 2\,\delta N}, \quad \text{(A2.1.10)}$$

which suffice to determine ρ_1 and ρ_2 completely. While, of course, these results are very restrictive, and true for a symmetric (boson) ground-state wave function, whereas the main interest is, of course, N-fermion problems here, the results of Cohen and Lee suffice to allow a valuable test of the Colle–Salvetti (CS) results (see Chapter 3, Section 3.7). However, other studies on real systems show that the CS correlation energy is often only semiquantitative in molecules (compare Appendix 4.3).

A2.2. ATOMIC ENERGIES FROM RENORMALIZATION OF LARGE-DIMENSIONALITY RESULTS

The purpose of this appendix is to supplement the account in Chapter 2 of the dimensionality dependence of atomic energies. This will be done by summarizing the main findings of Kais *et al.* (1993).

The key feature that these workers exploit is that by working in dimensionality D the many-electron effects tend to have only a relatively mild dependence on dimensionality.

Renormalization using Hartree-Fock Data

As the electronic energy of atoms is a smooth function of dimensionality D and nuclear charge Ze, Kais *et al.* (1993) argue that an effective value, say Z_∞, of

TABLE A2.2.1 Renormalized Charges and Energies[a] (in Atomic Units) from Hartree–Fock Data for Two-Electron Atomic Ions. Adapted from Kais *et al.*, 1993 for ground-state $1s^2\ {}^1S$ $(D = 3)$

Z	ΔZ_∞	$\Delta Z_\infty^{\mathrm{HF}}$	$E_\infty(Z)$	$E_\infty(Z_\infty^{\mathrm{HF}})$	% error
4	0.0449	0.0429	−13.326	−13.640	0.1105
10	0.0425	0.0417	−93.084	−93.891	0.0162
18	0.0419	0.0414	−311.428	−312.89	0.0049
20	0.0418	0.0414	−386.014	−387.64	0.0039

[a]Adapted from Kais *et al.*, 1993. Renormalized charges such that $Z_\infty = Z + \Delta Z_\infty$ and $Z_\infty^{\mathrm{HF}} = Z + \Delta Z_\infty^{\mathrm{HF}}$. Renormalized energy calculated by Kais *et al.* from $D \to \infty$ limit. Percentage error recorded = $100\ [E_D(Z) - E_\infty(Z_\infty^{\mathrm{HF}})]/E_D(Z)$.

the nuclear charge exists such that the large-D limit, which is analytically tractable, becomes equal to some scaled energy for $D = 3$ with the actual value of Z. These workers then approximate Z_∞ by Z_∞^{HF}. The results of calculations on a few light atoms with two electrons only are summarized in Table A2.2.1. Fig. A2.2.1 shows the variation of the percentage error in the ground-state density of such two-electron atoms with the square of the difference $Z_\infty - Z_\infty^{\mathrm{HF}}$ in renormalized nuclear charges as derived, respectively, from the exact energy and from the HF approximation.

Kais *et al.* perform similar studies of neutral atoms for $Z \leq 20$ from HF data. Their conclusion again is that by augmenting approximations such as HF or $1/Z$ expansions with knowledge gained from the large-D limit, one can substantially improve the accuracy of neutral-atom energies.

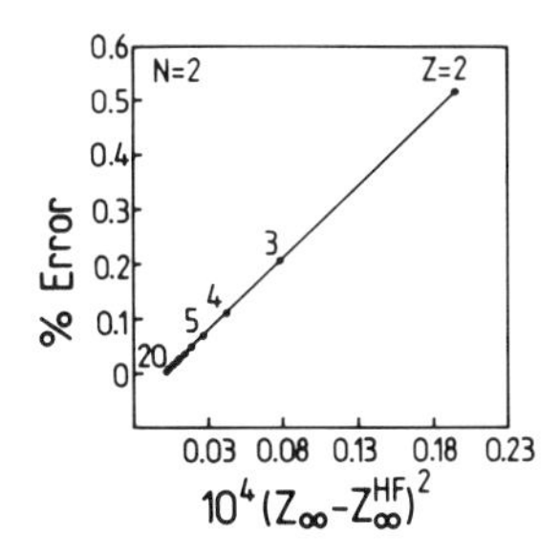

FIGURE A2.2.1. Percentage error in ground-state energy of two-electron atoms vs. square of difference in renormalized nuclear charges (Redrawn from Kais *et al.*, 1993).

A2.3. SIZE-EXTENSIVITY, CUMULANTS, AND COUPLED-CLUSTER EQUATIONS

The energy is a size-extensive quantity, i.e., the energy of two well-separated, but otherwise identical systems equals twice the energy of a single system. Approximations that are applied in calculations of, e.g., the ground-state energy of a system must preserve this property. The argument below follows Schork and Fulde (1992) closely.

From classical statistical mechanics, it is known that size consistency is attained by expressing the size-extensive quantities in terms of cumulants (Ursell, 1927). Kubo (1962) showed how the quantum mechanical expectation value of an operator can be expressed in terms of them. Consider the expectation value with respect to a state $|\phi\rangle$of the product of operators $A_1 \ldots, A_n$. The cumulant (denoted by the superscript c) of this expectation value is defined by

$$\langle\phi|\prod_{i=1}^{n} A_i|\phi\rangle^c = (\prod_i \partial_i)\ln\langle\phi|\prod_{i=1}^{n} e^{\lambda_i A_i}|\phi\rangle|_{\lambda_i=0} \tag{A2.3.1}$$

From this definition, one immediately finds that, e.g.,

$$\begin{aligned}\langle A_1\rangle^c &= \langle A_1\rangle \\ \langle A_1A_2\rangle^c &= \langle A_1A_2\rangle - \langle A_1\rangle\langle A_2\rangle, \\ \langle A_1A_2A_3\rangle^c &= \langle A_1A_2A_3\rangle - \langle A_1\rangle\langle A_2A_3\rangle - \langle A_2\rangle\langle A_1A_3\rangle \\ &\quad - \langle A_3\rangle\langle A_1A_2\rangle + 2\langle A_1\rangle\langle A_2\rangle\langle A_3\rangle,\end{aligned} \tag{A2.3.2}$$

etc. Here $\langle\cdot\cdot\cdot\rangle$ stands for $\langle\phi|\cdot\cdot\cdot|\phi\rangle$.

The ground-state energy E_0 of a system described by a Hamiltonian H can be expressed conveniently in terms of cumulants, implying that all approximations subsequently made are *a priori* size consistent. Usually the ground-state energy can be computed exactly for only a part H_0 of H, while the contribution of the remaining part $H_1 = H - H_0$ must be treated by approximations. Let $|\Phi_0\rangle$ denote the ground state of H_0. Following Becker and Fulde (1988, 1989) the ground-state energy is given by

$$E_0 = (H|\Omega), \tag{A2.3.3}$$

where the abbreviation

$$(A|B) = \langle\Phi_0|A^\dagger B|\Phi_0\rangle^c, \tag{A2.3.4}$$

has been introduced so that E_0 is indeed expressed by a cumulant. The operator Ω is defined by

$$\Omega = \lim_{z\to 0}\left(1 + \frac{1}{z - L_0 - H_1} H_1\right), \tag{A2.3.5}$$

where L_0 is the Liouvillian belonging to H_0. It acts on any operator B according to

$$L_0 B = [H_0, B]_- . \tag{A2.3.6}$$

Ω specifies the exact ground state $|\Psi_0\rangle$ of the system in the sense that the result

$$\langle\Psi_0|A|\Psi_0\rangle = \langle\Phi_0|\Omega^\dagger A\Omega|\Phi_0\rangle^c \tag{A2.3.7}$$

holds for all operators A (Becker and Brenig, 1990). Thus Ω is related closely to the wave operator $\tilde{\Omega}$, which transforms $|\Phi_0\rangle$ into the exact ground state $|\Psi_0\rangle$, i.e.,

$$|\Psi_0\rangle = \tilde{\Omega}|\Phi_0\rangle. \tag{A2.3.8}$$

The difference is that Ω appears only when expectation values are calculated, i.e., only within a cumulant.

It is important to notice that the above considerations are independent of the separation of H into $H = H_0 + H_1$. In particular, H_0 need not be a one-electron Hamiltonian, as in a self-consistent field (SCF) calculation. Instead, it may also contain two-particle operators. Therefore Eqs. (A2.3.3)–(A2.3.5) are well suited for a treatment of systems with strongly correlated electrons. Consider, e.g., the treatment of a spin system. The Heisenberg Hamiltonian for the antiferromagnetic case ($J>0$) reads

$$H = J \sum_{\langle ij\rangle} S_i S_j, \tag{A2.3.9}$$

where $\langle ij\rangle$ are nearest-neighbor sites and S_i are the usual spin operators. In this case, one may choose the Ising part as starting Hamiltonian H_0,

$$H_0 = J \sum_{\langle ij\rangle} S_i^z S_j^z. \tag{A2.3.10}$$

Its ground state $|\Phi_0\rangle$ is the Néel state.

A different way to obtain size-consistent results for the ground-state energy is provided by the coupled cluster (CC) method. A detailed account of the method has been given by Bishop (1991).

Within the CC method, one chooses a model state $|\Phi_0\rangle$ and the following ansatz is made for the exact ground state:

$$|\Psi_0\rangle = e^S|\Phi_0\rangle, \tag{A2.3.11}$$

where $|\Psi_0\rangle$ satisfies the Schrödinger equation $H|\Psi_0\rangle = E_0|\Psi_0\rangle$.

A2.3.1. Different Representations of Ω

When Ω, as defined by Eq. (A2.3.5), is expanded in powers of $1/z$, the resulting products of operators $(L_0 + H_1)^m H_1$ are *composite* as regards cumulant formation. They are not to be treated as an entity when a cumulant involving Ω is evaluated. The question arises whether the composite operators must be treated as independent dynamic variables when $|\Omega)$ is searched for, or whether their contributions to $|\Omega)$ are bound tightly to those operators that are treated as an entity when the cumulant is calculated. We will call the latter operators prime. The notation $|\Omega)$ indicates that Ω always acts on the ground state $|\Phi_0\rangle$ and appears only within a cumulant.

In the following, let us assume that $|\Omega)$ can be represented by an universal analytic function f of a prime operator S,

$$|\Omega) = |f(S)), \tag{A2.3.12}$$

which implies that composite operators are not considered as independent variables. Since f is a universal function, it does not depend on the Hamiltonian H. Its Taylor series is

$$f(x) = \sum_{\mu} \alpha_\mu x^\mu \tag{A2.3.13}$$

with real coefficients α. One notes that composite operators enter $|\Omega)$ only in the form of powers of S. Their weights are fixed once the universal coefficients α_μ are known.

Next it can be shown that $f(S)$ must be of the form $\exp S$. One starts by noting that the expectation value of an operator product AB with respect to the exact ground state can be expressed in two different ways. One has by applying Eq. (A2.3.7),

$$\langle\Psi_0|AB|\Psi_0\rangle = (\Omega|(AB)\bullet\Omega). \tag{A2.3.14}$$

The dot indicates that the operator product must be treated as an entity when the cumulant is evaluated. The second is by the relation

$$(\Psi_0|\delta A\delta B|\Psi_0\rangle = (\Omega|AB\Omega), \tag{A2.3.15}$$

where $\delta A = A - \langle\Psi_0|A|\Psi_0)$ (Becker and Brenig, 1990). Here the operators are treated separately when cumulants are formed. Equations (A2.3.14) and (A2.3.15) imply that

$$(\Omega|(AB)\bullet\Omega) = (\Omega|AB\Omega) + (\Omega|A\Omega)(\Omega|B\Omega). \tag{A2.3.16}$$

Inserting the ansatz [Eq. (A2.3.12)] and the definition of cumulants, the terms in Eq. (A2.3.16) read

$$\begin{aligned}(f(S)|Af(S)) &= \sum_{\mu\nu} \alpha_\mu\alpha_\nu (S^\mu|AS^\nu) \\ &= \sum_{\mu\nu} \alpha_\mu\alpha_\nu\, \partial_1^\mu\, \partial_3^\nu \left.\frac{\langle e^{\lambda_1 S^\dagger} A e^{\lambda_3 S}\rangle}{\langle e^{\lambda_1 S^\dagger} e^{\lambda_3 S}\rangle}\right|_{\lambda_i=0},\end{aligned} \tag{A2.3.17}$$

$$(f(S)|(AB)\bullet f(S)) = \sum_{\mu\nu} \alpha_\mu\alpha_\nu\, \partial_1^\mu\, \partial_3^\nu \left.\frac{\langle e^{\lambda_1 S^\dagger} AB e^{\lambda_3 S}\rangle}{\langle e^{\lambda_1 S^\dagger} e^{\lambda_3 S}\rangle}\right|_{\lambda_i=0},$$

and

$$(f(S)|ABf(S)) = \sum_{\mu\nu} \alpha_\mu \alpha_\nu \partial_1^\mu \partial_3^\nu \left[\frac{\langle e^{\lambda_1 S^\dagger} A B e^{\lambda_3 S} \rangle}{\langle e^{\lambda_1 S^\dagger} e^{\lambda_3 S} \rangle} - \frac{\langle e^{\lambda_1 S^\dagger} A e^{\lambda_3 S} \rangle \langle e^{\lambda_1 S^\dagger} B e^{\lambda_3 S} \rangle}{\langle e^{\lambda_1 S^\dagger} e^{\lambda_3 S} \rangle^2} \right] \Bigg|_{\lambda_i = 0} . \tag{A2.3.18}$$

With the abbreviation

$$g_A(\lambda_1,\lambda_2) = \frac{\langle e^{\lambda_1 S^\dagger} A e^{\lambda_3 S} \rangle}{\langle e^{\lambda_1 S^\dagger} e^{\lambda_3 S} \rangle}, \tag{A2.3.19}$$

Eq. (A2.3.16) yields the following condition for the coefficients α:

$$\sum_{\mu\nu} \alpha_\mu \alpha_\nu \partial_1^\mu \, \partial_2^\nu [g_A(\lambda_1,\lambda_2) g_B(\lambda_1,\lambda_2)] |_{\lambda_1=\lambda_2=0}$$

$$= \sum_{\mu\nu\rho\sigma} \alpha_\mu \alpha_\nu \alpha_\rho \alpha_\sigma \; \partial_1^\mu \partial_2^\nu \, g_A(\lambda_1,\lambda_2) \partial_3^\rho \partial_4^\sigma \, g_B(\lambda_3,\lambda_4) |_{\lambda_1=\lambda_2=\lambda_3=\lambda_4=0.} \tag{A2.3.20}$$

Differentiating the left-hand side according to the product rule, one can obtain

$$0 = \sum_{\mu\nu\rho\sigma} \left[\alpha_\mu \alpha_\nu \alpha_\rho \alpha_\sigma - \alpha_{\mu+\rho} \alpha_{\nu+\sigma} \binom{\mu+\rho}{\rho} \binom{\nu+\sigma}{\sigma} \right] (S^\rho | A S^\sigma)(S^\mu | B S^\nu), \tag{A2.3.21}$$

which has to be fulfilled for all operators A and B, as well as for all Hamiltonians, the parameters of which enter S. On the other hand, f is desired to be a universal function. Therefore the relation

$$\alpha_\mu \alpha_\nu \alpha_\rho \alpha_\sigma = \alpha_{\mu+\rho} \alpha_{\nu+\sigma} \binom{\mu+\rho}{\rho} \binom{\nu+\sigma}{\sigma} \tag{A2.3.22}$$

must hold for all μ, ν, ρ, σ, i.e.,

$$\alpha_{\mu+\rho}(\mu + \rho)! = \alpha_\mu \mu! \alpha_\rho \rho! \; . \tag{A2.3.23}$$

Setting $\alpha_0 = 1$ (which corresponds to $\Omega = 1$ for $H_1 = 0$), one readily obtains

$$\alpha_\mu = \frac{\kappa^\mu}{\mu!} \tag{A2.3.24}$$

and

$$|\Omega) = |f(S)) = |e^{\kappa S}). \tag{A2.3.25}$$

The parameter κ (= α_1) may be absorbed in the operator S. Therefore one sets κ = 1 (Schork and Fulde, 1992).

A2.3.2. Calculating the Ground-State Energy

The aim of this section is to show how ground-state energies are calculated using the results of the previous section. In order to determine the operator S, one first sets up equations fulfilled by Ω as defined in Eq. (A2.3.5). This shows (Schork and Fulde, 1992)

$$(A|H\Omega) = 0 \tag{A2.3.26}$$

holds, where A is an arbitrary prime operator and Ω is defined by Eq. (A2.3.5). In fact, this relation is intuitively obvious. As pointed out before, Ω characterizes the exact ground state that is an eigenstate of H. Therefore the expectation value corresponding to Eq. (A2.3.26) factorizes for any operator A, and the cumulant in Eq. (A2.3.26) vanishes. Only if A is replaced by a c-number then Eq. (A2.3.26) does not vanish. If A is replaced by the c-number 1, the ground-state energy in Eq. (A2.3.3) is recovered. From Eq. (A2.3.26) one expects that

$$(AB|H\Omega) = 0 \tag{A2.3.27}$$

holds and that similarly constructed products involving more than two prime operators vanish, too. This can be proved by analogy with Eq. (A2.3.26).

Eqs. (A2.3.26) and (A2.3.27) must hold also when the special form $|\Omega) = |\exp S)$ is used. In this case Eq. (A2.3.27) follows directly from Eq. (A2.3.26). This is seen by applying the definition in Eq. (A2.3.1) of cumulants to obtain

$$(AB|He^S) = [(AB)\bullet|He^S)] - (A|e^S)(B|e^S)(B|e^S)(A|He^S). \tag{A2.3.28}$$

Due to Eq. (A2.3.26), each term on the right-hand side vanishes.

With these results one has a method at hand to calculate the ground-state energy when the form $|\Omega) = |\exp S)$ is used. Let $\{A_\mu\}$ denote a basis of the prime operators in the Liouville space. Let us expand S in terms of this basis:

$$S = \sum_\mu \alpha_\mu A_\mu. \tag{A2.3.29}$$

From Eq. (A2.3.26) one expects that

$$(A_\mu|He^S) = 0 \tag{A2.3.30}$$

yields a set of equations for the coefficients α_μ. But operators for which $A_\mu|\Phi_0\rangle = 0$ need no further consideration: They fulfill Eq. (A2.3.30) trivially, leaving α_μ undetermined. The same holds true for operators that generate linearly dependent states $A_\mu|\Phi_0\rangle$, $A_\nu|\Phi_0\rangle$. Therefore only a subset $\{S_\nu\}$ of the basis $\{A_\mu\}$ is needed to construct S. It is given by the minimal set of operators A_μ that generates a basis of the Hilbert space when applied to $|\Phi_0\rangle$. Stated differently, $\{|\Phi_\nu\rangle = S_\nu|\Phi_0\rangle\}$ is a basis covering at least the irreducible vector space that contains the exact ground state. These operators are similar to multiconfigurational excitation operators (see, for example, Bishop, 1991). Thus, one sets

$$S = \sum_\nu \beta_\nu S_\nu \tag{A2.3.31}$$

and determines the coefficients η_ν from Eq. (A2.3.26), i.e., from

$$(S_\nu|He^S) = 0. \tag{A2.3.32}$$

It is to be noted that this ensures that Eq. (A2.3.26) holds for all operators A.

The ground-state energy is given by

$$E_0 = (H|e^S). \tag{A2.3.33}$$

All approximations that are subsequently made to evaluate E_0 preserve the important condition of size consistency due to the use of cumulants (cf. Schork and Fulde, 1992).

A2.3.3. Derivation of the Coupled-Cluster Equations

In this subsection, following Schork and Fulde, the CC equations will be derived from Eqs. (A2.3.31–33). Equation (A2.3.32) yields a set of equations for the coefficients η_ν in the expansion Eq. (A2.3.31):

$$(S_\nu|He^S) = 0. \tag{A2.3.34}$$

From the previous section one knows that this implies that Eq. (A2.3.27) holds, and thus

$$0 = (e^{-S\dagger}\, S_\nu|He^S) = (S_\nu|e^{-S}He^S) \tag{A2.3.35}$$

is true as well. Inserting the definition in Eq. (A2.3.1) of cumulants, one finds that the cumulant agrees with the conventional expectation value because the unconnected terms cancel one another. Thus the set of Eq. (A2.3.34) is equivalent to

$$0 = \langle\Phi_0|S_\nu^\dagger e^{-S}\, He^S|\Phi_0\rangle^c = \langle\Phi_0|S_\nu^\dagger e^{-S}He^S|\Phi_0\rangle \tag{A2.3.36}$$

and Eq. (A2.3.11) is effectively recovered.

Similarly,

$$E_0 = (H|e^S) = (1|He^S) = (e^{-S\dagger}|He^S) = \langle\Phi_0|e^{-S}\, He^S|\Phi_0\rangle^c \tag{A2.3.37}$$

since only the c-number 1 of the expansion of $\exp(-S^\dagger)$ contributes, because of Eq. (A2.3.26,27). Again, the cumulant agrees with the conventional expectation value, and Eq. (A2.3.37) reduces to the CC equation for the ground-state energy.

$$E_0 = \langle\Phi_0|e^{-S}He^S|\Phi_0\rangle. \tag{A2.3.38}$$

It can therefore be shown that Eq. (A2.3.38) can be derived from the many-body equation [Eq. (A2.3.3)] which is based on cumulants. It is not clear at the time of writing whether one can also derive Eq. (A2.3.3) by starting from Eqs. (A2.3.8,11). What is apparent is that different size-consistent methods are linked with one another. For example, the relation of Eq. (A2.3.3) to various coupled electron-pair approximations (CEPA) was previously established by Becker and Fulde (1988, 1989).

The virtues and the broad range of applications of CC theory are well established. But it may also prove useful to start instead from Eq. (A2.3.3,5), because different approximations may be made than in the CC theory. For example, the Schrödinger perturbation expansion follows immediately by expanding Ω in powers of $(z - L_0)^{-1} H_1$. A very powerful approximation scheme is the projection technique, which can be applied to $(H|\Omega)$ (see, e.g., Becker and Fulde, 1988; 1989). Let us also draw attention to another type of approximation, which Schork and Fulde (1992) term the *independent mode* approximation.

A2.3.4. Independent Mode Approximation

Suppose that the perturbing Hamiltonian H_1 has been expanded in terms of eigenoperators A_ν of the Liouvillian L_0:

$$H_1 = \sum_\nu A_\nu, \tag{A2.3.39}$$

where

$$L_0 A_\nu = a_\nu A_\nu. \tag{A2.3.40}$$

It sometimes proves advantageous to group the operators A_ν, e.g., according to an induced momentum transfer (Schork and Fulde, 1992). This grouping is indicated by an additional index i: $A_\nu, a_\nu \to A_{i,\nu}, a_{i,\nu}$. Let us define new operators $B_i = \sum_\nu A_{i,\nu}$, so that $H_1 = \sum_i B_i$. In general, B_i is not an eigenoperator of L_0 because the $a_{i,\nu}$ usually differ for different ν. One approximates the different eigenvalues $a_{i,\nu}$ by a single one, ω_i, so that

$$L_0 B_i = \omega_i B_i. \tag{A2.3.41}$$

In this approximation B_i acts like a single mode with respect to L_0. Next, one assumes that these modes are independent, i.e., that the B_i commute:

$$[B_i, B_j]_- = 0. \tag{A2.3.42}$$

Thus, H_1 acts like a sum of independent excitation modes. Summing the geometric series in Eq. (A2.3.5), one obtains

$$\Omega = \lim_{z \to 0} \sum_{\nu=0}^{\infty} \left(\frac{1}{z - L_0} H_1 \right)^\nu = \exp\left(- \sum_i \frac{1}{\omega_i} B_i \right). \tag{A2.3.43}$$

In the case when all ω_i are equal, i.e., that H_1 acts like a single mode with an eigenvalue ω_0, one finds that

$$\Omega = \exp\left(- \frac{H_1}{\omega_0} \right). \tag{A2.3.44}$$

It is to be noted that now the perturbing Hamiltonian H_1 enters the exponent of Ω, in contrast to the CC theory, where the operator S is exponentiated. The unknown prefactors ω_i in Eq. (A2.3.43) describe the eigenfrequencies of the different modes. They can be determined by minimizing the ground state energy $(H|\Omega)$. Wave functions corresponding to Eq. (A2.3.43) have been employed successfully in a number of cases. One example is the treatment of the electron gas within the random-phase approximation (RPA; see also Chapter 3). The Hamiltonian reads $H = H_0 + H_1$, with

$$H_0 = \sum_{k\sigma} \varepsilon_k c^\dagger_{k\sigma} c_{k\sigma}$$

$$H_1 = \sum_{q\neq 0} \frac{4\pi e^2}{q^2} \rho_q \rho_{-q}, \qquad \text{(A2.3.45)}$$

where $c^\dagger_{k\sigma}$ creates an electron with momentum k and spin σ and $\rho_q = \sum_{k\sigma} c^\dagger_{k+q\sigma} c_{k\sigma}$ denotes a density fluctuation having momentum q. The operators $\rho_q\rho_{-q}$ are identified with the operators B_i above. They indeed commute with one another. Let $|\Phi_0\rangle$ denote the ground state of H_0. The variational wave function introduced by Gaskell (1958) reads

$$|\Psi_0\rangle = \exp(-\sum_q \tau_q \rho_q \rho_{-q})|\Phi_0\rangle \qquad \text{(A2.3.46)}$$

and corresponds to Eq. (A2.3.44).

Another example is the Gutzwiller (1963) wave function (see also Chapter 4):

$$|\Psi_0\rangle = \exp(-\eta \sum_i n_i{\uparrow} n_i{\downarrow})|\Phi_0\rangle, \qquad \text{(A2.3.47)}$$

which refers to the Hubbard Hamiltonian $H = H_0 + H_1$ with

$$H_0 = \sum_{ij\sigma} t_{ij} c_{i\sigma}{}^\dagger c_{j\sigma}$$

$$H_1 = U \sum_i n_i{\uparrow} n_i{\downarrow}, \qquad \text{(A2.3.48)}$$

where $n_{i\sigma} = c^\dagger_{i\sigma} c_{i\sigma}$. The index i is a site index and $|\Phi_0\rangle$ is the ground state of H_0. In this case the perturbing Hamiltonian H_1 is treated as describing a single mode.

A2.4. THE HILLER–SUCHER–FEINBERG IDENTITY AND IMPROVEMENT OF CUSP CONDITION

In the main text, it is noted that an important property of the exact electron density $\rho(\mathbf{r})$ is that it satisfies the cusp condition of Kato (1957; see also Kryaschko and Ludeña, 1990). Unfortunately (see, for example Appendix 4.3) many *ab initio* electronic structure calculations employ cuspless Gaussian functions, and the cusp condition is often satisfied only quite poorly by the computed electron densities (cf. Cioslowski and Challacombe, 1992).

It is therefore of interest to record here an identity proposed by Hiller *et al.* (1978) that can sometimes be used to improve the values of $\rho(\mathbf{r})$ at nuclei. Some difficulties that arise at other spatial points are set out in the work of Cioslowski and Challacombe referred to already. The Hiller–Sucher–Feinberg identity is merely quoted below:

$$\tilde{\rho}(\mathbf{R}) = (2\pi)^{-1} \langle \Psi | \sum_i \hat{D}_i(\mathbf{R}) | \Psi \rangle, \tag{A2.4.1}$$

where

$$\hat{\mathbf{D}}_i(\mathbf{R}) = |\mathbf{r}_i - \mathbf{R}|^{-3}[(\mathbf{r}_i - \mathbf{R}) \times \nabla_i] \cdot [(\mathbf{r}_i - \mathbf{R}) \times \nabla_i] + |\mathbf{r}_i - \mathbf{R}|^{-1} (\mathbf{r}_i - \mathbf{R}) \cdot (\nabla_i V). \tag{A2.4.2}$$

Equation (A2.4.1) is useful at nuclei in a molecule, even when basis functions without cusps have been used. For exact wave functions, the densities $\rho(\mathbf{R})$ and $\tilde{\rho}(\mathbf{R})$ must, of course, be identical. The potential energy V in Eq. (A2.4.2) is that for a Coulombic system described by a nonrelativistic Born–Oppenheimer Hamiltonian.

Example of 1s State of Hydrogen-like Atom

To exemplify Eqs. (A2.4.1) and (A2.4.2), let us write down $\tilde{\rho}(\mathbf{R})$ for the 1*s* state of the hydrogen-like atom with

$$V = -Z/r. \tag{A2.4.3}$$

Let us merely assume a general trial function $\Psi(\mathbf{r})$, to insert with Eq. (A2.4.3) into Eq. (A2.4.1), that decays at least exponentially at large r. Then direct use of the above equations yields:

$$\tilde{\rho}(\mathbf{R}) = (4/3R) \int_0^R dr\, r\Psi^* (Z\Psi + 2\Psi' + r\Psi'') + (2/3)$$
$$\times \int_R^\infty dr\, r^{-2}\, \Psi^* \, [-Z\,(R^2 - 3r^2)\Psi - 2R^2\,(\Psi' - r\Psi'')]. \tag{A2.4.4}$$

One now can note that, for large R, the second integral falls off at least exponentially and the asymptotic behavior of Eq. (A2.4.4) is given by the first term. This can be evaluated in the limit of large R to yield

$$\tilde{\rho}(R) \to -\left(\frac{1}{3\pi R}\right)\langle \Psi | 2\hat{T} + V | \Psi \rangle, \tag{A2.4.5}$$

where $\hat{T}$ as usual denotes the exact kinetic energy operator. Unless the virial theorem is exactly satisfied by the chosen trial wave function, the resulting density $\tilde{\rho}(\mathbf{R})$ is evidently not integrable (see Cioslowski and Challacombe, 1992).

Some final comments are called for. First, the computation of $\hat{\rho}(\mathbf{R})$ is more costly than that of $\rho(\mathbf{R})$ because it involves complicated molecular integrals. Also, as was demonstrated by Larsson (1981), $\tilde{\rho}(\mathbf{R})$ calculated from the Hartree–Fock limit single-determinant wave function only approaches the exact Hartree–Fock electron density. Nevertheless because Eq. (A2.4.1) involves a global operator, $\tilde{\rho}(\mathbf{R})$ at special points (and in particular at nuclei) can be expected to approximate the exact electron density better than $\rho(\mathbf{R})$. In relation to Chapter 3, it is of interest to note that the Hiller–Sucher–Feinberg identity has been re-expressed in density functional form by Holas and March (1994).

A2.5. TWO ELECTRONS WITH COULOMB INTERACTION MOVING IN AN EXTERNAL OSCILLATOR POTENTIAL

Since few problems with electron correlation are exactly soluble, the purpose of this appendix is to record the solution of one example: two electrons subject to Coulomb repulsion but moving in an external oscillator potential. Early work by White and Byers-Brown (1970) [see also Benson and Byers-Brown (1970) and Tuan (1969)] is noteworthy. The account below follows that of Taut (1993).

A2.5.1. Solution of Schrödinger Equation

The Hamiltonian of the above example reads (using units in which $\hbar = m = e = 1$ throughout)

$$H = -\frac{1}{2}\nabla_1^2 + \frac{1}{2}\omega^2 r_1^2 - \frac{1}{2}\nabla_2^2 + \frac{1}{2}\omega^2 r_2^2 + \frac{1}{|\mathbf{r}_1 - \mathbf{r}_2|}. \tag{A2.5.1}$$

Using the difference vector $\mathbf{r}_1 - \mathbf{r}_2$ and the center of mass $\mathbf{R} = \frac{1}{2}(\mathbf{r}_1+\mathbf{r}_2)$ as new variables, the Hamiltonian decouples into the form

$$H = -\nabla_\mathbf{r}^2 + \frac{1}{4}\omega^2 r^2 + \frac{1}{r} - \frac{1}{4}\nabla_\mathbf{R}^2 + \omega^2 R^2 \equiv H_\mathbf{r} + H_\mathbf{R}. \tag{A2.5.2}$$

Since H is spin-independent, the total wave function can be factorized into the form (s_1 and s_2 being spin coordinates)

$$\Psi(1, 2) = \phi(\mathbf{r})\,\zeta(\mathbf{R})\,\chi(s_1, s_2) \tag{A2.5.3}$$

The center-of-mass equation is then the well-known three-dimensional oscillator solution and needs no further consideration. The relative motion of the two electrons is characterized by

$$\left[-\frac{1}{2}\nabla_\mathbf{r}^2 + \frac{1}{2}\omega_\mathbf{r}^2 r^2 + \frac{1}{2r}\right]\phi(r) = \varepsilon'\phi(\mathbf{r}), \tag{A2.5.4}$$

where $\omega_r = \frac{1}{2}\omega$. Introducing spherical coordinates, which separate the modulus r from the angular coordinates $\hat{\mathbf{r}} = \mathbf{r}/r$, one can write

$$\phi(r) = \frac{u(r)}{r} Y_{lm}(\hat{\mathbf{r}}). \tag{A2.5.5}$$

Here Y_{lm} are the usual spherical harmonics, while $u(r)$ satisfies

$$\left[-\frac{1}{2}\frac{d^2}{dr^2} + \frac{1}{2}\omega_r^2 r^2 + \frac{1}{2r} + \frac{l(l+1)}{2r^2} \right] u(r) = \varepsilon' \, u(r) \tag{A2.5.6}$$

As Taut (1993) notes, to satisfy the Pauli principle the solutions corresponding to even (odd) l belong to the singlet (triplet) state.

A2.5.2. An Exact Solution

Introducing reduced variables into Eq. (A2.5.6), $\rho = \sqrt{\omega_r}\, r$, $\varepsilon'' = 2\varepsilon'/\omega_r$ and splitting off the asymptotic solution for $r \to \infty$,

$$u(\rho) = e^{-(1/2)\rho^2} \cdot t(\rho), \tag{A2.5.7}$$

leads to

$$\rho^2 t'' - 2\rho^3 t' + \left[(\varepsilon'' - 1)\rho^2 - \frac{1}{\sqrt{\omega_r}}\rho - l(l+1)\right] t = 0. \tag{A2.5.8}$$

A power series expansion

$$t(\rho) = \rho^m \sum_{\nu=0}^{\infty} a_\nu \rho^\nu \tag{A2.5.9}$$

where $m = l + 1$ (the irregular solution $m = -l$ is dropped) transforms Eq. (A2.5.8) into a recurrence relation for the coefficients a_ν:

$$a_0 \neq 0 \tag{A2.5.10a}$$

$$a_1 = \frac{1}{2(l+1)\sqrt{\omega_r}} a_0 \tag{A2.5.10b}$$

$$a_\nu = \frac{1}{\nu(\nu + 2l + 1)} \left\{ \frac{1}{\sqrt{\omega_r}} a_{\nu-1} + [2(l+\nu) - 1 - \varepsilon''] a_{\nu-2} \right\} \quad \text{for } \nu \geq 2. \tag{A2.5.10c}$$

Now let us seek the condition that defines the discrete eigenvalue spectrum. This is not as straightforward as for the harmonic oscillator or the hydrogen problem, because Eq. (A2.5.10) is a three-step recurrence relation. In the present problem, termination of the power series, which guarantees normalization, can be reached in the following way (Taut, 1993). Using Eq. (A2.5.10), one can determine a_ν for arbitrary ν:

$$a_\nu = F(l,\nu,\varepsilon'',\omega_r) \cdot a_0. \tag{A2.5.11}$$

Now one can assume that the series of a_ν terminates at a certain $\nu = n$:

$$\ldots a_{n-1} \neq 0, \quad a_n = 0, \quad a_{n+1} = 0 \ldots,$$

so that the order of the polynomial $t(\rho)$ is $n + l$.

To reach this situation one must guarantee that $a_n = 0$ and $a_{n+1} = 0$. The first condition is fulfilled if

$$F(l, n, \varepsilon'', \omega_r) = 0, \tag{A2.5.12}$$

and for the second, one rewrites Eq. (A2.5.10c) as

$$a_{n+1} = \frac{1}{(n+1)(n+2l+2)}\left\{\frac{1}{\sqrt{\omega_r}}a_n + [2(l+n+1) - 1 - \varepsilon'']a_{n-1}\right\} \tag{A2.5.13}$$

It is then found that the bracket in Eq. (A2.5.13) must vanish:

$$\varepsilon'' = 2(l + n) + 1. \tag{A2.5.14}$$

Equations (A2.5.14) and (A2.5.12) are formulas for ε'' and ω_r, which must be fulfilled simultaneously and which define the energy spectrum. Unfortunately, one cannot find the energy spectrum for a given oscillator frequency ω_r straightforwardly. Equation (A2.5.14) allows one to calculate the reduced energies ε'' for given n and l. On the other hand, insertion of Eq. (A2.5.14) into Eq. (A2.5.12) provides an equation for ω_r:

$$F(l,n,2(l + n) + 1,\omega_r) = 0. \tag{A2.5.15}$$

Thus, Eq. (A2.5.15) determines which oscillator frequency ω_r belongs to the reduced energy given by Eq. (A2.5.14).

It is of interest to mention that additional information about the function $\varepsilon'(\omega_r)$ can be obtained by application of the Hellmann–Feynman theorem to Eq. (A2.5.6), giving (Taut, 1993)

$$\frac{d\varepsilon'}{d\omega_r} = \omega_r \int dr[u(r)]^2 r^2.$$

This can be calculated exactly for those ω_r for which analytical solutions have been found.

A2.5.3. Results

Following Taut (1993), let us consider some special cases. For $n = 2$ and arbitrary l, Eq. (A2.5.14) provides the reduced energies

$$\varepsilon'' = 2l + 5, \tag{A2.5.16}$$

and from Eqs. (A2.5.10), (A2.5.11), and (A2.5.14) it follows that

$$F(\omega_r) = \frac{1}{2(2l+3)}\left[\frac{1}{2(l+1)}\,\frac{1}{\omega_r} - 2\right] \tag{A2.5.17}$$

which has the zero

$$\omega_r = \frac{1}{4(l+1)}. \tag{A2.5.18a}$$

Thus, the energies for $n = 2$ and arbitrary l are

$$\varepsilon' = \frac{\omega_r}{2}\,\varepsilon'' = \frac{2l+5}{8(l+1)} \xrightarrow[l\to\infty]{} \frac{1}{4}. \tag{A2.5.18b}$$

The corresponding radial wave functions $u(r)$ read (apart from a normalization factor)

$$u(r) = r^{(l+1)} e^{\frac{r^2}{8(l+1)}}\left[1 + \frac{r}{2(l+1)}\right]. \tag{A2.5.18c}$$

For $n = 3$ and arbitrary l, the solutions are

$$\omega_r = \frac{1}{4(4l+5)}. \tag{A2.5.19a}$$

$$\varepsilon' = \frac{2l+7}{8(4l+5)} \tag{A2.5.19b}$$

$$u(r) = r^{(l+1)} e^{\frac{-r^2}{8(4l+1)}}\left[1 + \frac{r}{2(l+1)} + \frac{r^2}{4(l+1)(4l+5)}\right]. \tag{A2.5.19c}$$

Let us now consider the case $l = 0$ and arbitrary n. In this case, n is the order of the polynomial $t(\rho)$. The corresponding F up to $n = 11$ is readily calculated, and the solutions have been tabulated by Taut (1993). (The results up to $n = 5$ can be obtained in closed form.) It is to be stressed that solutions found in this way are not necessarily ground states. To which excitation a solution belongs depends on the number of nodes of the polynomial $t(\rho)$ for $\rho > 0$. Generally, one finds that for a given n, the number of real positive roots of the equation $F(\omega_r) = 0$ is $N_\omega =$

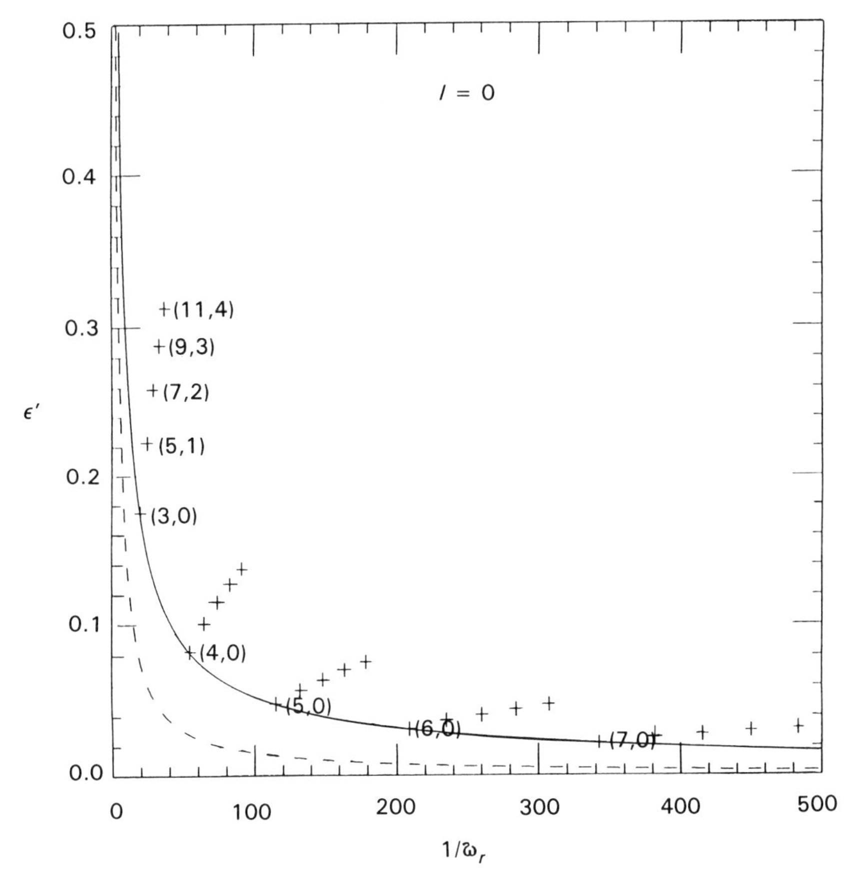

FIGURE A2.5.1. Exact solutions (crosses) for $l = 0$ and $n < 15$, which lie in the chosen ε' and ω_r range. Some solutions are supplemented by two numbers describing the order of the corresponding polynomial n and its number of zeros N_r. The full line is the result for the ground state in the Taylor expansion approximation and the dashed line refers to the independent particle picture (electron–electron interaction neglected) (redrawn from Taut, 1993).

int(n/2) and that among the N_ω eigensolutions for $t(\rho)$, the one with the smallest ω_r has no nodes (ground state), while that with the second-largest ω_r has one node (first excited state), and so on.

Some more solutions for $l = 0$ are shown in Fig. A2.5.1. A practical way for finding the solution for a given ω_r and l is to calculate the solutions for some ω_r in the range of interest and then interpolate between the energy values. An improved closed-form expression also for larger ω_r is obtained by interpolation between small-ω_r and large-ω_r approximations (see Taut, 1993):

$$\varepsilon'_{int} = \frac{3}{2^{5/3}}\,\omega_{\mathbf{r}}^{2/3} + \frac{C(2N_r + \frac{3}{2})\omega_{\mathbf{r}}^{P} + \sqrt{3}\,(N_r + \frac{1}{2})\;\omega_{\mathbf{r}}^{-P}}{C\omega_{\mathbf{r}}^{P} + \omega_{\mathbf{r}}^{-P}}\;\omega_{\mathbf{r}}, \qquad \text{(A2.5.20)}$$

where the constant C is fitted to reproduce the exact energy value for the lowest available $1/\omega_{\mathbf{r}}$ and the exponent is optimized to provide the lowest rms error for the remaining exact energy values. In this way Taut finds for $l = 0$ and N_r from 0 to 4 the values for C of 0.521245, 0.833802, 1.16225, 1.51111, and 1.91576 and for P the values 0.24, 0.22, 0.20, 0.19, and 0.19. The accuracy of this interpolation formula is better than 0.1%. Of course, any approximation in finding the solution for a given $\omega_{\mathbf{r}}$ can be avoided by resorting to numerical solution of the ordinary differential equation [Eq. (A2.5.6)] by standard methods.

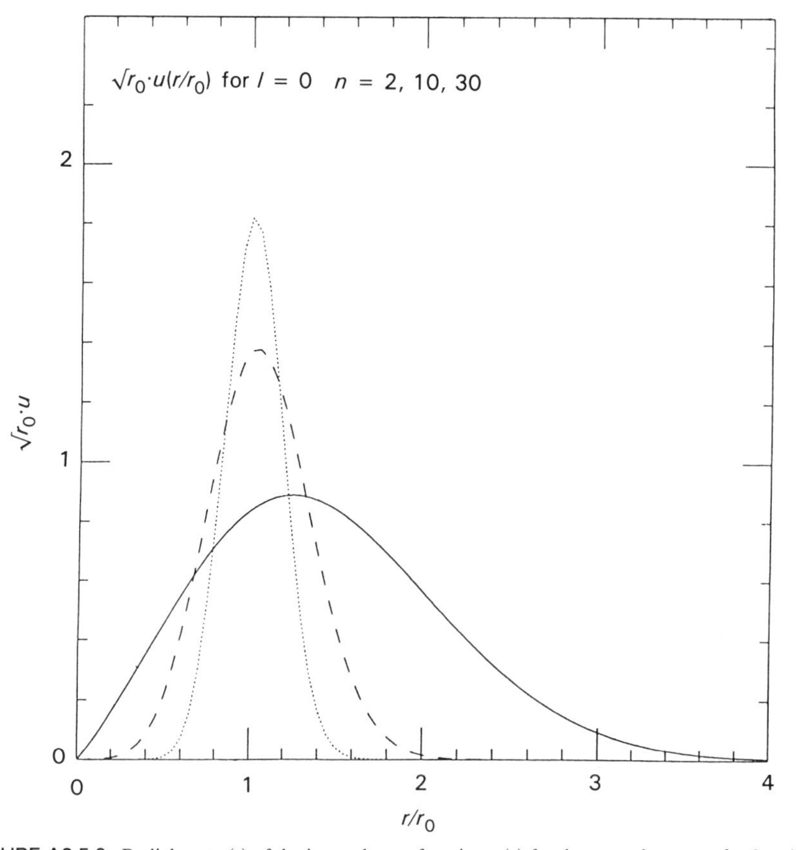

FIGURE A2.5.2. Radial part $u(r)$ of the internal wavefunction $\varphi(\mathbf{r})$ for the ground states to $l = 0$ and $n = 2$ (full), 10 (dashed), 30 (dotted), which belong to the solutions $\omega_{\mathbf{r}} = 0.25$, $9.4828{\cdot}10^{-4}$ and $3.2429{\cdot}10^{-5}$. r_0 is the classical electron distance as defined in Section 3.1 (redrawn from Taut, 1993).

A2.5.4. Some Physical Consequences

Let us briefly discuss some physical consequences. Correlations are most appropriately examined by means of the pair-correlation function (Taut, 1993)

$$\begin{aligned} G(\mathbf{r}) &= \langle \psi | \sum_{i \langle j} \delta(\mathbf{r}_i - \mathbf{r}_j - \mathbf{r}) | \psi \rangle \\ &= \int d^3\mathbf{r}' |\Psi(\mathbf{r} + \mathbf{r}', \mathbf{r}')|^2 \\ &= |\phi(\mathbf{r})|^2, \end{aligned}$$

which is determined by the internal wave function $\varphi(\mathbf{r})$ only. In Fig. A2.5.2 the radial part $u(r)$ of $\varphi(\mathbf{r})$ is given for $l = 0$ and three ω_r values. One sees that the probability for the electrons to have a certain distance r bunches around the classical distance r_0. The variance of the distribution is smaller than the smaller

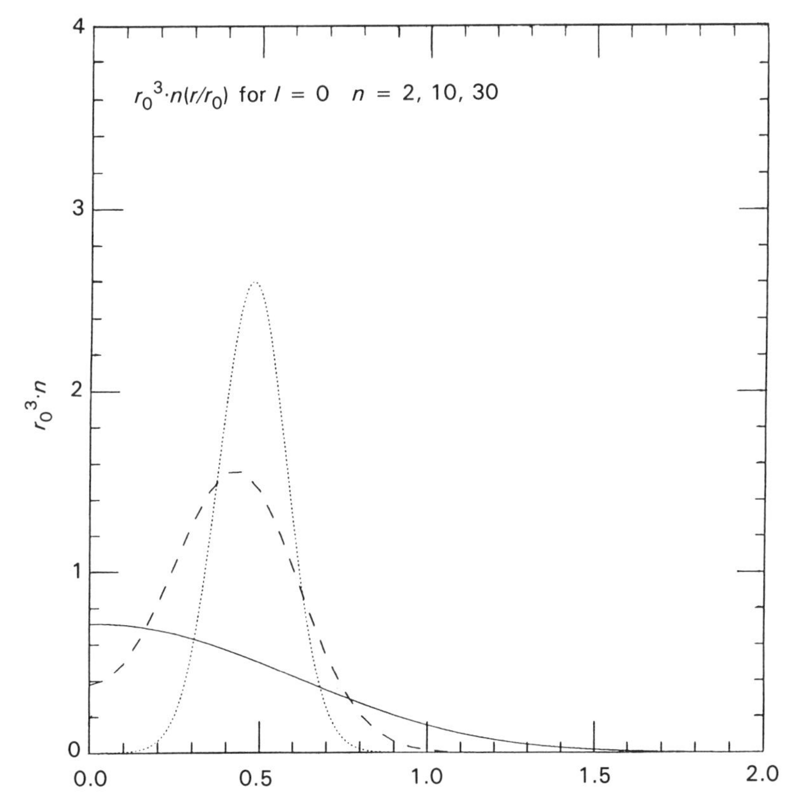

FIGURE A2.5.3. Electron density $n(\mathbf{r})$ vs r/r_0 corresponding to the solution shown in Fig. A2.5.2. (redrawn from Taut, 1993).

ω_r, and consequently the density, is. This is because for decreasing density the interaction energy dominates and governs the properties (as in the Wigner crystal; see Chapters 5 and 9). The same tendency is seen in the charge density

$$n(\mathbf{r}) = \langle\psi|\sum_i \delta(\mathbf{r} - \mathbf{r}_i)|\psi\rangle$$

$$= 2\int d^3\mathbf{r}'|\phi(\mathbf{r}')^2|\xi(\mathbf{r} + \frac{\mathbf{r}'}{2})|^2,$$

which is a convolution integral between the internal and center-of-mass motion and which is depicted in Fig. A2.5.3. For high densities, correlations are unimportant and both electrons are seeking independently from each other to minimize the potential energy in the external potentials. This results in a density centered at the origin. In this limit a one-particle picture applies. But with decreasing ω_r, the importance of correlations increases and the electrons are approaching a model in which they are distributed on a spherical shell of diameter r_0. Thereby they are located always at antipodal points. This last fact follows immediately from the pair-correlation function.

Appendix to Chapter 3

A3.1. EXAMPLE OF SPIN DENSITY DESCRIPTION: BLOCH'S HARTREE–FOCK TREATMENT

Let us take as the simplest possible example of spin density theory the Hartree–Fock approximation for a uniform electron assembly. Thus we consider the generalization of the Thomas–Fermi–Dirac theory for uniform electrons described by plane waves. If $\rho_\uparrow$ and $\rho_\downarrow$ denote the densities (assumed constant) for the upward and downward spins, the single-particle kinetic energy density of the Thomas–Fermi theory is readily generalized to read

$$t[\rho_\uparrow, \rho_\downarrow] = \text{constant } [\rho_\uparrow^{5/3} + \rho_\downarrow^{5/3}]. \tag{A3.1.1}$$

Similarly the exchange energy density takes the form

$$\varepsilon_x[\rho_\uparrow, \rho_\downarrow] = \text{constant } [\rho_\uparrow^{4/3} + \rho_\downarrow^{4/3}]. \tag{A3.1.2}$$

As Bloch (1930) was the first to show, by minimizing $t + \varepsilon_x$ with respect to $\rho_\uparrow$ and $\rho_\downarrow$, subject to $\rho_\uparrow + \rho_\downarrow = N/\Omega$ for N electrons in volume Ω, the ferromagnetic state becomes stable at sufficiently low densities, relative to the paramagnetic state. In terms of the total density $\rho = \rho_\uparrow + \rho_\downarrow$ and the exchange constant c_e in Eq. (3.36), the condition for ferromagnetism in the ground state in this model is found to be

$$c_e\rho^{1/3}/E_f > 4(2^{1/3} + 1)/5, \tag{A3.1.3}$$

where as usual E_f is the Fermi energy.

This example is merely illustrative, and the reader is cautioned that the Hartree–Fock model overestimates the tendency to ferromagnetism (grossly in this case, as one knows precisely now from the quantum Monte Carlo simulations of jellium in Section 5.3) because in this theory there are Fermi-statistical correlations between all parallel-spin electrons but no correlations between antiparallel spins. In fact, in the uniform electron assembly the introduction of correlation stabilizes the antiferromagnetic phase relative to the paramagnetic phase at sufficiently low densities.

If in the uniform gas theory one introduces the relative magnetization defined by

$$\zeta = \frac{\rho_\uparrow - \rho_\downarrow}{\rho_\uparrow - \rho_\downarrow} = \frac{\rho_\uparrow + \rho_\downarrow}{\rho} \tag{A3.1.4}$$

the above-ground-state energy can be written in the form

$$\frac{E(\zeta)}{NE_\mathrm{f}} = \frac{3}{10}\,[(1+\zeta)^{5/3} + (1-\zeta)^{5/3}] - \frac{3}{8}\frac{\varepsilon_\mathrm{x}}{E_\mathrm{f}}[(1+\zeta)^{4/3} + (1-\zeta)^{4/3}], \tag{A3.1.5}$$

and $\varepsilon_\mathrm{x} = (4/3)\, c_e\, \rho^{1/3}$ in terms of the constant c_e introduced above.

Here let us simply refer to the curves of Lidiard (1951) who has plotted (his Fig. 3) $[E(\zeta) - E(O)]/N$ against the relative magnetization for particular values of $\varepsilon_\mathrm{x}/E_\mathrm{f}$. Let us merely quote the result that in the Hartree–Fock theory of jellium the simple ferromagnetic state $\zeta = 1$ is stable, from Eq. (A3.1.3), when $\varepsilon_\mathrm{X}/E_\mathrm{f} > \mathrm{e_c}$ while the paramagnetic Hartree–Fock state $\zeta = 0$ is stable relative to the ferromagnetic state for $\varepsilon_\mathrm{x}/E_f < \mathrm{e_c}$ where $\mathrm{e_c} = 4(2^{1/3} + 1)/5$.

No intermediate values of ζ are possible in this model, in contrast to the more realistic collective electron model of Stoner (1938, 1939). His model, in fact, for uniform electronics, could be regarded as related to Eq. (A3.1.5) if one argues, motivated by the (relative) success of the Weiss theory of ferromagnetism, that the effect of including correlation is to restore an interaction energy in Eq. (A3.1.5) whose dependence on the relative magnetization ζ is simply quadratic. Then the character of the solution is altered drastically and intermediate values of ζ are possible.

Criterion for Ferromagnetism. In addition to this merit of the Stoner model, one can insert realistic energy bands, and the criterion that ferromagnetism will occur takes the form

$$IN(E_\mathrm{f}) > 1, \tag{A3.1.6}$$

where I measures the strength of the exchange and correlation interaction and $N(E_\mathrm{f})$ is the one-electron density of states at the Fermi surface.

Returning to Eq. (A3.1.5), one can easily calculate the spin susceptibility χ in the paramagnetic region both at $T = 0$ and also (approximately) at elevated temperatures (March and Donovan, 1954; Lidiard et al., 1956). If N is the total number of electrons, then one finds

$$\frac{\mu_b^2 N}{\chi E_\mathrm{f}} = \frac{2}{3} - \frac{1}{3}\frac{\varepsilon_\mathrm{x}}{E_\mathrm{f}} \tag{A3.1.7}$$

where $\varepsilon_\mathrm{x} = 4/3\, c_e\, \rho_0^{1/3}$. This reduces to the usual Pauli result for free electrons if one sets $\varepsilon_x = 0$. The interaction (exchange) term enhances the susceptibility as usual.

The above spin density theory framework is very suitable for discussing the incorporation of correlation effects in the calculation of the spin paramagnetic susceptibility in metals, early work in this area being due to Callaway and his co-workers (See Callaway and March, 1984, for references). Actually, even earlier Sampson and Seitz (1940) calculated the spin susceptibility by a density method. They essentially corrected Eq. (A3.1.5) by adding correlation energy, which they noted to be principally dependent on the number of electrons with opposite spin. Their treatment employed the Wigner (1938) formula for the correlation energy and their computed susceptibility for sodium was close to the observed value.

In connection with the Stoner theory, the same enhancement effect noted above occurs in the spin susceptibility, the noninteracting value, say, χ_0 being increased according to

$$\chi = \frac{\chi_0}{1 - IN(E_f)}, \tag{A3.1.8}$$

with I measuring the strength of the electron–electron interaction, while as usual $N(E_f)$ is the density of electronic states at the Fermi level. The connection of this equation with the criterion of Eq. (A3.1.6) is evident. Stoddart and March (1971) discussed modifications of this criterion for the existence of ferromagnetism from their formulation of spin density functional theory; however the detail will not be given here. Rather, in the following chapter (Section 4.5) there will be a discussion of their work leading to the condition for local moment formation (see also Appendix 3.3) which has been used in subsequent theories of ferromagnetism (Hubbard 1979a, Roth 1978, Hasegawa 1980a: Heine *et al.*). However, in connection with Eq. (A3.1.6) the discussion of Gunnarsson (1976), who has used the same spin density functional theory to calculate magnetic properties of the transition metals should be emphasized. Gunnarsson's results for $IN(E_f)$ are shown in Table A3.1.1.

It will be seen that the ferromagnetic metals Fe, Co, and Ni do indeed satisfy the criterion of Eq. (A3.1.6) while the other three paramagnetic metals do not (see also Sigalas and Papaconstantopoulos, 1994). This circumstance would not obtain with a straightforward application of Dirac–Slater exchange theory which (cf. Section 3.5) gives larger values of $IN(E_f)$ than in the local spin density theory used by Gunnarsson.

TABLE A3.1.1. Gunnarsson's Results for $IN(E_f)$ of Transition Metals

Transition metal	V	Fe	Co	Ni	Pd	Pt
$IN(E_f)$	0.8–0.9	1.5–1.7	1.6–1.8	2.1	0.8	0.5

The spin density theory can also be used to formulate the effect of correlation on the wave-number dependent spin susceptibility, as discussed in Appendix 3.2.

A3.2. WAVE-NUMBER DEPENDENT MAGNETIC SUSCEPTIBILITY

Reference will be made here to the work of Stoddart (1975), in which the wave-number dependent magnetic susceptibility is expressed in terms of the spin density formalism.

Stoddart's work essentially effects the generalization of the relation between perturbative methods and gradient expansions referred to in section 3.4 (see Stoddart *et al.*, 1971) to the spin-polarized case. Let us confine ourselves below to the essential steps in the argument. Working in terms of $\rho(\mathbf{r})$ plus the magnetization

$$m(\mathbf{r}) = -\tfrac{1}{2}\,[\rho_\uparrow(\mathbf{r}) - \rho_\downarrow(\mathbf{r})] \tag{A3.2.1}$$

(for convenience, the factor $-\frac{1}{2}$ is included in the definition of m here), Stoddart writes formally the total ground-state energy as

$$E[\rho, m] = G[\rho, m] + \frac{1}{2}\int \frac{\rho(\mathbf{r})\rho(\mathbf{r}')}{|\mathbf{r} - \mathbf{r}'|}\, d\mathbf{r}d\mathbf{r}' + E_{\text{ext}}, \tag{A3.2.2}$$

where E_{ext} represents the energy of interaction with an external magnetic field, say. Then one can write

$$E_{\text{ext}} = -\int m(\mathbf{r})\, H(\mathbf{r})\, d\mathbf{r}.$$

Using density gradient expansions, Stoddart finds

$$G[\rho, m] = G_0[\rho, m] + \int g_1(\mathbf{r}, m(\mathbf{r}))[\nabla\rho(\mathbf{r})]^2 d\mathbf{r}$$

$$+ \int g_2(\rho,m)[\nabla\rho(\mathbf{r})\cdot\nabla m(\mathbf{r})]d\mathbf{r} + \int g_3(\rho, m)[\nabla_r m(\mathbf{r})]^2 d\mathbf{r} + \ldots, \tag{A3.2.3}$$

where the spin-polarized system is assumed throughout to have a single magnetization direction. In Eq. (A3.2.3),

$$G_0(\rho, m) = \int d\mathbf{r}\; g_0(\rho(\mathbf{r}), m(\mathbf{r})), \tag{A3.2.4}$$

where $g_0(\rho, m)$ is the sum of kinetic, exchange, and correlation energies per unit volume for a uniform electron gas with density ρ and magnetization m. [Compare Eq. (3.20) for this result in the Dirac–Slater exchange approximation. More general energy functionals and a corresponding spin-dependent one-body potential have been proposed by von Barth and Hedin (1972)].

Stoddart then effects the generalization of the perturbation expansion of $G[\rho, m]$ to read, for the magnetic case

$$G[\rho, m] = G_0(\rho_0, m_0) + \int ds\, G_1(\mathbf{s}) \int d\mathbf{r}\Delta^2(\mathbf{r}) + \int ds\, G_2(\mathbf{s}) \int d\mathbf{r}\Delta(\mathbf{r})\delta(\mathbf{r})$$

$$+ \int ds\, G_3(\mathbf{s}) \int d\mathbf{r}\delta^2(\mathbf{r}) - \tfrac{1}{2} \int\int G_1(\mathbf{r}')[\Delta(\mathbf{r} + \tfrac{1}{2}\mathbf{r}') - \Delta(\mathbf{r} - \tfrac{1}{2}\mathbf{r}')]^2 d\mathbf{r}d\mathbf{r}'$$

$$- \tfrac{1}{2} \int\int G_2(\mathbf{r}')[\Delta(\mathbf{r} + \tfrac{1}{2}\mathbf{r}') - \Delta(\mathbf{r} - \tfrac{1}{2}\mathbf{r}')][\delta(\mathbf{r} + \tfrac{1}{2}\mathbf{r}') - \delta(\mathbf{r} - \tfrac{1}{2}\mathbf{r}')\, d\mathbf{r}d\mathbf{r}'$$

$$- \tfrac{1}{2} \int\int G_3(\mathbf{r}')[\delta(\mathbf{r} + \tfrac{1}{2}\mathbf{r}') - \delta(\mathbf{r} - \frac{1}{2}\mathbf{r}')]^2 d\mathbf{r}d\mathbf{r}' \qquad \text{(A3.2.5)}$$

where Δ ($\mathbf{r}$) is the displaced charge given by

$$\rho(\mathbf{r}) = \rho_0 + \Delta(\mathbf{r}), \qquad \text{(A3.2.6)}$$

while δ is defined by

$$m(\mathbf{r}) = m_0 + \delta(\mathbf{r}). \qquad \text{(A3.2.7)}$$

Following the method used by Stoddart *et al.* (1971) for the nonmagnetic case, the g_i in the gradient development of Eq. (A3.2.3) can be related to the kernels G_i in the perturbative expansion of Eq. (A3.2.5).

Finally, defining the spin response $\chi(q)$ for the external field H through

$$\chi(q) = -\delta(\mathbf{q})/H(q), \qquad \text{(A3.2.8)}$$

Stoddart then obtains the result

$$\chi(q) = \frac{1}{2}\left[\tilde{G}_i(q, \rho_0, m_o) - \frac{G_{i2}^2(q, \rho_o, m_0)}{4[(2\pi/q)^2 + \tilde{G}_1]}\right]^{-1}, \qquad \text{(A3.2.9)}$$

where $\tilde{G}_i$ is defined in terms of Eq. (A3.2.5) as

$$\tilde{G}_i(q, \rho_o, m_0) = \int d\mathbf{r}G_i(\mathbf{r}) \exp(i\mathbf{q}\cdot\mathbf{r}). \qquad \text{(A3.2.10)}$$

If one (somewhat too drastically) illustrates Eq. (A3.2.9) by neglecting exchange and correlation, one finds, with $m_o = 0$,

$$G_1\,(q, \rho_0, 0) = -\tfrac{1}{4}\,K$$

$$G_2 = 0$$

$$G_3 = -K, \qquad \text{(A3.2.11)}$$

where $K = K(q, \rho_0)$ and $K(q)$ is the Lindhard-type response function

$$K = \frac{-2\pi^2}{k_f}\left[\frac{1}{2} + \frac{1-\eta^2}{4\eta}\ \ln\left|\frac{1+\eta}{1-\eta}\right|\right]^{-1} \qquad \text{(A3.2.12)}$$

with $\eta = q/2K_f$ and $\rho_0 = K_f^3/3\pi^2$ as usual.

Then $\chi(q)$ reduces to the free-electron result [cf. Hebborn and March, 1970)]

$$\chi(q) = \chi_{\text{Pauli}}\left[\frac{1}{2} + \left[\frac{1-\eta^2}{4\eta}\right]\ln\left|\frac{1+\eta}{1-\eta}\right|\right], \tag{A3.2.13}$$

which is closely connected also with the charge susceptibility $\chi_0(q)$ (see Appendix A3.5).

Extensive band-structure calculations, based on the functionals and corresponding spin-dependent one-body potentials discussed in the main text, have been carried out by Langlinais and Callaway (1972; see also the later work of Callaway and Chatterjee, 1978).

A3.3. KOSTER–SLATER-LIKE MODEL OF CRITERION FOR LOCAL MOMENT FORMATION

A summary is given below of the argument of Stoddart and March (1971), who obtained the local moment criterion of Eq. (4.4) from a Koster–Slater-like model, using spin density functional theory.

Having summarized the basic formulation of spin-dependent one-body potentials in the body of the text, we start by finding the eigenfunctions and the Green function when the Hartree–Fock interactions are switched on at one site, say $\mathbf{t}_\alpha$.

The Green function for a spin-dependent one-body potential satisfies the usual integral equation, with

$$V_\sigma(\mathbf{r}) = \bar{V}(\mathbf{r}) + v_\sigma(\mathbf{r}), \tag{A3.3.1}$$

where $v_\sigma(\mathbf{r})$ is a functional of $\Delta_\sigma(\mathbf{r}) = \rho_\sigma(\mathbf{r}) - \rho_p(\mathbf{r})$, which represents the spin density deviation from the nonmagnetic state, which is assumed to have electron density $\rho_p(\mathbf{r})$ for both + and − spin electrons. The integral equation is

$$G_\sigma(\mathbf{r}, \mathbf{r}_o E) = G_p(\mathbf{r}, \mathbf{r}_0 E) + \int d\mathbf{r}_1 G_p(\mathbf{r}, \mathbf{r}_1 E)\, v_\sigma(\mathbf{r}_1)\, G_0(\mathbf{r}_1\mathbf{r}_0 E), \tag{A3.3.2}$$

where G_p is evidently the nonmagnetic lattice Green function. Writing the general Bloch wave ($\psi_\mathbf{k}(\mathbf{r})$) expansion of $G_\sigma(\mathbf{r}\ \mathbf{r}_0 E)$ as

$$G_\sigma(\mathbf{r}, \mathbf{r}_o, E) = \sum_{\mathbf{k}_1\mathbf{k}_2} \psi_{\mathbf{k}_1}(\mathbf{r})\, \psi_{\mathbf{k}_2}(\mathbf{r}_0)\, \tilde{G}_\sigma(\mathbf{k}_1, \mathbf{k}_2, E). \tag{A3.3.3}$$

it follows that the above integral equation takes the form

$$\tilde{G}_\sigma(\mathbf{q}_1\mathbf{q}_2 E) = \frac{\Delta\mathbf{q}_1\mathbf{q}_2}{E - \varepsilon(\mathbf{q}_1) + i\eta} + \sum_{\mathbf{k}} \frac{\alpha_\sigma(\mathbf{q}_1\mathbf{k})\tilde{G}_\sigma(\mathbf{k}\mathbf{q}_2 E)}{E - \varepsilon(\mathbf{q}_1) + i\eta}, \tag{A3.3.4}$$

where

$$\alpha_\sigma(\mathbf{q}, \mathbf{k}) = \int d\mathbf{r}\ \psi_\mathbf{q}{}^* v_\sigma(\mathbf{r})\psi_\mathbf{k}(\mathbf{r})$$

$$= N^{-1} \sum_{\mathbf{nm}} e^{-i\mathbf{q}\cdot t_\mathbf{n} + i\mathbf{k}\cdot t_\mathbf{m}} \lambda_\sigma{}^{\mathbf{nm}} \qquad \text{(A3.3.5)}$$

and

$$\lambda_\sigma^{\mathbf{nm}} = \int d\mathbf{r}\ w^*(\mathbf{r} - \mathbf{t}_\mathrm{n})\ v_\sigma w(\mathbf{r} - \mathbf{t}_\mathrm{m}) \qquad \text{(A3.3.6)}$$

where the w represent the Wannier states.

With the one-site assumption, it is clear from Eq. (A5.3.5) that we must write

$$\alpha_\sigma(\mathbf{q}, \mathbf{k}) = N^{-1}\ e^{i\mathbf{t}_a\cdot(\mathbf{k}-\mathbf{q})} U\delta\mathbf{n}_{-\sigma}^{(a)}, \qquad \text{(A3.3.7)}$$

since

$$\lambda_\sigma^{\mathbf{m}l} = U\ \delta\mathbf{n}_{-\sigma}^{(a)}\ \Delta_{\mathbf{m}a}\Delta_{\mathbf{l}a}. \qquad \text{(A3.3.8)}$$

Making use of the expansion

$$\phi_{k\sigma}(\mathbf{r}) = \sum_\mathbf{n} b_\sigma(\mathbf{k}, \mathbf{t}_\mathrm{n})\ w(\mathbf{r} - \mathbf{t}_\mathbf{n}) \qquad \text{(A3.3.9)}$$

in terms of the Wannier functions, one finds

$$b_\sigma(\mathbf{k}, \mathbf{t}_\mathbf{n}) = e^{i\,\mathbf{k}\cdot\mathbf{t}}\mathbf{n} + \sum_{\mathbf{m,l}} F(\mathrm{n}, \mathrm{m}, E)\ \lambda_\sigma^{\mathbf{m}\ell} b_\sigma(\mathbf{k}, \mathbf{t}_\mathbf{l}) \qquad \text{(A3.3.10)}$$

with

$$F(n, m, E) = N^{-1} \sum_\mathbf{q} \frac{e^{i\,\mathbf{q}\cdot(t_n - t_m)}}{E - \varepsilon(\mathbf{q}) + i\eta}. \qquad \text{(A3.3.11)}$$

In particular, the diagonal form is given by

$$F(n, n, E) = F(E)$$

$$= N^{-1} \sum_\mathbf{q} \frac{1}{E - \varepsilon(\mathbf{q}) + i\eta}$$

$$= P(E) - i\pi\ n(E), \qquad \text{(A3.3.12)}$$

which is the usual Koster–Slater function, P denoting the principal part:

$$P(E) = P \int dE' \frac{n(E')}{E - E'} \qquad \text{(A3.3.13)}$$

with $n(E)$ the density of states per atom.

Also, it follows that at the site a at which the interactions are switched on,

$$b_\sigma(\mathbf{k}, \mathbf{t}_a) = \frac{e^{i\mathbf{k}\cdot\mathbf{t}_a}}{1 - U\delta n_{-\sigma}^{(a)} F(E)}. \qquad \text{(A3.3.14)}$$

In terms of the eigenfunctions the change in the number of σ electrons on site *a* in going from the nonmagnetic to the magnetic state is given by

$$\delta\, n_\sigma^{(a)} = N^{-1} \int_{-\infty}^{\mu'} dE\, n_\sigma(E)\, |b_\sigma(E, \mathbf{t}_a)|^2 - N^{-1} \int_{-\infty}^{\mu} dE\, n(E), \quad \text{(A3.3.15)}$$

where the Fermi energy μ' is defined by

$$\sum_\sigma \int_{-\infty}^{\mu'} dE\, n_\sigma(E) = 2 \int_{-\infty}^{\mu} dE\, n(E), \quad \text{(A3.3.16)}$$

$n(E)$ being the nonmagnetic lattice density of states, with μ the corresponding Fermi energy. As n_σ and b_σ depend only on the parameters $\delta n_\sigma^{(a)}$ that represent the density dependence, Eq. (A3.1.16) is the fundamental Euler equation in the density-functional formalism.

All that remains to investigate whether a magnetic solution corresponding to $\delta n_\sigma(a) = 0$ is possible is to use the expansions of $|b_\sigma|^2$ and n_σ for small $\delta n_\sigma^{(a)}$ (Stoddart and March, 1971). The desired local moment criterion in Eq. (4.10) then follows readily.

It is relevant here to mention the all-electron local density theory of local moments in metals by Delley *et al.* (1983a and b). Also elaborating a little the local moment work of Roth (1978), she obtained the same local moment criterion as Stoddart and March (1971) in a coherent potential approximation calculation of the high-temperature state of itinerant magnets. The same result also arises in the spin fluctuation theory of Usami and Moriya (1980; for some contact of such theories with experiment, see Lowde *et al.*, 1983). One must also mention in the present context the further studies of magnetism in transition metals at elevated temperatures by You *et al.* (1980), You and Heine (1982), and Holden and You (1982).

A3.4. DYNAMIC PROPERTIES OF JELLIUM

While plasmons will be discussed at length in Chapter 6, it is of interest to supplement the discussion of static correlation functions in jellium given in Chapter 3 by a brief account of some dynamic properties of jellium. The account below follows closely that of Holas (1993).

A3.4.1. Response Function and Screening

In order to study such dynamic properties of jellium, let us perturb it at time $t = t_0$ by switching on a weak external potential $\phi_{\text{ext}}(\mathbf{r}, t)$. The corresponding perturbing Hamiltonian is

$$H_{\text{ext}}(t) = \int d\mathbf{r}\, \hat{n}(\mathbf{r}, t)(-e)\phi_{\text{ext}}(\mathbf{r}, t), \quad \text{(A3.4.1)}$$

where $\hat{n}(\mathbf{r}, t)$ is the particle density field operator of the unperturbed system (the jellium). Using perturbation theory, one calculates the change in the electron density due to the perturbation (note that the unperturbed density n is uniform:

$$\langle \hat{n}(\mathbf{k}, \omega) \rangle - n = \chi^R(\mathbf{k}, \omega)(-e)\phi_{ext}(\mathbf{k}, \omega) + O(\phi_{ext}^2). \qquad (A3.4.2)$$

Here the ($\mathbf{r}$,t) dependence has been Fourier transformed to ($\mathbf{k}$, ω); $\chi^R(\mathbf{k}, \omega)$ defines the linear part of the response to the perturbation and is termed the response function. In order to distinguish it from responses to other possible perturbations, it is usually referred to as the density–density response function, as it represents a retarded (R) correlation function of two density–deviation operators. It should be noted that the Feynman diagrammatic technique (see e.g. March *et al.*, 1995) applies to the time-ordered operators only. So let us define χ^{TO} ($\mathbf{k}$, ω), an associated time-ordered (TO) correlation function of the same operators as occur in $\chi^R(\mathbf{k}, \omega)$. The two functions are closely related; in particular, for real ω,

$$\mathrm{Re}\, \chi^R(\mathbf{k}, \omega) = \mathrm{Re}\, \chi^{TO}(\mathbf{k}, \omega),$$
$$\mathrm{Im}\, \chi^R(\mathbf{k}, \omega) = \mathrm{Im}\, \chi^{TO}(\mathbf{k}, \omega)\, \mathrm{sgn}(\omega). \qquad (A3.4.3)$$

If χ^{TO} ($\mathbf{k}$, ω) is obtained in some approximation, by evaluation of the appropriate diagrams, then, using Eq. (A3.4.3), one has χ^R ($\mathbf{k}$, ω) in the same approximation. Because of the isotropy of the jellium, χ depends on $k = |\mathbf{k}|$ rather than on $\mathbf{k}$. The interpretation of the diagrammatic representation of $\chi(k, \omega)$ suggests another name for this function—the (full) polarizability. An important partial summation of the infinite series of diagrams representing χ allows it to be written in the form

$$\chi(k, \omega) = \frac{\pi(k, \omega)}{1 - v(k)\pi(k, \omega)}, \qquad (A3.4.4)$$

where

$$v(k) = 4\pi e^2/k^2. \qquad (A3.4.5)$$

is the Fourier transform (FT) of the Coulombic electron–electron interaction, while $\pi(k, \omega)$, the proper (or irreducible) polarizability, representing a sum of such polarization diagrams that cannot be disconnected by breaking one interaction line only [i.e., a line representing $v(k)$].

The expression in the denominator of Eq. (A3.4.4),

$$\varepsilon(k, \omega) = 1 - v(k)\Pi(k, \omega), \qquad (A3.4.6)$$

is the so-called dielectric function; it also defines an effective (or screened) electron–electron interaction as

$$v_{eff}(k, \omega) = v(k) / \varepsilon(k, \omega). \qquad (A3.4.7)$$

It can be shown that, for small k,

$$v_{\text{eff}}(k, \omega) = 4\pi e^2 / (k^2 + k_{\text{TF}}^2), \tag{A3.4.8}$$

where k_{TF}, the so called Thomas–Fermi wave number (see also Appendix 6.2), equals

$$k_{\text{TF}} = 1.563\ r_s^{-1/2}\ a_B^{-1}. \tag{A3.4.9}$$

One sees from Eq. (A3.4.8) that the screening present in the electron gas removes the singular behavior of the bare interaction (A3.4.5), for $k \to 0$.

A3.4.2. Random Phase Approximation and Beyond

The sum of the diagrams representing the proper polarizability can be written as the following series:

$$\Pi(k, \omega) = \Pi_0(k, \omega) + \Pi_1(k, \omega) + \ldots, \tag{A3.4.10}$$

where Π_n represents a sum of all irreducible diagrams possessing n lines of interaction. While Π_0 and Π_1 can be evaluated separately, any separate higher-order term is infinite in the thermodynamic limit, but their sum remains finite, of course.

The diagram of the leading term $\Pi_0(k, \omega)$—a "loop" of two Green functions—can be evaluated in terms of elementary functions. The result is the so-called Lindhard function. An approximate theory, in which Π is represented only by its leading term,

$$\Pi(k, \omega) \approx \Pi_0(k, \omega), \tag{A3.4.11}$$

is called, for historical reasons, the random phase approximation (RPA). After replacing Π by Π_0 in Eq. (A3.4.6), one obtains the RPA dielectric function. Doing the same in Eq. (A3.4.4) gives the RPA response function. In this way, the RPA version may be obtained for any jellium property derivable from the response function. Some examples are

- The dynamic structure factor (giving the cross section of the scattering of any probe of the electron system):

$$S(k, \omega) = \begin{cases} -\hbar (\pi n)^{-1}\ \text{Im}\ \chi(k, \omega) & \text{for } \omega > 0 \\ 0 & \text{for } \omega < 0 \end{cases}; \tag{A3.4.12}$$

  it should be noted that this function can also be written in terms of $\varepsilon(k, \omega)$, because, from Eqs. (A3.4.4) and (A3.4.6), one has

$$v(k)\ \text{Jm}\ \chi(k, \omega) = \text{Jm}\ [1/\varepsilon(k, \omega)]; \tag{A3.4.13}$$

- The static structure factor (see also March, 1981):

$$S(k) = \int d\omega\, S(k, \omega); \tag{A3.4.14}$$

- The pair correlation function

$$g(r) = 1 + n^{-1}(2\pi)^{-3} \int d\mathbf{k}\, [S(k)-1] \exp(i\mathbf{kr}); \tag{A3.4.15}$$

- And the plasmon, a collective excitation in the electron system (see especially Chapter 6). Its dispersion $\omega = \omega_{\rm pl}(k)$ and the reciprocal lifetime $\gamma(k)$ represent a complex solution of the equation

$$\varepsilon(k, \omega_{\rm pl}(k) - i\gamma(k)) = 0. \tag{A3.4.16}$$

From Eqs. (A3.4.12), (A3.4.13), and (A3.6.16), it is clear that a sharp peak must be observed for $S(k, \omega)$ at the plasmon frequency (for details, see Chapter 6). In the RPA, this peak is given by a delta function; the plasmon is undamped.

Although, from its definition in Eq. (A3.4.11), the RPA seems to be a very crude approximation, it nevertheless describes most of the jellium properties quite well over a wide range of r_s. This is due to the fact that χ in the form of Eq. (A3.4.4) still represents a sum of an infinite number of diagrams, even if Π is taken in the lowest order as in Eq. (A3.4.11).

Since the RPA results are well known, it is often convenient to write the exact χ in a RPA-like manner, allowing, however, all the necessary corrections to be represented as a modification of the interaction

$$\chi(k, \omega) = \frac{\Pi_0(k, \omega)}{1 - v(k)[1 - G(k, \omega)]\Pi_0(k, \omega)}. \tag{A3.4.17}$$

The function $G(k, \omega)$ is called the local-field correction. Equating χ in its two forms, Eqs. (A3.4.4) and (A3.4.17), one obtains for G the expression

$$G(k, \omega) = (1/v(k))\, [1/\Pi(k, \omega) - 1/\Pi_0(k, \omega)]. \tag{A3.4.18}$$

This equation can be used to express Π in terms of Π_0 and G:

$$\Pi(k, \omega) = \frac{\Pi_0(k, \omega)}{1 + V(k)G(k, \omega)\Pi_0(k, \omega)}. \tag{A3.4.19}$$

Although improvement over the RPA can be achieved by taking into account the next term in the expansion (A3.4.10), this becomes insufficient for r_s exceeding 2. Therefore nonperturbative approaches to evaluation of Π (or G) are of interest. An important contribution in this respect is the STLS theory (see following Appendix) of semiclassical origin (due to Singwi, Tosi, Land, and Sjölander,*

*Referred to below as STLS: see Appendix A3.5.

1968) and its subsequent development and quantum generalization. In this approach the local-field correction, simplified to a static, real function $G(k)$ in the earlier papers and treated as a complex, frequency-dependent $G(k, \omega)$ in some generalizations, is calculated in a self-consistent scheme [see especially the review of Singwi and Tosi (1981)].

A3.5. Static Local-Field and Dielectric Function Applied to Screened Interactions in Metals

The local field correction $G(k, \omega)$ in Appendix A3.4 will first be set down in the static STLS approximation. Then the related static dielectric function $\varepsilon(k)$ will be applied to calculate screened interactions in simple metals (e.g., Na: see Perrot and March, for DFT calculations of the screened interaction). The account below follows closely that of Tosi (1994: see also March and Tosi, 1995).

Whereas in Hartree–Fock theory the radial distribution function g(r) gives the local structure as merely due to exchange between electrons with parallel spins ("Fermi hole"), the Coulomb repulsions induce additional local structure mainly through correlations between electrons with antiparallel spins ("Coulomb hole"). This effect is overestimated in the RPA (see Appendix A3.4): it yields values of g(r) that are increasingly negative at small r with increasing r_s from $r_s \approx 1$. As a consequence the RPA ground-state energy is a poor approximation for electrons at metallic densities.

Having recalled the above main points about local structure in the electron plasma, let us stress that there is a physically evident connection between local structure and dielectric response: as a consequence of the Fermi and Coulomb hole the local density of polarizable electron fluid around an electron is lower than the average density n. An analogous problem is well known in the theory of dielectrics: the Lorentz local field is introduced there in order to account for the role of the local microscopic structure of the dielectric medium around any given atom in determining the effective field that acts on the atom to polarize it.

From this analogy one infers that, in a weak external potential, the effective potential seen by an average electron consists of three terms: (a) the external potential $V_e(\mathbf{q}, \omega)$, (b) the potential $(4\pi e^2/q^2)n(\mathbf{q}, \omega)$ of the polarization charges (which added to (a) gives the Hartree potential), and (c) an additional "local field" potential associated with the local structure. Let us write the latter as $-(4\pi e^2/q^2)G(q,\omega)n(\mathbf{q},\omega)$ where $G(q,\omega)$ is the "local field factor". One expects $G(q,\omega)$ to be related (in an as yet unknown manner) to the (as yet unknown) dynamical local structure. The frequency dependence of the local field is due to the inertia of the Fermi and Coulomb hole.

STLS approximation. One assumes that the electron fluid responds as an ideal Fermi gas to the effective potential introduced just above (in the static case this assumption is fully justified within density functional theory):

$$n(\mathbf{q}, \omega) = \chi_0(q,\omega)\left[V_e(\mathbf{q},\omega) + \frac{4\pi e^2}{q^2}\, n(\mathbf{q},\omega) - \frac{4\pi e^2}{q^2}\, G(\mathbf{q}, \omega)\, n(\mathbf{q},\omega)\right] \tag{A3.5.1}$$

namely

$$\chi(q,\omega) = \frac{\chi_0(\mathbf{q},\omega)}{1 - \dfrac{4\pi e^2}{q^2}\,[1 - G(q,\omega)]\chi_0(q,\omega)}. \tag{A3.5.2}$$

The STLS scheme introduces two approximations in order to determine the local field factor self-consistently with the local structure: (1) the frequency dependence of $G(q,\omega)$ is neglected, and (2) $G(q)$ is explicitly related to the radial distribution function by

$$G(q) = -\frac{1}{n}\int \frac{d\mathbf{k}}{(2\pi)^3}\,\frac{\mathbf{q}\cdot\mathbf{k}}{k^2}\,[S(|\mathbf{q} - \mathbf{k}|) - 1]. \tag{A3.5.3}$$

This expression was justified by STLS using an analysis of kinetic equations for the plasma. At any rate, since at this point $\chi(q,\omega)$ is an explicit functional of the local structure described by $S(q)$, one may use the fluctuation-dissipation theorem

$$S(q) = \int_{-\infty}^{\infty} \frac{d\omega}{2\pi}\, S(q,\omega) = -\,\frac{n}{\pi}\int_0^{\infty} d\omega\, \mathrm{Im}\,\chi(q,\omega) \tag{A3.5.4}$$

to simultaneously determine both $G(q)$ and $S(q)$ in a self-consistent manner.

As an illustration we report below the calculated values for the correlation energy (in Ryd) compared with the results from quantum Monte Carlo (see Chapter 5 below):

r_s	1	2	5	10	20
STLS	−0.124	−0.092	−0.056	−0.036	−0.022
QMC	−0.120	−0.090	−0.056	−0.037	−0.023

(by definition the ground state energy is the sum of the correlation energy and of the Hartree–Fock energy).

The coefficient of the q^2 term in the plasmon dispersion relation (see especially Chapter 6) is substantially smaller than in RPA and becomes negative with increasing r_s (as was observed in electron-energy-loss experiments on alkali metals by Fink *et al.*, 1989). Subsequent developments of the theory have mainly been aimed at reaching self-consistency with the compressibility sum rule and at accounting for a frequency dependence of the local field factor (Singwi and Tosi,

1981; Ichimaru, 1982) and work on two-dimensional electron plasmas by Neilson *et al.*). The local field factor G(q,0) in the static case is related to the exchange and correlation energy and the dielectric function $\varepsilon(k)$ within the framework of density functional theory (DFT) (a dynamic extension within time-dependent DFT is immediate).

Screened Interactions in Metals

Below let us merely sketch the consequences of these results for the electron theory of simple metals. Consider for simplicity an ion inside the fluid of interacting electrons as a static point charge Z_1e which is linearly screened by the electrons. A second ion with charge Z_2e sees the Hartree potential created by the first, and hence the effective interaction between the two ions is (see also Corless and March, 1961):

$$V_{ii}(k) = \frac{4\pi Z_1 Z_2 e^2}{k^2 \varepsilon(k)}, \ \varepsilon(k) = 1 - \frac{4\pi e^2}{k^2} \chi_0(k)\left[1 - \frac{4\pi e^2}{k^2} G(k, 0)\chi_0(k)\right]^{-1}. \tag{A3.5.5}$$

On the other hand, the effective ion-electron interaction is given by the self-consistent Slater–Kohn–Sham potential, and may in this case be written as

$$V_{ie}(k) = -\frac{4\pi Z_1 e^2}{k^2 \varepsilon_p(k)}, \ \varepsilon_p(k) = 1 - \frac{4\pi e^2}{k^2}[1 - G(k, 0)\chi_0(k)]^{-1}. \tag{A3.5.6}$$

$\chi_0(k)$ being the static charge susceptibility (compare Eq. (A3.2.12). The two dielectric functions giving the screening of the bare interactions are different, because of exchange and correlation with the electronic screening cloud.

In practice, linear response theory gives useful results for simple metals because the divergence of the bare Coulomb potential generated by an ion is probed neither by the other ions nor by the conduction electrons. The orthogonality of the conduction-electron states to the occupied core states may be embedded into a *pseudopotential* which keeps the conduction electrons mostly outside the ionic cores (for a review of pseudopotentials and their applications, see Seitz and Turnbull, 1970). The screened ion-ion interaction with specific reference to liquid metals will be considered in more detail in Chapter 8.

Appendix to Chapter 4

A4.1. MODEL OF TWO-ELECTRON HOMOPOLAR MOLECULE

Here a summary of the results of the model of Falicov and Harris (1969) of a two-electron homopolar molecule will be presented. To define their one-band Hamiltonian, it is convenient to use second quantization operators and to restrict the model to four orbitals, one of each spin on each of the two centers. The Hamiltonian then takes the form

$$H = H_\alpha + H_\beta + H_u + H_K, \tag{A4.1.1}$$

where

$$H_\alpha = \alpha\,(n_{1\uparrow} + n_{1\downarrow} + n_{2\uparrow} + n_{2\downarrow}) \tag{A4.1.2}$$

$$H_\beta = =\beta(c^\dagger_{1\uparrow}\, c^\dagger_{2\uparrow} + c^\dagger_{1\downarrow}\, c_{2\downarrow} + c^\dagger_{2\uparrow}\, c_{1\uparrow} + c^\dagger_{2\downarrow}\, c_{1\downarrow}) \tag{A4.1.3}$$

$$H_u = U(n_{1\uparrow}\, n_{1\downarrow} + n_{2\uparrow}\, n_{2\downarrow}) \tag{A4.1.4}$$

$$H_K = K(n_{1\uparrow}\, n_{2\downarrow} + n_{1\uparrow}\, n_{2\downarrow} + n_{1\downarrow}\, n_{2\uparrow} + n_{1\downarrow}\, n_{2\downarrow}). \tag{A4.1.5}$$

Here $c^\dagger_{i\sigma}$, $c_{i\sigma}$, and $n_{i\sigma} = c^\dagger_{i\sigma}\, c_{i\sigma}$ are, respectively, creation, annihilation, and number operators for the orbital of spin σ centered on nucleus i. H_α and H_β are, respectively, the single-particle diagonal and off-diagonal terms, H_u is the intraatomic Coulomb repulsion, while H_k is the corresponding interatomic term. The parameters β, U, and K are positive definite quantities such that $U > K$.

Since one is considering only two-electron states, the following results can be exploited:

(i) The electron-number operator

$$N = \sum_i n_{i\sigma} \tag{A4.1.6}$$

is completely diagonal and can everywhere be replaced by the number 2, e.g.,

$$H_\alpha = 2\alpha. \tag{A4.1.7}$$

(ii) The operator

$$N^2 = \sum_{ij\sigma\sigma'} n_{i\sigma} n_{j\sigma'} \tag{A4.1.8}$$

is also completely diagonal and can everywhere be replaced by the number 4.

(iii) Recalling that

$$n_{i\sigma} n_{i\sigma} \equiv n_{i\sigma} \tag{A4.1.9}$$

for any i and σ, the results (i) and (ii) immediately above can be used to write the identity

$$4 = 2 + 2\,(n_{1\uparrow}n_{1\downarrow} + n_{2\uparrow}n_{2\downarrow}) + 2(n_{1\uparrow}n_{2\uparrow} + n_{1\uparrow}n_{2\downarrow} + n_{1\downarrow}n_{2\uparrow} + n_{1\downarrow}n_{2\downarrow}) \tag{A4.1.10}$$

(iv) The Hamiltonian can be rewritten

$$H = 2\alpha + K + \mathcal{H}, \tag{A4.1.11}$$

with

$$\mathcal{H} = H_\beta + H_u + (H_k - K), \tag{A4.1.12}$$

where the final term can be written

$$H_K - K = -K(n_{1\uparrow}n_{1\downarrow} + n_{2\uparrow}n_{2\downarrow}) = -(K/U)H_u. \tag{A4.1.13}$$

From the above expressions, or alternatively from simple physical arguments, it can readily be seen that the eigenvalues of H can depend on U and K only through the difference U–K. This permits one to write the ground-state energy E as

$$E = \langle H \rangle = 2\alpha + K + \varepsilon, \tag{A4.1.14}$$

where

$$\varepsilon = \langle \mathcal{H} \rangle = \beta\varepsilon(x), \tag{A4.1.15}$$

with ε dependent only on the variable $x = (U - K)/\beta$.

Exact Ground State. Any eigenstate of Eq. (A4.1.12) with two electrons should be a linear combination of the six states

$$|1\uparrow 2\uparrow\rangle,\quad |1\downarrow 2\downarrow\rangle,\quad |1\uparrow 1\downarrow\rangle,\quad |2\uparrow 2\downarrow\rangle,\quad |1\uparrow 2\downarrow\rangle, \text{ and } |2\uparrow 1\downarrow\rangle,$$

where

$$|i\sigma j\sigma'\rangle = c^\dagger_{i\sigma} c^\dagger_{j\sigma'} |0\rangle, \tag{A4.1.16}$$

TABLE A4.1.1. Matrix elements of H

	$\|1\uparrow 2\uparrow\rangle$	$\|1\downarrow 2\downarrow\rangle$	$\|1\uparrow 1\downarrow\rangle$	$\|2\uparrow 2\downarrow\rangle$	$\|1\uparrow 2\downarrow\rangle$	$\|2\uparrow 1\downarrow\rangle$
$\langle 1\uparrow 2\uparrow\|$	0	0	0	0	0	0
$\langle 1\downarrow 2\downarrow\|$	0	0	0	0	0	0
$\langle 1\uparrow 1\downarrow\|$	0	0	U–K	0	–β	–β
$\langle 2\uparrow 2\downarrow\|$	0	0	0	U–K	–β	–β
$\langle 1\uparrow 2\downarrow\|$	0	0	–β	–β	0	0
$\langle 2\uparrow 1\downarrow\|$	0	0	–β	–β	0	0

with $|0\rangle$ as the vacuum state. The matrix elements of H in this manifold are given in Table A4.1.1, taken from Falicov and Harris (1969).

An exact diagonalization of the Hamiltonian matrix yields, for the ground-state $|G\rangle$:

$$|G\rangle = 2[16 + x^2(x^2 + 16)^{1/2}]^{-1/2}(|1\uparrow 1\downarrow\rangle + |2\uparrow 2\downarrow\rangle) + 0.5[x + (x^2 + 16)^{1/2}]$$
$$[16 + x^2 + x(x^2 + 16)^{1/2}]^{-1/2} \times (|1\uparrow 2\downarrow\rangle + |2\uparrow 1\downarrow\rangle), \qquad \text{(A4.1.17)}$$

while the energy, expressed in the variables discussed above, is characterized by

$$\varepsilon_G = 0.5[x-(x^2 + 16)^{1/2}]. \qquad \text{(A4.1.18)}$$

Falicov and Harris have used these exact results to assess the accuracy of various approximate solutions such as the Heitler–London and molecular orbital solutions discussed in Section 4.1, but the details will be omitted here, except to say that the most successful approximate solutions of their model are in the form of symmetrized spin density wave trial functions.

The generalization of this model to more complicated, many-electron chains has also been discussed (Harris and Falicov, 1969; Fenton, 1968).

A4.2. DEPENDENCE ON ATOMIC NUMBER OF CORRELATION ENERGIES IN NEUTRAL ATOMS

The correlation energy of neutral atoms was extracted from experiment by Clementi (1963) for nuclear charge Z up to 18. His results are shown by the continuous line in Fig. A4.2.1. It can be seen that there is a grossly linear dependence of the correlation energy E_c on Z for $Z < 18$. The aim of this Appendix is to propose an interpretation of this result (March and Wind, 1992).

As a starting point, let us write the total correlation energy E_c in terms of the

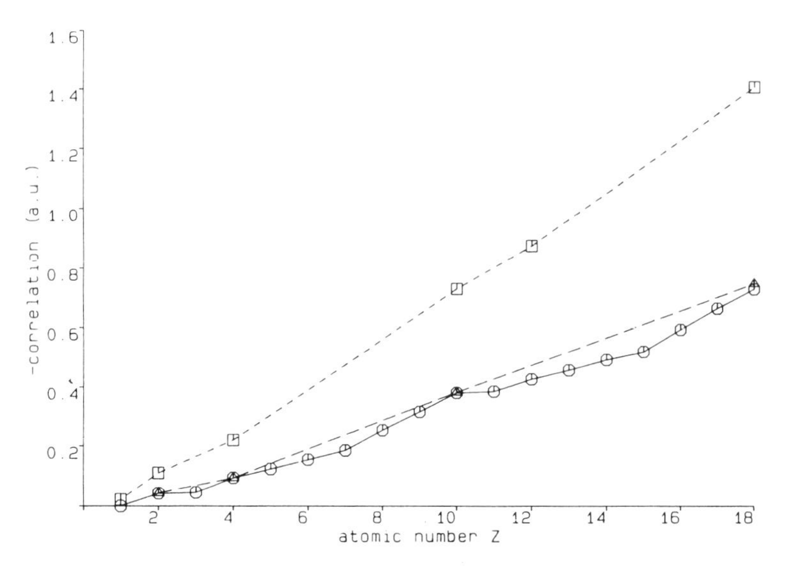

FIGURE A4.2.1. Magnitudes of total correlation energies in light atoms for $Z \leq 18$. 'Experiment' (o) corresponds to the results of Clementi (1963). Also shown for comparison are results of local density approximation (upper curve) and of the Colle–Slavetti (1975) correlation energy (– –Δ– –) after Lee *et al.* (1988). (Dashed lines are merely guides for the eye). (From March and Wind, 1992).

local correlation energy per electron ε_c (r) at distance r from the nucleus and the ground state density $\rho(r)$ as

$$E_c = \int \rho(r)\varepsilon_c(r)d\mathbf{r} = \int_0^\infty D(r)\varepsilon_c(r)dr \tag{A4.2.1}$$

where $D(r)$ is the radial electron density $4\pi r^2$ $\rho(r)$.

If one can assume that $\varepsilon_c(r)$ is slowly varying over the range of r, centered around some value r, to be specified below, for which the magnitude of $D(r)$ is significant, then one can write

$$E_c \simeq \varepsilon_c(\bar{r}) \int_0^\infty D(r)\mathrm{dr} = \varepsilon_c(r)Z, \tag{A4.2.2}$$

the last step following from normalization of $\rho(r)$ for neutral atoms. Returning to Fig. A4.2.1, the approximate linearity with Z would then evidently follow, provided $\varepsilon_c(r)$ varies only slowly with atomic number Z. With these two assumptions presented at the outset to provide a focus for the ensuing discussion, let us turn to the approximate routes that currently allow estimates of $\varepsilon_c(r)$ entering Eq. (A4.2.1) to be made.

The simplest of these routes, motivated by the Thomas–Fermi–Dirac statistical theory (see Chapter 3), is to replace in Eq. (A4.2.1) the exact ε_c by the correlation energy per electron of a homogeneous electron assembly, but with the constant density of that system replaced by the inhomogeneous atomic ground-state density $\rho(r)$. This is the local density approximation (LDA) to the correlation energy. Since the correlation energy of the homogeneous electron system is known from the quantum Monte Carlo simulations of Ceperley and Alder (see Chapter 5), this approximation is readily evaluated for neutral atoms. Such results for E_c have been obtained, for example, by Perdew and Zunger (1981), and are shown in Fig. A4.2.1. Though the approximate linearity with Z is already apparent, it is also clear that this local density approximation is too crude to give more than the general trend for atoms, because of its neglect of electron density gradients.

Fortunately, an alternative proposal is available from the work of Colle and Salvetti (1975), whose results have been cast into density functional form by Lee *et al.* (1988). Their numerical findings are also shown in Fig. A4.2.1 and are seen to reflect not only the correct trend with Z but also magnitudes which accord quite well with Clementi's results.

To examine further the two assumptions made in relation to Eq. (A4.2.2), Fig. A4.2.2 has been constructed using the two approximate approaches detailed above, where $\varepsilon_c(r)$ is plotted for the heavier closed-shell atoms Kr and Xe using the almost Hartree–Fock quality densities given by Clementi and Roetti (1974). In this figure, the radial distance r is scaled as essentially $rZ^{1/3}/a_0$, with a_0 the Bohr radius $\hbar^2/me^2$ or more precisely the quantity x in Fig. A4.2.2 is the scale of the Thomas–Fermi atom (see Chapter 2)

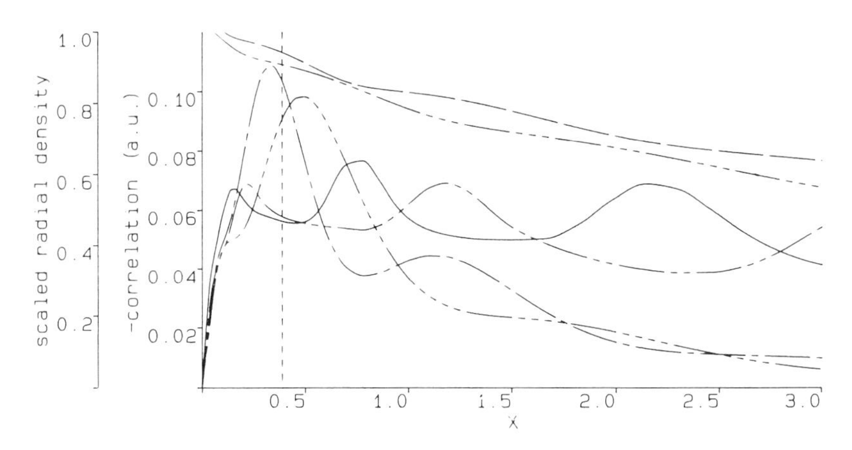

FIGURE A4.2.2. Spatial variation of both $D(r)$ and $\varepsilon_c(r)$ in Eq. (A4.2.1) for closed-shell atoms Kr and Xe. Clementi–Roetti densities of almost Hartree–Fock quality were employed. Upper two curves show $\varepsilon_c(r)$ from local density approximation, while the two other curves of $\varepsilon_c(r)$ are calculated from the Colle–Salvetti (1975) approximation. $D(r)/Z^{4/3}$ is the other quantity plotted, using Clementi–Roetti densities for Kr and Xe. (From March and Wind, 1992).

$$x = r/b; \quad b = 0.88534a_0/Z^{1/3}. \tag{A4.2.3}$$

The scaled radial densities $D(r)/Z^{4/3}$ for Kr and Xe are also depicted in Fig. A4.2.2. Though there is some (minor) structure in $\varepsilon_c(r)$ for these two atoms in the local density approximation, it will be seen that $\varepsilon_c(r)$ is indeed slowly varying over the range of r for which $D(r)$ has significant magnitude, for both Kr and Xe, justifying therefore the approximate formula in Eq. (A4.2.2). At the maximum of $D(r)$, occurring at position r_m, say, $\varepsilon_c(r_m)$ is also insensitive to whether Kr or Xe is considered. Though this local density approximation therefore lends support to the two assumptions proposed above in relation to Eq. (A4.2.2), it is too crude to explain Clementi's results for the light atoms considered in Fig. A4.2.1. Therefore in Fig. A4.2.2 the results for $\varepsilon_c(r)$ for Kr and Xe are plotted from the Colle–Salvetti (1975) formula (see also Lee *et al.*, 1988) which involves explicitly not only $\rho(r)$ but also the first and second derivatives of the density. Though there is now more structure in $\varepsilon_c(r)$ than for the local density approximation, the prime conclusions are again that (*i*) $\varepsilon_c(r)$ has a characteristic scale that is large compared with the half-width of $D(r)$, and (ii) $\varepsilon_c(r_m)$ is not appreciably different for Kr and Xe. Furthermore, as already noted from Fig. A4.2.1 (see also Lee *et al.*, 1988), the total correlation energies E_c are in much better agreement (see Fig. A4.2.1) with Clementi's results for $Z \leq 18$ than are the local density values of heavy neutral atoms. First of all this universal treatment gives for the position r_m of the maximum in $D(r)$

$$r_m = bx_m; \quad x_m = 0.386 \tag{A4.2.4}$$

which is shown by the vertical dashed line in Fig. A4.2.2. This is already a useful prediction for the position of the maxima in $D(r)$ for both Kr and Xe from the Hartree–Fock quality densities. Second, $\varepsilon_c(r_m)$ for the Colle–Salvetti approximation using this universal Thomas–Fermi density description is also plotted for neutral atoms versus atomic number Z in Fig. A4.2.3. The anticipated weak dependence of $\varepsilon_c(r_m)$ on Z is evident for the larger values of Z in the plot, at which the Thomas–Fermi approximation becomes useful. It can be shown analytically in connection with Fig. A4.2.3 that this curve tends to a constant as Z tends to infinity (of course, all the time within the nonrelativistic theory used in this volume), the correction term for large Z being of order $Z^{-2/3}$ (March and Wind, 1992).

This is now the point at which to take up the determination of the characteristic distance in $\varepsilon_c(r)$, over which $\varepsilon_c(r)$ varies by an appreciable fraction of itself. This is most readily discussed by referring again to the physics of the homogeneous electron liquid. In this system (see Chapters 3 and 6) the long-range Coulomb interaction e^2/r_{ij} between electrons i and j at separation r_{ij} gives rise to collective modes; the plasma oscillations with characteristic frequency $\omega_p = (4\pi\rho_0 e^2/m)^{1/2}$, with ρ_0 the now-homogeneous electron density. The important length, l_c say, in this system is then the product of the characteristic time $2\pi/\omega_p$ and

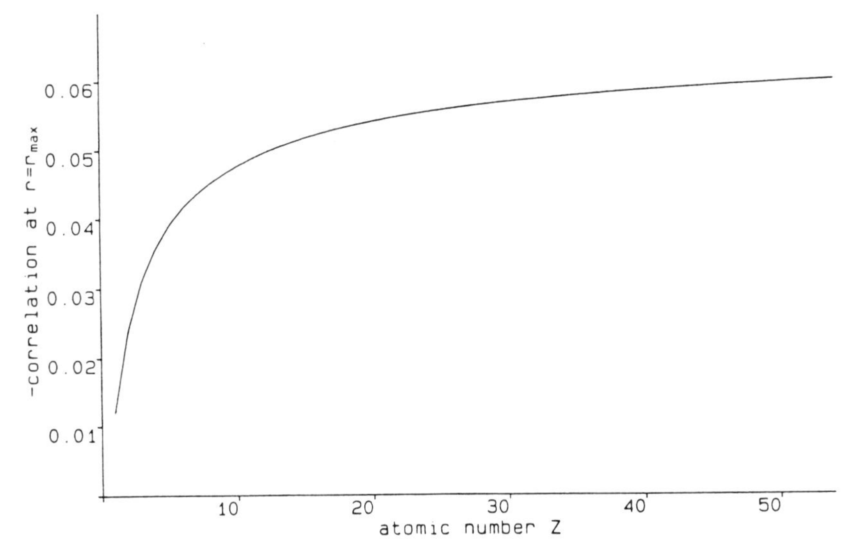

FIGURE A4.2.3. Local correlation energy per electron $\varepsilon_c(r_m)$ at maximum of radial density *D(r)* plotted vs atomic number *Z* using a Colle–Salvetti form of correlation energy, after Lee *et al.* (1988), and approximating *D(r)* by the universal Thomas–Fermi self-consistent field density for heavy neutral atoms. The important conclusion here is the insensitivity of $\varepsilon_c(r_m)$ (of Eq. A4.2.2)) to atomic number *Z* for *Z* sufficiently large to validate the Thomas–Fermi density (From March and Wind, 1992).

the Fermi velocity $v_f = \hbar k_f/m$, with k_f, the Fermi wave number, related to ρ_0 by $k_f = (3\pi^2)^{1/3}\rho_0^{1/3}$. After accounting for this collective behavior, the residual interaction left between electrons is of the screened form $(e^2/r_W)\exp(-r_{ij}/l_c)$ where l_c is plainly the degenerate analog of the classical Debeye–Hückel length, namely the Thomas–Fermi screening length (cf. Appendices 3.5 and 6.2). In the inhomogeneous atomic charge cloud, one can write, essentially in the local density approximation,

$$l_c^{\text{atomic}} = \frac{a_0^{1/2}}{2}\left(\frac{\pi}{3\rho(r_m)}\right)^{1/6}, \tag{A4.2.5}$$

where ρ_0 in the homogeneous liquid treated above has now been replaced by the characteristic atomic density $\rho(r_m)$, that is, the atomic density at the distance defined by the maximum in $D(r)$. Using Eq. (A4.2.5), the ratio of the length r_m to l_c^{atomic} in atoms is evidently

$$\frac{r_m}{l_c^{\text{atomic}}} = \frac{2r_m}{a_0^{1/2}}\left[\frac{3D(r_m)}{4\pi^2 r_m^2}\right]^{1/6}. \tag{A4.2.6}$$

However, $D(r_m)/Z^{4/3}$ in the Thomas–Fermi theory has the universal magnitude

$0.383a_0^{-1} = d$, say, and hence using $r_m = bx_m$ in Eq. (A4.2.6), one reaches the desired form

$$\frac{r_m}{l_c^{\text{atomic}}} = \frac{2}{a_0^{1/2}} \left(\frac{3d}{4\pi^2}\right)^{1/6} (bZ^{1/3}x_m)^{2/3} = 0.542, \qquad \text{(A4.2.7)}$$

which is seen to be independent of Z from Eq. (A4.2.3). Having established that, in nonrelativistic theory and for heavy atoms, r_m and l_c^{atomic} are about the same in magnitude, the final step is to estimate the derivative $d\varepsilon_c/dr$ at r_m, which is plainly a measure of the spatial variation of $\varepsilon_c(r)$ around the maximum of $D(r)$. This can again be done most readily in the local density approximation; Fig. A4.2.2 shows that the characteristic scale for the local density and the Colle–Salvetti approximations are not fundamentally different. For the homogeneous electron liquid, ε_c is solely a function of l_c in Eq. (A4.2.5), the Perdew–Zunger form being

$$\varepsilon_c^{-1} = A_1 + A_2 l_c + A_3 l_c^2, \qquad \text{(A4.2.8)}$$

where the A_i are known constants. One can readily calculate $d\varepsilon_c/dl_c$, and it remains therefore to relate this to $d\varepsilon_c/dr$ at the atomic distance r_m. Since l_c is proportional to the density to the power − from Eqs. (A4.2.5) and (A4.2.6), one readily calculates $d\varepsilon_c/d\rho$ from $d\varepsilon_c/dl_c$. The remaining step is to write

$$\frac{d\varepsilon_c}{dr} = \frac{d\varepsilon_c}{d\rho}\frac{d\rho}{dr} \qquad \text{(A4.2.9)}$$

and hence, using the property $dD(r)/dr = 0$ at $r = r_m$, one finds the desired result:

$$\left(\frac{d\varepsilon_c}{d\rho}\right)_{r_m} = \frac{l_c^{\text{atomic}}}{3r_m}\left(\frac{d\varepsilon_c}{dl_c}\right)_{l_c^{\text{atomic}}} \qquad \text{(A4.2.10)}$$

This atomic formula reproduces the slope of the local density curves at r_m in Fig. A4.2.2, when combined with the Perdew–Zunger result in Eq. (A4.2.8) and Eq. (A4.2.7). The slow variation of $\varepsilon_c(r)$ is due almost entirely to the form of dependence of Eq. (A4.2.8) on the Thomas–Fermi screening length l_c.

To summarize, evidence has been presented to support the two assumptions that enable Eq. (A4.2.1) to be written in the approximate form of Eq. (A4.2.2). These are

(i) the slow variation of $\varepsilon_c(r)$ over a range of r around r_m (identified now by $\bar{r}$ in Eq. (A4.2.2)) over which $D(r)$ has appreciable magnitude; this is amply borne out by both Fig. A4.2.2 for Kr and Xe and by Eq. (A4.2.8) to (A4.2.10), and

(ii) the relative insensitivity of $\varepsilon_c(r_m)$ to variation of Z, which is substantiated by Fig. A4.2.3 for the larger values of Z in that plot, for which the Thomas–Fermi approximation becomes a useful one.

Needless to say, to make a comparison between theory and 'experiment' as in Fig.

A4.2.1, for heavier atoms the relativistic 'corrections' made by Clementi in arriving at the results shown in Fig. A4.2.1 will naturally become more important with increasing Z. Also, for large Z, Kais *et al.* (1994) argue that $Z^{1.33}$ is a refinement on a grossly linear behavior.

A4.3. CORRELATION ENERGY OF DIATOMIC MOLECULES VERSUS NUMBER OF ELECTRONS

In Appendix 4.2, density functional theory has been used to explain the approximately linear variation of light-atom correlation energies with atomic number Z. This Appendix extends the discussion to treat some neutral diatomic molecules (Grassi *et al.*, 1994).

Let us again write the correlation energy E_c in the form

$$E_c = \int \varepsilon_c(\mathbf{r})\rho(\mathbf{r})\,d\mathbf{r} \tag{A4.3.1}$$

where $\varepsilon_c(\mathbf{r})$ is the correlation energy per electron at position $\mathbf{r}$ while $\rho(\mathbf{r})$ is the electron density. Following conventional quantum-chemical treatments of diatomic molecules, one expands the molecular wave function in a suitable atomic basis set, say, $\phi_\mu(\mathbf{r})$. Then, in standard quantum-chemical notation,

$$\begin{aligned} E_c &= \int \varepsilon_c(\mathbf{r})\rho(\mathbf{r})\,d\mathbf{r} \\ &= \sum_\mu^{A} P_{\mu\mu} \int \varepsilon_c(\mathbf{r})\phi_\mu^2(\mathbf{r})\,d\mathbf{r} + \sum_\mu^{B} P_{\mu\mu} \int \varepsilon_c(\mathbf{r})\phi_\mu^2(\mathbf{r})\,d\mathbf{r} \\ &\quad + \sum_\mu \sum_{\mu\neq\nu} P_{\mu\nu} \int \varepsilon_c(\mathbf{r})\phi_\mu(\mathbf{r})\phi_\nu(\mathbf{r})\,d\mathbf{r}, \end{aligned} \tag{A4.3.2}$$

where the partitioning is into atomic (A and B) and overlap parts. Equation (A4.3.2) is the starting point of the model proposed to treat the correlation energy E_c of neutral diatomic molecules (Grassi *et al.*, 1994).

The next step, following March and Wind (1992), is to assume $\varepsilon_c(\mathbf{r})$ is a slowly varying function of $\mathbf{r}$, compared with the $\phi_\mu^2(\mathbf{r})$ pieces in the atomic part of Eq. (A4.3.2).

Then one can write, for the atomic contribution in $E_c = E_{\text{atomic}} + E_{\text{over}}$ in Eq. (A4.3.2),

$$E_{\text{atomic}} = \varepsilon_A \sum_\mu^{A} P_{\mu\mu} + \varepsilon_B \sum_\mu^{B} P_{\mu\mu}, \tag{A4.3.3}$$

where ε_A and ε_B are suitably chosen averages of $\varepsilon_x(\mathbf{r})$ corresponding to atoms A and B, respectively. The term in $\Sigma_\mu P_{\mu\mu}$ in Eq. (A4.3.3) represents the effective charge on atom A, say n^a_{eff} (A), so that Eq. (A4.3.3) becomes

$$E_{\text{atomic}} = \varepsilon_A n^a_{\text{eff}}(A) + \varepsilon_B n^a_{\text{eff}}(B). \tag{A4.3.4}$$

The new step, transcending the problem discussed by March and Wind (1992; see Appendix 4.2) is to treat the overlap term E_{over} in Eq. (A4.3.2) by introducing a suitable average, say, ε_0, than rewriting E_{over} as

$$E_{over} = \varepsilon_c \sum_{\mu} \sum_{\nu \neq \mu} P_{\mu\nu} S_{\mu\nu}. \tag{A4.3.5}$$

Recalling the orthonormality of the atomic functions, one can finally write

$$E_c = \varepsilon_A n^a_{eff}(A) + \varepsilon_B n^a_{eff}(B) + \varepsilon_0 n^0(AB), \tag{A4.3.6}$$

where $n^0(AB)$ represents the total overlap population $\Sigma^A_\mu \Sigma^B_\nu P_{\mu\nu} S_{\mu\nu}$.

For ε_A and ε_B, Grassi *et al.* (1994) take tabulated experimental values. Two routes to estimating ε_0 have been utilized:

(i) $\varepsilon_0 = k N$, where N is the total number of electrons in the molecule and the constant k has been calculated by χ^2 minimization of the experimental values of the correlation energies of LiH, BH, Li_2, NH, HF, C_2, N_2, and F_2. The best value found for k is (12.5 a.u.)e^{-2}.

(ii) $\varepsilon_0 \cong (\varepsilon_A n^a_{eff}(A) + \varepsilon_B n^a_{eff}(B))$.

At this point, to test routes (i) and (ii) proposed above, one can compare the calculated correlation energies. For this purpose, one can use the experimental values for molecules (and atoms) reported by Savin *et al.* (1986). The values of the total atomic charges and total overlap population for each molecule have been taken from the work of Fraga and Ransil (1961).

The final results thereby obtained by Grassi *et al.* are depicted in Fig. A4.3.1. Table A4.3.1 also records the values for each individual molecule. For completeness, the values obtained by the Method II of Hollister and Sinanoglu (1965), to which the above approach is somewhat related, are also reported.

Though it is too much to expect such a simple model to reflect all the quantitative details, Fig. A4.3.1 leaves no doubt that both methods (i) and (ii) reproduce the correctly general trends of the correlation energy in diatomic molecules.

In summary, a generalization of the approach adopted by March and Wind (1992: see Appendix 4.2) can be made for diatomic molecules. Equation (A4.3.5) is then the main result. Two methods of estimating the overlap energy ε_0 lead to similar trends of the total correlation energy with the number of electrons.

A4.4. EFFECT OF CORRELATION ON VON WEIZSÄCKER INHOMOGENEITY KINETIC ENERGY: SCALING AND MOLECULAR DISSOCIATION

The purpose of this Appendix is to record the effect of including correlation on the von Weizsäcker inhomogeneity kinetic energy of density functional theory (see Section 3.4). This contribution to the kinetic energy is defined by

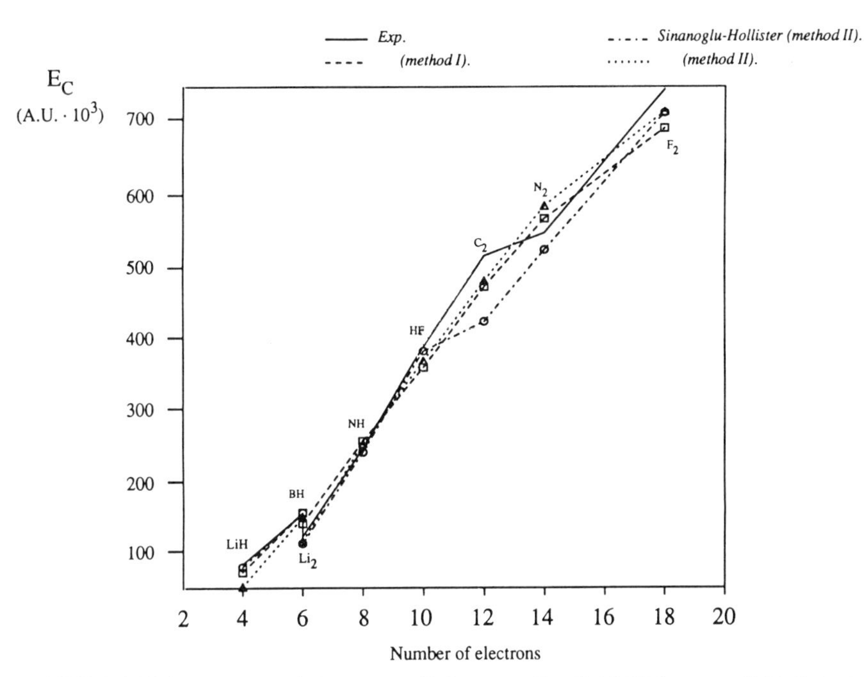

FIGURE A4.3.1. Shows correlation energy E_c [compare Eq. (A4.3.6)] for some light diatomic molecules. 'Experimental' values taken from Savin *et al.* (1986). Values obtained by Hollister and Sinanoglu (1965) are also shown. While the simple models used cannot reflect all the quantitative details, the point to be stressed is that they reproduce correct general trends of correlation energy in diatomics. (From Grassi *et al.*, 1994).

TABLE A4.3.1. Correlation Energies E_c of Some Light Diatomic Molecules[a]

		Energies (mhartree)			
	N	E_c(exp)[b]	Method I	Method II	Hollister–Sinanoglu (Method II)
LiH	4	83.0	71.8	51.8	79.1
BH	6	153.0	154.6	148.1	154.4
Li_2	6	122.0	139.8	115.0	112.0
NH	8	243.0	253.9	250.7	238.6
HF	10	387.0	356.4	366.0	380.0
C_2	12	514.0	470.7	479.0	421.9
N_2	14	546.0	566.1	584.0	522.4
F_2	18	746.0	690.9	714.4	712.9

[a]After Grassi *et al.* (1994).
[b]From Savin *et al.* (1986).

$$T_2 = (\hbar^2/72m) \int (\nabla\rho)^2/\rho \, d\mathbf{r}, \tag{A4.4.1}$$

and Laming *et al.* (1994) have calculated T_2 for a number of diatomics. Their results are set out in Table A4.4.1, where Hartree–Fock calculations are compared with the results obtained by incorporating correlation in two approximate versions of density functional theory.

It will be useful to summarize first the procedure by which the results in Table A4.4.1 were obtained (Laming *et al.*, 1994). The density $\rho(\mathbf{r})$ and its gradient $\nabla\rho(\mathbf{r})$, needed to evaluate T_2 in Eq. (A4.4.1), were found from the self-consistent solution of the Slater–Kohn–Sham equations:

$$(-\tfrac{1}{2}\nabla^2 + V_{\text{eff}})\phi_i = \varepsilon_i\phi_i \tag{A4.4.2}$$

where

$$V_{\text{eff}} = V_{\text{ext}} + V_{\text{Hartree}} + V_{\text{XC}}. \tag{A4.4.3}$$

In Eq. (A4.4.3) the external potential is merely that created by the bare nuclei in the diatomics investigated, while V_{Hartree} is written for the total electrostatic potential (energy) created by $\rho(\mathbf{r})$. The final, many-electron, contribution to the one-body potential, $V_{\text{eff}}(\mathbf{r})$, comes from exchange plus correlation interactions. The density $\rho(\mathbf{r})$ is now constructed in the usual way by summing the squares of the Slater–Kohn–Sham orbitals over the lowest occupied energy levels ε_i; i.e.,

TABLE A4.4.1. Inhomogeneity Kinetic Energy T_2 in au

System	Method	T_2	$t = T_2/N^2$
N_2	Hartree–Fock	9.593208	0.0489
	BLYP	9.554933	0.0487
	S–VWN	9.428059	0.0481
O_2	Unrestricted Hartree–Fock	12.636134	0.0494
	BLYP	12.613529	0.0493
	S–VWN	12.458345	0.0487
F_2	Hartree–Fock	16.202429	0.0500
	BLYP	16.160067	0.0498
	S–VWN	15.980084	0.0493
P_2	Hartree–Fock	46.870369	0.0520
	BLYP	46.757730	0.0520
	S–VWN	46.459168	0.0516
Cl_2	Hartree–Fock	60.824434	0.0526
	BLYP	60.685530	0.0525
	S–VWN	60.336579	0.0522

$$\rho(\mathbf{r}) = \sum_{\text{occupied}_i} |\phi_i|^2. \tag{A4.4.4}$$

In the work of Laming *et al.* (1994), a Gaussian-type orbital approach has been adopted for the density functional calculations. Thus, the Slater–Kohn–Sham orbitals $\phi_i(\mathbf{r})$ are formed from a linear superposition of Gaussian functions, say, η_α.

$$\phi_i = \sum_\alpha c_{i\alpha}\eta_\alpha \tag{A4.4.5}$$

It follows, using Eq. (A4.4.4), that

$$\nabla\rho = \sum_i \nabla|\phi_i|^2 = 2\sum_i \phi_i \nabla \phi_i, \tag{A4.4.6}$$

and substituting Eq. (A4.4.5) into Eq. (A4.4.6) then yields

$$\nabla\rho = 2\sum_i\sum_\alpha\sum_\beta c_{\alpha i}c_{\beta i}\eta_\alpha \nabla\eta_\beta. \tag{A4.4.7}$$

Returning now to the term V_{XC} in Eq. (A4.4.3), we note that Eq. (A4.4.2) can be regarded as the analogue of Hartree–Fock equations, but with the exchange operator replaced by the density functional theory (DFT) exchange-correlation potential V_{XC}. This means that, in the first instance, a DFT code can be created by making a minimal number of changes to a standard quantum-chemistry code such as cadpac5 or Gaussian. A little additional coding is required to implement Eqs. (A4.4.4) and (A4.4.7). These modifications were made by Laming *et al.* (1994) to the code, enabling one to choose whether to perform a Hartree–Fock or a DFT calculation.

The integral appearing in Eq. (A4.4.1) was then evaluated, using the input from Eqs. (A4.4.4) and (A4.4.7), by numerical quadrature as described in Murray *et al.* (1992, 1993). All the calculations were performed using a TZ2P + f basis set and very accurate numerical quadrature: 64 radial points, 24 theta points, and 48 phi points per atom. This quadrature guarantees that the integrated density is accurate to at least 10^{-7}. The above implementation can be readily applied to other molecular systems.

The acronym S-VWN corresponds to a self-consistent field (SCF) Slater–Kohn–Sham calculation performed using the Dirac (1930) form for the exchange energy and the Vosko *et al.* (1980) form for the correlation energy. The acronym BLYP signifies a SCF Slater–Kohn–Sham calculation using the Dirac form for the exchange energy with Becke's (1988) gradient correction. The LYP part corresponds to the correlation energy calculated using the correlation functional of Lee *et al.* (1988).

Having established how Table A4.4.1 was constructed, let us turn to examine its scaling properties following earlier work of Pucci and March (1987). They argue that one should expect for homonuclear diatomic molecules that

$$T_2 = N^2 f(e), \tag{A4.4.8}$$

where

$$e = R_e N^{1/3}. \tag{A4.4.9}$$

N and R_e are the number of electrons and the equilibrium distance, respectively. With the data available, we have represented the as yet unknown function $f(x)$ in Eq. (A4.4.8) as a Taylor expansion

$$f(x) = f(o) + xf'(o) + \ldots \tag{A4.4.10}$$

This expansion then motivates the plot shown in Fig. A4.4.1. Keeping only the first two terms of the Taylor expansion, Eq. (A4.4.10), T_2 can be approximately written

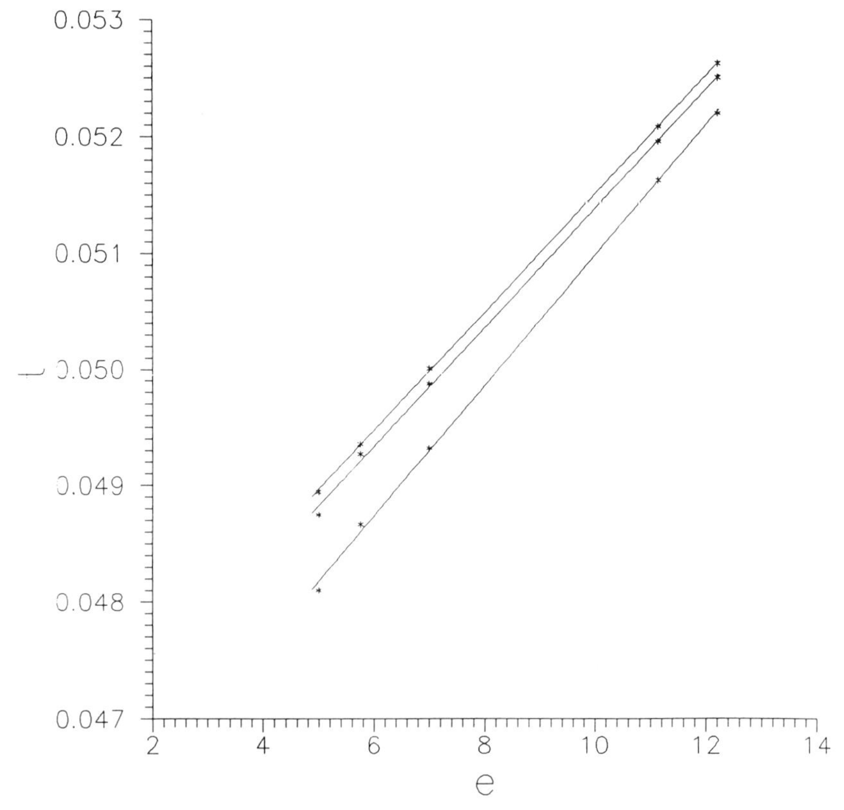

FIGURE A4.4.1. The scaled inhomogeneity kinetic energy $t = T_2/N^2$ as a function of $e = R_e N^{1/3}$ for HF (upper line), BLYP (middle line), and the S–VWN (lower line) approximations. (after Laming *et al.*, 1994). It is to be noted that electron correlation lowers the HF line, but the LDA line (S–VWN) overestimates effect of electron correlation (see also Appendix A4.2 on this point for atoms).

TABLE A4.4.2. Constants C_1 and C_2 in Eq. (A4.4.12)

Method	C_1 (a.u.)	C_2 (a.u.)
Hartree–Fock	0.04642	0.00051
BLYP	0.04626	0.00051
S–VWN	0.04537	0.00056

$$T_2 = N^2 t, \tag{A4.4.11}$$

where

$$t = C_1 + C_2\, e, \tag{A4.4.12}$$

with $C_1 = f(o)$ and $C_2 = f'(o)$. The constants can be obtained by a least-squares fit using the values of T_2 in Table A4.4.1. The values of C_1 and C_2 are also collected in Table A4.4.2 for all the three approximations for T_2. Figure A4.4.1. shows the lines corresponding to Eq. (A4.4.12) for the three calculations of t. In all three approximations the five diatomics fit nicely on a line. Electron correlation is seen to lower the HF-line somewhat.

Having discussed the scaling properties associated with the T_2 values in Table A4.4.1, let us turn finally to make some brief comments on the relation to dissociation energy D. That T_2 might be used to characterize molecular binding was first proposed by Mucci and March (1983). These workers argued that one could make a merit out of the theorem of Teller (1962), which states the molecules do not bind in a completely local DFT; i.e., treating single-particle kinetic energy locally (avoided when solving Slater–Kohn–Sham equations). Evidence has accumulated in support of the proposal of Mucci and March (see Allan *et al.*, 1985; Lee and Ghosh, 1986) on conventional quantum-chemical systems and the subsequent work of Cordero *et al.* (1993) on alkali metal clusters. These workers all emphasize that the dissociation energy should correlate with T_2. Moreover, it has been shown (March, 1991) that there is a one-sixth power law relating molecular dissociation energy D and the inhomogeneity kinetic energy T_2 (March and Nagy, 1992).

Appendix to Chapter 5

POSITIVITY OF STATIC GREEN FUNCTION G(R,R′)

Here, a proof will be given, following Senatore and March (1994), that the Green function $G(\mathbf{R},\mathbf{R}')$ obtained by solving Eq. (5.11) is positive and never vanishing for finite $\mathbf{R},\mathbf{R}'$. Let us start by remarking that with a suitable choice of the constant V_0 all the eigenvalues of the Schrödinger operator $H + V_0$ are positive, and in particular the lowest one, $\varepsilon_0 > 0$. Of course the ground-state wave function $\phi_0(\mathbf{R})$ is positive and vanishes only at infinity. One next notices that $G(\mathbf{R},\mathbf{R})$ is clearly positive, as can easily be seen in the energy representation, Eq. (5.15). Thus, for given $\mathbf{R}'$, the function $\tilde{\phi}_0(\mathbf{R})$ is defined by

$$\phi_0(\mathbf{R}) \equiv G(\mathbf{R},\mathbf{R}'), \quad \text{fixed } \mathbf{R}', \tag{A5.1.1}$$

will be positive in some region around $\mathbf{R}'$ for reasons of continuity. The strategy is to show that such a region must extend to infinity, so that $G(\mathbf{R},\mathbf{R}')$ may vanish only when either $\mathbf{R}$ or $\mathbf{R}'$ goes to infinity. To this purpose, let us assume *per absurdum* that $\tilde{\phi}_0(\mathbf{R})$ vanishes on some surface in the $3N$-dimensional space of particle configurations. By continuity, it is apparent that such a surface must divide the space in a domain D_1 containing $\mathbf{R}'$ and its complement D_2. It is not difficult to convince oneself that it is always possible to find a subdomain of D_2 that will be denoted by D', possibly coinciding with D_2, such that $\tilde{\phi}_0(\mathbf{R})$ does not change sign in D' and vanishes on its boundary. It follows that in the closed domain D' the function $\tilde{\phi}_0(\mathbf{R})$ satisfies the equation

$$(H + V_0)\tilde{\phi}_0(\mathbf{R}) = 0, \tag{A5.1.2}$$

with homogeneous boundary conditions. In other words the Schrödinger operator $H + V_0$ restricted to D', and with the conditions that the eigenfunctions vanish at the boundary, possesses the ground-state eigenfunction $\tilde{\phi}_0$ with eigenvalue $\tilde{\varepsilon}_0 = 0$. Clearly, one has $\varepsilon_0 > \tilde{\varepsilon}_0$. On the other hand, it can be shown using the calculus of variations (see, for instance, Courant and Hilbert, 1953; in particular, Chap. VI, Sec. 2, theorem 3) that under the present circumstances, i.e., D' is a proper subdomain of the full space, it must hold instead the condition $\tilde{\varepsilon}_0 > \varepsilon_0$. Thus, one is led to a contradiction. The conclusion is that $G(\mathbf{R},\mathbf{R}')$ is positive and can vanish only for $\mathbf{R}$ or $\mathbf{R}'$ at infinity.

Appendix to Chapter 6

A6.1. k SPACE TREATMENT OF ELECTRON ENERGY LOSS BY PLASMON EXCITATION

In order to calculate the energy dissipated by a fast charge traversing a metallic medium, one can have recourse to the following formula of classical electrodynamics:

$$P = \frac{1}{4\pi} \mathcal{E} \cdot \frac{\partial \mathbf{D}}{\partial t}, \qquad \text{(A6.1.1)}$$

which relates the power dissipated in unit volume to the electric and displacement fields. In the present case, **D** is the field due to the external charge e and is given by

$$\mathbf{D} = -\nabla \frac{e}{|\mathbf{r} - \mathbf{v}t|}. \qquad \text{(A6.1.2)}$$

Here **v** is the velocity vector of the charge, and this is assumed to be so large that all recoil effects can be neglected. The next step is to Fourier analyze **D** in the form

$$\mathbf{D}(\mathbf{k}, t) = \int \frac{d\mathbf{k}\,d\omega}{(2\pi)^4} \exp(i\mathbf{k}\cdot\mathbf{r} - \omega t)\mathbf{D}(\mathbf{k}, \omega) \qquad \text{(A6.1.3)}$$

from which, using Eq. (A6.1.2), one obtains the explicit result

$$\mathbf{D}(\mathbf{k}, \omega) = -i\mathbf{k} \frac{4\pi e}{\mathbf{k}^2} \delta(\omega - \mathbf{k} \cdot \mathbf{v}). \qquad \text{(A6.1.4)}$$

Each Fourier component is related to the corresponding component of the internal field by

$$\mathcal{E}(\mathbf{k}, \omega) = \frac{D(\mathbf{k}, \omega)}{\varepsilon(\mathbf{k}, \omega),}, \qquad \text{(A6.1.5)}$$

where $\varepsilon(\mathbf{k},\omega)$ is the frequency- and wave-number-dependent dielectric function of the metallic medium, which is assumed to be uniform and isotropic.

The corresponding contribution to the power dissipation can now be calculated using Eq. (A6.1.1) as

$$\mathbf{P}(\mathbf{k},\omega) = \frac{1}{4\pi}\{\mathrm{Re}[\mathcal{E}(\mathbf{k},\ \omega)e^{-i\omega t}]\,\mathrm{Re}[-i\omega\mathbf{D}(\mathbf{k},\ \omega)e^{-i\omega t}]\}$$

$$= \frac{1}{4\pi}\mathbf{D}(\mathbf{k},\ \omega)\left\{\mathrm{Re}\left[\frac{1}{\varepsilon}\cos(\omega t)\right] + \mathrm{Im}\left[\frac{1}{\varepsilon}\sin(\omega t)\right]\right\} \times \mathbf{D}(\mathbf{k},\ \omega)[-\omega\sin(\omega t)], \tag{A6.1.6}$$

and averaging over a cycle one finds

$$\mathbf{P}(\mathbf{k},\omega) = -\frac{\omega}{8\pi}\ \mathrm{Im}\left[\frac{1}{\varepsilon(\mathbf{k},\omega)}\right][\mathbf{D}(\mathbf{k},\omega)]^2, \tag{A6.1.7}$$

which is the desired result. Clearly there is a peak in the power dissipation whenever collective modes are excited, which is equivalent to saying that $\varepsilon(\mathbf{k},\omega) = 0$. This last equation gives the condition for plasmon excitation, since it follows from Eq. (A6.1.5) that this latter equation can then be satisfied for $\mathbf{D}(\mathbf{k},\omega) \neq 0$, i.e., for no external charges. It follows that one can then have $\mathcal{E}(\mathbf{k},\omega) \equiv 0$ even in the absence of external charges. This condition is indeed realized when a plasma oscillation is set up in a metallic medium.

For an r-space formulation of the energy loss problem (discussed above in **k**-space) generalized to include a weak lattice potential, the reader is referred to the work of March and Tosi (1973).

A6.2. THOMAS–FERMI MODEL OF STATIC DIELECTRIC FUNCTION OF SEMICONDUCTOR

The dynamic model dielectric constant of Inkson (1972) was employed in Section 6.3. Here, the Thomas–Fermi model of static dielectric screening in a semiconductor will be discussed, following Resta (1977). Whereas in a metal the dielectric function $\varepsilon(k)$ diverges as k^{-2} as k tends to zero, in the semiconductor the value must tend to the static dielectric constant.

Essentially, Resta solves the linearized Thomas–Fermi equation, namely,

$$\nabla^2 V = q^2\,(V - A). \tag{A6.2.1}$$

The model semiconductor is thereby being treated as an electron gas, the unperturbed electron density being denoted by ρ_0.

Resta now argues for the existence of a finite screening radius R around a test charge Ze such that the density $\rho(r)$ becomes equal to ρ_0 at this distance from the test charge, that is,

$$\rho(R) = \rho_0, \tag{A6.2.2}$$

which gives for the constant A in Eq. (A6.1) the value $V(R)$. Beyond this screening distance R, the screened potential energy $V(r)$ of the point charge Ze is given by

$$V(r) = -Ze^2/\varepsilon(0)\, r, \qquad r>R, \tag{A6.2.3}$$

$\varepsilon(0)$ being the static dielectric constant of the semiconductor. Writing the solution of the linear equation [Eq. (A6.2.1)] as

$$V(r) = -(Ze^2/r)\,[\propto \exp(qr) + \beta \exp(-qr)] + A, \qquad r \leq R, \tag{A6.2.4}$$

one finds by imposing continuity at $r = R$. using $V \to -(Ze^2/r)$ as $r \to 0$, and putting $A = V(R)$ as discussed above, Eq. (A6.2.4) takes the form

$$V(r) = -(Ze^2/r)\,\frac{\sinh q(R-r)}{\sinh qR} - \frac{Ze^2}{\varepsilon(0)R}, \qquad r \leq R. \tag{A6.2.5}$$

The screening distance R is found by requiring continuity of the electric field at $r = R$. This is related to $\varepsilon(0)$ through

$$\frac{\sinh qR}{qR} = \varepsilon(0). \tag{A6.2.6}$$

Equation (A6.2.6) yields a finite solution for R for any $\varepsilon(0) > 1$. The metallic limit already referred to above corresponds to $\varepsilon(0) \to \infty$, when we find $R \to \infty$, as expected.

The wave-number-dependent dielectric function $\varepsilon(k)$ can be readily calculated from the above treatment and takes the form (Resta, 1977)

$$\varepsilon(k) = \frac{k^2 + q^2}{[q^2/\varepsilon(0)] \sin kR/kR + k^2}. \tag{A6.2.7}$$

A merit of Eq. (A6.2.7) in common with other usable models of $\varepsilon(k)$, is that the only input data required to evaluate $\varepsilon(k)$ are the static dielectric constant and the Fermi momentum k_f of the valence electrons. Comparison of Eq. (A6.2.7) with other models is made by Resta (1977), to whose paper the interested reader is referred.

The final point to be made, which adds to one's confidence in Resta's Thomas–Fermi treatment, is that the numerical values of R found from Eq. (A6.2.6) are very close to the nearest-neighbor distances in the crystals in the cases of the valence semiconductors diamond, silicon, and germanium.

A6.3. SEMIEMPIRICAL SELF-ENERGY CORRECTIONS TO DENSITY FUNCTIONAL (LDA) BANDS OF SEMICONDUCTORS

In the main text (see also Appendix 6.8), the GW approximation for semiconductors has been discussed at some length. While such GW calculations have allowed impressive progress to be made, it is clear from the treatment in Chapter 6 that they are elaborate and time-consuming. Therefore, in this appendix, a link will be forged, following Fiorentini and Baldereschi (1992; hereafter FB), with density functional theory (at LDA level) by considering semiempirical self-energy corrections to LDA energy bands of semiconductors.

The focus of the work of FB is on correcting the LDA bands in a class of materials by a procedure that incorporates as much as possible of the relevant physics (long-range dielectric screening, energy dependence), with the goal of precise eigenvalues and related properties such as masses as well as transferability to more complex systems (e.g., alloys and superlattices).

As in the main text, one starts from the leading term in the many-body expansion of the self-energy Σ in powers of the screened interaction. This is the GW form of the self-energy Σ:

$$\overset{\mathrm{GW}}{\Sigma}(\mathbf{r}, \mathbf{r}', E) = \frac{i}{4\pi} \int \exp(i\omega\delta) G(\mathbf{r}, \mathbf{r}', E + E') W(\mathbf{r}, \mathbf{r}', E') dE'. \tag{A6.3.1}$$

Here δ is an infinitesimal, G the one-electron Green function, and W the effective screened interaction. While in metals the Coulomb interaction is well screened and the potential W decays rapidly as $|\mathbf{r} - \mathbf{r}'|$ tends to infinity (in RPA, $W \simeq r^{-3} \times$ Friedel oscillations: see especially Chapter 8) and the self-energy is short-ranged, in semiconductors W tends to $1/\varepsilon_\infty\ |\mathbf{r} - \mathbf{r}'|$ in the same limit, due to the incomplete screening of the Coulomb interaction (see Gygi and Baldereschi, 1989). In the case of insulators, one may then decompose W into a short-range metal-like part [when one recovers the local exchange-correlation potential of the inhomogeneous electron gas, V_{xc} (see Chapter 3), if its weak energy dependence is neglected] and a long-range part, small as $|\mathbf{r} - \mathbf{r}'|$ tends to zero and with the above long-range behavior. The expectation value of the self-energy part related to this term, $\delta\Sigma$, is known to be discontinuous across the energy gap (Godby *et al.*, 1988; Hybertson and Louie, 1986). The achievement of the work of FB is to reproduce $\delta\Sigma$, the slowly varying long-range correction to $V_{\mathrm{xc}}^{\mathrm{LDA}}$, by means of smooth nonlocal potential.

A6.3.1. Summary of Technique

The equilibrium structure and volume of the bulk crystal are determined *ab initio* by a converged LDA pseudopotential (Bachelet *et al.*, 1982) plane-wave calculation, and the electronic eigensystem is computed at a set M of **k**-points (Γ, X, and L in the work of FB). The eigenvalue for the 'empirical quasiparticle' (EQP) Hamiltonian,

$$H = H_{\mathrm{LDA}} + V_E, \tag{A6.3.2}$$

is then solved, expanding the EQP eigenstates on the basis of the LDA eigenstates. In Eq. (A6.3.2), H_{LDA} is the self-consistent LDA Hamiltonian, and the nonlocal potential V_E is a sum of local potentials projected onto occupied (valence) and virtual (conduction-band) states:

$$V_E = V_E^v P_v + V_E^c(1-P_v) \tag{A6.3.3}$$

where $P_v = \Sigma_v |v\mathbf{k}\rangle\langle v\mathbf{k}|$ denotes the projector on the LDA valence manifold. V_E^c is expanded in Fourier space as

$$V_E^c(\mathbf{r}) = \sum_{\mathbf{G}} V_E^c(\mathbf{G})S(\mathbf{G})\exp(i\mathbf{G}\cdot\mathbf{r}) = \sum_{\mathbf{G}} V_E^{Sc}(\mathbf{G})\exp(i\mathbf{G}\cdot\mathbf{r}) \tag{A6.3.4}$$

(S is the crystal basis structure factor $\Sigma_{\mathbf{G}} \exp(i\mathbf{G}\cdot\mathbf{r}_i)$, the $\mathbf{G}$ are reciprocal-lattice vectors, and V_E^{Sc} is defined by the second equality), and analogously, for V_E^v.

The matrix elements of H between the LDA eigenstates $(m\mathbf{k})$ and $(n\mathbf{k})$ are

$$A_{m,n} = \sum_{\mathbf{G},\mathbf{G}'} C^*_{m,\mathbf{k}}(\mathbf{G})C_{m,\mathbf{k}}(\mathbf{G}')V_E^S(\mathbf{G},\mathbf{G}') + E^{\mathrm{LDA}}_{n,\mathbf{k}}\delta_{m,n}, \tag{A6.3.5}$$

where m, $n = 1, \ldots, N$ are band indices, C the coefficients of the plane-wave expansion of the LDA wave function at a specific **k**-point, and $\mathbf{G},\mathbf{G}'$ are again reciprocal-lattice vectors. Matrix elements between conduction and valence states are zero by construction, so in Eq. (A6.3.5) the label i stands for either c or v.

While N is fixed for valence levels, when dealing with conduction states an appropriate truncation $N = N_c$ of the expansion of EQP functions on LDA states has to be selected; since the potential proves to be weak, couplings of the states of interest with bands much higher in energy rapidly become negligible, and $N = N_c \cong 10$ turns out to be adequate. The value of N_c needs to be chosen carefully to avoid artificial splittings of degenerate multiplets. In the applications under discussion, $N_c \cong 25$–30 proves suitable, depending on **k** values and degeneracies.

The Fourier components $V_E(\mathbf{G})$ are to be determined by requiring that the computed eigenvalues fit experiment within the measurement errors. This involves a multidimensional nonlinear optimization process, which can be recast to a minimization problem, as indicated below. The latter problem can then be tackled by the robust simplex method of Nelder and Mead (1965). To estimate the difference of the experimental band structure from the calculated form (which is evidently a functional of V_E) and to produce the potential V_E minimizing this difference, FB use a cost function given by the infinity norm of the calculation-to-experiment error, i.e., the maximum weighted deviation of the calculated eigenvalues from the pertaining experimental results:

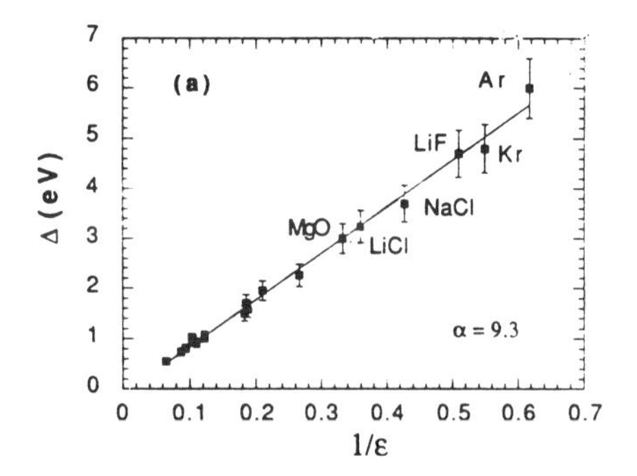

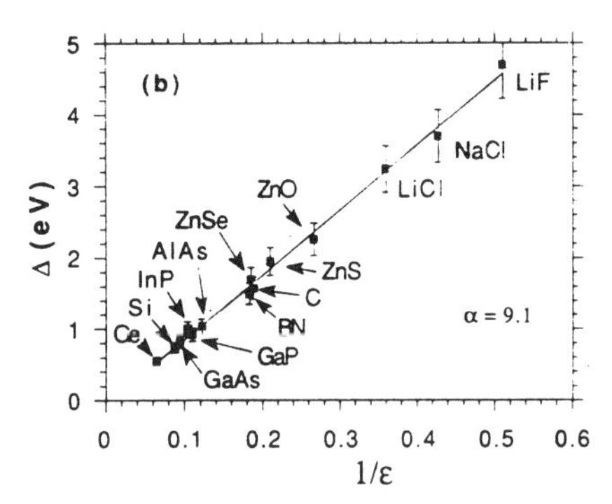

FIGURE A6.3.1. The scaling rule for the scissor operator (a) and an expansion of the high-ε region (b). The data points indicate the scissor operator as determined in the work of Fiorentini and Baldereschi (1992); for C, InP, ZnSe, ZnS, ZnO, MgO, LiCl, NaCl, LiF, Kr, and Ar the differences between the experimental and computed gaps is plotted instead. Error bars are 10% of the data. (Reproduced from Fiorentini and Baldereschi, 1992.)

$$F_c = |e|_\infty = \max_{i \varepsilon m} w_i \, (E_i^{\text{th}} - E_i^{\text{ex}}). \tag{A6.3.6}$$

The weight for the ith state is the ratio of a reference experimental error $e_r = \min_{(i\varepsilon m)} e_i$ to the ith experimental error

$$w_i = (e_r/e_i)p_i, \tag{A6.3.7}$$

where i runs over the set M of the relevant states included in the minimization procedure. Very precise experimental values ($w \cong 1$) can be overweighted by choosing $p > 1$; FB take values of p ranging from 1 to 1.1. The calculation of the cost function amounts to computing the eigensystem of the Hamiltonian H on the basis of LDA eigenfunctions at each of the points of the set M. The eigenenergies are

$$E_i^{\text{th}} = \frac{E_{i,\,\text{EQP}}}{E_{i,\text{LDA}}} + \langle i_{\text{EQP}}\mathbf{k}|V_E|i_{\text{EQP}}\mathbf{k}\rangle \simeq E_{i,\,\text{LDA}} + \langle i_{\text{QP}}\mathbf{k}|\delta\Sigma|i_{\text{QP}}\mathbf{k}\rangle. \tag{A6.3.8}$$

It follows that these energies depend sensitively on the LDA eigenvalues. The eigenfunctions of $H + V_E$ are the EQP wave functions $|i_{\text{EQP}}\mathbf{k}\rangle = \Sigma_j\, q_j|j_{\text{LDA}}\mathbf{k}\rangle$, and these turn out in the study of FB to closely resemble the LDA states, i.e.,

$$|i_{\text{EQP}}\mathbf{k}\rangle \approx |i_{\text{LDA}}\mathbf{k}\rangle \approx |i_{\text{QP}}\mathbf{k}\rangle. \tag{A6.3.9}$$

For the cases examined, the minimization is found to converge to a single point in the parameter space of the Fourier components of V_E in a limited number of simplex iterations, independently of the starting point.

A6.3.2. Some Specific Examples

The materials focused on below are GaAs, AlAs, and Ge. In view of the reasonable accord between LDA results (see FB) and GW calculation, plus experiment, for valence states, one can set the valence self-energy correction to zero and then focus on that for the conduction states.

As noted above, the dependence on the LDA energies is crucial. It is clear then that care must be taken to fully converge on the eigenvalues. (FB note that the maximum deviation is less than 10 meV for their study of the conduction minimum in GaAs.)

FB also observed that, as a general strategy, given the limited experimental data and the expected long-range nature of the potential, one should consider Fourier coefficients of V_E up to the $\langle 200\rangle$ FCC reciprocal-lattice shell.

For zincblende and diamond structures, one then has four or two free parameters per material, respectively. The total of ten free parameters for the three materials is well below the number of experimental data; it is nevertheless important, as FB stress, to seek possible transferability rules for the potential (or parts of it) among different materials, a practice which will then permit a reduction in the number of free parameters.

FB find that independent minimizations for each material well reflect the experimental data and show that

1. The Fourier form of V_E is dominated by the $\mathbf{G} = (000)$ component, the scissor operator Δ, which is ~0.5–1 eV;
2. in Ge the $\mathbf{G} \neq 0$ component vanishes (a computational result, not a symmetry restriction); and
3. the scissor operator Δ scales as the ratio of the high-frequency dielectric constants, i.e., $\varepsilon \times \Delta$ is a constant,

$$\Delta = \alpha/\varepsilon, \tag{A6.3.10}$$

with $\alpha \cong 9$ eV.

Thus, according to FB, the scissor operators of the various materials are not independent parameters, thereby reducing to seven the number of free parameters for the three materials under consideration.

FB also emphasize that the scaling law of the scissor operator, Eq. (A6.3.10), is a useful rule of thumb that can be used to obtain estimates of gaps. In many cases (e.g., for wide gaps), the LDA estimate alone is not useful and this rule provides a "correction" restoring some agreement with experiment. It also turns out that the validity of this rule extends beyond the semiconductors considered above. In Fig. A6.3.1, the difference Δ between LDA calculated and experimental gaps is depicted for a number of materials. It can be seen that the rule of Eq. (A6.3.10) is well obeyed up to very large gaps.

A6.4. BIORTHONORMAL REPRESENTATION OF r-SPACE INVERSE DIELECTRIC FUNCTION

By Fourier transforming Eq. (6.26) with respect to time according to van Haeringen *et al.* (1987) one obtains (see Engel *et al.*, 1991a)

$$\int d^3r_3 \varepsilon(\mathbf{r}_1\mathbf{r}_3;\, \varepsilon)\varepsilon^{-1}(\mathbf{r}_3,\, \mathbf{r}_2;\, \varepsilon) = \delta(\mathbf{r}_1 - \mathbf{r}_2). \tag{A6.4.1}$$

It is not difficult to verify that the function $\varepsilon^{-1}(\mathbf{r}_1,\mathbf{r}_2;\, \varepsilon)$, like the Green function $G(\mathbf{r}_1,\, \mathbf{r}_2;\, \varepsilon)$, may be expressed in the form of a biorthonormal representation:

$$\varepsilon^{-1}(\mathbf{r}_1,\mathbf{r}_2;\, \varepsilon) = \sum_n \frac{\xi_n(\mathbf{r}_1;\, \varepsilon)\zeta_n^*(\mathbf{r}_2;\, \varepsilon)}{D_n(\varepsilon)}, \tag{A6.4.2}$$

in which the functions ξ_n and ζ_n satisfy the equation [cf. van Haeringen *et al.* (1987)]:

$$[D_n(\varepsilon) - \mathcal{K}(\varepsilon)]\xi_n(\mathbf{r};\, \varepsilon) = 0, \tag{A6.4.3}$$

$$[D_n^*(\varepsilon) - \mathcal{K}^\dagger(\varepsilon)]\zeta_n(\mathbf{r};\, \varepsilon) = 0. \tag{A6.4.4}$$

Here the operator $\mathcal{K}(\varepsilon)$ has been defined by

$$\mathcal{K}(\varepsilon)f(\mathbf{r}) = \int d^3r' \varepsilon(\mathbf{r},\mathbf{r}';\, \varepsilon)f(\mathbf{r}'), \tag{A6.4.5}$$

where $f(\mathbf{r})$ is an arbitrary function. The adjoint operator $\mathcal{K}^\dagger(\varepsilon)$ has been defined through the relation $\langle \mathcal{K}(\varepsilon)f, g\rangle = \langle f, \mathcal{K}^\dagger(\varepsilon)g\rangle$, where $\langle , \rangle$ stands for the inner product, defined according to the expression $\langle f, g\rangle = \int d^3r\, f^*(\mathbf{r})g(\mathbf{r})$. As shown by van Haeringen *et al.* (1987) the functions ξ_n and ζ_n can be constructed to be biorthonormal, in the sense that

$$\langle \xi_m, \zeta_n\rangle = \delta_{m,n}. \tag{A6.4.6}$$

Moreover, these functions are supposed to satisfy the closure relation [cf. van Haeringen *et al.* (1987)]

$$\sum_n \xi_n(\mathbf{r}_1;\, \varepsilon)\zeta_n^*(\mathbf{r}_2;\, \varepsilon) = \delta(\mathbf{r}_1 - \mathbf{r}_2). \tag{A6.4.7}$$

In view of the fact that $\varepsilon(\mathbf{r}_1,\mathbf{r}_2;\, \varepsilon)$ has the dimension m^{-3}, one observes from Eq. (A6.4.5) that $\mathcal{K}(\varepsilon)$ and therefore also $D_n(\varepsilon)$ are dimensionless. The dimension of the product $\xi_n(\mathbf{r}_1;\, \varepsilon)\zeta_n^*(\mathbf{r}_2;\, \varepsilon)$ in Eq. (A6.4.2) is m^{-3}, which allows one to postpone the choice of dimension for the functions ζ_n and ζ_n themselves, as long as their product has dimension m^{-3}. It follows readily from Eq. (A6.4.5) that

$$\mathcal{K}^\dagger(\varepsilon)f(\mathbf{r}) = \int d^3r'\varepsilon^*(\mathbf{r}',\mathbf{r};\, \varepsilon)f(\mathbf{r}'), \tag{A6.4.8}$$

again for arbitrary functions $f(\mathbf{r})$.

Substituting Eq. (A6.4.2) in the right-hand side of Eq. (6.26), one obtains (Engel *et al.*, 1991a):

$$W(\mathbf{r}_1,\mathbf{r}_2;\, \varepsilon) = \sum_n \frac{\xi_n(\mathbf{r}_1;\, \varepsilon)\zeta'^*_n(\mathbf{r}_2;\, \varepsilon)}{D_n(\varepsilon)}, \tag{A6.4.9}$$

in which

$$\zeta'_n(\mathbf{r};\, \varepsilon) = \int d^3r' v(\mathbf{r} - \mathbf{r}')\zeta_n(\mathbf{r}';\, \varepsilon). \tag{A6.4.10}$$

From Eqs. (A6.4.7) and (A6.4.10) it follows that

$$\sum_n \xi_n(\mathbf{r}_1;\, \varepsilon)\zeta'^*_n(\mathbf{r}_2;\, \varepsilon) = v(\mathbf{r}_1 - \mathbf{r}_2). \tag{A6.4.11}$$

By using Eqs. (6.26), (A6.4.4), (A6.4.8), and (A6.4.10) one can readily show that $\zeta'_n(\mathbf{r};\, \varepsilon)$ satisfies the equation

$$\int d^3r' \left[\delta(\mathbf{r}'-\mathbf{r}) - \int d^3r'' \mathrm{P}^*(\mathbf{r}', \mathbf{r}'';\, \varepsilon)v(\mathbf{r}'' - \mathbf{r})\right] \zeta'_n(\mathbf{r}';\, \varepsilon) = D_n^*(\varepsilon)\zeta'_n(\mathbf{r};\, \varepsilon). \tag{A6.4.12}$$

In the system under consideration (i.e., no spins and magnetic fields), $P(\mathbf{r}_1, \mathbf{r}_2;\, \varepsilon)$ is equal to $P(\mathbf{r}_2;\, \mathbf{r}_1;\, \varepsilon)$, so that one has, in fact,

$$[D_n^*(\varepsilon) - \mathcal{K}^*(\varepsilon)]\zeta'_n(\mathbf{r};\, \varepsilon) = 0. \tag{A6.4.13}$$

Therefore the functions ξ_n and ζ'^*_n actually satisfy the same equation [compare Eqs. (A6.4.13) and A6.4.3)].

In a crystal, where $\varepsilon(\mathbf{r}_1, \mathbf{r}_2;\, \varepsilon) = \varepsilon(\mathbf{r}_1 + \mathbf{R}, \mathbf{r}_2 + \mathbf{R};\, \varepsilon)$, in which $\mathbf{R}$ stands for a lattice vector, the functions $\xi_n(\mathbf{r};\, \varepsilon)$ and $\zeta'_n(\mathbf{r};\, \varepsilon)$ can be chosen as Bloch functions and may be denoted as $\xi_{l,\mathbf{k}}(\mathbf{r};\, \varepsilon)$ and $\zeta'_{l,\mathbf{k}}(\mathbf{r},\varepsilon)$. Here l is a band index and $\mathbf{k}$ is a vector in the first Brillouin zone (1BZ). As the functions $\zeta'^*_{l,\mathbf{k}}(\mathbf{r};\, \varepsilon)$ have to be linear combinations of the $\xi_{l,\mathbf{k}}(\mathbf{r};\, \varepsilon)$ functions, and as $\langle \xi^*_{l,\mathbf{k}}, \xi_{l,\mathbf{k}}\rangle$, and $\langle \zeta'_{l,\mathbf{k}}, \xi_{l,\mathbf{k}}\rangle$ are equal to zero for any $\mathbf{k}' \neq -\mathbf{k}$ (due to the Bloch property), it follows from Eq. (A6.4.6) that $\zeta'^*_{l,\mathbf{k}}(\mathbf{r};\, \varepsilon)$ has to be identified, apart from a possible numerical factor, with $\xi_{l,-\mathbf{k}}(\mathbf{r};\, \varepsilon)$ (compare the related discussion in van Haeringen *et al.* (1987).

Choosing a factor equal to unity, i.e., by writing $\zeta'^{*}_{l,\mathbf{k}}(\mathbf{r};\ \varepsilon) = \xi_{l\text{-}\mathbf{k}}(\mathbf{r};\ \varepsilon)$, one simultaneously fixes the dimension of the eigenfunctions ζ' and ξ as $(Jm^{-3})^{1/2}$. Denoting $D_n(\varepsilon)$ as $D_{l,\mathbf{k}}(\varepsilon)$, one may now write Eq. (A6.4.9) as

$$W(\mathbf{r}_1,\ \mathbf{r}_2;\ \varepsilon) = \sum_{l,\mathbf{k}} \frac{\xi_{l,\mathbf{k}}(\mathbf{r}_l;\ \varepsilon)\xi_{l,-\mathbf{k}}(\mathbf{r}_2;\ \varepsilon)}{D_{l,\mathbf{k}}(\varepsilon)}. \tag{A6.4.14}$$

As $\xi_{l,\mathbf{k}}$ and $\xi_{l\text{-}\mathbf{k}}$ are eigenfunctions belonging to the same eigenvalue, one finds that $D_{l,\mathbf{k}}(\varepsilon)$ and $D_{l,-\mathbf{k}}(\varepsilon)$ are in fact identical. The static values $D_{l,\mathbf{k}}(\varepsilon = 0)$ correspond to the eigenvalues defining the concept of dielectric band structure (see, for example, Baldereschi and Tosatti, 1979: see also Car *et al.*, 1981). It should be noted that $\xi_{l,-\mathbf{k}}(\mathbf{r};\ \varepsilon)$ is equal to $\xi^{*}_{l,\mathbf{k}}(\mathbf{r};\ \varepsilon)$ only if $P(\mathbf{r}_1,\ \mathbf{r}_2;\ \varepsilon)$ is real-valued. By expanding the functions ξ in Eq. (A6.4.14) in plane waves $\exp|i(\mathbf{k} + \mathbf{G})\cdot\mathbf{r}|$, $\mathbf{G}$ denoting a reciprocal-lattice vector, according to the relation

$$\xi_{l,\mathbf{k}}(\mathbf{r};\ \varepsilon) = \frac{1}{\sqrt{\Omega}} \sum_{G} \Xi_{l,\mathbf{k}}(\mathbf{G};\ \varepsilon)e^{i(\mathbf{k}+\mathbf{G})\cdot\mathbf{r}}, \tag{A6.4.15}$$

with Ω the volume of the crystal, and Fourier transforming $W(\mathbf{r}_1,\mathbf{r}_2;\ \varepsilon)$ according to van Haeringen *et al.* (1987), one obtains the plane-wave matrix elements of $W(\mathbf{r}_1,\ \mathbf{r}_2;\ \varepsilon)$ as follows ($\mathbf{K}$ and $\mathbf{K}'$ are reciprocal-lattice vectors):

$$W_{\mathbf{K},\mathbf{K}'}(\mathbf{k};\ \varepsilon) = \sum_{l} \frac{\Xi_{l,\mathbf{k}}(\mathbf{K};\ \varepsilon)\Xi_{l,-\mathbf{k}}(-\mathbf{K}';\ \varepsilon)}{D_{l,\mathbf{k}}(\varepsilon)}, \tag{A6.4.16}$$

where the $\Xi_{l,\mathbf{k}}$ functions have the dimension $J^{1/2}$. From Eqs. (A6.4.3), (A6.4.4), and (A6.4.15) it follows directly that the plane-wave coefficients $\Xi_{l,\mathbf{k}}(\mathbf{K};\ \varepsilon)$ satisfy the following system of linear equations:

$$\sum_{\mathbf{K}'} \varepsilon_{\mathbf{K},\mathbf{K}'}(\mathbf{k};\ \varepsilon)\Xi_{l,\mathbf{k}}(\mathbf{K}';\ \varepsilon) = D_{l,\mathbf{k}}(\varepsilon)\Xi_{l,\mathbf{k}}(\mathbf{K},\ \varepsilon). \tag{A6.4.17}$$

If the system under consideration is symmetric with respect to the origin, it can be shown that $\Xi_{l,\mathbf{k}}(\mathbf{K};\ \varepsilon)$ and $\Xi_{l,-\mathbf{k}}(-\mathbf{K};\ \varepsilon)$ are in fact identical.

The closure relation in Eq. (A6.4.11) becomes

$$\sum_{l} \Xi_{l,\mathbf{k}}(\mathbf{K};\ \varepsilon)\Xi_{l,\mathbf{k}}(-\mathbf{K}';\ \varepsilon) = v_{\mathbf{K},\mathbf{K}'}(\mathbf{k}), \tag{A6.4.18}$$

whereas the normalization is fixed by the Fourier transform of Eq. (A6.4.7), with ζ_n replaced by using Eq. (A6.4.10);

$$\sum_{\mathbf{k}} \frac{1}{v_{\mathbf{K},\mathbf{K}}(\mathbf{k})} \Xi_{l,\mathbf{k}}(\mathbf{K};\ \varepsilon)\Xi_{l',-\mathbf{k}}(-\mathbf{K};\ \varepsilon) = \delta_{l,l'}, \tag{A6.4.19}$$

where

$$v_{\mathbf{K},\mathbf{K}'}(\mathbf{k}) = \frac{e^2}{\varepsilon_0}\ \frac{1}{|\mathbf{k}+\mathbf{K}|^2}\ \delta_{\mathbf{K},\mathbf{K}'} \tag{A6.4.20}$$

denotes the Fourier transform of the bare Coulomb interaction function.

Let us next consider $W_{\mathbf{K},\mathbf{K}'}(\mathbf{k};\, \varepsilon)$ for large values of $|\varepsilon|$. The asymptotic behavior is known to be

$$W_{\mathbf{K},\mathbf{K}'}(\mathbf{k};\, \varepsilon) \to v_{\mathbf{K},\mathbf{K}'}(\mathbf{k}) + O(\varepsilon^{-2}) \qquad \text{as } |\varepsilon| \to \infty. \tag{A6.4.21}$$

Similarly, the matrix elements of the dielectric matrix have the property

$$\varepsilon_{\mathbf{K},\mathbf{K}'}(\mathbf{k};\, \varepsilon) \to \delta_{\mathbf{K},\mathbf{K}'} + O(\varepsilon^{-2}) \quad \text{as } |\varepsilon| \to \infty. \tag{A6.4.22}$$

As a consequence of this, it follows from Eq. (A6.4.17) that

$$D_{l,\mathbf{k}}(\varepsilon) \to 1, \quad \text{as } |\varepsilon| \to \infty. \tag{A6.4.23}$$

Therefore, it proves advantageous to introduce a function $\mathcal{D}_{l,\mathbf{k}}(\varepsilon)$ via the relation

$$\frac{1}{D_{l,\mathbf{k}}(\varepsilon)} = 1 + \frac{1}{\mathcal{D}_{l,\mathbf{k}}(\varepsilon)}, \tag{A6.4.24}$$

in which, owing to Eq. (A6.4.23), $1/\mathcal{D}_{l,\mathbf{k}}(\varepsilon) \to 0$, as $|\varepsilon| \to \infty$. Due to the fact that the plane-wave coefficients $\Xi_{l,\mathbf{k}}(\mathbf{K};\, \varepsilon)$ approach constant values for large $|\varepsilon|$, plus use of Eq. (A6.4.21), one must have $1/\mathcal{D}_{l,\mathbf{k}}(\varepsilon) = O(\varepsilon^{-2})$ as $|\varepsilon| \to \infty$. Introducing Eq. (A6.4.24) into Eq. (A6.4.16), while making use of Eq. (A6.4.18) one obtains

$$W_{\mathbf{K},\mathbf{K}'}(\mathbf{k};\, \varepsilon) = v_{\mathbf{k},\mathbf{k}'}(\mathbf{k}) + \widetilde{W}_{\mathbf{K},\mathbf{K}'}(\mathbf{k};\, \varepsilon). \tag{A6.4.25}$$

The function

$$\widetilde{W}_{\mathbf{K},\mathbf{K}}(\mathbf{k};\, \varepsilon) = \sum_{l} \frac{\Xi_{l,\mathbf{k}}(\mathbf{K};\, \varepsilon)\, \Xi_{l,-\mathbf{k}}(-\mathbf{K}';\, \varepsilon)}{\mathcal{D}_{l,\mathbf{k}}(\varepsilon)} \tag{A6.4.26}$$

is in fact the screening part of the electron–electron interaction.

As mentioned above, the biorthonormal representation in Eqs. (A6.4.14) and (A6.4.26) is formally equivalent to the representation of the one-particle Green function in terms of eigenfunctions and eigenvalues of an energy-dependent Hamiltonian. Therefore it seems natural to make an approximation for W similar to the quasiparticle approximation of the Green function. In such an approximation, the eigenvalues $\mathcal{D}_{l\mathbf{k}}(\varepsilon)$ are assumed to have M_l simple zeroes $e_l^m(\mathbf{k})$ for which

$$\mathcal{D}_{l,\mathbf{k}}(e_l^m(\mathbf{k})) = 0 \leftrightarrow D_{l,\mathbf{k}}(e_l^m(\mathbf{k})) = 0. \tag{A6.4.27}$$

Equation (A6.4.26) is then written as a sum over residues:

$$\widetilde{W}_{\mathbf{K},\mathbf{K}}(\mathbf{k};\, \varepsilon) \simeq \sum_{l} \sum_{m=1}^{M_l} f_{l,\mathbf{k}}^m \frac{\Xi_{l,\mathbf{k}}^m(\mathbf{K})\, \Xi_{l,-\mathbf{k}}^m(-\mathbf{K}')}{\varepsilon - e_l^m(\mathbf{k})}, \tag{A6.4.28}$$

in which the function $\Xi_{l,\mathbf{k}}^m(\mathbf{K})$ stands for $\Xi_{l,\mathbf{k}}(\mathbf{K};\, e_l^m(\mathbf{k}))$ and

$$f_{l,\mathbf{k}}^m \equiv \left[\frac{\partial \mathcal{D}_{l,\mathbf{k}}(\varepsilon)}{\partial \varepsilon}\right]^{-1}_{\varepsilon = e_l^m(\mathbf{k})} = \left[\frac{\partial D_{l,\mathbf{k}}(\varepsilon)}{\partial \varepsilon}\right]^{-1}_{\varepsilon = e_l^m(\mathbf{k})}. \tag{A6.4.29}$$

It should be noted that $e_l^m(\mathbf{k}) = e_l^m(-\mathbf{k})$. Obviously, at $\varepsilon = e_l^m(\mathbf{k})$ the solution of Eq. (A6.4.17) with vanishing right-hand side [cf. Eq. (A6.4.27)] is nontrivial if and only if

$$\det[\varepsilon_{\mathbf{K},\mathbf{K}}(\mathbf{k};\, e_l^m(\mathbf{k}))] = 0, \tag{A6.4.30}$$

which is just the dispersion relation for plasmon energies (see also Appendix A6.1). Therefore one has shown that the poles $e_l^m(\mathbf{k})$ of the biorthonormal type of representation can be identified with plasmon excitation energies. Note that only those solutions of Eqs. (A6.4.27) and (A6.4.30) are physically acceptable for which either $\mathrm{Re}[e_l^m(\mathbf{k})] > 0$ and $\mathrm{Im}\,[e_l^m(\mathbf{k})] < 0$ *or* $\mathrm{Re}[e_l^m(\mathbf{k})] < 0$ and $\mathrm{Im}\,[e_l^m(\mathbf{k})] > 0$. For practical purposes, it is advantageous to look for the poles of $\mathrm{Tr}[\varepsilon^{-1}_{\mathbf{K},\mathbf{K}'}(\mathbf{k};\, \varepsilon)]$ rather than for the solutions of Eq. (A6.4.30), as one has

$$\sum_{\mathbf{K}} \varepsilon^{-1}_{\mathbf{K},\mathbf{K}}(\mathbf{k};\, \varepsilon) = \sum_{l} \frac{1}{D_{l,\mathbf{k}}(\varepsilon)}\,. \tag{A6.4.31}$$

In this connection, it must be recognized that the function $\det[\varepsilon_{\mathbf{K},\mathbf{K}'}(\mathbf{k};\, \varepsilon)]$, for semiconductors, is real-valued for real energy values ε within a finite interval around $\varepsilon = 0$. Furthermore, the function is analytic everywhere in the complex ε plane, except for branch cuts on the real axis. Consequently, the reflection principle of Schwarz (see e.g. Titchmarsh (1985)) implies that the function takes complex conjugate values at complex conjugate energies. Therfore the zeros of the above function, if any, should be symmetrically located with respect to the real axis. This would imply poles of W in all four quadrants of the complex energy plane, which contradicts the above assertion about the location of poles of W, unless they lie precisely on the real axis. The latter possibility, however, is in conflict with the observation that, for three-dimensional systems, the function W has to be continuous and therefore bounded. The solution to this seemingly contradictory situation is that one has to search for the possible zeros of the *analytic continuation* (see Knopp, 1945, 1947) of $\det[\varepsilon_{\mathbf{K},\mathbf{K}'}(\mathbf{k};\, \varepsilon)]$ across the branch cut on the real axis, i.e., on the nonphysical Riemann sheets.

Engel *et al.* (1991a,b) derive a formalism that circumvents the practical difficulties in solving Eqs. (A6.4.27), (A6.4.29), and (A6.4.30) directly, which arise from the necessity of analytically continuing $\varepsilon_{\mathbf{K},\mathbf{K}'}(\mathbf{k};\, \varepsilon)$ into the nonphysical Riemann sheet. This is achieved by making an estimate of the pole positions e_l^m and using the weights $f_{l,k}^m$ as fitting parameters. If one is interested merely in a good numerical approximation for W, it is even possible to choose pole positions independent of the band index l, i.e.,

$$\widetilde{W}_{\mathbf{K},\mathbf{K}'}(\mathbf{k};\, \varepsilon) \simeq \sum_{m} \left[\sum_{l} f^m_{l,\mathbf{k}}\, \Xi^m_{l,\mathbf{k}}(\mathbf{K})\, \Xi^m_{l,-\mathbf{k}}(-\mathbf{K}') \right] \left[\frac{1}{\varepsilon - e_m(\mathbf{k})} - \frac{1}{\varepsilon + e_m(\mathbf{k})} \right], \tag{A6.4.32}$$

where, concerning the minus sign between the two terms inside the last set of large parentheses, use has been made of the fact that the dielectric matrix $\varepsilon_{\mathbf{K},\mathbf{K}'}(\mathbf{k};\ \varepsilon)$ is an *even* function of ε.

If, however, one is interested in explaining the main features of the plasmon spectrum in terms of as few collective excitations as possible, direct solution of Eqs. (A6.4.27), (A6.4.29), and (A6.4.30) yields a physically more appealing picture, especially in view of calculating plasmon band structures. As this requires knowledge of the analytic continuation of W across the branch cut along the real energy axis, the answers one obtains for the excitation energies and lifetimes become more ambiguous the further away from the real axis the corresponding pole lies, and it becomes more and more difficult to distinguish between individual excitations. It should be noted that whatever choice of parameters in Eq. (A6.4.28) has been made, Eqs. (A6.4.27), (A6.4.29), and (A6.4.30) will always be satisfied exactly if one replaces the exact functions $\varepsilon_{\mathbf{K},\mathbf{K}'}$ and $D_{l,\mathbf{k}}$ by those which would be obtained from the approximation to W in Eq. (A6.4.28). The technique of analytic continuation has been used successfully to determine plasmon bands along the Δ and Λ directions in silicon (see also Appendix 6.6).

In order to find the connection between the expansion functions $\xi_{l,\mathbf{k}}(\mathbf{r};\ \varepsilon)$ and the charge-density functions participating in plasma oscillations, the reader is reminded that the *total* potential $\delta\chi(\mathbf{r};\ \varepsilon)$ due to an *external* potential $\delta\phi(\mathbf{r};\ \varepsilon)$ is equal to

$$\delta\chi(\mathbf{r};\ \varepsilon) = \delta\phi(\mathbf{r};\ \varepsilon) + \int d^3r' v(\mathbf{r}-\mathbf{r}')\delta\rho(\mathbf{r}';\ \varepsilon), \tag{A6.4.33}$$

in which $\delta\rho(\mathbf{r}';\ \varepsilon)$ stands for the charge-density deviation from the charge-density distribution in the ground state. Moreover, $\delta\phi(\mathbf{r};\ \varepsilon)$ is related to $\delta\chi(\mathbf{r};\ \varepsilon)$ through the relation (Hedin and Lundqvist, 1969)

$$\delta\phi(\mathbf{r};\ \varepsilon) = \int d^3r' \varepsilon(\mathbf{r},\ \mathbf{r}';\ \varepsilon)\delta\chi(\mathbf{r}';\ \varepsilon). \tag{A6.4.34}$$

In the absence of the external potential, making use of Eq. (A6.4.33), this expression reduces to

$$\int d^3r_1 d^3r_2 \varepsilon(\mathbf{r},\mathbf{r}_1;\ \varepsilon)v(\mathbf{r}_1 - \mathbf{r}_2)\delta\rho(\mathbf{r}_2;\ \varepsilon) = 0, \tag{A6.4.35}$$

or, in the plane-wave representation,

$$\sum_{\mathbf{K}'} \varepsilon_{\mathbf{K},\mathbf{K}'}(\mathbf{k};\ \varepsilon)v_{\mathbf{K},\mathbf{K}'}(\mathbf{k})\delta\rho(\mathbf{k}+\mathbf{K}';\ \varepsilon) = 0. \tag{A6.4.36}$$

Eq. (A6.4.36) allows the dispersion relation for plasmon band structures to be extracted; due to the fact that $\det[v_{\mathbf{K},\mathbf{K}'}(\mathbf{k})] \neq 0$, Eq. (A6.4.36) can have nontrivial solutions if and only if $\det[\varepsilon_{\mathbf{K},\mathbf{K}'}(\mathbf{k};\ \varepsilon)] = 0$. It immediately follows that the eigenfunctions $\Xi^m_{l,\mathbf{k}}$ can be identified, apart from a multiplicative constant α, with the amplitudes of charge-density fluctuations:

$$\delta\rho(\mathbf{k} + \mathbf{K};\ e^m_l(\mathbf{k})) = \alpha\, \frac{\varepsilon_0}{e^2} |\mathbf{k} + \mathbf{K}|^2\, \Xi^m_{l,\mathbf{k}}(\mathbf{K}). \tag{A6.4.37}$$

A6.4.1. Calculations of the Electron Self-Energy in the GW Scheme

Most of the calculations of the electron self-energy operators Σ in crystalline materials thus far have employed the *GW* approximation (see also Appendix 6.8), in which only contributions to lowest order in the screening potential W are accounted for. If one replaces the Green function G by its unperturbed version G^0, the plane-wave matrix elements of Σ can be written as

$$\sum_{\mathbf{G}\mathbf{G}'}(\mathbf{k};\,\varepsilon) = \frac{i}{2\pi\hbar^2\Omega}\sum_{\mathbf{k}'}\sum_{\mathbf{K},\mathbf{K}'} F_{\mathbf{K},\mathbf{K}'}(\varepsilon;\,\mathbf{G},\,\mathbf{G}',\,\mathbf{k},\,\mathbf{k}'), \tag{A6.4.38}$$

where

$$F_{\mathbf{K},\mathbf{K}'}(\varepsilon;\,\mathbf{G},\,\mathbf{G}',\,\mathbf{k},\,\mathbf{k}') = \hbar\sum_n d_{n,\mathbf{k}-\mathbf{k}'}(\mathbf{K})d^*_{n,\mathbf{k}-\mathbf{k}'}(\mathbf{K}')$$

$$\int_{-\infty}^{\infty} d\varepsilon' \frac{W_{\mathbf{G}-\mathbf{K},\mathbf{G}'-\mathbf{K}'}(\mathbf{k}',\,\varepsilon')e^{-i\varepsilon'\eta_0/h}}{\varepsilon-\varepsilon'-\varepsilon_n(\mathbf{k}-\mathbf{k}')-i\eta_1\,\mathrm{sgn}\,[\mu-\varepsilon_n(\mathbf{k}-\mathbf{k}')]}. \tag{A6.4.39}$$

In Eq. (A6.4.39) the $d_{n,\mathbf{k}}(\mathbf{K})$ coefficients are plane-wave components of the Bloch wave functions of the unperturbed system with band index n and wave vector $\mathbf{k}$; $\varepsilon_n(\mathbf{k})$ is the related (real) energy; μ separates occupied from unoccupied bands; and η_0,η_1 are infinitesimally small positive numbers. Making use of Eqs. (A6.4.25) and (A6.4.26) one obtains from Eq. (A6.4.39) the result of Engel *et al.* (1991a):

$$\begin{aligned}F_{\mathbf{K},\mathbf{K}'}(\varepsilon;\,\mathbf{G},\,\mathbf{G}',\,\mathbf{k},\,\mathbf{k}') = {}& 2\pi i\hbar\sum_n d_{n,\mathbf{k}-\mathbf{k}'}(\mathbf{K})d^*_{n,\mathbf{k}-\mathbf{k}'}(\mathbf{K}')\,[\theta(\mu-\varepsilon_n(\mathbf{k}-\mathbf{k}'))\\ &\times v_{\mathbf{G}-\mathbf{K},\mathbf{G}'-\mathbf{K}'}(\mathbf{k}') + H_{\mathbf{G}-\mathbf{K},\mathbf{G}'-\mathbf{K}'}(\mathbf{k}';\,\varepsilon,\,\varepsilon_n(\mathbf{k}-\mathbf{k}'),\,\mu)],\end{aligned} \tag{A6.4.40}$$

in which

$$\mathbf{H}_{\mathbf{G}-\mathbf{K},\mathbf{G}'-\mathbf{K}'}(\mathbf{k}';\,\varepsilon,\,\varepsilon_n(\mathbf{k}-\mathbf{k}'),\,\mu) \equiv \frac{1}{2\pi i}$$

$$\times\int_{\infty}^{\infty} d\varepsilon' \frac{\widetilde{W}_{\mathbf{G}-\mathbf{K},\mathbf{G}'-\mathbf{K}'}(\mathbf{k}',\,\varepsilon')}{\varepsilon-\varepsilon'-\varepsilon_n(\mathbf{k}-\mathbf{k}')-i\eta_1\,\mathrm{sgn}\,[\mu-\varepsilon_n(\mathbf{k}-\mathbf{k}')]}. \tag{A6.4.41}$$

In arriving at Eq. (A6.4.41), use has been made of the residue theorem to obtain the term containing the θ function. By substituting Eq. (A6.4.32) in Eq. (A6.4.41) and again using the residue theorem, one finds

$$\mathbf{H}_{\mathbf{G}-\mathbf{K},\mathbf{G}'-\mathbf{K}'}(\mathbf{k}',\,\varepsilon,\,\varepsilon_n(\mathbf{k}-\mathbf{k}'),\,\mu) =$$

$$-\sum_{l,m} f^m_{l,\mathbf{k}'}\frac{\Xi^m_{l,\mathbf{k}'}(\mathbf{G}-\mathbf{K})\Xi^m_{l,-\mathbf{k}'}(-\mathbf{G}'+\mathbf{K}')}{\varepsilon-\varepsilon_n(\mathbf{k}-\mathbf{k}')+e_m(\mathbf{k}')\mathrm{sgn}\,[\mu-\varepsilon_n(\mathbf{k}-\mathbf{k}')]}. \tag{A6.4.42}$$

The practical advantage of Eq. (A6.4.42) along with Eq. (A6.4.40) over Eq.

(A6.4.39) is obvious: there is no energy integration left to be carried out numerically. The price to be paid is an additional summation over the plasmon band index l and over the pole positions m. In general, an efficient method for obtaining the coefficients $f^m_{l,\mathbf{k}'}$, can be found (Engel *et al.*, 1991a,b) which makes evaluation of Σ^{GW} using Eqs. (A6.4.38) (A.6.4.40) and (A6.4.42) comparable in terms of numerical cost with schemes employing crude approximations for W, like the plasmon-pole model in which each matrix element of W is represented in terms of two simple poles. In the main text further development of the theory presented is given along with numerical results of calculations on a model semiconductor [Engel *et al.* (1991b)].

A6.5. COEFFICIENTS IN THE CONTINUED FRACTION EXPANSION

The coefficients a_i and b_i in Eq. (6.19) can be evaluated recursively from the associated orthogonal polynomials (Stieltjes theorem; see Engel *et al.*, 1991b):

$$\left.\begin{aligned} P_{-1}(x) &= 0, \\ P_0(x) &= 1, \\ P_{n+1}(x) &= (x - a_n)P_n(x) - b_n^2 P_{n-1}(x). \end{aligned}\right\} \tag{A6.5.1}$$

where

$$a_n \equiv \frac{\langle xP_n, P_n\rangle}{\langle P_n, P_n\rangle}, \tag{A6.5.2}$$

$$b_n^2 \equiv \frac{\langle xP_n, P_{n-1}\rangle}{\langle P_{n-1}, P_{n-1}\rangle} - \frac{\langle P_n, P_n\rangle}{\langle P_{n-1}, P_{n-1}\rangle}, \tag{A6.5.3}$$

Here one has defined the inner product $\langle P, Q\rangle$ by

$$\langle P, Q\rangle \equiv \int_a^b dx\, P(x)g(x)Q(x) \tag{A6.5.4}$$

for arbitrary real functions P and Q. The $\{P_n\}_n$ are orthonormal, i.e., $\langle P_n, P_m\rangle = \delta_{n,m}$, by construction.

A6.6. PLASMON PROPERTIES AND EFFECTIVE SCREENED INTERACTION

Plasmon energies are closely related to the complex zeros of the determinant of the dielectric matrix or, equivalently, to the complex poles of the dynamically screened interaction matrix of electronic systems. In this Appendix, following

Farid *et al.* (1994), the true positions of these zeros, or poles, in the complex energy plane will be clarified. Concluding that these points are on the nonphysical Riemann sheets associated with the abovementioned matrices, two methods which enable one to determine not only the plasmon energy bands but also their lifetimes throughout the whole Brillouin zone of reciprocal space are considered.

Then the plasmon energy bands of silicon in the diamond structure (within the norm-conserving pseudopotential framework) along a symmetry axis will be studied. Finally, it will be demonstrated how the electronic self-energy operator within the above *GW* framework can be accurately and relatively easily calculated, following Farid *et al.* (1994).

A6.6.1. Screened Interaction

In an electronic system, interaction and screening are inseparable notions. The static and approximate theory of Thomas and Fermi (see, e.g., Alonso and March, 1989), for instance, shows that within a uniform system the $1/r$ behavior of the free-space electronic interaction, with r the distance between two electrons, transforms into $e^{-q_{TF}r}/r$. Here q_{TF} stands for the Thomas–Fermi wave number (see also Appendix 6.2) that takes values between 1.2 and 2.1 Å^{-1} for metallic densities.

The relationship between the dynamically screened interaction W and the bare Coulomb interaction υ is a linear one characterized by the dielectric function ε via (in symbolic notation) $W = \varepsilon^{-1}v$. In this Appendix, the main tool (see Holas, 1993; also Appendix 4.3) will be the random-phase approximation (RPA) of ε and, where appropriate, corrections to this in terms of the local-field-correction function (Ichimaru, 1982; Farid *et al.*, 1993). This local-field correction has to be distinguished from "local-field effects" as discussed by Farid *et al.* (1994).

Besides relating W to v, ε relates, in the linear-response regime, the charge fluctuations in the system to an externally applied (time-dependent) potential. To be explicit, let us first define $F(\varepsilon) \equiv \int_{-\infty}^{+\infty} dt\, e^{i\varepsilon t/h} f(t)$ as the time-Fourier transform of $f(t)$. Symbolically, one can write $\Phi_{ext}(\varepsilon) = \in(\varepsilon)\Phi_{int}(\varepsilon)$, in which Φ_{ext} and Φ_{int} denote, respectively, an external and the resulting internal potential in the system; Φ_{int} is equal to the sum of Φ_{ext} and an induced potential Φ_{ind}. Φ_{ind} is directly related to the number-density fluctuations $\delta\rho(\varepsilon)$ in the system via $\Phi_{ind}(\varepsilon) = \upsilon\delta\rho(\varepsilon)$.

In uniform systems the abovementioned relation between Φ_{int} and Φ_{ext} is diagonal in the momentum representation, and nonvanishing charge fluctuations in the absence of Φ_{ext} are possible at energies $\varepsilon = e(q)$ for which $\varepsilon[(q;e(q)] = 0$, where q stands for a linear momentum. For nonuniform systems, of which here only semiconducting or insulating crystals will be considered, the momentum-representation relationship between Φ_{int} and Φ_{ext} is not diagonal, and nonvanishing $\delta\rho(\varepsilon)$ for $\Phi_{ext} \equiv 0$ are expected at $\varepsilon = e_l(\mathbf{q})$ for which $\det\{\varepsilon(\mathbf{q};e_l(\mathbf{q})\} = 0$, holds in which $(\in_{\mathbf{G},\mathbf{G}'}(\mathbf{q};\, \varepsilon)) = \in_{\mathbf{G},\mathbf{G}'}(\mathbf{q};\, \varepsilon)$, with $\mathbf{G}$ and $\mathbf{G}'$ reciprocal lattice vectors and $\mathbf{q}$ a vector inside the first Brillouin zone (1BZ). The energies $e(q)$ and $e_l(q)$, with l a band index, are plasmon energies (if sufficiently low, they are also called "exciton" energies), and the above equations from which these are obtained are plasmon dispersion relations. It should be noted that the off-diagonal elements of $\varepsilon(\mathbf{q};\, \varepsilon)$ are a consequence of Umklapp or local-field effects.

As in a many-electron system electrons interact via W rather than v, it seems natural to employ W, which is "weaker" than v, in the perturbation expansion of, e.g., the single-particle Green function G of the interacting system in terms of that of a noninteracting system. Many properties of interacting systems, such as quasiparticle energies (Hybertsen and Louie, 1986; Godby, *et al.*, 1988) or ground-state total energies (Farid, *et al.*, 1990), can be obtained from G. Using the Dyson equation, G can be obtained from the self-energy operator Σ. Hedin (1965) has derived a systematic expansion scheme for Σ in "powers" of W of which the first term is the GW approximation; Σ consists of a combination of G and W: This is discussed in more detail in the main text of Chapter 6.

Briefly, Farid *et al.* (1994) present an analysis of the dielectric function and of the plasmon dispersion relation, in the complex energy domain. They then consider three techniques of analytic continuation and use these to demonstrate the way the energy dependence of W can be very accurately described in terms of plasmon-like excitations and how Σ^{GW} can be evaluated accurately and relatively easily in real crystals. In the remainder of this Appendix, some numerical results for the plasmon band energies in silicon will then be summarized. These results will be discussed in comparison with experiment and some possible areas for future research indicated (see Engel *et al.*, 1994).

6.6.2. Plasmon Properties

The results presented below are for silicon in the diamond structure. They originate from two independent calculations, both based on LDA band structures within a nonlocal pseudopotential framework (Engel *et al.*, 1994).

In Fig. A6.6.1 both the solutions of $\det[\varepsilon(\mathbf{q};\, z)] = 0$ and $\varepsilon_{0,0}(\mathbf{q};z) = 0$ for $\mathbf{q}$ along the λ direction of the 1BZ on the fourth quadrant of an NPRS* obtained by analytically continuing $\varepsilon(\mathbf{q};\, z)$ (see Daling *et al.*, 1992) are shown. Note that solutions of $\varepsilon_{0,0}(\mathbf{q};z) = 0$ would be contained in the solutions of $\det[\in(\mathbf{q};\, z)] = 0$ if the local-field effects were vanishing. In this picture one clearly observes how the off-diagonal elements of $\in(\mathbf{q};\, z)$, e.g., give rise to a gap in $e_l(q)$ at the zone boundary.

As in experiments, Engel *et al.* (1994) obtain the plasmon energies and lifetimes by taking the positions and widths of peaks in the loss function $-\mathrm{Im}\ \varepsilon^{-1}_{0,0}(\mathbf{q};\, \varepsilon)$. Their computational results, along with the experimental values for the plasmon energies along the λ direction, are presented in Fig. A6.6.2. In this figure

*LDA denotes local density approximation, while NPRS refers to non-physical Riemann sheets.

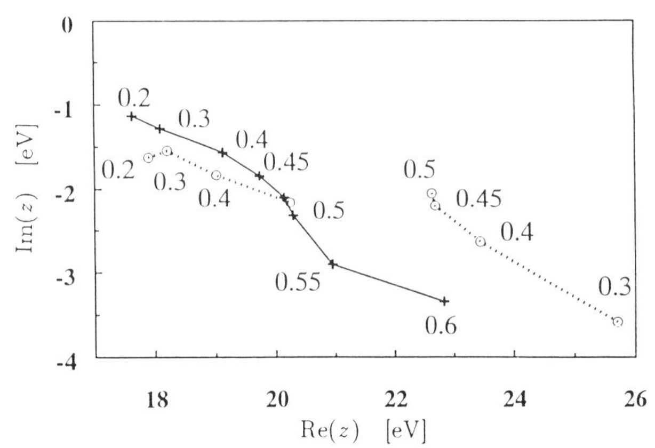

FIGURE A6.6.1. Complex plasmon energies from an LDA calculation (Daling *et al.*, 1992) for $\mathbf{q}$ along the λ direction of the 1BZ. The solid lines join data points obtained by neglecting local-field effects; that is, they are solutions of $\varepsilon_{\mathbf{G},\mathbf{G}}(\mathbf{q}; z) = 0$. The numbers attached to the + symbols give the values of q in $\mathbf{q} + \mathbf{G} = (q, q, q)$; $q = 0.5$ corresponds to the L point. The data points joined by dotted lines are the solutions of $\det[\varepsilon(\mathbf{q}; z)] = 0$. The size of $\varepsilon(\mathbf{q}; z)$ is 27×27. The numbers attached to the O symbols give the values of q in $\mathbf{q} = (q, q, q)$.

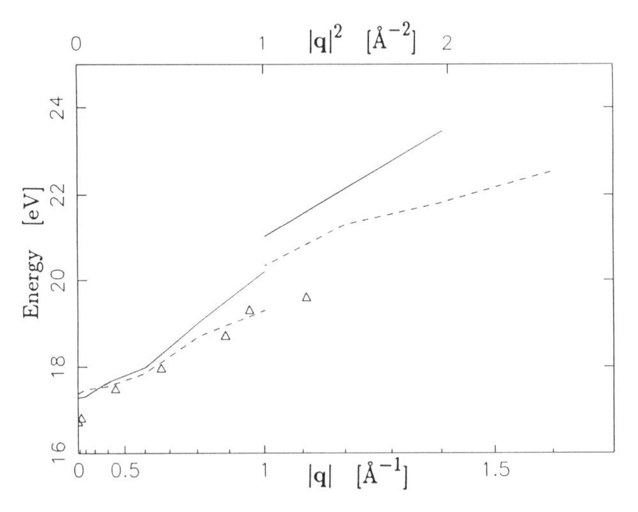

FIGURE A6.6.2. Dispersion of the peak position in the energy-loss function $-\mathrm{Im}\{\varepsilon_{0,0}^{-1}(\mathbf{q}; \varepsilon)\}$ (i.e., plasmon dispersion) along the Λ direction of the BZ. The solid line corresponds to an RPA calculation, whereas the dashed line is obtained by including an LDA local-field-correction function (Engel and Farid, 1992). Δ: experiment, according to Stiebling and Raether (1978).

it is also shown how the calculated RPA curves are improved by including a local-field-correction function, calculated within the LDA, in the calculations. Such an improvement is also observed in calculations on a uniform electron gas (Utsumi and Ichimaru, 1981). For details the reader should consult also Engel and Farid (1992).

In Fig. A6.6.3 their results for the negative of the energy-loss function are compared with experiment. To make the calculated result comparable with experiment, they have convoluted the direct result with a Lorentzian of the same width as used in obtaining the experimental curve.

Comparison of the calculated plasmon bandstructure for silicon with experiment indicates that for wave vectors not too near the Brillouin zone center and boundary, very satisfactory results can be obtained when a local-field-correction function is included in the calculations.

As for disagreements with experiment, two prime reasons can be indicated. First, the basis of the calculations has been the density functional theory of ground-state properties, which is not specifically suited for the calculation of the excited-state properties. Second, leaving that unsuitability of the DFT aside, the LDA can be a major source of error. This is apparent, e.g., from the fact that although the macroscopic dielectric constant, $1/\varepsilon^{-1}_{0,0}(\mathbf{q} \rightarrow 0; \varepsilon = 0)$, is a ground-state property, it is overestimated by about 17% by the LDA [see, e.g., (Engel and Farid, 1992)]. The overestimation of the plasmon-band dispersion, as shown in

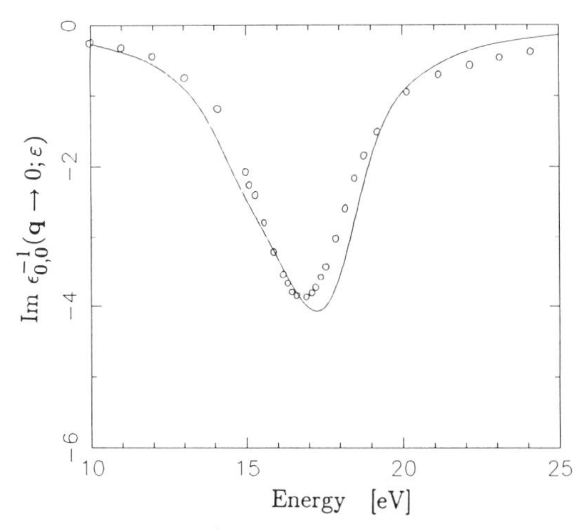

FIGURE A6.6.3. Comparison of Im $\{\varepsilon^{-1}_{0,0}(\mathbf{q} \rightarrow 0; \varepsilon)\}$ (solid line), convoluted with a Lorentzian $\eta/\pi(\varepsilon^2 + \eta^2)$ of halfwidth $\eta = 0.4$ eV, with experiment (○○○○○) (Stiebling, 1978). For numerical parameters see Engel and Farid (1992).

Fig. A6.6.2, in the vicinity of the zone boundary is probably an LDA problem, as within the LDA the contribution of the local-field-correction function is non-dispersive. The same can be said about the overestimation of the plasmon energies in the vicinity of the zone center. In addition, here Engel and Farid (1994) find several peaks in the calculated loss function (due to the local-field effects), making identification of a single plasmon energy very difficult. This is undoubtedly a matter which also complicates experimental observations.

Therefore, two directions for development are to

1. use a theory that is suited for describing excited states of the many-electron systems, and to
2. improve on the LDA.

In fact, a density functional theory for describing energy-dependent effects does exist (Runge and Gross, 1984) and an LDA within this formalism has been developed by Gross and Kohn (1985). However, using such a theory in the above context confronts the fact that the eigenstates of the independent-particle, time-dependent Slater–Kohn–Sham Hamiltonian are not stationary. Concerning the nonlocal effects neglected in LDA, there are now formalisms available that take these effects approximately into account (Langreth and Mehl, 1983).

A6.7. SOME DIFFICULTIES WITH PLASMON-POLE APPROXIMATIONS AND PROPOSED REMEDIES

As seen in the main text, it proves possible to extend the plasmon-pole approximation to include moments of higher order. However, certain problems associated with such plasmon-pole approximations can arise: these are pointed out below and some possible remedies are then referred to (see also Engel *et al.*, 1991b).

The first drawback is the numerical effort needed for their construction. Evaluating the inner products directly in Eq. (A6.5.2) and (A6.5.3), for example, is impractical, as it would require the knowledge of the imaginary part of $W(G,\Sigma)$ at a large number of real energies (this was, however, needed to construct Figs. 6.3 and 6.4). Sum rules, like the f-sum rule, allow a fairly ready evaluation of the lowest moments; similarly, the asymptotic behavior can be used to determine these [Eq. (6.45)]. For some functions, it is also possible to carry out the energy integrations in Eqs. (A6.5.2) and (A6.5.3) analytically. The polarization function P in the random-phase approximation (from which the dielectric function and its inverse can be readily found; Engel *et al.*, 1991b) in the plane wave representation, for example, can be written as

$$p_{\mathbf{K},\mathbf{K}'}(\mathbf{q};\ \varepsilon) = p_{\mathbf{K},\mathbf{K}'}(\mathbf{q};\ \varepsilon) + p_{\mathbf{K},\mathbf{K}'}(\mathbf{q};\ -\varepsilon), \tag{A6.7.1}$$

where

$$p_{\mathbf{K},\mathbf{K}'}(\mathbf{q};\,\varepsilon) = \sum_{l_v,l_c} \int d^3k\, F^{l_v,l_c}_{\mathbf{K},\mathbf{K}'}(\mathbf{k},\,\mathbf{q}) \left[\frac{1}{\varepsilon - \varepsilon_{l_c}(\mathbf{k}+\mathbf{q}) + \varepsilon_{l_v}(\mathbf{k}) + i\eta} \right]. \qquad \text{(A6.7.2)}$$

Here the $\varepsilon_{l_v}(\mathbf{k})$ and $\varepsilon_{l_c}(\mathbf{k})$ are valence- and conduction-band energies, respectively. The $F^{l_v,l_c}_{\mathbf{K},\mathbf{K}'}$ are combinations of plane-wave coefficients [see also Eq. (A6) of Engel *et al.*, 1991b). From Eq. (A6.7.2), one finds (assuming, for the sake of simplicity, $F^{l_v l_c}_{\mathbf{K},\mathbf{K}'}$ to be real),

$$\int_0^\infty d\varepsilon\; \mathrm{Im}\, p_{\mathbf{K},\mathbf{K}'}(\mathbf{q};\,\varepsilon) Q_n(\varepsilon) = -\pi \sum_{l_v,l_c} \int d^3k\, F^{l_v,l_c}_{\mathbf{K},\mathbf{K}'}(\mathbf{k},\mathbf{q}) Q_n[\varepsilon_{l_c}(\mathbf{k}+\mathbf{q}) - \varepsilon_{l_v}(\mathbf{k})], \qquad \text{(A6.7.3)}$$

for an arbitrary polynomial $Q_n(\varepsilon)$

For the plasmon-pole model and its extensions discussed in this Appendix, the poles depend on the matrix element under consideration, which in conjunction with their k-dependence makes computation difficult for the GW self-energy operator.

It is also to be noted that, in the formulation given earlier, the discontinuity $g(x)$ has to be positive or negative definite over the entire energy range. While this is often the case for the diagonal elements of operators, the off-diagonal elements do not in general satisfy this requirement (for some discussion, see the GW calculations by Hybertson and Louie, 1985).

AN ALTERNATIVE REPRESENTATION

One wants to approximate an analytic function $f(z)$, defined as in Eq. (6.36) with a branch cut along some part of the positive (negative) real axis, by a function $h(z)$ of the form

$$h(z) \equiv \sum_{i=1}^{N} \frac{w_i}{z - \xi_i}, \qquad \text{(A6.7.4)}$$

where the N poles ξ_i are restricted to lie in the lower (upper) half of the complex z-plane. Let us therefore minimize (an asterisk indicates complex conjugation)

$$I \equiv \int_{-\infty}^{\infty} dx \left| \sum_i \frac{w_i}{x - \xi_i} - f(x) \right|^2$$

$$= \int_{-\infty}^{\infty} dx \left(\sum_{i,j} \frac{w_i^* w_j}{(x - \xi_i^*)(x - \xi_j)} - \sum_i \frac{w_i^*}{x - \xi_i^*} f(x) - \sum_i \frac{w_i}{x - \xi_i} f^*(x) + |f(x)|^2 \right). \qquad \text{(A6.7.5)}$$

Using contour integration and the fact that f does not have any poles in the upper (lower) half-plane (see above) and falls off like $1/|z|$ as $|z| \to \infty$, this yields

$$I = 2\pi i \left(\sum_{i,j} \frac{w_i^* w_j}{\xi_i^* - \xi_j} - \sum_i w_i^* f(\xi_i^*) - \sum_i w_i f^*(\xi_i^*) \right) + \int_{-\infty}^{\infty} dx |f(x)|^2. \tag{A6.7.6}$$

Minimizing this expression with respect to w_i^* and ξ_i^*, one obtains, respectively,

$$\sum_j \frac{1}{\xi_i^* - \xi_j} w_j = f(\xi_i^*), \tag{A6.7.7a}$$

$$\sum_j \frac{1}{(\xi_i^* - \xi_j)} w_j = - \frac{df}{dz} (\xi_i^*). \tag{A6.7.7b}$$

In principle, this system of equations completely determines the optimal pole positions ξ_i and the weights w_i. However, numerically solving Eq. (A6.7.7) for the pole positions ξ_i is difficult to date. In practice, it is possible to fix the positions of the poles ξ_i *a priori*, using Gaussian integrations as a guideline. Then Eq. (A6.7.7a) shows that minimization of the functional I in Eq. (A6.7.5) with respect to the weights w_i is equivalent to interpolating the function $f(z)$ by the function $h(z)$ [see Eq. (A6.7.4)] at the points ξ_i^*.

Equation (6.47) of the main text shows that an exact analytic continuation of f through the branch cut will, mathematically, always give rise to the branch cut being located to another position; any attempt to replace the branch cut by a finite number of simple poles is necessarily an approximation.

One way of evaluating the contour integral in Eq. (6.47) would be to employ some Gaussian-quadrature rule. This results in an approximation for $\tilde{f}(z)$ that is formally equivalent to the expression for $h(z)$ in Eq. (A6.7.4). These considerations motivate the following choice of pole positions i in Eqs. (A6.7.4)–(A6.7.7):

$$\xi_i = \left(y_i, \frac{\mp \sqrt{(y_i - a)(b - y_i)}}{b - a} (y_{N/2+1} - y_{N/2}) \right), \qquad i = 1, 2, \ldots, N. \tag{A6.7.8}$$

Here the $\{y_i\}_i$ can, for example, be the zeros of a Chebyshev polynomial of order N rescaled to cover the interval $[a, b]$; the sign of the imaginary part depends on whether the branch cut in f lies below (– sign) or above (+ sign) the real axis. The imaginary part of ξ_i in the above expression is chosen so as to ensure that the resulting function $h(z)$ in Eq. (A6.7.4) is smooth along the real axis. This is the case if the distance between two poles is approximately equal to their imaginary parts (the imaginary part of $1/(x - i\gamma)$ is a Lorentzian with halfwidth γ). In Eq. (A6.7.8) use has been made of the fact that the (rescaled) density of zeros of the Chebyshev polynomials $T_N(x)$ can be shown to be proportional to $1/\sqrt{(x - a)(b - x)}$ in the limit of $N \to \infty$, where a and b are the branch points. Of course, it is possible to make different choices for the pole positions ξ_i, the main restriction being that they should be chosen rather dense in the vicinity of the branch points a and b. In the calculations of Engel *et al.* (1991b), the form Eq. (A6.7.8) was adopted throughout, with N ranging from 10 to 40.

A considerable advantage of the strategy suggested above, when applied to the matrix elements of $\widetilde{W}$, over the plasmon-pole-type expressions discussed in the earlier sections is that the pole positions ξ_i can be chosen independent of the matrix element under consideration. Their dependence on the wave vector $\mathbf{q}$ is extremely simple and is determined by the dependence of the branch points a and b on $\mathbf{q}$. The final expression for $\widetilde{W}$ is

$$\widetilde{W}_{\mathbf{K},\mathbf{K}'}(\mathbf{q};\,\varepsilon) = \sum_m \widetilde{W}^m_{\mathbf{K},\mathbf{K}'}(\mathbf{q})\left\{\frac{1}{\varepsilon - e_m(\mathbf{q})} - \frac{1}{\varepsilon + e_m(\mathbf{q})}\right\}, \tag{A6.7.9}$$

where the symbols $\widetilde{W}^m_{\mathbf{K},\mathbf{K}'}(\mathbf{q})$ and $e_m(\mathbf{q})$ denote the weights and poles.

The approximate expression in Eq. (A6.7.9) is found to be indistinguishable from the exact function for 40 poles and still quite accurate for as few as 10 poles (see Fig. 6.3). Thus the complete function $\widetilde{W}_{\mathbf{K},\mathbf{K}'}(\mathbf{q};\,\varepsilon)$ can be recovered by evaluating it for ten energy values only. Moreover, none of these energies needs to be real. This is important, as for real energies the integrand in the expression for the polarization function P[Eq. (A6.7.1)]—which is closely related to W—becomes singular, making its numerical evaluation difficult. Figure A6.7.1 was obtained by minimizing

$$I \equiv \int_{-\infty}^{\infty} dx \left| \sum_m \frac{2e_m(\mathbf{q})\,\widetilde{W}^m_{\mathbf{K},\mathbf{K}'}(\mathbf{q})}{x - e_m(\mathbf{q})^2} - \widetilde{W}_{\mathbf{K},\mathbf{K}'}(\mathbf{q};\,\sqrt{x}) \right|^2, \tag{A6.7.10}$$

which is of the same form as Eq. (A6.7.6) and use has been made of the symmetry

$$\widetilde{W}_{\mathbf{K},\mathbf{K}'}(\mathbf{q};\,\varepsilon) = \widetilde{W}_{\mathbf{K},\mathbf{K}'}(\mathbf{q};\,-\varepsilon). \tag{A6.7.11}$$

The integral over x from $-\infty$ to 0 then corresponds in the energy domain ($\varepsilon^2 = x$) to an integration along the imaginary energy axis, and the fit thus ensures optimal agreement along both the real and the imaginary energy axis.

It is apparent that the function h in Eq. (A6.7.4) with the poles chosen according to Eq. (A6.7.8) will reproduce only those features of the imaginary part $g(z)$ at the position x accurately that vary over a range larger than the imaginary part of the pole ξ_i closest to x. This means that, if $g(x)$ is sharply peaked at a particular position x_0, the contour in Fig. 5 of Engel *et al.* (1991b) would have to be deformed and the density of poles increased in the vicinity of that peak accordingly. The same holds if there are gaps or kinks present in the function g (van Hove singularities, etc.). In the latter case, even without such adjustments, the expression in Eq. (A6.7.4) with the poles chosen according to Eq. (A6.7.8) and the weights determined by Eq. (A6.7.7a) reproduces the overall features of f extremely well. This can be seen in Fig. A6.7.1b for one of the off-diagonal elements of W in the simple two-band test model, where (unphysical) kinks are present in the exact

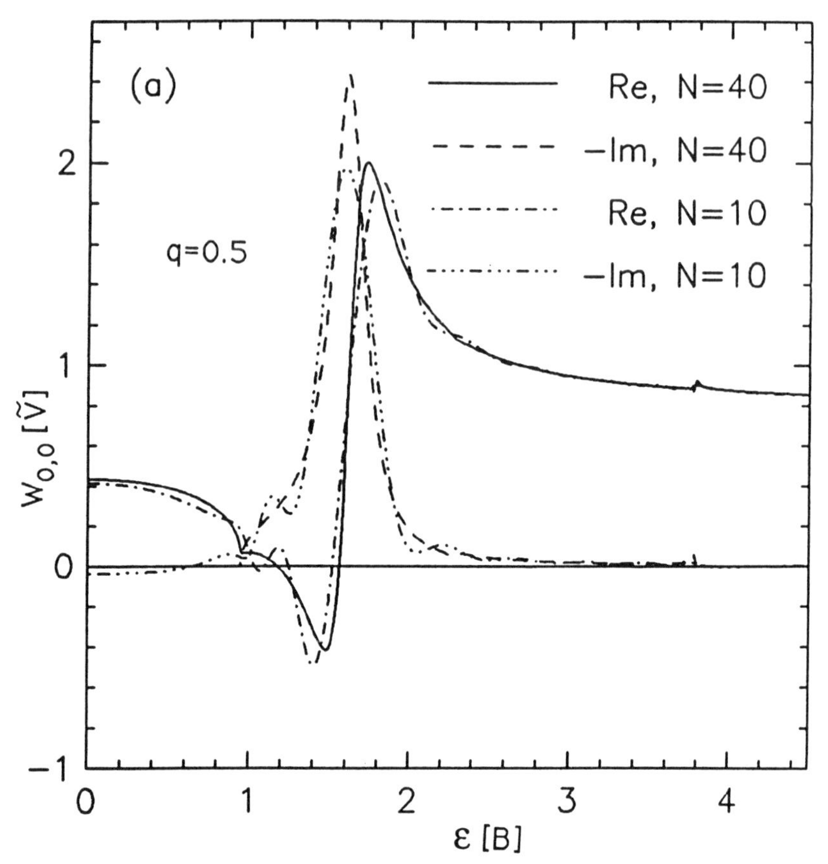

FIGURE A6.7.1.(a) Head matrix element $W_{0,0}$ of screened interaction for different number of poles N.

function W due to the restriction of two plane-wave coefficients. These kinks have been smoothed by the fitting procedure without affecting the overall agreement for the rest of the spectrum.

It was found, however, that sharp and strong peaks in $g(x)$ on the real axis may cause problems for the accuracy of the above representation if the density of poles in their vicinity has been chosen too small. This is not surprising, as it is clear from Fig. 5 of Engel *et al.* (1991b) that one should choose poles along a contour that excludes possible poles in $\tilde{g}(z)$, the analytic continuation of $g(x)$, unless additional residue contributions are taken into account [Eq. (6.47)]. In general, neglecting such contributions can lead to strong oscillations of the approximating function around the exact values over a large energy range. It is not difficult to understand this feature in terms of the least-squares integral [Eq. (A6.7.5)]; sharp

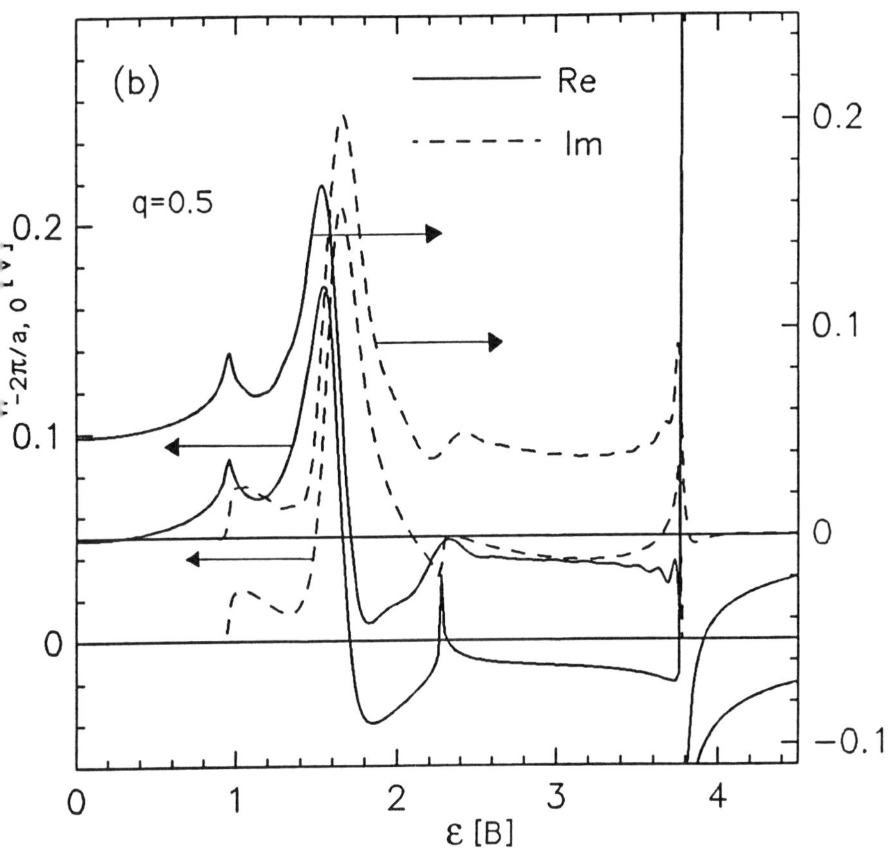

FIGURE A6.7.1.(b) An off-diagonal matrix element of screened interaction.

peaks make a large contribution to this integral (for δ-function peaks it diverges) and in order to minimize it, solving Eq. (A6.7.7a) ensures as good a fit as possible in the vicinity of such peaks. If, however, the imaginary part of the poles in the vicinity of such a peak has been chosen too large, this closest-possible fit can be achieved only by a strongly oscillatory behavior of the fitting function. It is to be noted, incidentally, that the evidence available does not suggest the occurrence of extremely sharp peaks in the imaginary part of W for real three-dimensional systems; for example, the calculations of the dielectric function of silicon by Walter and Cohen (1972) show a very smooth behavior with fairly broad peaks in the imaginary part.

As an artifact of the one-dimensional model system, it turns out that W in this model has a pole on the real axis at an energy slightly above the high-energy cutoff

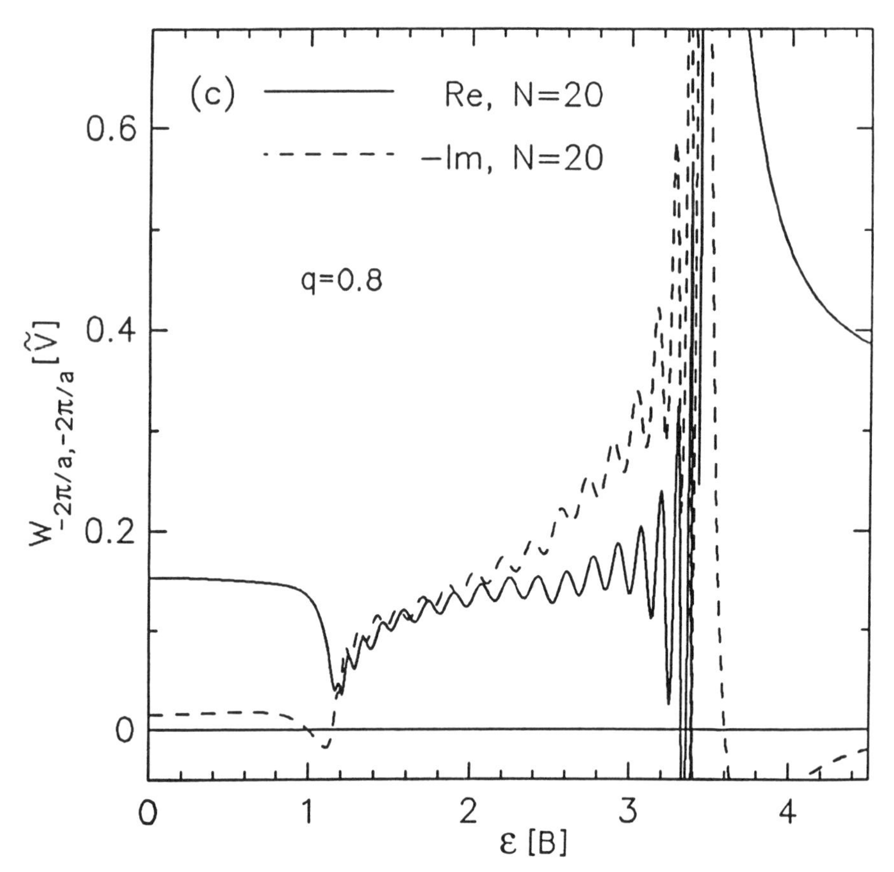

FIGURE A6.7.1.(c) Result of choosing poles according to Eq. (A6.7.8).

for the electron–hole excitation region. Though this is a pathological feature of this particular model, it is instructive to see its effect on the accuracy of the present representation. Figure A6.7.1(c) shows the result of neglecting the existence of this pole and naively choosing poles according to Eq. (6.7.8); because these poles cannot mimic the behavior of the function near the pole on the real axis, the resulting expression shows large oscillations. These are almost completely removed if one chooses an extra pole on the real axis at its proper position (Fig. A6.7.1d).

It may seem disturbing that the procedure does not automatically guarantee satisfaction of the f-sum rule, nor is the value of $\widetilde{W}_{\mathbf{K},\mathbf{K}'}(\mathbf{q}; \varepsilon = 0)$ fixed explicitly. Strictly speaking, using poles ξ_i with a finite imaginary part [as in Eq. (A6.7.8}], the imaginary part of the function $h(z)$ defined in Eq. (A6.7.4) will in general fall off not faster than $O(\varepsilon^{-2})$ as $|e|$ tends to infinity, and the first frequency moment

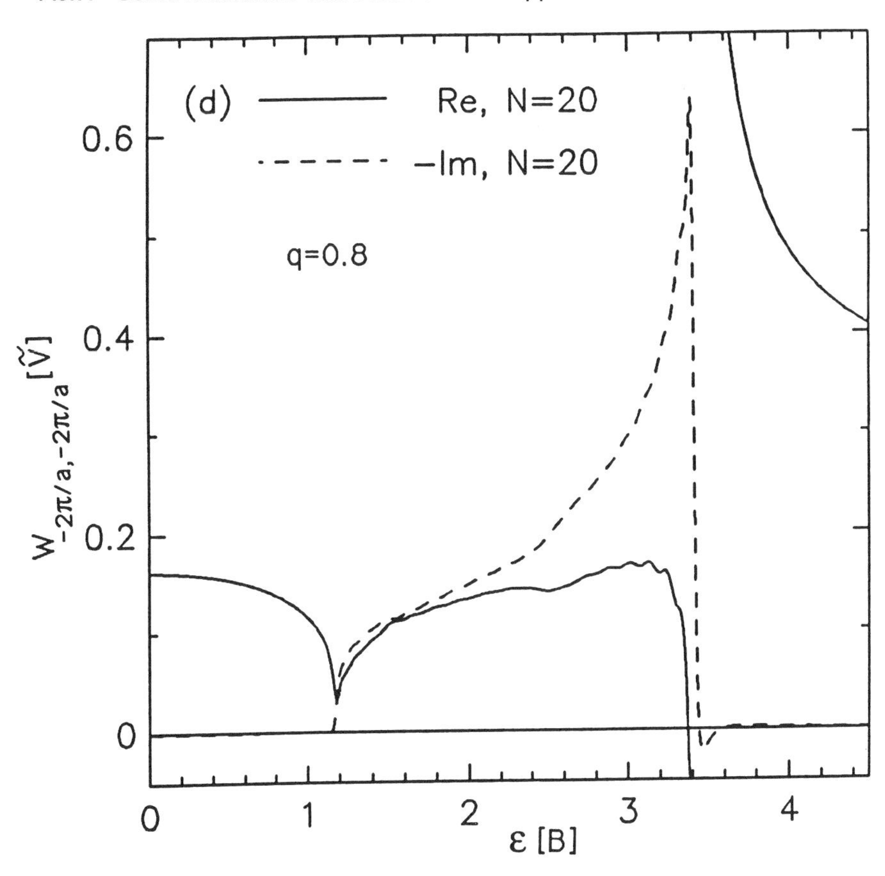

FIGURE A6.7.1.(d) Effect of extra pole on real axis (after Engel *et al.*, 1991b).

over the imaginary part diverges, whereas the imaginary part of the exact function $f(z)$, for which $h(z)$ is an approximation, was supposed to fall off at least exponentially, ensuring the existence of moments of all orders. This seemingly contradictory result can, however, be understood if one recalls that the moments of Im $f(z)$ are merely the expansion coefficients of an expansion of $f(z)$ in powers of $1/z$ [see Eq. (A6.45)], where this was shown for the case in which f represents the screening part of the screened interaction function $\tilde{W}$). Expanding $h(z)$ in Eq. (A6.7.4) in powers of $1/z$, one finds, for large $|z|$,

$$h(z) = \sum_{n=0}^{\infty} \left(\sum_i w_i\, \xi_i^n\right) 1_z n+1. \qquad \text{(A6.7.12)}$$

The nonexistence of the frequency moments of the imaginary part of $h(z)$ simply means that the expansion coefficients $\Sigma i\ W_i\ \xi_i^n$ in this expression are not purely real, as they would be for the exact function f. For the numerical calculation of $\widetilde{W}$ in the two-band model, the difference between the expansion coefficients in front of $(1/\varepsilon^2)$ (corresponding to the first frequency moment) of the approximate and exact $\widetilde{W}$ was found to be negligibly small (of the order of 10^{-4} of its absolute value for the approximation with 40 poles). The same excellent agreement was found for the static value of W at $\varepsilon = 0$. Moreover, both conditions characterizing the plasmon-pole approximation can easily be incorporated into the above formalism as additional constraints, most trivially by introducing a penalty function for their violation into Eq. (A6.7.10) or by incorporating them via a Lagrange multiplier. By doing so, the imaginary part of $\widetilde{W}$ can be made to fall off like ε^{-3}, which is in fact physically more sensible than having a finite energy cutoff for the imaginary part.

A6.8. DIFFERENT FORMS OF GW APPROXIMATIONS

Most of the ensuing discussion will be about the calculation of the self-energy of an electron. The self-energy is expressed as an integral over the wave vectors of the excitations. The argument of this integral has terms which as in the main text are represented by the symbols "GW." The discussion below will be confined to the model Hamiltonian of the *homogeneous electron gas* (see Mahan, 1994)

$$H = H_0 + V, \tag{A6.8.1}$$

$$H_0 = \sum_{p\sigma} E_p C^+_{p\sigma} C_{p\,5}, \tag{A6.8.2}$$

$$V = \frac{1}{2\Omega} \sum_{pkqs\sigma} \nu_q C^+_{p+q,s} C^+_{k-q,\sigma} C_{k\sigma} C_{ps}, \tag{A6.8.3}$$

$$\nu_q = \frac{4\pi e^2}{\varepsilon_x q^2}. \tag{A6.8.4}$$

The form for ν_q is for three dimensions. For 2D one must use area A instead of volume Ω, and $\nu_q = 2\pi/\varepsilon_x q$. Most of the discussion is for three dimensions. Electron density is denoted as usual by r_s, which is the radius in atomic units which encloses on unit of charge: a sphere in 3D and a circle in 2D. Low density metals in 3D have $r_s \sim 5$–6 and high density metals have $r_s \leq 2$.

Quinn and Ferrell (1958) were the first to calculate the self-energy of an electron for the Hamiltonian in A6.8.1–A6.8.4. An equation from their paper is

$$\Sigma(p) = \int \frac{d^4 q}{(2\pi)^4} \frac{\nu_q}{\varepsilon(q)} \frac{1}{E_p - \omega - E_{p-q} \pm i\delta}. \tag{A6.8.5}$$

This seems to be the first use of the GW Approximation, although they did not use that phrase. The modern terminology is to call

$$W(q) = \frac{\nu_q}{\varepsilon(q)}, \tag{A6.8.6}$$

$$G_0(p) = \frac{1}{E_p - \omega - E_{p-q} \pm i\delta}, \tag{A6.8.7}$$

where $W(q)$ is the screened interaction between electrons. They used the RPA form for the dielectric function $\varepsilon_{\text{RPA}}(q)$, which will be defined below. The symbol $G_0(p)$ stands for the Green's function. Note that their Green's function is "on the mass shell," which means the electron energy variable has been set equal to E_p. They used a noninteracting Green's function.

Following Mahan (1994) let us call this GW (RPA), which indicates that the electron–electron interaction ν_q is screened by the RPA dielectric function, and the electron Green's function is noninteracting. Usually the energy of the electron in the Green's function is not on the mass shell, but is set equal to some variable E. Then the self-energy $\Sigma(p,E)$ is a function of two variables: momentum and energy. One can also calculate its real and imaginary parts. The most complete calculations were done by Hedin (1965): see Hedin and Lundqvist (1969).

A further important paper for the GW Approximation was by DuBois(1959). He also calculated the self-energy of an electron. Most of his results were for small values of $r_s < 1$. However, he was the first to notice the significant cancellation between vertex functions and self-energy diagrams. Vertex functions make correlation functions larger, while self-energy effects make them smaller. If one includes the right combination of vertex functions and self-energy diagrams, often the sum of these two contributes only a small net residue. That observation raises the second question: how does one know the right combination of self-energy and vertex functions?

A6.8.1. Vertices and Self-Energies

The vertex function is defined as all contributions which are not self-energies. So let us start with the definition of a self-energy contribution. Dyson's equation states that the Green's function for an interacting electron has the form (see, e.g., Mahan, 1990):

$$G(p) = \frac{1}{E - E_p - \Sigma(p)}. \tag{A6.8.8}$$

Here a four-variable notation $p = (\mathbf{p}, E)$ is used. The extra term $\Sigma(p, E)$ in the denominator is the self-energy of an electron. It is calculated from all of the interactions between electrons and other particles; ion vibrations (phonons), impurities, defects, and other electrons.

Let us discuss various contributions using Feynman diagrams. In these diagrams, a non-interacting Green's function $G_0(p)$, which is defined as one which lacks a self-energy, is given by a straight, solid line. Self-energies are Feynman diagrams which have a single solid line entering one side, and another single solid line leaving the other side. Figure A6.8.1 shows some Feynman diagrams of an electron. The dashed lines are the interactions. They can either be Coulomb interactions or phonons. Note that all of the dashed lines begin and end at a solid line which denotes the electron (see also Mahan, 1994).

The term "self-energy" means that the electron interacts with its surroundings, which may either be ions or other electrons. These surroundings become polarized by this interaction. This polarization, in turn, acts on the original electron. A classical analogy is a boat moving on the surface of water. It makes a bow wave, and this bow wave acts on the boat and can actually limit the boat's velocity. An electron also makes microscopic polarization waves which act on the electron (Mahan, 1994).

An important set of Feynman diagrams are those which define the polarization of the electron gas. The exact dielectric function is defined as

$$\varepsilon(q) = 1 - \nu_q P(q). \tag{A6.8.9}$$

The function $P(q)$ is the exact polarization of the electron gas. The calculation of this quantity plays a central role in the theory. RPA is where this polarization is evaluated in the lowest order of perturbation theory:

$$P_0(q) = \frac{1}{\Omega} \sum_{p,\sigma} G_0(p)G_0(p + q) \tag{A6.8.10}$$

$$= 2 \int \frac{d^3p}{(2\pi)^3} \frac{n_F(E_p) - n_F(E_{p+q})}{\omega + E_p - E_{p+q}}. \tag{A6.8.11}$$

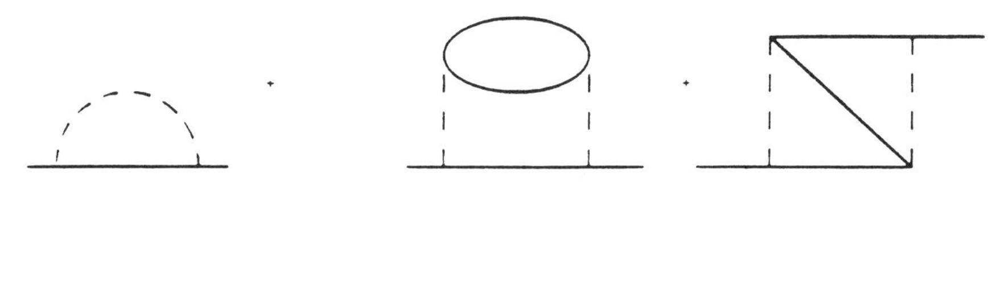

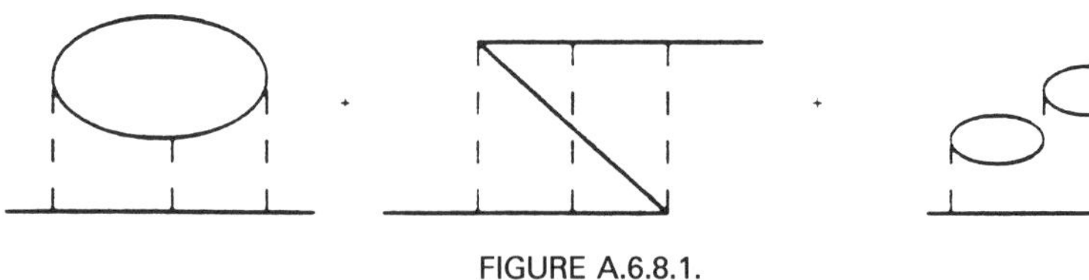

FIGURE A.6.8.1.

The wave vector integrals are easily evaluated at zero temperature. The central issue below is how to improve upon RPA as a first approximation.

Figure A6.8.2a depicts a bubble diagram which represents $P_0(q)$. The two solid lines are the two Green's functions in Eq. (A6.8.10): This diagram represents RPA for the polarization operator $P(q)$. Figure A6.8.2b shows three diagrams which also contribute to $P(q)$ and represent corrections to RPA. The first two are self-energy diagrams while the third is a vertex diagram. There are various arguments (see Mahan, 1990, 1994) which state that if one is going to include this vertex correction, then one must also include these two self-energy diagrams. Explicit calculations (Geldart and Taylor, 1970; Engel and Vosko, 1990) show that the contributions from these three terms largely cancel. The vertex diagram alone makes $P(q)$ larger, while the self-energy diagrams make it smaller. The correct answer is obtained only by having the right combination of vertex and self-energy diagrams.

Several methods have been proposed for deciding the correct combination of vertex and self-energy diagrams. One is the "conserving approximation" of Baym

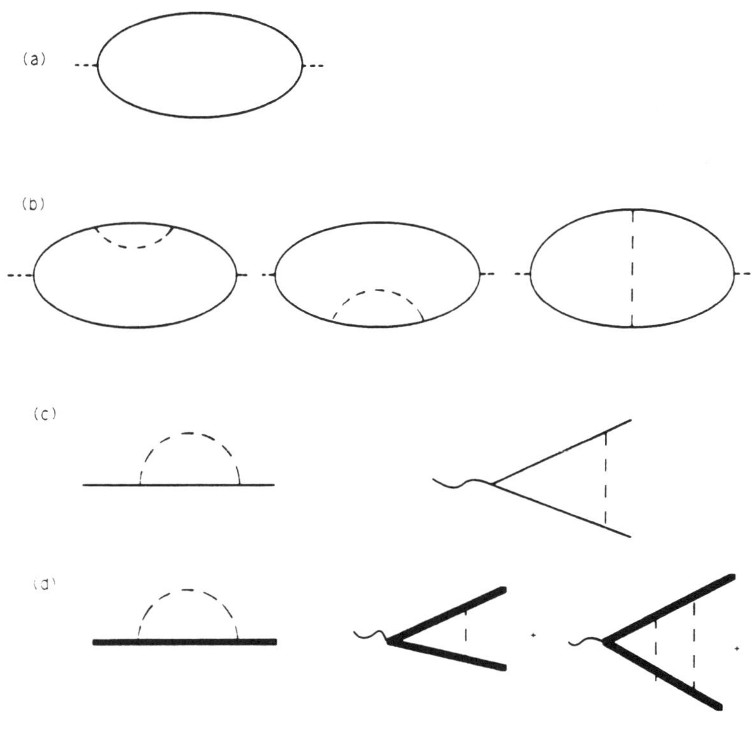

FIGURE A.6.8.2.

and Kadanoff (1961) while another is Ward (1950; see also Mahan, 1990) identities. The latter are easier to describe using Feynman diagrams, and will be discussed below. The vertex diagrams associated with a self-energy operator are found by attaching an external line to the internal Green's function lines of the self-energy. If the self-energy has more than one electron line, then a different vertex diagram results as the external line is attached, in turn, to each internal electron line.

Two examples will be given following Mahan (1994). The first is the simple self-energy shown in Fig. A6.8.2c, which has one internal electron Green's function:

$$\Sigma(p) = \sum_q W(q) \frac{1}{E + \mu + \omega - E_{p+q}}. \tag{A6.8.12}$$

Attaching the external line to this line results in the vertex diagram which is shown. These two components are the basis for the three terms in Fig. A6.8.2b. The Ward identities demonstrate that these vertices and self-energies go together.

The second example is based on the self-energy

$$\Sigma(p) = \sum_q W(q) \frac{1}{E + \mu + \omega - E_{p+q} - \Sigma(p+q)}. \tag{A6.8.13}$$

Note that this is a self-consistent equation for the self-energy, which must be solved numerically by some iteration procedure. The Feynman diagram is depicted in Fig. A6.8.2d. The double solid line indicates a full Green's function, including the self-energy. This line has an infinite number of noninteracting Green's functions, and thus generates an infinite number of vertex functions. They are also shown in the figure as a summation of ladder diagrams. The conclusion to be drawn from this is that if one wants to include a set of ladder diagrams in the evaluation of $P(q)$, then the internal lines must be interacting Green's functions with the self-energy given above. Note that the self-energy does not have a vertex correction. If it did, the resulting vertex corrections would have additional contributions. Also note that it has been assumed that the interaction term, given by the dashed line, does not have any electron lines as internal lines. If it does, then there are additional vertex contributions resulting from this self-energy. Actual interactions are screened, which implies that they do have internal electron lines in the interaction.

Three early studies of Dubois, Ward, and Baym-Kadanoff concluded that an accurate theory required that vertex functions should only be added to correlation functions when the corresponding self-energy diagrams were used to dress the Green's functions in the internal lines. Geldart and Taylor (1970) constructed a theory of electron–electron interactions based on this principle.

Near the mass shell ($E \sim E_p$) the interactions tended to renormalize the

Green's function by a constant Z. Near the Fermi surface, the self-energy can be expanded around this point:

$$\Sigma(p,E) \approx \Sigma(p_F, 0) + a(E_p - \mu) + bE, \tag{A6.8.14}$$

$$G(p,E) \approx \frac{Z}{E + \mu' - E_p'}, \tag{A6.8.15}$$

$$\mu' = \frac{\mu(1 + a) - \Sigma(p_F, 0)}{1 - b}, \tag{A6.8.16}$$

$$E'_p = E_p \frac{1 + a}{1 - b}, \tag{A6.8.17}$$

and

$$Z = \frac{1}{1 - b}. \tag{A6.8.18}$$

Here a is the kinetic energy derivative of the self-energy, while b is the derivative with respect to the energy E. The effect of the interactions, near the Fermi surface, is to (i) renormalize the chemical potential, (ii) renormalize the band width, and (iii) multiply the Green's function by Z. Calculations for actual metals indicate that $Z \approx 0.7$–0.8.

A6.8.2. Summary of Effects of Exchange and Correlation

Hubbard (1957) was the first to include exchange effects in the theory of dielectric screening. If the screening is provided partly by an electron of spin up, then other electrons of spin up cannot get too close, and are less likely to help with the screening. This intuitive picture is obviously correct. Later the picture was expanded to include correlation, so all electrons will tend to stay away from the screening electrons. This effect does alter the form for the dielectric function. It weakens the screening and makes the interaction larger. Hubbard showed that it tended to provide a modified form for the dielectric function, which is called the Hubbard form:

$$\varepsilon_H(q) = 1 - \frac{v_q P_0(q)}{1 + G_H(q) v_q P_0(q)}, \tag{A6.8.19}$$

$$G_H(q) = \frac{1}{2} \frac{q^2}{q^2 + k_s^2}. \tag{A6.8.20}$$

The factor $G_H(q)$ is not a Green's function, but is the local field factor (see Singwi and Tosi, 1981). Hubbard suggested the form given above: first with the screening wave vector equal to the Fermi wave vector ($k_s^2 = k_F^2$), and later with other versions such as including the Thomas-Fermi screening length $k_s^2 = k_F^2 + q_{TF}^2$. The latter choice works well (Rice, 1965). The Hubbard factor is actually frequency dependent $G_H(q,\omega)$ but this dependence is seldom incorporated in numerical studies. The problem with implementation of the exchange theory is that the Hubbard correction is equivalent to including the ladder diagrams. Numerous authors have shown that one can derive the Hubbard factor by summing the ladder diagrams shown in Fig. A6.8.3. This equivalence is exact for a contact potential, and is accurate for a screened Coulomb potential. In this case the dashed lines are the screened electron–electron interaction. The Hubbard result is obtained only when one uses noninteracting Green's functions during the summation of ladders. This procedure violates the early ideas that one must include self-energies in the Green's functions since these contributions go with the vertex diagrams, and largely cancel them. Instead, the usual Hubbard result is obtained by including vertices without self-energies. This procedure violates the requirements of self-consistency. Nevertheless it is the standard approximation at the time of writing. Geldart and Taylor (1970) made the only attempt to construct a Hubbard factor which included self-energies (see Mahan, 1994).

The first GW Approximation is RPA: the interaction $W(q) = \nu_q/\varepsilon_{\mathrm{RPA}}(q)$. The second GW Approximation Mahan (1994) calls GW_{II}. It is obtained by using the Hubbard form of the dielectric function in screening the Coulomb interaction. In a four-vector notation

$$\Sigma(p) = \int d_q W_H(q) G_0(p+q). \tag{A6.8.21}$$

$$W_H(q) = \frac{\nu_q}{\varepsilon_H(q)}. \tag{A6.8.22}$$

This expression for the screened interaction is also called W_{tt}, which stands for "testparticle–testparticle interaction." A testparticle is any charged particle which

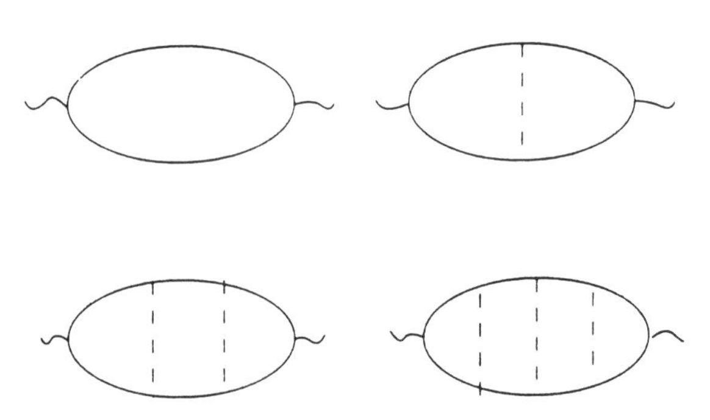

FIGURE A.6.8.3.

is not an electron, such as a proton or muon. The above expression gets it self-energy from the screened interactions with the surrounding electrons (Hedin and Lundqvist, 1969).

A different expression is needed for the self-energy of an electron because it has additional exchange interactions with the other electrons which are absent for a testparticle. These extra contributions are vertex corrections $\Gamma(p, p + q)$. For the Hubbard dielectric function, the Hubbard vertex correction $\Gamma_H(q)$ is (Rice, 1965)

$$\Gamma_H(q) = \frac{1}{1 + \nu_q G_H(q) P_0(q)}, \tag{A6.8.23}$$

$$\Sigma(p) = \int dq W_H(q) \Gamma_H(q) G_0(p + q), \tag{A6.8.24}$$

$$\widetilde{W}_H(q) = W_H(q) \Gamma_H(q) = \frac{\nu_q}{1 - \nu_q(1 - G_H(q)) P_0(q)}. \tag{A6.8.25}$$

This expression Mahan (1994) terms the third GW Approximation. Including the vertex correction results in an effective dielectric function of the form $\tilde{\varepsilon}_H(q) = 1 - \nu_q(1 - G_H)P_0$.

This form of the GW Approximation was first used by Rice (1965). The single particle energy of the electron can be obtained from the ground state energy by a functional derivative with respect to the occupation number

$$\varepsilon_k + \Sigma_k = \frac{\delta E_G}{\delta n_k}. \tag{A6.8.26}$$

There is an expression for the ground state energy which involves an integral over the dielectric function (Mahan, 1990). When using the Hubbard form for the dielectric function in the expression for the ground state energy, the functional derivative yields the third expression for the GW Approximation. The Hubbard vertex expression goes with the Hubbard dielectric function. It is not a conserving approximation, in the sense of Baym and Kadanoff, since both G_H and Γ_H result only from vertex corrections which are ladder diagrams. There are no self-energy corrections in G_H or in Γ_H.

The electron self-energy calculations of Hedin (1965) and Rice (1965) agreed well. Since Hedin used the first GW Approximation (RPA) while Rice used the third in Eqs. (A6.8.24)-(A6.8.25), the factor of G_H seems to have little effect if included in the third GW Approximation. This was confirmed by Mahan (1994) whose group did extensive calculations using both approximations and found little difference in the numerical results.

The compressibility sum rule is a statement that the long wavelength limit of polarization can be related to the compressibility K of the electron gas:

$$\lim_{q \to 0} = P_0(0) \frac{K}{K_f}, \tag{A6.8.27}$$

where K_f is the compressibility of the noninteracting electron gas—this term derives from the kinetic energy. From Eq. (A6.8.20) one finds

$$\frac{K_f}{K} = 1 + \frac{2\pi e^2}{\varepsilon_x k_s^2} P(0) \tag{A6.8.28}$$

which relates the compressibility to the Hubbard local field correction. The right-hand side vanishes as the electron density $r_s \rightarrow 5.1$: a behavior expected from other arguments. Thus it seems that the Hubbard factor gives the right form for the compressibility. There are two difficulties with this interpretation. The first, as noted by Hedin, is that if one uses dressed Green's functions by adding a factor of Z^2, the agreement vanishes. The second difficulty is that the divergence in the compressibility is caused by an entirely different set of diagrams than these simple ladders. These two difficulties are consistent. The ladder diagrams, with dressed Green's functions, are not supposed to cause a divergence in the compressibility, and do not. Mahan (1994) also discussed the local density approximation in the present context, as well as spin and charge susceptibilities. The interested reader is referred to his account of these topics.

Appendix to Chapter 7

A7.1. RELATION OF RESONATING VALENCE BOND AND GUTZWILLER METHODS

Motivated by Eq. (4.49), Anderson makes up a genuine RVB state first of all by the straightforward Gutzwiller method (see Section 4.8). One forms the projection operator

$$P_d = \prod_i (1 - n_{i\uparrow} n_{i\downarrow}), \tag{A7.1.1}$$

where n_i is the occupation number, and

$$\psi_{\mathrm{RVB}} = P_d (b^+)^{N/2} \psi_0 \tag{A7.1.2}$$

is then the trial wave function of Anderson.

He notes that one may also use quasifermion operators with the double occupancy projected out:

$$c_{i\uparrow} = c^+_{i\downarrow}\,(1 - n_{i\downarrow}) \tag{A7.1.3}$$

to write

$$b^+ = \sum_k a(k) c_{k\uparrow} c_{-k\downarrow} \tag{A7.1.4}$$

(compare Eq. (4.47)).

At this stage, he resorts to the Dyson transformation between product states and Bardeen–Cooper–Schrieffer (BCS) states. As shown by Dyson (see Anderson, 1987), if one notes that $(b^+_k)^2 = 0$, with $b^+_k = c^+_{k\uparrow} c^+_{-k\downarrow}$, then

$$1 + a_k b^+_k = \exp(a_k b_k^+), \tag{A7.1.5}$$

and hence

$$\Psi_{\text{BCS}} = \frac{\prod_k [\sqrt{1-h_k} + \sqrt{h_k} b_k^+] \psi_0}{\alpha \exp\left[\sum_k \frac{\sqrt{h_k}}{1-h_k} b_k^+\right] \psi_0}, \tag{A7.1.6}$$

where b_k is the variational parameter in the standard BCS treatment.

If one projects ψ_{BCS} on the state with just $N/2$ pairs, one then obtains, following Anderson (1987),

$$P_{N/2}\Psi_{\text{BCS}} = \left[\sum_k \left[\frac{\sqrt{h_k}}{\sqrt{1-h_k}}\right] b_k^+\right]^{N/2} \psi_0, \tag{A7.1.7}$$

or, in other words, just a projected BCS function.

The approximation ψ_{RVB} is also related to the appropriate projection of a BCS function:

$$\psi_{\text{RVB}} = P_{N/2} P_d \left[\frac{1}{\sqrt{1+a_k^2}} + \frac{a_k}{\sqrt{1+a_k^2}} c_{k\uparrow}^+ c_{-k\downarrow}^+\right] \psi_0, \tag{A7.1.8}$$

or

$$\psi_{\text{RVB}} = P_{N/2} \prod_k \left[\frac{1}{\sqrt{1+a_k^2}} + \frac{a_k}{\sqrt{1+a_k^2}} c_{k\uparrow}^+ c_{-k\downarrow}^+\right] \psi_0. \tag{A7.1.9}$$

In the insulating state, every site i is filled once, so that $|a_k|$ might as well be constant, $a_k = \pm|$, and the wave function contains a "pseudo-Fermi surface," at which a_k changes sign. Anderson believes that the "pseudo-Fermi surface" is physical and that the spin excitations may resemble those of a real Fermi liquid.

On the other hand, he points out that, considered as a solution to the Hubbard model, there is a gap for any charged excitation. By assumption, the Hubbard U is so large that adding the $(N + 1)$th electron costs an extra $\tilde{U}$ in energy, relative to adding the Nth.

Anderson (1987) then considers the state obtained by doping the system to remove the "half-filled" Hubbard-model criterion and make it metallic, but the reader is referred to his original paper for the details.

A7.2. REDUCTION OF HUBBARD AND EMERY MODELS TO EFFECTIVE SPIN HAMILTONIANS

In this Appendix, by employing the well-known Dirac identity, the problem of the reduction of some strongly correlated electronic models to effective spin Hamiltonians will be considered, following closely the work of Cheranovskii (1992). Such strong correlation models, such as those utilized in Chapter 3 for ferromagnetism (see also Chapter 8) and in Chapter 7 for the description of quasi-1D systems (see also March and Tosi, 1995), have come back into the center of the stage in relation to high-T_c superconductors (see Appendices 7.4 and 7.5). Let us note again, as in Chapter 4, that a valuable model for such interacting electrons in a crystal lattice is described by the one-band Hubbard Hamiltonian (see also, Klein, March and Alexander, 1995):

$$H = \sum t_{ij}(a^+_{i\sigma}a_{j\sigma} + a^+_{j\sigma}a_{i\sigma}) + U\sum a^+_{i\sigma}a_{i\sigma}a^+_{i\sigma}a^+_{i-\sigma}a_{i-\sigma}, \qquad \text{(A7.2.1)}$$

where t_{ij} is a matrix element of the electron transfer between the states of ith and jth sites and U is the repulsive potential of the δ-function type.

It is well known that the Hubbard model with a strong repulsion can be reduced to a Heisenberg spin Hamiltonian in the case of the half-filled band. However, for the non-half-filled band, an analogous reduction has been found in the case of the one-dimensional lattice only.

In this Appendix it will be shown that a number of models with strong electron correlation, such as the above one-band Hubbard Hamiltonian and the Emery (1989) model with infinite repulsion of holes at the copper band, can be rewritten using spin-free fermions and operators of cyclic spin permutations. In some cases, such as the Emery model with the one additional hole in the oxygen band, this factorized representation allows one to transform the lattice Hamiltonian into a spin model with a structure transcending that of the Heisenberg Hamiltonian. The spectra of the corresponding spin Hamiltonian can be studied with the help of various methods taken from the theory of spin systems, and one can draw some conclusions about exact spectral properties and can carry out numerical calculations on lattice clusters of larger size than in the case of direct use of the Hubbard Hamiltonian.

Cyclic spin permutations have been used to study the Hubbard model with infinite repulsion (see Cheranovskii, 1992) and the one-dimensional Emery model with one additional hole in the oxygen band. In particular, using cyclic permutations allows one to calculate the exact spectra of perovskite lattice clusters with 14 atoms of Cu (see Cheranovskii, 1992, and references therein). Here attention is focused on methodological aspects and the algebraic structure of the spin Hamiltonians.

A7.2.1. Hubbard Hamiltonian with Strong Electron Repulsion Energy

Let us consider a rectangular lattice described by the Hubbard Hamiltonian with infinite repulsion energy and any filling of the band. Finding the spectra of this lattice is a complicated kinematic task. The exact solution of it is known in the one-dimensional case only (Lieb and Wu, 1980). A wave function of the lattice can be expressed as the following superposition:

$$\psi = \sum \Phi(n_1, n_2, \ldots, n_s)\Theta(\sigma_1, \sigma_2, \ldots \sigma_s), \tag{A7.2.2}$$

where Φ is a coordinate function that describes the electron distribution over the lattice, n_i labels the ith singly occupied lattice point and Θ is a function of spin variables σ_i, which describe the spin configuration of the lattice.

Let us enumerate all variables of Eq. (A7.2.2) in succession along rows of the lattice beginning from the upper one (Fig. A7.2.1). When the Hamiltonian (A7.2.1) acts upon the function Ψ, the electrons hop to neighboring unfilled lattice points. For example, the first electron can hop to the second site or to the fifth site. With the above enumeration over the lattice rows, two new electron configurations appear as the result of these processes. It is easily shown that all electron numbers are fixed in the case of electron transfer between the states at the first and second sites. Therefore, this process leads to a change of the variable n_i in the coordinate function only. If the electron hops to the fifth site, the first electron becomes the third one, the second electron becomes the first one, and the third electron becomes the second one. Thus, to retain the chosen numeration, one needs to perform the cyclic permutation of numbers of electrons situated between the first and fifth lattice sites. This procedure leads to cyclic permutation of the three spin variables in the function Θ in Eq. (A7.2.2) and to an obvious change of the coordinate function. The desired spin permutation can be written in the standard form

$$Q_{13} = \begin{pmatrix} 1 & 2 & 3 \\ 3 & 1 & 2 \end{pmatrix}, \qquad Q_{13}|\sigma_1\sigma_2\sigma_3\rangle = |\sigma_2\sigma_3\sigma_1\rangle, \tag{A7.2.3}$$

where the upper row determines the initial spin configuration and the lower row corresponds to the final one.

It should be noted that the invariant character of Θ for the case of the electron transfer along the lattice row corresponds to the well-known spin degeneracy of the spectra of the one-dimensional Hubbard model with free ends and infinite repulsion.

Making use of this treatment for all electrons and sites, one obtains the Hubbard Hamiltonian with infinite repulsion in the following form:

$$H = \sum_{i<j} t_{ij}(c_i^+ c_j Q_{kl} + c_j^+ c_i Q_{kl}), \tag{A7.2.4}$$

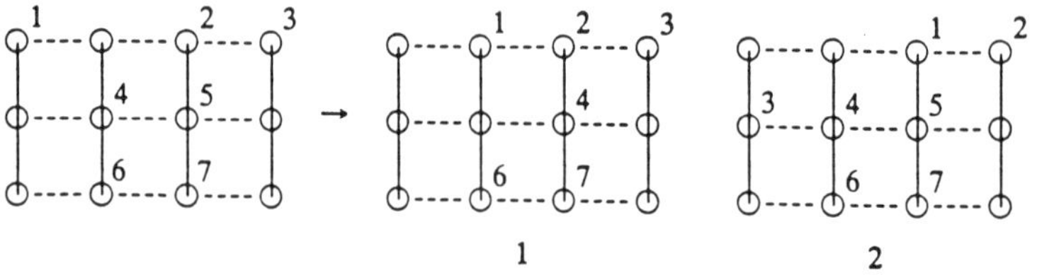

FIGURE A7.2.1. Electron configurations appearing as the result of electron hops to the neighbors of the first site.

where c_i^+ is a spin-free fermion operator and Q_{kl} is the cyclic spin permutation of spin variables of electrons situated between neighbor sites with ith and jth numbers.

The proposed representation allows one to construct a simple algorithm for numerical calculation of the spectra of small lattice clusters. Most calculations with the help of Eq. (A7.2.4) are performed in spin space, with evidently a smaller size than the full spin-coordinate space. It should be noted that the formula in Eq. (A7.2.4) is valid for other types of lattices such as the triangular form.

Assume now that the repulsion energy U is large but finite. It allows the use of perturbation theory (PT) in $\delta = U^{-1}$ for the construction of effective lattice Hamiltonians H. In the case of $\delta = 0$, such a Hamiltonian is as defined above. Below, the method of the construction of H through the first PT order in U^{-1} will be described. For higher orders in δ, H is constructed in a similar way. However, concrete calculations get rather cumbersome.

Let us first consider the one-dimensional lattice. In this case the effective Hamiltonian can be found without using the formalism of cyclic spin permutations. As noted, for $\delta = 0$ energy levels are degenerate with respect to spin. In the first order in δ, two processes of the electron transfer can be realized. The first is a simple exchange of electrons of neighbor sites. The second process is the electron hop to the nearest neighbor site with electron exchange. Consequently, the finite value of U leads to the mixing of different spin configurations and to resolving the spin degeneracy. In these terms it is described by the effective Hamiltonian

$$\left.\begin{aligned} H &= H_1 + H_2 \\ H_1 &= t \sum_{i=1}^{n} (c_i^+ c_{i+1} + c_{i+1}^+ c_i) \\ H_2 &= \delta t^2 \sum_{i=1}^{n} (Q_{kk+1} - 1)(2c_i^+ c_i - c_i^+ c_{i+2} - c_{i+2}^+ c_i) c_{i+1}^+ c_{i+1} \end{aligned}\right\}, \quad \text{(A7.2.5)}$$

where n is the full number of sites and k is the number of electrons labeled on the ith site.

Averaging over the ground state of H_1, one obtains an effective Hamiltonian that describes low-lying states of the exact spectra of the one-dimensional lattice containing s electrons and n sites:

$$H_{\text{eff}} = -n|t| \sin(\pi\rho)\pi^{-1} + \delta t^2[1 - (2\pi\rho)^{-1} \sin(2\pi\rho)]\rho \sum_{i=1}^{n} (S_i * S_{i+1} - \tfrac{1}{4}), \quad \text{(A7.2.6)}$$

where $\rho = s/n$ is the electron density, S_i is the one-electron spin operator for an ith electron (all spin transpositions occurring in Eq. (A7.2.6) are expressed in terms of scalar products of spin operators S_i by the Dirac identity $Q_{ii+1} = 2S_i * S_{i+1} + \frac{1}{2}$).

It is to be noted that one can analogously obtain an effective spin Hamiltonian for the one-dimensional lattice with alternating values of hopping integrals of

$i-1$	i	$i+1$
$j-1$	j	$j+1$
	j'	

FIGURE A7.2.2. Square lattice with Hubbard $U \gg |t|$.

neighbor chemical bonds. The exact spectrum of that lattice can be obtained in the case of $\delta = 0$. After a similar treatment for the square lattice (Fig. A7.2.2), one finds the following Hamiltonian to first order in δ (Cheranovskii, 1992)

$$\begin{aligned} H = \delta t^2\{2\ &\sum[(P_{kk+1} - 1)n_i n_{i+1} + (P_{km} - 1)n_i n_j] - \sum[(P_{kk+1} - 1)n_{i+1} c_i^{\dagger} c_{i+2} \\ &+ c_{j+1}^{\dagger} c_i n_{i+1} Q_{k+1m}(P_{kk+1} - 1) + c_{j-1}^{\dagger} c_i n_{i-1} Q_{km-1}(P_{k-1k} - 1) \\ &+ (c_{j+1}^{\dagger} Q_{km-1} + c_{j-1}^{\dagger} Q_{km-1} + c_j^{\dagger} Q_{mm'} Q_{km-1}) c_i n_j P_{km} - 1) + \text{h.c.}]\}, \end{aligned} \qquad (A7.2.7)$$

where $P_{km} = \binom{k\ m}{m\ k}$; $n_i = c_i^{\dagger} c_i$; subscripts j and j' are associated with nearest neighbors and next-nearest neighbors of ith site, respectively; $k(m)$ is the number of electrons labeled on ith (jth) site. [To obtain Eq. (A7.2.7) one uses the identity $P_{kk+1} Q_{k+1m}^{-1} = Q_{k+1m}^{-1} P_{km}$.]

When studying strongly anisotropic lattices, the cyclic spin permutation technique has a rather clear advantage. Let us consider a rectangular lattice that consists of weakly interacting segments (Fig. A7.2.3). First treat the interaction of two n-site segments each containing s electrons. Eigenfunctions of noninteracting segments have the form

$$\begin{aligned} \Phi_r(s) &= \sum A_r(n_1, n_2, \ldots, n_s) a_{n_1}^{+}\sigma_1 a_{n_2}^{+}\sigma_2 \ldots a_{n_s}^{+}\sigma_s|0\rangle, \\ A_r(n_1, n_2, \ldots, n_s) &= \det[f(r_1, n_1) f(r_2, n_2) \ldots f(r_s, n_s)], \\ f(r_i, n_j)] &= \left(\frac{2}{(n+1)}\right)^{1/2} \sin(\pi r_i n_j / n + 1), \\ &1 \le n_1 < n_2 < \cdots < n_s \le n; \qquad 1 \le r_1 < r_2 < \cdots < r_s \le n. \end{aligned} \qquad (A7.2.8)$$

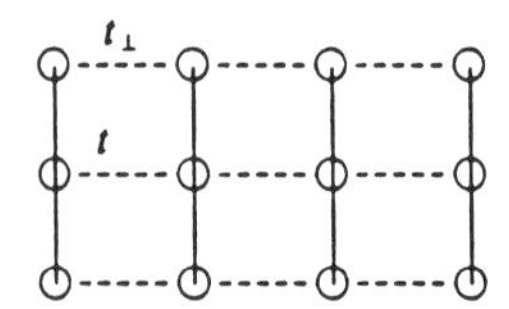

FIGURE A7.2.3. Anisotropic lattice consisting of weakly interacting three-site segment.

These states are spin degenerate and have energies

$$E_r^i(s) = 2t_i \sum \cos\left(\frac{\pi r_j}{n+1}\right), \qquad i = 1, 2. \tag{A7.2.9}$$

If $t_i < 0$, the set of numbers $r = j$ and energy

$$E_0 = \frac{2t_i \sin(\beta s) \cos[\beta(s+1)]}{\sin(\beta)}, \qquad \beta = \pi(2n+2)^{-1},$$

are associated with the ground state of the segment.

Assume that t_1 is associated with the first segment and t_2 is associated with the second. The interaction between segments leads to electron hops, If electrons are numbered in succession over segments, these hops lead to cyclic spin permutations just as in the previous considerations. As a result of these permutations the mixing of spin configurations and consequently splitting of the degenerate energy levels takes place. This splitting can be considered by means of the PT with a value of hopping integral $t_\perp$ describing electron transfer between segments. In the second order of PT using the cyclic spin permutations technique, one may carry out summations over the lattice variables to obtain the lattice Hamiltonian in a "pure" spin form (in the first order in PT the interaction between segments is absent):

$$H = \sum_{kp}^{s} \sum_{lq}^{s+1} J_1(klpq) Q^+_{\mathrm{kl}+s-1} Q_{pq+s-1} + J_2(klpq) Q_{lk+s} Q^+_{qp+s}\,. \tag{A7.2.10}$$

Here $J_1(klpq)$ are effective exchange integrals determined by the following expression (Cheranovskii, 1992):

$$J_1(klpq) = \frac{\sum_{rr'} B_{rr'}^{ss+1}(kl) B_{rr'}^{ss+1}(pq)}{E_0^1(s) + E_0^2(s) - E_r^1(s-1) - E_{r'}^2(s+1)},$$

$$B_{rr'}^{ss'}(kl) = t_\perp \sum_{i=1}^{n} G_{0r}(i, k, s) G_{r'0}(i, l, s')(-1)^{k+l},$$

$$G_{rr'}(i, k, s) = \sum_{1 \le n_1 < n_2 < \ldots < n_s \le n} A_r(n_1, n_2, \ldots, n_{k-1}, n_{k+1}, \ldots, n_s)$$
$$\times A_{r'}(n_1, n_2, \ldots, n_{k-1}, n_{k+1}, \ldots, n_s). \tag{A7.2.11}$$

Integrals $J_2(klpq)$ are obtained from $J_1(klpq)$ by the transposition of superscripts for energies $E(s)$.

As an example, let us consider the case of $n = 2$ and $s = 1$. In accordance with Eq. (A7.2.10), the Hamiltonian describing the interaction of these segments has the form

$$H = \frac{t_\perp^2}{|t_1 + t_2|}\left[\begin{pmatrix} 1 & 2 \\ 2 & 1 \end{pmatrix} - 1\right]. \tag{A7.2.12}$$

Rewriting the spin permutations with the help of the Dirac identity, one finds the following spin Hamiltonian:

$$H = J(S_1 * S_2 - \tfrac{1}{4}), \qquad J = \frac{2t_\perp^2}{|t_1 + t_2|}. \tag{A7.2.13}$$

When n is increased, only the form of the exchange integral J is changed. For example, in the case of $n = 3$,

$$J = \frac{3t_\perp^2|t_1 + t_2|[3(t_1t_2)^{-1} + (t_1 + 2t_2)^{-1}(t_1 + 2t_2)^{-1}]}{8\sqrt{2}}. \tag{A7.2.14}$$

Increasing s leads to more complicated spin Hamiltonians. For example, in the case of $s = 2$, $n = 3$, and $t_1 = t_2 = t$, this Hamiltonian takes the form

$$H_{12} = 2J\left[\begin{pmatrix} 1 & 3 \\ 3 & 1 \end{pmatrix} + \begin{pmatrix} 2 & 4 \\ 4 & 2 \end{pmatrix}\right] + \frac{J}{9}\left[\begin{pmatrix} 1 & 2 \\ 2 & 1 \end{pmatrix} + \begin{pmatrix} 3 & 4 \\ 4 & 3 \end{pmatrix}\right] - \frac{44J}{9J} + H'$$

$$H' = J\begin{pmatrix} 1 & 2 & 3 & 4 \\ 2 & 4 & 1 & 3 \end{pmatrix} - \frac{J}{6}\left[\begin{pmatrix} 1 & 2 & 4 \\ 2 & 4 & 1 \end{pmatrix} + \begin{pmatrix} 1 & 2 & 3 \\ 3 & 1 & 2 \end{pmatrix}\right.$$

$$\left. + \begin{pmatrix} 2 & 3 & 4 \\ 3 & 4 & 2 \end{pmatrix} + \begin{pmatrix} 1 & 3 & 4 \\ 4 & 1 & 3 \end{pmatrix}\right] + \text{h.c.}$$

$$J = 3t_\perp^2(16\sqrt{2}\,|t|)^{-1}. \tag{A7.2.15}$$

If the lattice consists of L n-site segments, the corresponding spin Hamiltonian has the additive form

$$H_{\text{eff}} = \sum_m H_{mm+1}, \tag{A7.2.16}$$

where H_{mm+1} is the Hamiltonian that is similar to Eq. (A7.2.10) and describes the interaction of neighboring segments. For example, the lattice consisting of n-site segments with one electron per segment and alternating integrals t_1 and t_2 (Fig. A7.2.4) is described by the Hamiltonian of the uniform Heisenberg spin chain. Thus this model can be exactly solved.

For $s > 1$, the analogous reduction to exactly solvable models cannot be performed. Nevertheless, using properties of cyclic permutations one can draw some general conclusions about exact spectra of the resulting Hamiltonians. Thus

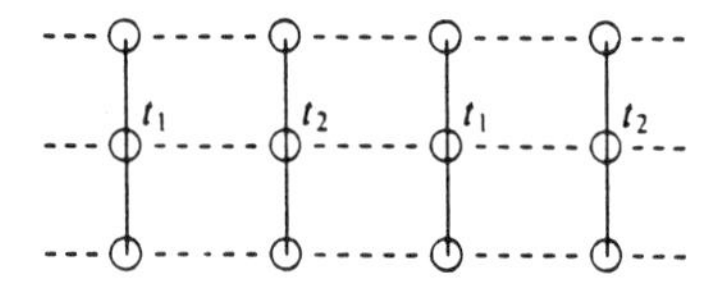

FIGURE A7.2.4. Anisotropic lattice with alternating hopping integrals of segments.

one can show that Σ_{klpq} J($klpq$) = 0. Therefore, as a consequence of the fully symmetric form of spin functions with a maximal spin, the ferromagnetic state energy of the lattice must vanish. It can also be shown that in the limit of an infinite number of segments the lattice with an odd number of electrons per segment and nondegenerate ground state has no energy gap in the excitation spectrum. A proof of this conclusion is similar to the consideration of Heisenberg spin lattices in Klein and Seitz (1980).

All cyclic permutations as in Eq. (A7.2.4) can be rewritten in the form of a product of spin transportations. So, using the Dirac identity, it can be easily shown that a cyclic permutation is expressed in the following form:

$$\begin{pmatrix} 1 & 2 & 3 & \ldots & n \\ n & 1 & 2 & \ldots & n-1 \end{pmatrix} = \prod_{i=1}^{n-1} (2S_i * S_n + \tfrac{1}{2}). \qquad \text{(A7.2.17)}$$

This representation allows the study of the spectrum of the lattice Hamiltonian using different approximations from the theory of spin systems. There are the spin-wave approximation, the localized-site cluster technique, the renormalization group approach, and others: see Cheranovskii (1992). By making use of the Jordan–Wigner (JW) transformation one can transform from spin operators to spin-zero fermions.

That representation permits the use of different variants of Hartree–Fock (HF) approximations that are similar to quantum chemical methods. Cyclic spin permutations can be expressed in the form of product or spin transpositions in various ways. Therefore the JW transformation of the same lattice Hamiltonian leads to a set of isospectral models with different numbers of interacting Fermions. The HF approximation destroys this isospectrality, and the problem of choice of the most adequate model arises.

For example, let us consider the uniform Heisenberg spin chain

$$H = \sum_{i=1}^{n} (S_i * S_{i+1} - \tfrac{1}{4}). \qquad \text{(A7.2.18)}$$

In this case the JW transformation gives a Hamiltonian with four-Fermion interactions. At the same time this spin chain can be also described by the following spin Hamiltonian:

$$H = \sum_{i=1}^{n/2} (S_{2i-1} * S_{2i} - \tfrac{1}{4}) + \sum_{i=1}^{n/4} (S_{4i-2} * S_{4i} - \tfrac{1}{4}) + \sum_{i=1}^{n/4} (S_{4i-1} * S_{4i+1} - \tfrac{1}{4}). \qquad \text{(A7.2.19)}$$

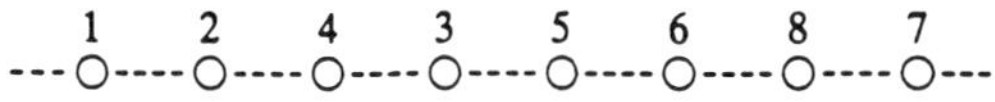

FIGURE A7.2.5. One-dimensional lattice with a special type of labelling of sites.

(This form of the Hamiltonian is associated with the numeration of one-dimensional lattice sites, which is shown in Fig. A7.2.5.) In this case the JW transformation leads to a Hamiltonian with six-Fermion interaction. Omitting details, one notes that the HF approximation leads to a more adequate estimate of the exact spectrum in the case of the Hamiltonian in Eq. (A7.2.18).

The representation in Eq. (A7.2.17) allows the construction of matrix elements of Hamiltonian of Eq. (A7.2.10) in a spin symmetry–adapted basis if a genealogical scheme is used. It simplifies calculations of the exact spectrum of small lattice clusters. The results of these calculations are discussed in Klein, March, and Alexander (1995). In all cases, when a number of electrons is significantly less than the number of sites, the lattice ground state has a minimal spin.

Next assume that hopping integrals describing electron transfer within segments are equal to each other, and that the number of electrons is not commensurate with the number of segments. Then fillings of some neighboring segments differ by one, and spin degeneracy is resolved in the first PT order in $t_\perp$. By making use of the cyclic spin permutations formalism, it can be shown that the Hamiltonian of two neighboring segments is determined as

$$H = \sum_{kl=1}^{s+1} J(kl)\,\{Q_{kl+s}(s_1 - s) + Q^{+}_{kl+s}(s_2 - s)\}R_{12}, \qquad s = \min(s_1,s_2). \tag{A7.2.20}$$

Here s_1 and s_2 are occupation numbers of the first and second segments, respectively, and R_{12} is the operator that transposes the wave functions of the first and second segments:

$$J(kl) = (-1)^{k+l+s} \sum_{i=1}^{n} G_{00}(i,k, s+1)G_{00}(i,l,s+1).$$

In particular, the Hamiltonian of interacting n-site segments with one and two electrons has the form

$$H = [A(s_1 - s) + A^{+}(s_2 - s)\}R_{12,}$$

$$A = J_1(n)(P_{12} + P_{23}) + J_2(n)(1 + P_{23}P_{12}), \tag{A7.2.21}$$

where

$$J_1(n) + J_2(n) = -\tfrac{1}{2}t_\perp; \qquad J_1(2) = -\tfrac{1}{2}t_\perp, \qquad \mathrm{J}_1(3) = -\tfrac{1}{16}t_\perp$$

$$\begin{aligned} J_1(n) = \{&4[\cot 3\alpha - 3\cot\alpha + 4[\cot 2\alpha - 2\cot\alpha)\cos\alpha(\cos\alpha + 1)] \\ &\times(\sin\alpha + \sin 2\alpha)^{-1} + \tfrac{8}{3}\sin^{-2} 2\alpha - 8(2n^2 + 4n + 3) \\ &- [3 + 16\cos\alpha(\cos\alpha + 1) + 48\cos{}^2\alpha(\cos\alpha + 1)^2] \\ &\times(\sin\alpha + \sin 2\alpha)^{-2}\}(4(n+1))^{-2}, \end{aligned}$$

$$\alpha = \pi/(n+1), \qquad n > 3.$$

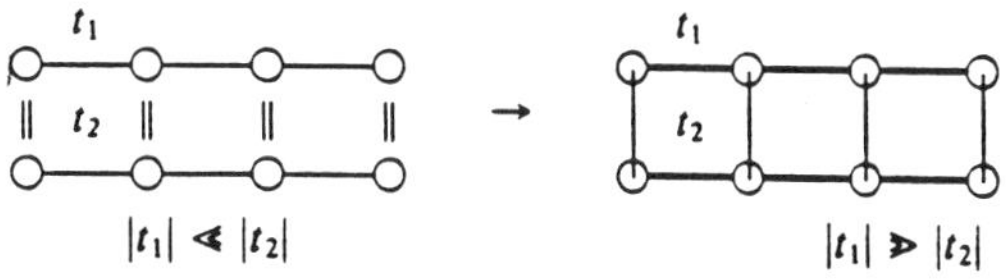

FIGURE A7.2.6. Structural transition with change of spin multiplicity.

It is readily shown that eigenvalues of the Hamiltonian of Eq. (A7.2.21) coincide with singular values of the matrix A. This connection allows one to obtain an exact lattice spectrum in analytical form and to show that the ground state has maximal spin for any segment length (Cheranovskii, 1992).

On the other hand, the lattice consisting of two n-site segments can be constructed from n two-site segments (Fig. A7.2.6). At the same time, if these two-site segments are weakly interacting ($|t_1| \ll |t_2|$), the lattice ground state has a minimal spin. Hence the multiplicity of the lattice ground state is a function of the ratio of hopping integrals t_1 and t_2. Using singular values simplifies calculations on more complicated lattices. For two weakly interacting n-site segments with $n < 7$ and odd numbers of electrons, exact calculations show that the ground state has a maximal spin. Therefore the structural transition with a change of spin multiplicity also takes place for these lattices.

Now consider a lattice that obeys the condition of Nagaoka's (1966) theorem (see also Klein, March, and Alexander, 1995):

$$N_e = N - 1,$$

where N_e and N are the numbers of electrons and sites, respectively. By means of Eq. (A7.2.20) the ferromagnetic character of the lattice ground state can be established (i.e., one can establish a generalization of Nagaoka's theorem). Thus all nonvanishing exchange integrals contained in Eq. (A7.2.20) have coinciding indexes in this example:

$$J(kl) = \frac{\delta_{kl}(-1)^s 2 \sin^2 [\pi k/(n + 1)]}{n + 1}.$$

Therefore, the Hamiltonian matrix can be chosen so that all matrix elements have nonnegative values in the space of spin configurations. At the same time Eq. (A7.2.20) contains $s + 1$ cyclic permutations Q_{kl}, which act on $s + 1$ spins. It can be shown that these permutations mix all spin configurations; or, put another way, it is always possible to rearrange spins in the lattice from one configuration to another. Therefore the Hamiltonian matrix cannot be decomposed into two or more disconnected block parts. Hence, according to the Perron–Frobenius theorem, its maximum eigenvalue corresponds to a positive eigenvector while the state of the regular spin multiplicity, having only positive components in the space of spin configurations, is a state with the maximum spin.

It can be also shown that $\Sigma_{kl}J(kl) = (-1)^s t_\perp$ and consequently ferromagnetic energy of the lattice containing two weakly interacting segments and having filling numbers that differ by one equals $-|t_\perp|$ independently of the numbers of electrons and sites.

Thus two kinds of interactions of neighboring segments take place if N_e and N are not commensurable values. This leads to two types of spin orderings. The competition of these types of interactions leads to a formation of regions with ferromagnetic ordering (spin polaron) and a cascade of concentration transitions with regular oscillations of spin multiplicity between minimal and maximal values (see Cheranovskii, 1992).

An anisotropic lattice consisting of n-site segments has the ground state with the maximal spin if ρ obeys the following condition:

$$A_+(1,n) < \rho < A_-(l + 1,n), \tag{A7.2.22}$$

where $A_\pm(1,n) = 1/n \pm (n\pi)^{-1}(3\pi\varepsilon_l/2)^{1/3}$; $l = 1, 2 \ldots < n$, and ε_l is an average energy of the interaction of two segments each containing l electrons.

Consider an n_1 by n_2 lattice with hopping integrals t_1 and t_2 much as indicated in Figure A7.2.6. It can be shown that in the cases of $|t_1| \ll |t_2|$ and $|t_2| \ll |t_1|$ this lattice has different multiplicities provided ρ satisfies the condition

$$\max[A_+(l_1,n_1), A_-(l_2,n_2)] < \rho < \min[A_-(l_1 + 1, n_1), A_+(l_2, n_2)]. \tag{A7.2.23}$$

Assume now that the lattice consists of two n-site segments with different hopping integrals t_1 and t_2. If $|t_1 - t_2| \ll |t_\perp|$), these segments interact in second order in $t_\perp$ only. By making use of the cyclic spin permutation technique, it can be shown that the corresponding Hamiltonian has the form

$$H_{12} = \sum_{klpq=1}^{s+1} J_1(klpq) Q^+_{kl+s} Q_{pq+s} + \sum_{kp=1}^{s} \sum_{lq=1}^{s+2} J_2(klpq) Q_{lk+s+1} Q^+_{qp+s+1}, \tag{A7.2.24}$$

where the first segment contains $s + 1$ electrons and the second contains s electrons:

$$J_1(klpq) = \sum_{rr'} \frac{B^{s+1s+1}_{rr'}(kl) B^{s+1s+1}_{rr'}(pq)}{E^1_0(s + 1) + E^2_0(s) - E^1_r(s) - E^2_{r'}(s + 1)},$$

$$J_2(klpq) = \sum_{rr'} \frac{B^{ss+2}_{rr'}(kl) B^{ss+2}_{rr'}(pq)}{E^1_0(s + 1) + E^2_0(s) - E^1_r(s + 2) - E^2_{r'}(s - 1)}.$$

For example, in the case of $n = 2$, $s_1 = 2$, and $s_2 = 1$, this Hamiltonian is expressed in the following form:

$$H = J\left\{ t_2 \left[\begin{pmatrix} 1 & 2 & 3 \\ 2 & 3 & 1 \end{pmatrix} + \begin{pmatrix} 1 & 2 & 3 \\ 3 & 1 & 2 \end{pmatrix} \right] + 2t_1 \right\}, \qquad J = \frac{t_\perp^2}{2|t_1^2 - t_2^2|}. \tag{A7.2.25}$$

It can readily be shown that the ground state of this lattice has a maximal multiplicity independently of the nonvanishing value of $|t_1 - t_2|$.

A7.2.2. Emery Model

It seems at the time of writing that some high-T_c superconducting copper oxides are approximately modelled by the two-band Hubbard Hamiltonian:

$$H = t \sum \{p^+_{i\sigma} d_{j\sigma} + d^+_{j\sigma} p_{i\sigma}\} + \alpha \sum p^+_{i\sigma} p_{i\sigma} + U_p \sum p^+_{i\sigma} p_{i\sigma} p^+_{i-\sigma} p_{i-\sigma}. \qquad \text{(A7.2.26)}$$

Here t is the hopping integral, which represents the hybridization of oxygen and copper bands, α is the orbital energy of the electron at an oxygen site, U_p is the repulsive potential that acts only when two holes are at the same oxygen site, $p^+_{i\sigma}$ and $d^+_{i\sigma}$ are creation operators for a hole of spin σ in the atomic state at the ith oxygen and copper sites, respectively. The similar repulsive potential for the copper sites has infinite value. (With this type of parametrization the Hamiltonian of Eq. (A7.2.26) is referred to as the Emery model.)

This model has been intensively discussed in many places. It is usually considered that $|t| \ll \alpha$. Nevertheless, the Hamiltonian obtained as the result of that consideration is rather complicated, and an exact solution of this problem is known only for the one-dimensional lattice with one reversed spin. It will be shown (Cheranovskii, 1992) that the Emery model with one oxygen hole can be reduced to a "pure" spin problem in which the impulse of the hole is a parameter of the effective spin Hamiltonian.

Let us first consider the one-dimensional lattice in which spins of holes are numbered in succession over the sites independently from type one (Fig. A7.2.7). Therefore the one-dimensional Hamiltonian with cyclic boundary conditions and one hole in an oxygen band can be written in the form

$$H = J_1 \sum a^+_l a_l P_{l,l\pm1} + J_2 \sum (a^+_l a_{l+1} + a^+_{l+1} a_l) + J_2(a^+_1 a_{n+1} Q^+ + a^+_{n+1} a_1 Q)$$

$$J_1 = t^2/(\alpha + U_p), \qquad J_2 = t^2/\alpha, \qquad \text{(A7.2.27)}$$

where a^+_l is the creation operator for a hole at the lth oxygen atom, while Q is the cyclic spin permutation of all $n + 1$ spin variables of $2n$-site lattice.

Basis functions of the space in which H is determined can be chosen in the following form:

$$\Psi_l(\lambda, S, \mathcal{M}) = \varphi(1)\Lambda(\lambda, S, \mathcal{M}) \qquad \text{(A7.2.28)}$$

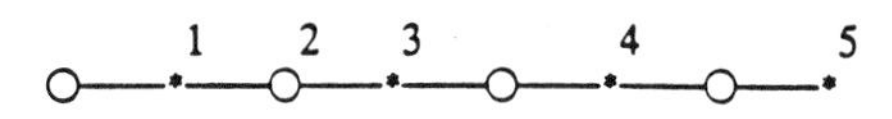

FIGURE A7.2.7. One-dimensional perovskite lattice (O is associated with oxygen sites and * with copper sites).

where φ is a coordinate part, which labels an oxygen hole with a number 1, and $\Lambda(\lambda, \mathcal{S}, \mathcal{M})$ is the spin-symmetry-adapted function of $n + 1$ spin variables

$$S^2\Lambda(\lambda, \mathcal{S}, \mathcal{M}) = \mathcal{S}(\mathcal{S} + 1)\Lambda(\lambda, \mathcal{S}, \mathcal{M}), \qquad S^z(\Lambda(\lambda, \mathcal{S}, \mathcal{M}) = \mathcal{M}\Lambda(\lambda, \mathcal{S}, \mathcal{M}).$$

The Hamiltonian in Eq. (A7.2.27) commutes with the operator that displaces all spins by one unit cell cyclically. Therefore, the eigenfunctions in Eq. (A7.2.28) must be characterized by hole impulse $k = 2\pi m/n$ ($m = 1,2, \ldots, n$). The symmetry-adapted basis functions corresponding to fixed values of impulse k can be constructed by the usual group theory techniques:

$$\Psi_{km}(\lambda, \mathcal{S}, \mathcal{M}) = (n)^{-1/2}\{\sum_{j=1}^{n} \exp(ikj)Q^j a^+_{m-j} + \sum \exp(ikj')Q^{j'+1}a^+_{n-j'+m}\}\,\Lambda(\lambda, \mathcal{S}, \mathcal{M})|0\rangle. \tag{A7.2.29}$$

Omitting cumbersome manipulations with the cyclic permutations, one notes that for the orthogonality of basis functions and reducibility of matrix H to block-diagonal form, it is necessary that the $\Lambda(\lambda, \mathcal{S}, \mathcal{M})$ should be eigenfunctions of Q:

$$Q\Lambda(\lambda, \mathcal{S}, \mathcal{M}) = \exp(ip)\Lambda(\lambda, \mathcal{S}, \mathcal{M}), \qquad p = \frac{2\pi m}{n+1}.$$

Thus

$$\Psi_{kl+a}(\lambda, \mathcal{S}, \mathcal{M}) = \exp[i(k + p)a]\Psi_{kl}(\lambda, \mathcal{S}, \mathcal{M}),$$

and the Hamiltonian H can be expressed in the pure spin form

$$H = J_1(P_{12} + P_{ln+1}) + 2J_2 \cos(k + p)\sigma_{kk'}\delta_{pp'}\delta_{\lambda\lambda'}. \tag{A7.2.30}$$

[To obtain the expression in Eq. (A7.2.30), use has been made of the identity $QP_{k,l}Q^{-1} = P_{k-1,l-1}$.]

For numerical calculations it is convenient to use a set of spin-symmetry-adapted functions that are constructed via a genealogical scheme and are not eigenfunctions of Q. In this representation the Hamiltonian, H may be rewritten in the following form (Cheranovskii, 1992):

$$H = J_1(P_{12} + P_{1n+1}) + J_2[\exp(ik)Q + \exp(-ik)Q^+]. \tag{A7.2.31}$$

If first copper spins are enumerated in a one-dimensional lattice, the corresponding Hamiltonian has the form

$$H = J_1 \sum_{m=1}^{n} a^+_m a_m(P_{m-1,n+1} + P_{m,n+1}) + J_2 \sum_{m=1}^{n-1} (a^+_m a_{m+1} + a^+_{m+1}a_m)P_{m+1,n+1}$$
$$+ J_2(a^+_1 a_n + a^+_n a_1)P_{1,n+1}. \tag{A7.2.32}$$

Carrying out similar manipulations, one obtains the following Hamiltonian:

$$H' = J_1(P_{nn+1} + P_{1n+1}) + J_2[\exp(ik)Q + \exp(-ik)Q^+]. \tag{A7.2.33}$$

The Hamiltonians in Eqs. (A7.2.31) and (A7.2.33) are isospectral as (Cheranovskii, 1992)

$$H' = QHQ^{-1}.$$

It should be noted that the Emery model is usually studied in the spin configuration space. In the case of a large number of the reversed spins, this does not allow one to obtain results in a convenient, compact form. Making use of the above technique permits investigation of the states with any multiplicity in operator form and essentially reduces the dimensionality of the Hamiltonian matrix for the lattice; for instance, in the case of 11 copper atoms, approximately by two orders of magnitude (Cheranovskii, 1992). For the two-dimensional perovskite lattice a simpler consideration is possible if first copper spins are enumerated similarly to the second type of enumeration for the one-dimensional lattice.

Let the (m by n) rectangular perovskite lattice have cyclic boundary conditions (Fig. A7.2.8). Using the cyclic spin permutation technique leads to the following Hamiltonian:

$$\begin{aligned} H = \sum_{i=1}^{n} \sum_{j=1}^{n} \{ & P_{(i-1)n+j,N+1}(a^{+}_{ij}a_{ij} + b^{+}_{ij}b_{ij}) + P_{(i-1)n+j-1,N+1}a^{+}_{ij}a_{ij} + P_{in+j,N+1}b^{+}_{ij}b_{ij} \\ & + [P_{(i-1)n+j,N+1}a^{+}_{ij+1}a_{ij} + b^{+}_{i+1j}b_{ij}P_{in+j,N+1} + (b^{+}_{ij} + b^{+}_{i-1j})a_{ij}P_{(i-1)n+j,N+1} \\ & + (b^{+}_{ij-1} + b^{+}_{i-1j-1})a_{ij}P_{(i-1)n+j-1,N+1} + \text{h.c.}], \end{aligned} \tag{A7.2.34}$$

where a^{+}_{ij} and b^{+}_{ij} are creation operators for a hole at an oxygen atom with type a and b, respectively, while N is the full number of copper sites. (For simplicity, one has taken $J_1 = J_2 = 1$.) This Hamiltonian commutes with operators that displace the enumeration of sites by one unit cell along and transverse to the lattice. These operators lead to the cyclic permutations of copper spins and have the following form:

$$R_1 = \prod_{i=1}^{m} Q_{[(i-1)n+1]in},$$

$$R_2 = \prod_{i=1}^{n} \begin{pmatrix} i & i+n & i+2n & \dots & i+(m-1)n \\ i+(m-1)n & i & i+n & \dots & i+(m-2)n \end{pmatrix}.$$

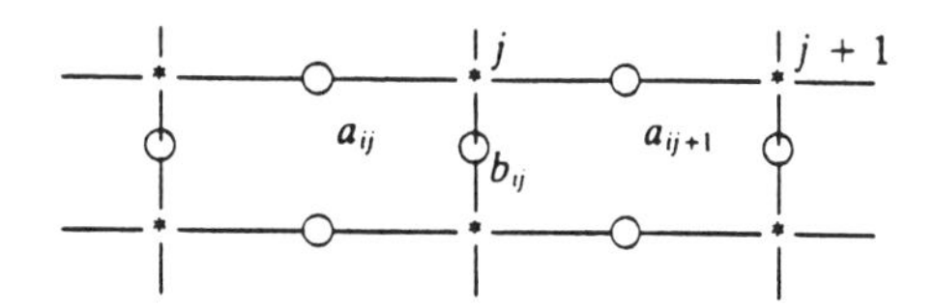

FIGURE A7.2.8. Two-dimensional perovskite lattice (O is associated with oxygen sites and * with copper sites).

By using a treatment similar to the abovementioned one-dimensional case, the lattice Hamiltonian takes the pure spin form:

$$H = \begin{pmatrix} H_1 & H_2^+ \\ H_2 & H_3 \end{pmatrix}.$$

$$H_1 = P_{l,N+1} + P_{N,N+1} + [R_1 \exp(ik_1) + \text{h.c.}]$$

$$H_2 = [P_{1,N+1} + P_{N,N+1} R_1 \exp(ik_1)][1 + R_2 \exp(ik_2)]$$

$$H_3 = P_{1,N+1} + P_{n+1,N+1} + [R_2 \exp(ik_2) P_{n+1,N+1} + \text{h.c.}].$$

In conclusion, using the Dirac identity, Hubbard and Emery models can be converted to effective spin Hamiltonians (see also Klein, March, and Alexander, 1995).

A7.3. LUTTINGER LIQUID: SPINONS AND HOLONS

Let us first clarify the distinction between a Luttinger liquid and a Landau Fermi liquid. In the latter, as in the homogeneous phase of jellium discussed at length in three dimensions in Chapter 3, the momentum distribution n(p) has a discontinuity at the Fermi momentum p_f. However, n(p) is qualitatively changed from the non-interacting Fermi gas (see later Appendices 8.1 and 8.2 and the quantal Monte Carlo results for 3D jellium in the range $1 \leq r_s \leq 10$: Ortiz and Ballone, 1994). In the interacting electron fluid n(p) has a high-momentum tail, asymptotically decaying as p^{-8} (Kimball, 1975), due to the creation of some holes inside the Fermi surface and correspondingly promotion of electrons outside. The magnitude q of the discontinuity at the Fermi surface depends, of course, on r_s. Such a system is a Landau Fermi liquid.

Now let us turn to the focus of this Appendix: the so-called Luttinger liquid (see also March and Tosi, 1995). This has attracted considerable interest since Anderson (1990a and b, 1991a and b) realized that there were analogies between the phenomenology of the high T_c cuprates, and of 1D metals. The latter are known to be Fermi liquids of a very special type (Luttinger liquids). Thus, the Fermi surface (defined by comparison with the 3D momentum distribution as a set of points where n(p) exhibits nonanalyticity) still exists and has the same 'volume' as for non-interacting electrons, in accord with Luttinger's theorem. But the elementary excitations are now totally different from those of a conventional metal. In particular, there are no quasiparticle excitations and the spectrum consists solely of collective modes.

To be quite specific, the two salient characteristic properties of Luttinger liquids are (i) the destruction of the Fermi discontinuity q in 3D and (ii) spin–charge separation. Let us turn almost immediately to point (ii) below. However it should be stressed that (i) and (ii) above are not necessarily the same issue (Houghton and Marston, 1993). Secondly, in spite of (i), it is useful to continue to use the language of Fermi surfaces (3D) and of Fermi points (1D). These points or surfaces are now to be defined more generally as manifolds of points in momentum space at which the zero-temperature occupancy $n(p)$ shows nonanalytic behavior characterized by an exponent α instead of a discontinuity. Specifically (compare also Holas and March, 1991), near the Fermi wave number $k_f = p_f/\hbar$ the occupancy varies as

$$n(k) \approx n(k_f) + C|k - k_f|^{2\alpha} \operatorname{sgn}(k - k_f) \tag{A7.3.1}$$

where C is a constant which is not universal but sets the momentum scale.

A7.3.1. Spin–Charge Separation

Let us turn then to point (ii) above, the second characteristic of a Luttinger liquid. It is now well established that spin–charge separation occurs automatically in one spatial dimension, at least in the weak coupling limit and at large length scales. Houghton and Marston, in particular, present an argument which will be briefly summarized below. They consider excitations near the two Fermi points k_f and $-k_f$. They write the action S in the noninteracting limit as

$$S_0 = \int dx\, dt \{\psi_L^{\dagger\alpha}\partial_-\psi_{L\alpha} + \psi_R^{\dagger\alpha}\partial_+\psi_{R\alpha}\}. \tag{A7.3.2}$$

In Eq. (A7.3.2), L and R refer to the left and right Fermi points; $\partial_\pm = \partial_t \pm iv_f\partial_x$ with $v_f = \hbar\, k_f/m$ which will, for convenience, be set equal to unity below. They are then led to the most general marginal interaction of the form

$$S_{\text{int}} = \int dx\, dt \left[\frac{\pi}{2}\delta v_c (J_L^2 + J_R^2) + \frac{\pi}{6}\delta v_s (J_{L\alpha}^{\beta} J_{L\beta}^{\alpha} + J_{R\alpha}^{\beta} J_{R\beta}^{\alpha}) + \lambda_c J_L J_R + \lambda_s J_{L\alpha}^{\beta} J_{R\beta}^{\alpha} \right]. \tag{A7.3.3}$$

Here J_L is the charge current at the left point, defined such that $\langle J_L(x,t)\rangle = 0$ by subtracting the constant background charge density from the current. The spin current appearing in Eq. (A7.3.3) has been conveniently expressed in matrix form,

$$J_{L\beta}^{\alpha}(x) = \Psi_L^{\dagger\alpha}(x)\Psi_{L\beta}(x) - \frac{1}{2}\delta_\beta^\alpha \Psi_L^{\dagger\gamma}(x)\Psi_{L\gamma}(x). \tag{A7.3.4}$$

The spin current has no charge current component because it is traceless, and it also has zero vacuum expectation.

The important outcome is that S_{int} involves only products of either pure spin or pure charge currents. The Gaussian part of the action S_0 can also be expressed solely in terms of separate products of charge and spin currents (see also Houghton and Marston, 1993). It has to be noted that omitted from the action are (a) terms that oscillate rapidly with wave vectors of order k_f, (b) interactions involving derivatives that arise from Taylor expansions of nonlocal interactions, and (c) terms with more than four fermion fields. Such terms can break charge-spin separation; however each is irrelevant in the renormalization-group sense and the coupling constants flow rapidly to zero in the low-energy limit.

Bosonization of the fermion fields then leads to the renormalized boson Hamiltonian H as the sum of two parts:

$$H = H_c + H_s. \tag{A7.3.5}$$

These pieces separately describe charge and spin excitations propagating at different velocities v_c and v_s, respectively, (compare Eq. (A7.3.3)):

$$H_c = \frac{\pi}{2} v_c \int dx \, \{J_{\rm L}^2(x) + J_{\rm R}^2(x)\} \tag{A7.3.6a}$$

and

$$H_s = \frac{\pi}{2} v_s \int dx \, \{J_{\rm Lz}^2(x) + J_{\rm Rz}^2(x)\}, \tag{A7.3.6b}$$

where $v_c = 1 + \delta v_c$ and $v_s = 1 + \delta v_s$. In writing Eqs. (A7.3.5) and (A7.3.6) use has been made of the Kac–Moody algebra:

$$[J_{\rm L\alpha}(x), J_{\rm L\beta}(y)] = -\frac{i}{2\pi} \delta_{\alpha\beta} \, \delta'(x-y) \tag{A7.3.7a}$$

and

$$[J_{\rm R\alpha}(x), J_{\rm R\beta}(y)] = \frac{i}{2\pi} \delta_{\alpha\beta} \, \delta'(x-y). \tag{A7.3.7b}$$

The charge current defined above can now be expressed in terms of these currents as

$$J_{\rm L}(x) = J_{\rm L\uparrow}(x) + J_{\rm L\downarrow}(x) \tag{A7.3.8}$$

and the z component of the spin current is simply

$$J_{\rm Lz}(x) = J_{\rm L\uparrow}(x) - J_{\rm L\downarrow}(x). \tag{A7.3.9}$$

Equations (A7.3.7)–(A7.3.9) suffice to complete the definitions of H_c and H_s in Eq. (A7.3.6).

It is to be stressed that Fermi liquid behavior is only recovered in the special case $v_c = v_s$ and exponent $\alpha = 0$ in Eq. (A7.3.1). However, the discontinuity at the Fermi surface remains even when $v_c \neq v_s$ if $\alpha = 0$.

A7.3.2. Can a 2D Luttinger Liquid Exist?

Let us turn next to consider the question posed above. Anderson (1991c) made a start on the problem by studying two coupled parallel Hubbard chains (see also Wen, 1990; Shultz, 1991; Nersesyan, Luther, and Kusmartsev, 1993). Anderson's analysis of this two-chain system led him to infer that weak interchain coupling does not change the 1D character significantly. But this conclusion has been challenged by Shultz and the later reference cited above, and it has been suggested that interchain coupling is relevant and that it destabilizes the Luttinger liquid.

It is of interest at this point to refer to the earlier study by Mattis (1987). In work of Jorgensen, Schütter, Hinks, Capone, Zhang, and Brodsky (1987) and of Matthiess (1987), a simple pseudo-2D band structure was proposed for the family of materials $La_{2-x}Ba_xCuO_4$, of which the composition $x = 0.15$ exhibits high-T_c behavior with $T_c \approx 35$ K, for the undoped (barium free) case $x = 0$. This band structure allows nesting in the [110] and $[\bar{1}10]$ directions and suggests that a distortion with a wave vector $\mathbf{Q} = \pi(1,1,0)$ or $\pi(-1,1,0)$ will occur in this material. This is, in fact, observed, leading to an energy gap and hence to semiconducting behavior.

Mattis studied interacting electrons in tight-binding (TB) band structures which readily nest, such as in the materials referred to above. One has then the well-known band dispersion relation

$$\varepsilon_{\mathrm{TB}}(\mathbf{k}) = -t[\cos(k_x) + \cos(k_y)]. \tag{A7.3.10}$$

For a half-filled band, the Fermi level lies along straight-line segments defined by $|k_x| + |k_y| = \pi$. The states for $\varepsilon < 0$ are all occupied while those with $\varepsilon > 0$ are unoccupied in the ground state. Mattis noticed then that the distortion instability with respect to the wave vectors $\mathbf{Q}$ listed above is not the only instability caused by this band structure and that there exist important infrared ($\mathbf{k}\to 0$) instabilities.

To demonstrate the above claim, Mattis applied a perturbation $V(\mathbf{k})$ with $\mathbf{k} = (k_x,k_y)$ and calculated the response to second order:

$$\delta E = -\tfrac{1}{2}\chi(\mathbf{k})|V(\mathbf{k})|^2. \tag{A7.3.11}$$

The susceptibility $\chi(\mathbf{k})$ at low temperatures can be calculated by expanding the Fermi function f to obtain

$$\begin{aligned}\chi(\mathbf{k}) &= -\sum_{\mathbf{k}} \partial f(\beta\varepsilon_{\mathrm{TB}}(\mathbf{k}))/\partial\varepsilon_{\mathrm{TB}}(\mathbf{k}) \\ &= (A/|t|)\ln(|t|)(k_B T + Bt|k|)^{-1}.\end{aligned} \tag{A7.3.12}$$

with A and B being numerical constants. This long-wavelength divergence at absolute zero reflects the logarithmic singularity of the density of states at the centre of the 2D tight-binding band structure. Mattis stressed that, once χ is singular, one can envisage higher-order terms being even more singular and perturbation theory must become suspect.

Mattis therefore simplified the form of the dispersion relation $\varepsilon_{\mathrm{TB}}(\mathbf{k})$ so as to achieve an exactly soluble model, thereby gaining some insight into the effects of electron–electron and electron–phonon interactions and pairing. His model was motivated by the separability of $\varepsilon_{TB}(\mathbf{k})$, i.e., of the form $\varepsilon_{TB}(\mathbf{k}) = \varepsilon(k_x) + \varepsilon(k_y)$. By use of bosonization techniques he was able, but now in 2D, to parallel the 1D pioneering studies of Tomonaga (1950) and to reformulate the original TB model in the presence of interactions to yield an exactly soluble model. As he emphasizes, while his model is an interesting linearized model in its own right, paralleling the work of Luttinger (1963) in 1D, the two-dimensionality is still reflected in the mixing of the boson creation and annihilation operators at every wave vector **k**. The study of Hlubina (1994) reaches the results of Mattis by a somewhat different route, but again nesting is a central feature.

To elaborate a little, Hlubina (see also March and Tosi, 1995) considers spinless electrons, again in 2D, with dispersion relation

$$\varepsilon(\mathbf{k}) = v_f(|k_x| + |k_y|). \tag{A7.3.13}$$

In momentum space the interactions among electrons have a finite range, which is small compared to the Fermi momentum. He then makes a "Golden Rule" calculation of the electron lifetime, which indicates a breakdown of Landau Fermi liquid theory in the model. At the one-loop level of perturbation theory, Hlubina demonstrates that the density wave and the superconducting instabilities cancel and that there is no symmetry breaking. Paralleling the earlier work of Mattis (1987), the model is solved via bosonization. The excitation spectrum is found to consist of gapless bosonic modes, as in a 1D Luttinger liquid.

It is important here to note that the above 2D model exhibits an analogue of spin-charge separation. The role of spin is now played by the momentum transverse to the direction of the quasi-1D motion and the electron Green function can be expected to have features beyond those in 1D. Hlubina points out the interest for further work of the influence of (i) the curvature of the Fermi surface, (ii) electron spin and (iii) large momentum scattering processes, on the stability of Luttinger liquids.

It is relevant in this context to note that Schulz (1987) and Dzyaloshinskii (1987) applied a complementary method to treat the TB spectrum. They showed that the superconducting (BCS) and charge-density wave (CDW) susceptibilities are dominated by electrons in the corners of the Fermi surface, which led to the replacement of the Fermi surface by four points. This simplification was then amenable to solution by renormalization group methods. As in the above studies, Houghton and Marston (1993) model a TB spectrum close to half filling by four points which, however, represent the branches of the spectrum and van Hove singularities are neglected. In this respect, the models of Hlubina and of Houghton and Marston are similar. For spinless fermions with repulsion, Houghton and Marston find, in the absence of large momentum scattering processes, no CDW instability and again, in this respect, their results are in accord with those of Hlubina. The final comment here is that the model studied by Hlubina can be viewed, essentially, as a problem of crossed 1D chains: also considered by Guinea and Zimanyi (1993).

Work by Feng (1994) uses a fermion-spin transformation to implement the charge-spin separation. He develops this to study the low-dimensional t-J model (see also Klein, March, and Alexander, 1995). In this approach, the charge and spin degrees of freedom are separated and the charge degree of freedom is represented by a spinless fermion while the spin degree of freedom is described via a hard-core boson. His approach is applicable to 1D and 2D systems. In the 1D case, Feng finds that the spinon as well as the physical electron behave like Luttinger liquids. He obtains a gapless charge and spin excitation spectrum, a good value for the ground-state energy, and a reasonable electronic momentum distribution within the mean-field approximation. The correct exponents of the correlation functions and momentum distribution are also found.

In the 2D case, within the mean-field approximation, the magnetized flux state with a gap in the spinon spectrum has the lowest energy at the half-filled limit. The antiferromagnetic long-range order is destroyed by hole doping of the order of 10–15% for t/J between 3 and 5 and a disordered flux state becomes stable. The calculated specific heat is roughly consistent with observed results on copper oxide high-T_c materials. The possibility of phase separation is also discussed at the mean field level.

To conclude this Appendix, a brief summary will be presented of related work by Schmeltzer and by Luther.

A7.3.3. Spinons, Holons and Gapless Spin Excitations

Schmeltzer (1992, 1993) has provided a discussion of a free model of spinons and holons with constraints. His conclusion is that they form a Luttinger liquid. The contrast with a Fermi liquid is that this latter has a discontinuity, say q, in the momentum distribution at the Fermi surface (cf. Appendices 8.1 and 8.2). In a Luttinger liquid, on the other hand, there is still nonanalyticity in $n(p)$ at the Fermi momentum $p = p_f$, and the discontinuity q has gone to zero.

Schmeltzer considers the physics of bosonization (see also Section A7.3.1) in one- and two-dimensional models for strongly correlated Fermions. In particular, within a Hubbard type of framework, he focusses on the strongly correlated limit as U/t tends to infinity for both $D = 1$ and $D = 2$.

His conclusions are that when $D = 1$, the only possible solution is a Luttinger liquid. For two dimensions, the situation is richer in behavior. For $T = T_{\text{K.T.}}$, one again has a Luttinger liquid. But in the two-dimensional case, as T tends to zero, there is a possible solution consisting of coupled Luttinger chains. Schmeltzer considers a renormalization group treatment of two coupled Luttinger chains in the Coulomb gas formulation.

The achievements of his work can be summarized as follows:

1. The bosonization for strongly correlated Fermions (exclusion of double occupancy) has been carried out in both one and two dimensions.
2. Constraints imposed (see Schmeltzer, 1992, 1993) lead in one dimension to a Luttinger liquid.
3. For $D = 2$, and again with constraints, Schmeltzer concludes that one is led to an "incoherent" Luttinger liquid.
4. For $D = 2$ at $T = 0$, one finds a situation that can be described as a set of coupled Luttinger chains.
5. The N coupled Luttinger chains have been investigated considering the Coulomb gas in the plasma phase.

Finally, as this Appendix was nearing completion, the important work of Luther (1994) appeared, treating interacting electrons on a square Fermi surface. In his work, electronic states near such a Fermi surface are mapped on to quantum chains. Using boson-fermion duality, Luther succeeds in (a) isolating and (b) diagonalizing the bosonic part of the interaction. He demonstrates that these interactions destroy Fermi-liquid behavior. Furthermore, he finds spin-charge separation and concludes that the square Fermi surface remains square under doping. At half-filling, there is a charge gap and insulating behavior together with gapless spin excitations (see also March and Tosi, 1995).

A7.4. NUMERICAL STUDY OF FERMI SURFACE AND LOW-ENERGY EXCITATIONS IN HIGH-T_c MATERIALS

The electronic properties of doped Hubbard–Mott insulators have been studied for many years. The discovery of high-T_c superconductivity in the cuprate materials and the assumption that simple, nearly half-filled one-band models retain the essential physics of these systems renewed interest in this problem.

A central question in the theory of high-T_c superconductivity concerns the nature of the Fermi surface and the low-energy excitations in these materials. One of the most controversial questions is whether spin and charge are deconfined, as in one-dimensional problems, or whether the excitations carry charge and spin together, being real electronlike quasiparticles (see Appendix 7.5).

Angle-resolved photoemission and inverse photoemission experiments in some of the high-T_c materials indicate the existence of a "large Fermi surface" consistent with that of a weakly interacting system. In fact, band-structure calculations, which clearly cannot be used to describe the undoped insulating materials, predict for the doped systems a Fermi surface consistent with the one obtained by photoemission. However, this successful prediction of one-electron-band calculations cannot be taken as evidence of weak interactions in these systems. If Luttinger's theorem is obeyed, the Fermi volume is invariant under interaction effects and a strongly interacting system should also have a large Fermi surface. It has been shown that in a two-dimensional (2D) square lattice, the Fermi surface of a *t-J* model is consistent with Luttinger's theorem. Monte Carlo simulations for a nearly half-filled 2D Hubbard model also support this result.

In this Appendix, a numerical study of the Fermi surface and the Hall resistivity of a 2D Hubbard model with strong on-site interactions will be reported. Consider the case of a nearly half-filled system (with the particle density $n \leq 1$): as will be shown below and in agreement with previous results, the one-particle spectral densities are consistent with a large Fermi surface. Moreover, the characteristic features of the spectral functions are strongly reminiscent of the experimental results. One of the important questions is whether this Fermi surface can be used to build a semiclassical theory for the dynamics of quasiparticles in an external magnetic field. If this is so, the Hall resistance should be negative for $n < 1$, indicating electronlike carriers. The de Haas–van Alphen effect should indicate the existence of such a large Fermi surface. If, on the other hand, charge and spin excitations are decoupled, the charge dynamics should be dominated by the existence of a pseudo-Fermi surface for charge excitations, which may be quite different from that observed in photo-emission (see Balseiro and Castillo, 1992).

In the Hubbard model for large U, the kinetic energy and the integrated low-energy optical conductivity are decreasing functions of n for a nearly half-filled band. Some authors have interpreted this result as an indication that charge carriers are holes, in accordance with the second point of view.

Ioffe *et al.* (1991) showed that a hole-doped Hubbard–Mott insulator has a positive Hall resistance and that its temperature dependence is consistent with experiments made in high-T_c materials. Their starting point is an effective Hamiltonian where charge and spin excitations are decoupled, and consequently the positive Hall resistance they obtained could be anticipated on physical grounds.

In order to build a consistent theory it is important to calculate in a good approximation and on the same footing the different properties of the system, i.e., the one-electron spectral densities, to obtain the Fermi surface and the Hall resistance of the system (Balseiro *et al.*, 1992).

To do so these workers have studied numerically a Hubbard Hamiltonian in the limit of strong on-site interaction U. All results presented below are obtained by exact diagonalization techniques. To reduce the size of the Hilbert space, they eliminated, through the usual canonical transformation, the doubly occupied states. The resulting Hamiltonian reads

$$H = \sum_{\langle ij\rangle}(t_{ij}c^{\dagger}_{i\sigma}c_{j\sigma} + \text{h.c.}) + \sum_{\langle ij\rangle} \frac{4|t_{ij}|^2}{U}\,(S_i\cdot S_j - \frac{1}{4}n_in_j)$$

$$+ \sum_{\langle ijk\rangle} \frac{t^*_{ij}t_{kj}}{U}\; c^{\dagger}_{k\sigma}c_{j\sigma}n_{j-\sigma'}\,c^{\dagger}_{j\sigma'}\,c_{i\sigma'}, \tag{A7.4.1}$$

where the operator $c_{i\sigma}$ destroys a Fermion of spin σ at site i, S_i is the spin operator for site i, $n_{i\sigma}$ is the Fermion number operator for site i and spin $\sigma(n_i = n_{i\uparrow} + n_{i\downarrow})$, the symbol $\langle ij \rangle$ denotes the sum over all pairs of nearest neighbors, $\langle ijk \rangle$ denotes the sum over all sets of three sites with i and k nearest neighbors of j, and t_{ij} is the hopping matrix element between sites i and j. All Fermion operators are subject to the constraint of having no double occupation. In what follows, t_x and t_y will denote the hopping matrix element along the x and y directions, respectively, and one defines $J_x = 4t_x^2/U$.

Balseiro *et al.* solved exactly the problem in two finite clusters of $N = 12$ sites each and all possible numbers of particles N_e $(1 \leq N_e \leq N)$. They considered rectangular clusters with periodic boundary conditions in the x direction and open boundary conditions in the y direction. The two types of clusters considered were 4×3 and 6×2. (The notation $m \times l$ indicates m sites in the x direction and l sites in the y direction.) These geometries were chosen in order to be able to include an external magnetic field B of arbitrary intensity perpendicular to the plane. They considered only the diamagnetic coupling of the external field with the charged particles. This is done through Peierls substitution, which in the present case can be effected by adding a phase to the hopping matrix elements in the x direction (this affects all terms except the spin–spin coupling). In what follows, their results will be presented for the one-particle spectral densities in the absence of magnetic field and the transverse conductivity σ_{xy} in the presence of a small magnetic field B.

For a rectangular $m \times l$ cluster with periodic boundary conditions only along the x direction, the band structure of uncorrelated particles consists of l bands. Within each band the quantum number k_x varies between $-\pi/a$ and π/a, where a is the lattice parameter. Each band is characterized by a quantum number a corresponding to a different wave function in the y direction. For the 4×3 cluster, they defined $a = -1, 0$, and 1 for the lower, medium, and upper bands, respectively. As the one-particle states are filled, each band has a different Fermi momentum k_{aF}. This collection of k_{aF} defines a "Fermi surface." Although this geometry is not the most appropriate for studying the Fermi surface, it is necessary for calculating the Hall conductivity and, as mentioned above, they calculate all the properties in the same cluster in order to minimize possible inconsistencies due to finite-size effects.

In the interacting system the expectation value for the occupation number n_{ak_x} of the state $|ak_x\rangle$ is greater than 0 and lower than 1. Balseiro *et al.* define k_{aF} as the momentum where this expectation value is equal to 0.5. With this criterion the Fermi surface in the interacting system is the same as in the noninteracting one. In Fig. A7.6.1 the expectation value of n_{ak_x} is shown; to present all the states in the same figure, they plotted $\langle n_{ak_x} \rangle$ as a function of the noninteracting one-particle energies ε_{ak_x}. As is evident from the Figure, for all densities $n = N_e/N$ the Fermi surface is consistent with Luttinger's (1961) theorem. In Fig. A7.4.1c, corresponding to $N_e = 10$ particles and $J_x = 0.1t_x$ in the 4×3 cluster, there are some oscillations in $\langle n_{ak_x} \rangle$. This is a consequence of plotting all bands on the same scale; within each

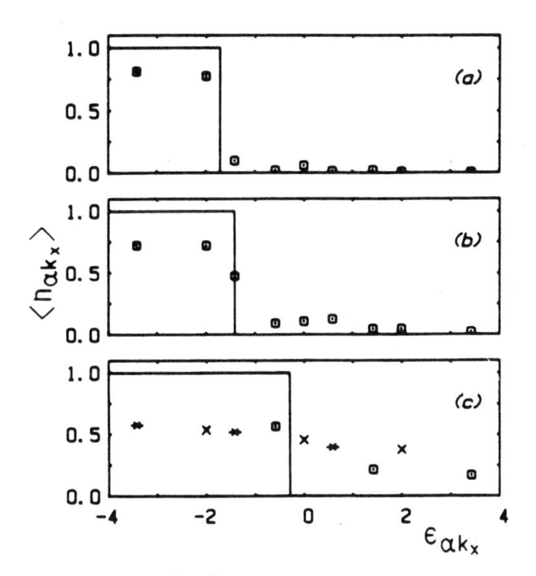

FIGURE A7.4.1. Occupation numbers $\langle n_{\alpha k_x} \rangle$ as a function of the noninteracting one-particle energies $\varepsilon_{\alpha k_x}$ for $J_x/t_x = 0.1$ in the 4×3 cluster. a) $n_e = 4$; b) $N_e = 6$; and c) $N_e = 10$. In part c) different symbols indicate different values of the quantum numbers α: $\alpha = -1$ (stars), $\alpha = 0$ (crosses), and $\alpha = 1$ (squares). Continuous lines indicate the occupation numbers for the noninteracting systems. (Reproduced from Balseiro and Castillo, 1992.)

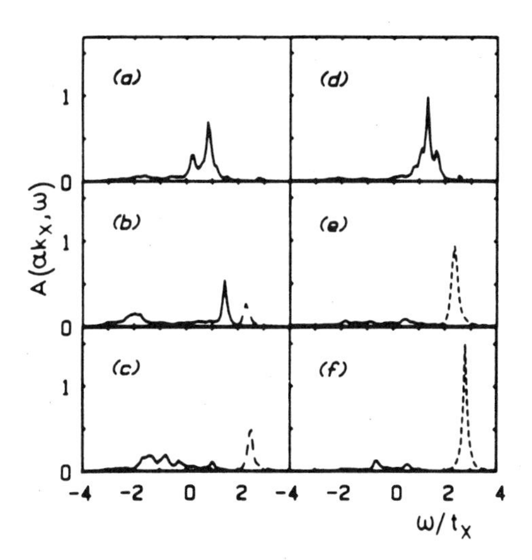

FIGURE A7.4.2. One-particle spectral densities for ten particles in the 4×3 cluster with $J_x/t_x = 0.1$. a)–c) correspond to $\alpha = 0$ with $k_x = 0$, $\pi/2$, and π, respectively; d)–f) correspond to $\alpha = 1$ with $k_x = 0$, $\pi/2$, and π. The solid and dashed lines correspond to the photoemission and inverse photoemission spectra, respectively. The spectral functions are plotted with a Lorentzian broadening of $\delta = 0.08t_x$. The Fermi energy is located at $\omega \simeq 2t_x$. (Reproduced from Balseiro and Castillo, 1992.)

band, however, the behavior of n_{ak_x} is monotonic. The same results are obtained for infinite U ($J_x = 0$).

The one-particle spectral densities obtained from the Green functions $\langle\langle c_{ak_x}, c^{\dagger}_{ak_x}\rangle\rangle$ are shown in Fig. A7.4.2. The spectral densities show a lot of structure that extends far from the Fermi energy. There is a strong peak that crosses the Fermi energy as k_x crosses the corresponding Fermi momentum k_{aF}. These spectral densities are in qualitative agreement with photoemission spectra obtained in high-T_c materials. Although the clusters studied are small and do not have x–y symmetry, these results are evidence that the Fermi surface in these strongly interacting systems is consistent with Luttinger's theorem if the number of holes is large enough. The case of a single hole may have a particular behavior.

Let us now present their results for the transverse conductivity σ_{xy}. The Hall resistance $R_H = R_{yx}/B$ or the Hall number n_H must be calculated by inverting the conductivity tensor. These quantities have the same sign as σ_{xy}. These workers calculated the transverse conductivity using linear response theory. A magnetic field perpendicular to the plane was included. These workers also considered the perturbation produced by a small transverse electric field. The coupling of the electrons to the electric field is given simply by $H_{\text{int}} = -eE\ \Sigma_i y_i$, where e is the electron charge, E is the electric field, and y_i is the y coordinate of the ith electron. The transverse conductivity is

$$\sigma_{xy} = -\ e^2a^2t_x\langle\langle j_x, y\rangle\rangle_{\text{ret}}\big|_{\omega=0}, \tag{A7.4.2}$$

where j_x and y are adimensional current and position operators. This equation

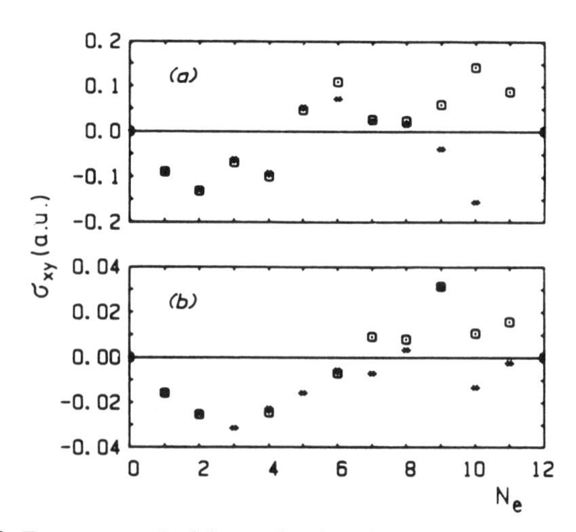

FIGURE A7.4.3. Transverse conductivity as a function of the total number of particles for (a) a 4×3 cluster with $t_y = t_x$, and (b) a 6×2 cluster with $t_y = 2t_x$. Squares and stars correspond to $j_x/t_x = 0$ and 0.1, respectively. (Reproduced from Balseiro and Castillo, 1992.)

yields, for a noninteracting system in the two clusters studied, a negative transverse conductivity for al! densities ($0 < n < 1$); this is in agreement with semiclassical theory.

In Fig. A7.4.3 the results obtained are shown as a function of N_e for both the 4×3 and 6×2 clusters and for different values of U. There are three points that are not included; for these points, because of quasidegeneracy of the ground state in these geometries, the computation of σ_{xy} is subject to large numerical uncertainties.

There are several features which Balseiro *et al.* (1992) stress: The transverse conductivity for $J_x = 0$ is electronlike for low n and holelike for larger n. The Hall number changes sign when the lower Hubbard band is approximately half filled ($n \simeq 0.5$). The single-hole problem constitutes a special point. The total spin of the system for one hole and $J_x = 0$ is maximum, in agreement with the Nagaoka theorem; consequently, the transverse conductivities for a single hole and a single electron are equal in absolute value and of different sign. For all other densities in the 4×3 cluster the total spin is zero for even N_e and $\frac{1}{2}$ or $\frac{3}{2}$ for odd N_e. For this system, the behavior of even and odd numbers of holes appears to be slightly different; this is probably due to finite-size effects. In the 6×2 cluster, both the eleven- and nine-particle ground states are completely polarized, but for all other densities the total spin is between 0 and $\frac{3}{2}$. It is to be noted that, although there is no electron–hole symmetry in the lower Hubbard band, the absolute value of the transverse conductivity for the case of a few electrons N_e is of the same order of magnitude as the conductivity for a few holes $N - N_e$. As U decreases, a tendency

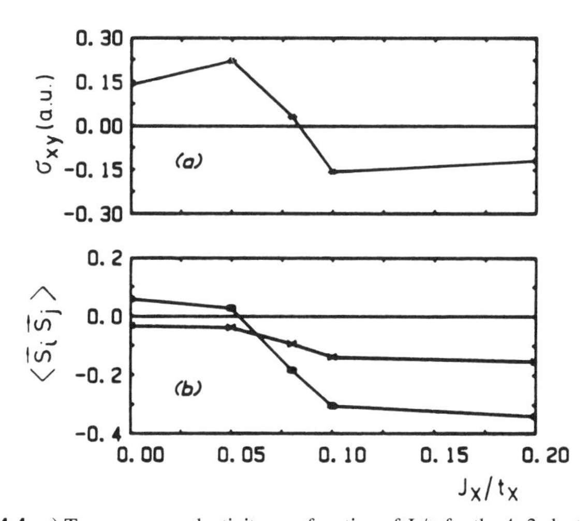

FIGURE A7.4.4. a) Transverse conductivity as a function of J_x/t_x for the 4×3 cluster with $N_e = 10$ and $t_y = 2t_x$. b) Nearest-neighbor spin-spin correlations. Parameters are the same as in part a). Squares indicate correlations along the upper and lower chains, and crosses indicate the correlations of the central chain. The lines are to guide the eye. (Reproduced from Balseiro and Castillo, 1992.)

toward a disappearance of holelike behavior for $n \cong 1$ is apparent. This could be a consequence of the antiferromagnetic correlations, which increase rapidly as U decreases. In fact, as shown in Fig. A7.4.4 the Hall conductivity for the 4×3 cluster with $N_e = 10$ changes sign when strong short-range antiferromagnetic correlations appear.

As a general behavior, these results show that for large U ($J_x \leq 0.1t_x$) the Hall number changes sign when the lower Hubbard band is approximately half filled, the carriers being electronlike for low density and holelike for intermediate densities. For larger densities ($n \simeq 1$) the transverse conductivity is dominated by the spin–spin correlations. If U is large enough, the Hall number remains positive up to $n = 1$; if U is smaller, however, the transverse conductivity becomes negative for low hole doping. For the system studied, there is a region of densities that depends on the cluster (typically $0.5 < n < 0.75$) where the Hall number is positive for a wide range of values of U. The interplay between antiferromagnetism and the Hall conductivity in the region $n \simeq 1$ cannot be studied in detail in these small clusters; the Hilbert spaces for larger clusters will require somewhat massive computing facilities.

The above results can be taken as indirect evidence that in the 2D Hubbard model, in the large-U limit, charge and spin excitations are decoupled, and, although one-particle band-structure calculations may correctly predict the shape of the Fermi surface, they should not be used to predict the dynamics of charge excitations.

One can enquire just what is the "Fermi surface" that a de Haas–van Alphen experiment probes: this problem is much more difficult to study in a small cluster.

To summarize, Balseiro *et al.* (1992) have calculated exactly the spectral densities and the Hall conductivity in the same strongly interacting system. Their results show that the Fermi surface calculated from the one-particle spectral functions is consistent with Luttinger's theorem. Furthermore, the Hall conductivity for a nearly half-filled system ($n \simeq 1$) indicates holelike carriers if the parameters are such that the antiferromagnetic correlations are weak. This makes evident the failure of the one-particle theories, which would predict the wrong sign for the conductivity. This apparent contradiction between the shape of the Fermi surface and the Hall conductivity can be taken as evidence of the deconfinement of charge and spin excitations (Balseiro *et al.*, 1992; see also Appendix 7.3).

A7.5. ELECTRON LIQUIDS FLOWING THROUGH ANTIFERROMAGNETIC ASSEMBLIES

The author has elsewhere (March, 1993) briefly reviewed some of the salient properties of electron (or hole) liquids flowing through antiferromagnetic assemblies, the motivation again being the high-T_c copper oxides.

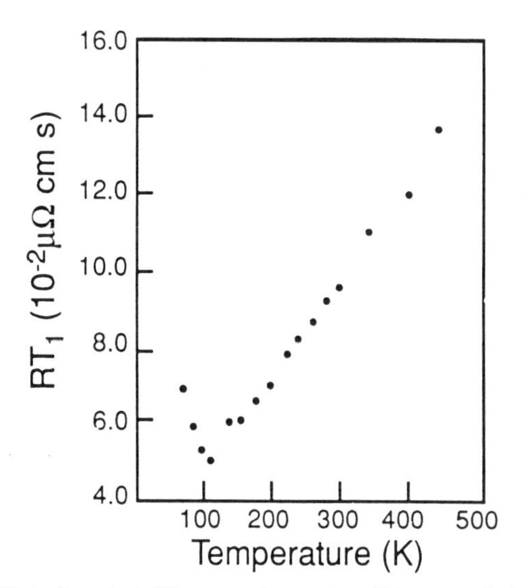

FIGURE A7.5.1. Plot of product RT_1 versus temperature T for normal state of high-T_c material YBa_2CuO_6. R is taken from experiment; T_1 is nuclear spin–lattice relaxation time obtained from Cu NMR data (after Egorov and March, 1994). Two-dimensional Fermi liquid theory predicts RT_1 proportional to T (cf. Egorov and March, 1994), and this theory is well obeyed above 120K. Departure from linearity at low temperature heralds a different phase of the electron liquid.

Without going over all the ground reviewed by March (1993), let us here briefly summarize the results of Egorov and March (1994). These workers start out from the two-dimensional Fermi liquid study of Kohno and Yamada (1991). These authors relate two apparently very different physical properties, the electrical resistivity R and the nuclear spin–lattice relaxation time T_1, to the magnetic susceptibility $\chi(\mathbf{Q})$, where $\mathbf{Q}$ is the antiferromagnetic wave vector. Then Egorov and March eliminate the susceptibility to obtain

$$RT_1 \propto T. \tag{A7.5.1}$$

Data for both R and T_1 have been taken from experiment for two materials for which both quantities have been measured over an appropriate temperature range. The results for one of these materials, $YBa_2Cu_4O_6$ are reproduced in Fig. A7.5.1. It can be seen that there is an appreciable range of temperature over which the result of Eq. (A7.5.1) is well borne out. However, a new and major feature appears at a temperature of around 100K, where the Hall coefficient and the thermoelectric power also show anomalous behavior (Egorov and March, 1994). This appears to be revealing a crossover from Fermi liquid behavior in Eq. (A7.5.1) to a new phase of the electron liquid in the lower temperature regime. Egorov and March have proposed that this manifestation of a Luttinger liquid or of coupled Luttinger chains (see, for example, Schmeltzer, 1992, 1993, and Appendix 7.3). It is too early to decide on either experimental or first-principles-theory grounds whether this proposal is indeed correct. One other possible reason for such a crossover is an effective change in dimensionality due to the (possible) role of interlayer coupling at low temperatures. The reader is also referred to the later study of Egorov and March (1995) in relation to plasmon experiments, biholons, and the transition temperature T_c for Bose-Einstein condensation.

Appendix to Chapter 9

A9.1. ELECTRON LIQUIDS AND THE QUANTIZED HALL STATE

The purpose of this Appendix is to indicate the way in which the wave function of a two-dimensional electron liquid in a magnetic field can be set up following Laughlin (1983). In turn, it will be stressed that his wave function can be intimately connected with the physics of the classical one-component plasma (OCP) model in two dimensions.

Specifically, Laughlin (1983) associated the $\nu = 1/m$ quantized Hall state, where m is an odd integer, with a liquid-like electron state based on a Jastrow (product of pair wave functions) trial total ground-state wave function. The importance of his wave function is that it can lead to a lower ground-state energy than the Wigner electron crystal and to a constant density (characteristic of a translationally invariant liquid). His wave function has the form

$$\Psi_m = \prod_{i<j}(z_i - z_j)^m \exp\left(-\tfrac{1}{4}\sum_{i=1}^{N} |z_i|^2\right), \tag{A9.1.1}$$

where $z_i = x_i + iy_i$ is the coordinate of the ith electron in units of the magnetic length l, $m = 1/\nu$ is an odd integer due to Fermi antisymmetry, and $\nu = n(2\pi l^2)$ is, as usual, the filling factor of the lowest Landau level.

The wave function (A9.1.1) consists, in fact, only of states in the lowest Landau level with angular momentum specified by m. The total angular momentum of N electrons is given by

$$L = \tfrac{1}{2}\, mN\,(N-1). \tag{A9.1.2}$$

It is important to stress here the relation to the classical OCP model.

Analogy with Two-Dimensional Classical OCP Model

Let us now make contact with the statistical mechanics of the classical $2d$ OCP model by writing (compare Ishihara, 1989)

$$|\Psi_m|^2 = \exp[-\beta\Phi], \tag{A9.1.3}$$

with $\beta = 1/m$ playing the role of inverse temperature $(k_B T)^{-1}$. One then finds by direct comparison with the form (A9.1.1) that

$$\Phi = -\sum_{i<j} 2m^2 \, ln\,|z_i - z_j| + \tfrac{1}{2}m \sum_i^N |z_i|^2. \tag{A9.1.4}$$

The interpretation of Φ defined in this way is then very simple. It is precisely the potential energy of the $2d$ OCP consisting of N identical particles with charge $\sqrt{2}\,m$. For $l = 1$, the first term in Eq. (A9.1.4) is the mutual repulsion, while the second term represents the interaction with an array of background neutralizing charges, uniformly distributed with density $1/\sqrt{2}\pi$. Such a plasma is "liquid-like" for $\Gamma = (\pi n)^{1/2} e^2 \beta = (2m)^{1/2} < 140$. It forms a hexagonal crystal above this Γ. Hence Laughlin's state can have a constant electron density, corresponding to a translationally invariant electron liquid in a certain parameter range. The probability density $|\Psi_m|^2$ then corresponds to a system having uniform density $1/(2\pi m)$.

Laughlin was able to calculate the overlap of his wave function with the true eigenfunction of angular momentum $3m$ as 0.99946 and 0.99468, for $m = 3$ and 5, respectively.

As discussed in Section 4.3, if the pair correlation function $g(r)$ of the OCP is given, then the potential energy per electron is immediately written on physical grounds as

$$\frac{V}{N} = \int_0^\infty \frac{ne^2}{r}[g(r)-1]\pi r\, dr. \tag{A9.1.5}$$

For large Γ, a model analogous to Onsager's ion sphere idea (ion disc in two dimensions) yields the approximation $V/N = (4/3\pi - 1)2e^2/R$. At the specific coupling strength $\Gamma = 2$, which corresponds to a full Landau level, Jancovici (1977) has found for the OCP the exact result

$$g(r) = 1 - \exp[-(r/R)^2], \tag{A9.1.6}$$

which yields from Eq. (A9.1.5) the result

$$\frac{V}{N} = -\frac{1}{2}\pi^{1/2}\frac{e^2}{R.}. \tag{A9.1.7}$$

For the $\nu = 1/m$ state, Laughlin's wave function tends to zero with r as r^m, and thus $g(r) \propto r^{2m}$ as r tends to zero. Such a strong "excluded-area" effect results in the low-ground-state energies given by his model.

A9.2. MELTING CRITERIA FOR THREE-DIMENSIONAL CLASSICAL AND QUANTAL WIGNER CRYSTALS

After a great deal of early work on the crystallization of a quantal electron plasma, at a critical coupling strength $a/a_0 = r_{sc}$, say, a measuring the mean interelectronic spacing, Ceperley and Alder (1980) established by quantal computer simulation that $r_{sc} \simeq 80$, as summarized in Chapter 5. The object of this appendix is to compare and contrast this strongly coupled quantal assembly near the Wigner transition with its classical counterpart (see March and Tosi, 1984). In the classical one-component plasma the coupling strength is conventionally measured by the dimensionless parameter Γ, defined by

$$\Gamma = e^2/ak_BT. \tag{A9.2.1}$$

Following the pioneering work of Brush *et al.* (1966), Hansen (1972) and later workers have established that the classical OCP freezes when Γ reaches a value, Γ_c say, of about 180.

March and Tosi (1985a) noted a marked qualitative similarity between the electron pair correlation function $g(r)$ of the strongly coupled electron liquid at $r_s = a/a_0$ equal to 100, as calculated by Lantto and Siemens (1979) and shown in Fig. A9.2.1a, and the pair function of the classical OCP at $\Gamma = 10$, given in Fig. A9.2.1b. To press this similarity and thereby to relate the two coupling strengths in Figs. A9.2.1a,b, let us consider the Fourier transform of $g(r) - 1$—that is, the liquid structure factor $S(k)$. Then in the long-wavelength limit the f-sum rule allows one to write the exact result (see, for example, March and Tosi, 1984).

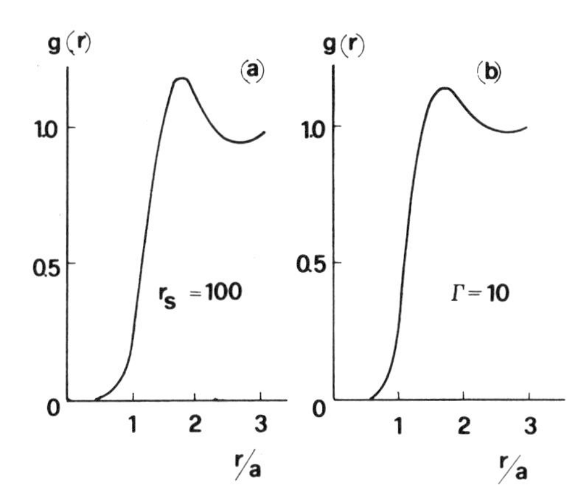

FIGURE A9.2.1. Pair distribution functions: (a) in quantal electron liquid at $r_s = 100$ (see Lantto and Siemens, 1979) and (b) in classical plasma at coupling constant = 10 (cf. Appendix 9.1 for two-dimensional analogue) (see Baus and Hansen, 1980) (after March and Tosi, 1985).

$$\lim_{k\to 0} S(k) = \frac{1}{2}\frac{\hbar\,\omega_p k^2}{4\pi\rho e^2} \tag{A9.2.2}$$

for the quantal system, ω_p being the electron plasma frequency and ρ the electron density $\frac{3}{4}\pi a^3$. The corresponding result for the classical plasma is

$$\lim_{k\to 0} S(k) = \frac{k_B T\, k^2}{4\pi\rho e^2}. \tag{A9.2.3}$$

Thus, this immediately suggests a correspondence between classical and quantal plasmas

$$k_B T \leftrightarrow \tfrac{1}{2}\hbar\omega_p, \tag{A9.2.4}$$

which is equivalent to

$$\Gamma \leftrightarrow (4r_s/3)^{1/2}. \tag{A9.2.5}$$

This correspondance, for $r_s = 100$ in the example in Fig. A9.2.1a, yields $\Gamma \simeq 12$, which is satisfactory in comparison with Fig. A9.2.1b.

From $g(r)$, the potential energy per electron is determined by the well-known formula

$$\frac{U}{N} = \frac{1}{2}\ \rho e^2 \int d\mathbf{r}\ \frac{g(r)-1}{r}. \tag{A9.2.6}$$

For classical and quantal systems, using the results of Slattery *et al.* (1980) and of Ceperley and Alder (1980), respectively, one finds again that the correspondence given in Eq. (A9.2.5) yields similar values for the potential energy per particle, when measured in units of e^2/a. Thus, although the correspondence in Eq. (A9.2.5) was derived from a long-wavelength argument, it turns out still to apply semi-quantitatively to liquid structure and potential energy.

However, when one turns to compare the coupling strengths for melting classical and quantal Wigner crystals, a wholly different situation obtains. Whereas the critical r_s value for the quantal case, as mentioned, is about 80, the critical Γ for the classical OCP is around 180. The conclusion from these numbers is that from liquid structure arguments the quantal crystal is more stable than its classical counterpart. The remainder of this appendix is therefore focused on how to reconcile these apparently quite different melting points, at least in a qualitative manner (cf. March and Tosi, 1985a).

It is helpful in this context to consider first the Lindemann criterion for melting the quantal Wigner crystal. Following early work by Pines and Nozières (1960) and by Coldwell–Horsfall and Maradudin (1963), extensive phonon calculations on the Wigner lattice were reported by Kugler (1981) and then utilized in the Lindemann criterion. His conclusion was that, because of disruption by vibrational energy, the Wigner crystal ought to melt at an r_s value that is an order of magnitude greater than the Ceperley and Alder (1980) finding. Though precise numerical values are difficult to obtain, the main point in the present context is that the kinetic energy of the localized electrons in a purely vibrational model must be too large, reduction clearly being essential to further stabilize the quantal crystal as required by the Ceperley–Adler results (March and Tosi, 1985a).

What is needed, therefore, is a mechanism whereby electron delocalization can occur, but which still retains the basic structural features of the insulating Wigner crystal. Though, as discussed, structure determines the potential energy, the virial theorem links potential with kinetic energy (March, 1958). The model advocated by March and Tosi (1985a) is to retain the insulating character of the Wigner crystal by allowing electron rotational degrees of freedom to play a role, through a mechanism somewhat analogous to that proposed long ago by Zener (1952). His idea was to consider a ring of atoms undergoing classical diffusive motions in unison. Of course, his process was a thermally activated one, and, to the writer's knowledge, the activation energy is sufficiently high in classical crystals so far examined that diffusion proceeds dominantly through other mechanisms. However, in the present case, such rotational motions can occur by quantum-mechanical tunneling at $T = 0$. This mechanism will delocalize electron wave functions from the localized Wigner Gaussian orbitals (see Section 4.6) and, equally important, will retain the insulating $T = 0$ character of the Wigner crystal by the use of only closed loops in the delocalization process. In making this model quantitative, it is important to consider carefully electron spin, the elementary picture of the Wigner crystal being one of a Néel antiferromagnet, with upward-spin electrons on the sites of one simple cubic lattice and downward spins on the other interpenetrating simple cubic lattice (see also Section 9.3). According to March and Tosi (1985a), this Néel-type antiferromagnetic array of electron spins on a body-centered-cubic lattice must be relaxed by rotational interchange of electrons with opposite spins.

A9.3. WIGNER OSCILLATOR IN MAGNETIC FIELD OF ARBITRARY STRENGTH

Because of the difficulty of studying electron assemblies analytically with their strong long-range Coulomb interactions, Lea and March (1989) have proposed an elementary model based on the Lindemann melting criterion, as well as a refined model set out in the following appendix. In a magnetic field the longitudinal plasmon mode $\omega_p(q)$ and the transverse phonon mode $\omega_t(q)$ are transformed to two magnetophonon modes $\omega_\pm(q)$, where

$$\omega_+^2 = \omega_c^2 + \omega_p^2 + \omega_t^2, \qquad \omega_- = \omega_p\omega_t/\omega_c. \tag{A9.3.1}$$

In high fields, $\omega_c \gg \omega_p, \omega_t$ the zero-point motion is contained in the ω_+ mode, close to the cyclotron frequency, ω_c. This can be described quantum mechanically by the theory of March and Tosi (1985) for a localized oscillator in a magnetic field with a vector potential $\mathbf{A} = (-\frac{1}{2}Hy, \frac{1}{2}Hx, 0)$. The canonical density matrix of such a Wigner electron oscillator is given by

$$C(\mathbf{r},\mathbf{r}_0,\beta) = f(\beta)\exp\{-i(x_0y + y_0x)\phi(\beta) - [(x - x_0)^2 + (y - y_0)^2]g(\beta) - [(x + x_0)^2 + (y + y_0)^2]h(\beta)\}, \tag{A9.3.2}$$

where the quantities in Eq. (A9.3.2) are given explicitly by March and Tosi and later workers. Defining the mean square displacement $\langle r^2 \rangle$ by

$$\langle r^2 \rangle = \frac{\int r^2 C(\mathbf{r}, \mathbf{r}, \beta)\, d\mathbf{r}}{\int C(\mathbf{r}, \mathbf{r}, \beta)\, d\mathbf{r}}. \tag{A9.3.3}$$

this is readily calculated to yield

$$\langle r^2 \rangle = \frac{\hbar}{m\omega_c b} \frac{1}{\coth(b\alpha) - \cosh\alpha/\sinh(b\alpha)} = \frac{\hbar}{m\omega_L bF} \tag{A9.3.4}$$

where $\alpha = \hbar\omega_L/kT$. The parameter b is given by

$$b = (1 + \omega_w^2/\omega_L^2)^{1/2}, \tag{A9.3.5}$$

where the frequency ω_w of the localized Wigner oscillator can be taken from Bonsall and Maradudin (1977) as $8e^2/ma^3$, where a is the lattice spacing. In terms of the parameters $r_s = (\pi n)^{-\frac{1}{2}}/a_0$ (a_0 is the Bohr radius) and the Landau filling factor $\nu = nhc/eH$, the melting criterion becomes, for $\alpha \gg 1$,

$$\frac{\langle r^2 \rangle}{a^2} = \frac{K\nu}{b} = \gamma_M, \qquad b = (1 + D\nu^2 r_s)^{1/2}, \tag{A9.3.6}$$

where $\gamma_M \simeq 0.08$ (Chui *et al.*, 1986). The numerical constants K and D (of order unity) should be calculated for a full theory but are used as fitting parameters in this simple model (Lea and March, 1989).

In the zero-temperature, high-field limit the electrons will form a classical two-dimensional solid at all densities. In the extreme quantum limit, $r_s = 0$, there is a critical filling factor $\nu_c^0 = \gamma_M/K$ at which the crystal will melt as the field is reduced. However, for $r_s > r_w = K^2/D\gamma_M$, the crystal will remain stable even in zero field (Ceperley, 1978), as shown on the phase diagram in Fig. 4 of Lea and March (1989).

At finite temperatures (but keeping $\alpha \gg 1$), the zero-point motion needs supplementing by contributions from phonon-like modes, but the reader should refer to Lea and March (1989) for further details.

A9.4. SHEAR MODULUS, ELECTRON DENSITY PROFILE, AND PHASE DIAGRAM FOR TWO-DIMENSIONAL WIGNER CRYSTALS

Defect energies in two- and three-dimensional classical crystals correlate with the shear modulus μ. In turn, this relates melting temperature T_m intimately to μ. Therefore, a model will first be summarized (Lea and March, 1992) that relates the shear modulus for two-dimensional Wigner crystals to the half-width σ of the localized electron density profile around a chosen Wigner lattice site. One then can solve self-consistently the classical limiting relation between σ^2 and $k_BT/\mu(\sigma)$ to obtain an approximate temperature dependence of the shear modulus. With the same model for $\mu(\sigma)$, this calculation will then be modified to include (a) the effect of a magnetic field, which has, of course, no influence on thermodynamics in the classical limit, and (b) zero-point motion, which is dominant in the extreme quantal limit.

For the high-field case, this modeling allows T_m/T_{mc}, the melting temperature measured in units of the classical limiting value T_{mc}, to be plotted against the Landau level filling factor ν. The predictions of the model can thereby be brought into contact with the experiments of Andrei *et al.* (1988) and Glattli *et al.* (1990) on the propagating shear modes in the electron assembly in a GaAs/GaAlAs heterojunction in strong magnetic fields (cf. Section 9.2). The model exhibits the main features found experimentally. Similarly, the zero-field quantal transition can be studied, this time the "experimental" contact being with the computer studies of Ceperley. The model is able to deal with the liquid-to-solid transition in the extreme quantal limit at $T = 0$.

In classical monatomic crystals, correlations have been known for a long time between melting temperature T_m, vacancy formation energy E_v, and elastic moduli. Some theoretical understanding of these empirical correlations has been afforded by (a) statistical mechanical models at elevated temperatures, appropriate to, say, condensed argon (Bhatia and March, 1984) and (b) current models of force fields including many-body contributions in metals like Cu (Johnson, 1989; see also March, 1989). In case (b), Johnson has stressed that the highest-quality correlation among elastic constants and E_v is via the shear modulus μ. While this is true in three dimensions, the well-known Kosterlitz–Thouless (1973, 1978) transition in two-dimensional classical crystals is driven by the shear modulus and the thermal unbinding of dislocation pairs.

Our concern is with two-dimensional Wigner electron crystallization, with and without magnetic fields. In zero magnetic field $B = 0$, two relevant studies exist. One is the classical limit of the phonon spectrum, worked out by Bonsall and Maradudin (1977). This provides a first-principles basis for scaling the model calculations presented in this appendix for the shear modulus (Lea and March,

1992). The second is the quantum Monte Carlo study of Ceperley (1978), in which he calculates the mean interelectronic separation, say r_W, at which the transition from electron liquid to two-dimensional Wigner electron crystal occurs, in the extreme quantal limit at $T = 0$ and in zero field $B = 0$. He finds the value $r_W = 33$; this one can take as an "experimental value," which again is a most valuable piece of information in the modeling set out here.

A9.4.1. Observation of Magnetically Induced Wigner Solid (MIWS)

Turning to nonzero magnetic fields, which is the prime motivation for the study of Lea and March (1991), magnetic-field-assisted Wigner electron crystallization, proposed by Durkan, Elliott, and March (1968), has been observed by Andrei *et al.* (1988) and Glattli *et al.* (1990). These workers demonstrated the onset of a propagating shear mode at low frequency in the electron assembly in a GaAs/GaAlAs heterojunction (see Section 9.2), which they took to be the fingerprint of a Wigner electron solid phase induced by an applied magnetic field. They were thus enabled to map out the melting curve of the Wigner solid as a function of the Landau level filling factor ν. Once such a plot was made, there remained only rather weak residual dependence on the carrier density n, related to the mean interelectronic separation r_s by

$$n = (\pi r_s^2)^{-1}. \tag{A9.4.1}$$

While there have been a number of attempts to calculate the melting curve of the Wigner crystal without an applied magnetic field (Ferraz *et al.*, 1978, 1979; Nagara *et al.*, 1987; see also March, 1988), it was only later that much attention was focused on the melting in the presence of an applied magnetic field (see Lea and March, 1991; Elliott and Kleppmann, 1975). This is therefore one of the focal points of the work of Lea and March (1992)—to propose a model that will lead to a prediction of the melting curve in a magnetic field.

The outline of the remainder of this appendix is as follows. In Section A9.4.2 a model calculation is reported of the shear modulus μ of a two-dimensional Wigner electron crystal from an assumed form of the localized density profile around a chosen Wigner lattice site (Lea and March, 1992). The idea behind the calculation is the extension of the approach of Bonsall and Maradudin (1977); which, however, applied in the limit of complete electron localization, i.e., for a delta function density profile. Their precise calculation will be used to scale the results of the present model to agree in this delta function limit of the density profile. Section A9.4.3 is then concerned with presenting self-consistent solutions of three limiting cases of the model: (i) the classical limit, in which both Einstein and Debye models lead to a relation between the half-width σ, squared, of the profile, and the ratio of thermal energy k_BT to shear modulus $\mu(\sigma)$; (ii) the high-magnetic-field limit; and (iii) the limit where the zero-point motion dominates. The contact with experiment—both the heterojunction data referred to for the high-magnetic-field case, and quantum Monte Carlo computer simulation for the zero-field case—is made in Section A9.4.4. A summary is given in Section A9.4.5, together with proposals for further experiments that would be of interest.

A9.4.2. Electron Density Profile and Shear Modulus

March and Tosi (1985) have considered the effective electron density of a localized Wigner oscillator in a magnetic field B of arbitrary strength (see the summary in Appendix 9.3). This can be thought of as the Einstein model of a Wigner crystal in an applied magnetic field. As well as for the ground state, it is remarkable that the effects of harmonic restoring force, magnetic field, and temperature T still contrive to leave a Gaussian profile for the site density $\rho(r)$, which we shall simply write in the form

$$\rho(r) = \text{constant} \exp(-r^2/\sigma^2). \quad \text{(A9.4.2)}$$

Here, one can see that the quantity σ is a measure of the half-width of the localized density profile. Thus, the distribution in space of the electron density can be characterized through the whole parameter range of magnetic field B, temperature, and carrier density, by this half-width σ. Though this can be calculated within the framework of the Einstein model, Lea and March (1992) model its dependence on the foregoing parameters in their approach, the Einstein results providing one useful guideline to achieve satisfactory modeling.

A9.4.2.1. Modeling of Shear Modulus μ as Function of Half-Width σ

An essential first step in setting up a model to treat the phase diagram of the two-dimensional Wigner crystal for a wide selection of parameter values is to relate the shear modulus μ to the half-width σ.

As starting point, one notes that Bonsall and Maradudin (1977) gave the Madelung energy of any two-dimensional crystalline array of localized electrons. Their results were later confirmed by Borwein *et al.* (1988). The ground state is a triangular crystal (one electron per lattice point on a 2D hexagonal Bravais lattice) with energy $-1.1061e^2/r_0$ per electron. Figure A9.4.1 displays the energies $E_1(\alpha)$ and $E_2(\alpha)$ for a simpler strain α applied along the $\langle 11\bar{2}\rangle$ and $\langle 10\bar{1}\rangle$ directions, respectively. For small strains, E_1 and E_2 are close to the harmonic energy

$$E_0(\alpha) = -1.1061e^2/r_0 + \tfrac{1}{2}\mu_0\pi r_0^2\alpha^2, \quad \text{(A9.4.3)}$$

where $\mu_0 = 0.044e^2/r_0^3$ is the shear modulus at $T = 0$ for a classical electron crystal.

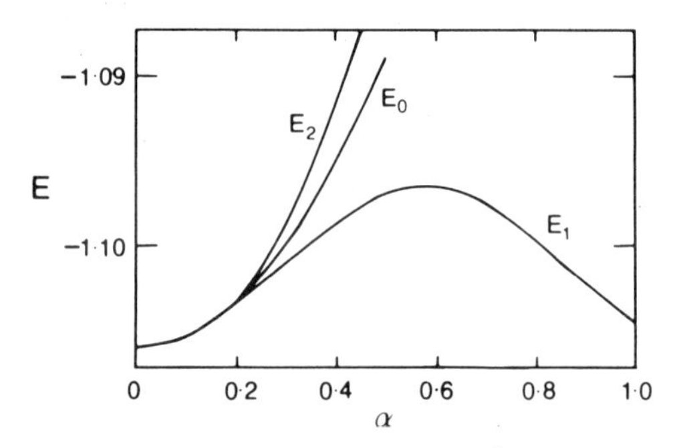

FIGURE A9.4.1. The Coulomb energy per electron, in units of e^2/r_0, for a two-dimensional triangular electron crystal as a function of a static shear strain α parallel to the $\langle 112 \rangle$ direction (energy E_1) and the $\langle 101 \rangle$ direction (energy E_2). The parabola E_0 shows the harmonic energy variation for the zero-temperature shear modulus. The calculations were made using the expressions given by Bonsall and Maradudin (1977).

For large strains, there is both anisotropy and anharmonicity. To a very good approximation

$$E_1(\alpha) = [-1.1013 - 0.0048 \cos(\sqrt{3}\pi\alpha)]e^2/r_0, \qquad (A9.4.4)$$

with $E_2(\alpha) > E_0(\alpha)$ with

$$E_1(\alpha) + E_2(\alpha) = 2E_0(\alpha). \qquad (A9.4.5)$$

The effect of this anharmonicity is to cause the shear modulus to depend on the thermal motion and zero-point motion of the electrons. It is assumed that this gives a Gaussian distribution $\exp(-\alpha^2/2\sigma^2)$ of resolved shear strain along any direction with variance σ^2. Hence the mean energy for an additional infinitesimal strain α_1 can be written as

$$\overline{E(\alpha_1)} = \int_{-\infty}^{\infty} E(\alpha + \alpha_1) \exp \frac{-\alpha^2}{2\sigma^2}\, d\alpha. \qquad (A9.4.6)$$

The effective shear modulus is then given by $\partial^2\bar{E}/\partial\alpha_1^2$, as α_1 tends to zero. Substituting the expressions for $\bar{E}_1$ and E_2 in (A9.4.5) we obtain

$$\mu_1^*(\sigma) = \mu_1(\sigma)/\mu_0 = \exp(-D\sigma^2), \qquad \mu_2^*(\sigma) = 2 - \mu_1^*(\sigma), \qquad (A9.4.7)$$

where $D = 3\pi^2/2 = 14.8$, while μ_1^* and μ_2^* are the normalized shear moduli for simple shear strains along the $\langle 11\bar{2} \rangle$ and $\langle 10\bar{1} \rangle$ directions, respectively. A static shear stress on the crystal produces both types of strain and a resultant modulus μ^*, isotropic by symmetry, is given by

$$1/\mu^* = 1/2\mu_1^* + 1/2\mu_2^*, \qquad \mu^*(\sigma) = \mu_1^*(\sigma)\mu_2^*(\sigma). \qquad (A9.4.8)$$

The next stage is to calculate the mean-square shear produced by the propagating shear modes, angular frequency $\omega(q)$ and amplitude U_q, in the crystal. The mean-square component of displacement of each electron, mass m, along any direction is given by

$$\langle u^2 \rangle = \frac{1}{2} \sum_q |U_q|^2 = \frac{1}{2} \sum_q \frac{\varepsilon}{mn\omega^2}. \tag{A9.4.9}$$

where

$$\varepsilon = (h\omega_q/2)\coth(h\omega_q/2k_B T) \tag{A9.4.10}$$

is the mean energy per mode. The mean-square shear strain, resolved along any direction, is

$$\sigma^2 = \langle \alpha^2 \rangle = \frac{1}{2} \sum_q q^2 |U_q|^2 = \frac{1}{2} \sum_q \frac{q^2 \varepsilon}{mn\omega^2} \tag{A9.4.11}$$

for pure shear modes, and will be isotropic by symmetry in a triangular crystal with contributions from all shear modes.

A9.4.3. Limiting Cases: Instabilities and Melting

The purpose of this section is to convert the result relating reduced shear modulus μ to half-width σ into potentially observable predictions. First let us consider the classical limit in zero field.

9.4.3.1. Classical Limit

Specifically, the calculation of the scaled shear modulus by Lea and March can be employed to construct the temperature dependence of the shear modulus μ and to exhibit instability of the lattice. Both the Einstein model referred to, as well as a phonon model, lead then to the relation

$$\sigma^2 = \frac{A\, k_B T}{\mu(\sigma)}. \tag{A9.4.12}$$

the constant A being, of course, model dependent.

A value of A can be obtained from a simplified Debye model. An unscreened 2D plasma is incompressible in the long-wavelength limit, so one includes only the shear modes (which have propagation and polarization directions in the plane of the electrons). The shear modes propagating along the two symmetry directions are $\omega_1 = c_1 q$, where $c_1^2 = \mu_1/mn$ for q_1 parallel to $\langle 10\bar{1} \rangle$ and polarization along $\langle 11\bar{2} \rangle$, and $\omega_2 = c_2 q$ with $c_2^2 = \mu_2/mn$ for q_2 parallel to $\langle 11\bar{2} \rangle$ and polarization along $\langle 10\bar{1} \rangle$. Integrating Eq. (A9.4.11) to the Debye wave vector $q_D = 2/r_0$ leads to

$$\sigma^2 = \frac{k_B T}{2\pi r_0^2}\left(\frac{1}{2\mu_1} + \frac{1}{2\mu_2}\right) = \frac{k_B T}{2\pi r_0^2 \mu} = \frac{0.0285t}{\mu^*(\sigma)}, \qquad \text{(A9.4.13)}$$

where t is the temperature normalized to the Kosterlitz–Thouless melting temperature $T_{mc} = e^2/r_0\Gamma_m k_B$, where Γ_m has been determined experimentally to be 127 (Deville, 1988) for electrons on liquid helium.

Equation (A9.4.13) has been solved self-consistently to obtain the shear modulus $\mu^*(t)$ as a function of temperature. The results are shown, along with plots of $\mu_1^*(t)$ and $\mu^*(t)$ in Fig. A9.4.2. Several interesting points emerge from this admittedly simple model. First, the shear mode with q_1 parallel to $\langle 10\bar{1}\rangle$, which corresponds to the close-packed lines of electrons sliding past each other, softens as t increases while the q_2 mode stiffens. Note that $\mu_1^*(t)$ decreases linearly with t at low temperatures. The total shear modulus $\mu^*(t)$ also decreases to an anharmonic instability at $\mu^* = 0.5$, $t = 1.46$ and $\sigma = 0.29$. For $\sigma > 0.29$, Eq. (A9.4.13) still has a solution (shown as a broken curve in Fig. A9.4.2) but the crystal will then be unstable. The Kosterlitz–Thouless transition occurs at a temperature T_m such that

$$T_m = \mu a^2/4\pi k_B, \qquad \text{(A9.4.14)}$$

where a is the lattice spacing. This transition will therefore occur when the reduced shear modulus $\mu_{KT} = 0.62t$ as shown by the full line in Fig. A9.4.2. This intersects the $\mu^*(t)$ graph for this model at $t = 1.22$. Also shown in Fig. A9.4.2 are measure-

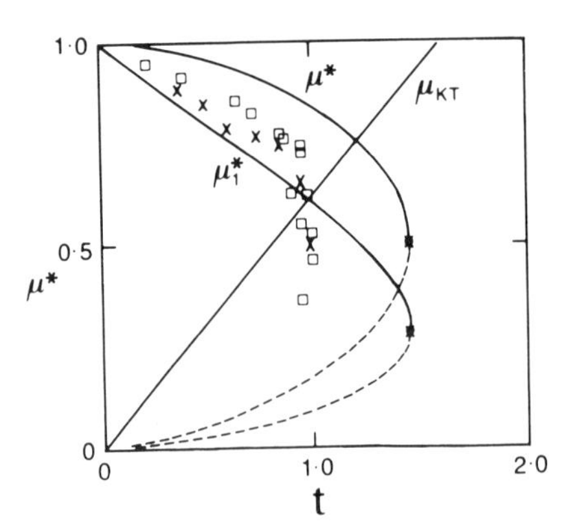

FIGURE A9.4.2. The reduced shear moduli μ^* and μ_1^*, calculated as a function of reduced temperature t. The points on the lines show the anharmonic stability limit. The broken curves show the unstable solutions. The full line μ_{KT} shows the theoretical locus of the Kosterlitz–Thouless transition. The crosses are the results from computer simulations by Morf (1979), while the squares are the experimental data of Deville *et al.* (1984) for electrons on liquid helium (after Lea and March, 1992).

ments of μ^* by Deville *et al.* (1984), together with the computer simulation results of Morf (1979). The shear modulus is found to decrease linearly at low temperature and this has been shown to be due to anharmonicity in detailed calculations by Chang and Maki (1983). The rapid decrease in μ near the transition found by Morf has been ascribed to renormalization due to thermally excited dislocations.

Note that the shear modulus and the anharmonic instability in the present model can be scaled in temperature by adjusting the absolute value of A. Hence the relative positions of the Kosterlitz–Thouless transition and the anharmonic instability can be varied and can occur very close together: with reference to Fig. A9.4.2 an increase in σ by only 20% above the Debye model would bring the upper curve into accord with the available data. The two transitions are closely linked in that dislocations are produced by slip along the $\langle 11\bar{2}\rangle$ direction which corresponds to the softened q_1 shear mode. It is tempting to associate the region between the Kosterlitz–Thouless transition and the anharmonic instability with the postulated hexatic phase (Nelson and Halperin, 1979).

A9.4.3.2. Quantum Limit

Equation (A9.4.6) can also be applied to the quantum crystal at $T = 0$ in zero field. In this case σ^2 is due to zero-point shear strain and the two-mode Debye model gives

$$\sigma_0^2 = \frac{1}{3\sqrt{\pi r_s}}\left(\frac{1}{2\sqrt{\mu_1}} + \frac{1}{2\sqrt{\mu_2}}\right) = \frac{0.45}{\sqrt{r_s}}\left(\frac{1}{\sqrt{\mu_1^*}} + \frac{1}{\sqrt{\mu_2^*}}\right). \tag{A9.4.15}$$

where $r_s = r_0/a_B$ and $a_B = \hbar^2/me^2$ is the Bohr radius. This can be solved for μ^* and μ_1^* as functions of $1/r_s$, the results being shown in Fig. A9.4.3. As zero-point motion increases, the shear modulus falls until the crystal becomes unstable at $\mu^* = 0.16$ and $r_s = r_W = 125$. There are no experimental results on this transition, which is the transition Wigner originally proposed (Wigner, 1934, 1938), though in two dimensions, but Ceperley has shown by computer simulation that $r_W = 33$. Hence, the quantum crystal is more stable than this simple model suggests. The notorious sensitivity of r_W to the model chosen is clear from the table of Care and March (1975). The Debye model probably overestimates the zero-point motion. Siringo *et al.* (1991) have also shown that the force constant between disks of electronic charge is greater than for point charges. Finally this model does not consider any specifically quantum effects which may result from the overlap of the individual electronic wave functions. Nonetheless a possible mechanism for two-dimensional Wigner quantum melting is clearly indicated, as an anharmonic instability.

It is interesting to follow this instability and the K–T transition at finite temperature, as r_s decreases from infinity in the classical limit, using the expres-

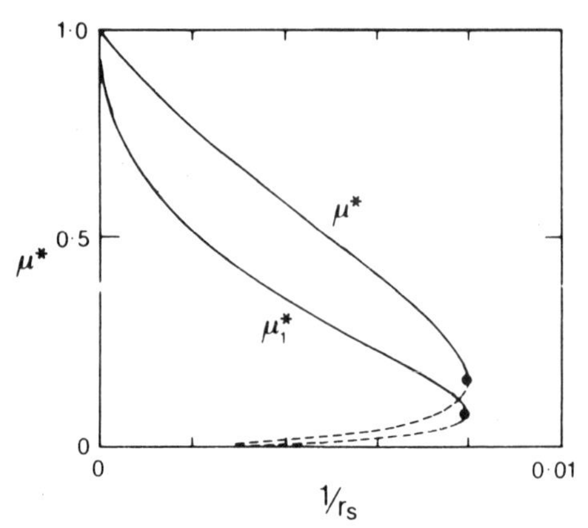

FIGURE A9.4.3. Reduced shear moduli μ^* and μ_1^*, calculated from Eq. (A9.4.4) as a function of $1/r_s$ at $T = 0$. Points on lines show anharmonic instability limit. Broken curves show unstable solutions of Eq. (A9.4.2) (after Lea and March, 1992).

sion for σ^2, supplemented by thermal contributions. The loci of these transitions on the t–$1/r_s$ plane is shown in Fig. A9.4.4. At $r_s = \infty$ the K–T transition occurs below the instability, as already discussed. But as r_s decreases the two transitions merge until, for $r_s < 400$, the K–T transition no longer occurs. In the present model the transition is an anharmonic instability for $125 < r_s < 400$.

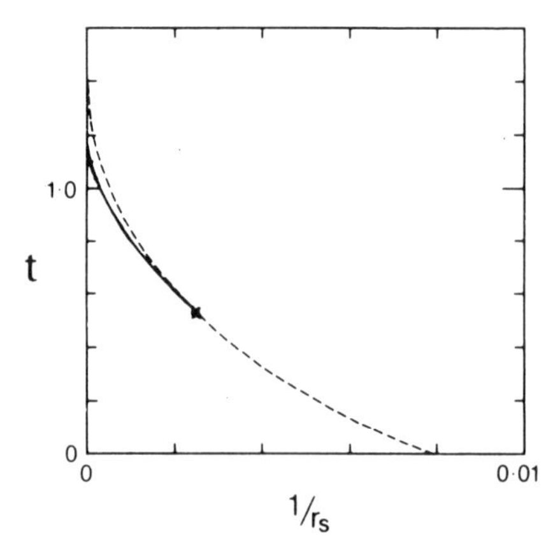

FIGURE A9.4.4. Model phase diagram of two-dimensional electron crystal in zero field on t versus $1/r_s$ plot. Full curve shows locus of Kosterlitz–Thouless transition, while broken curve is locus of anharmonic instability. Note that there is a predicted change in the nature of the transition as electron density increases (after Lea and March, 1992).

A9.4.4. The Magnetically Induced Wigner Solid (MIWS)

In the quantum limit of high density, when $r_s = 0$, it is well established theoretically that an infinite magnetic field can suppress the zero-point motion and induce a classical two-dimensional electron crystal. As the field is reduced, or as the Landau level filling factor ν increases, the cyclotron motion of individual electrons increases with a mean-square displacement along any direction $\langle u^2 \rangle = l_B^2$, where $l_B = (\hbar/eB)^{1/2}$ is the magnetic length. If we assume that this displacement produces both longitudinal and shear strains then the resolved shear strain can be written as

$$\sigma_B^2 = 0.0285t/\mu^*(\sigma) + G\nu, \qquad \text{(A9.4.16)}$$

where the first term is taken to be the same as for classical crystal in zero field. Fourier transforming the cyclotron motion into components of longitudinal and shear displacements and integrating to find the mean-square shear strain gives $G = 0.5$ in the limit $l_B < r_0$. Taking $G = 0.5$, Eq. (A9.4.16) is solved self-consistently. It is found that as ν increases the K–T transition and the instability temperature decrease as shown in Fig. A9.4.5. For the model parameters used, the instability lies above the K–T transition at all fields, but the relative position could well be field dependent.

This phase diagram has some of the features of the experimental data of Andrei *et al.* (1988) and Glattli *et al.* (1990). However, it now seems possible that there may be regions of two-dimensional electron liquid phases interspersed with

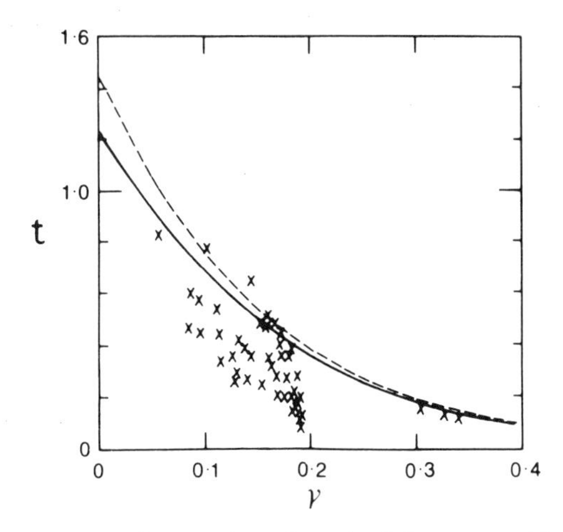

FIGURE A9.4.5. Model phase diagram of two-dimensional electron crystal in magnetic field on a t versus ν plot. Full curve shows locus of Kosterlitz–Thouless transition, while broken curve is locus of anharmonic instability. Data points are from Andrei *et al.* (1988) and Glattli *et al.* (1990). After Lea and March (1992).

solid phases (Jiang *et al.*, 1990; Buhmann *et al.*, 1991) and that the phase diagram calculated here could form an "envelope" for the solid regions. On this interpretation such a liquid phase is seen in the data close to a filling factor $\nu = 0.192$. Also the three points at $\nu > 0.3$, originally excluded by Andrei *et al.* from their analysis, could represent a solid phase in this region. It is known from the zero-magnetic-field treatments of melting, already referred to above, that near $T = 0$ the phase diagram will depend sensitively on the nature of the low-lying excitations in the two phases and these are not carefully treated in the present model. It is to be noted that Lozovik *et al.* (1985) have exposed an anharmonic instability as the Landau level factor increases beyond a critical value.

A9.4.5. Discussion and Summary

That there is an intimate relation between the dimensionless parameter σ characterizing the shear strain and the electron density profile (A9.4.2) has already been emphasized. Some work on deviations of this profile from Gaussian form is available from the study of Gann *et al.* (1979). Related to this, and for the ground state in zero magnetic field, the three-dimensional density functional calculations of Senatore and Pastore (1990) offer a way of checking the Gaussian structure factor $\rho(k)$, by extracting the Fourier components in the periodic density in the Wigner crystal at the reciprocal lattice vectors $\boldsymbol{K}_i$. Equally important would be a study of the way that shear affected the localized density profile. Underlying the present work is the model in which the "localized Gaussian blobs" move rigidly as the lattice sites are shifted by a "frozen phonon." This could be tested, at least in principle, by the approach of Senatore and Pastore (1990), who refer to both BCC and FCC lattices. As pointed out by Perrin *et al.* (1985), it is possible by continuous deformation via a body-centered tetragonal phase to pass from BCC to FCC; each periodic lattice could be explored to check the way in which the "localized" blobs on the Wigner sites have to be deformed as a crystal is sheared. In connection with atomic crystals, the model of localized blobs moving without deformation is the analogue of the rigid-ion model, which in turn is equivalent to a pair force field.

Such refinements will no doubt mean that the simple modeling of μ in terms of σ in (A9.4.7) will have to be transcended. Nevertheless, this modeling suggests that it will be of considerable interest to explore whether Bragg reflection studies of some sort are feasible on a heterojunction. If the phonon mode found by Andrei *et al.* (1988) and by Glattli *et al.* (1990) is the lower hybrid magnetophonon mode derived from the shear mode, this would confirm that one is dealing with a Wigner solid; not necessarily with long-range crystalline order. So this would be the first objective of Bragg reflection studies: to demonstrate crystallinity. The second would be to test the prediction of the above model of an intimate connection between shear modulus and scattering factors.

A further point of considerable interest is to recognize that a more general theoretical model will not merely cover a range of filling factors to $\nu > 0.4$, but will need to relate Wigner crystal theory to Laughlin-like electron liquid states as well as, possibly, the Hall crystal postulated by Halperin *et al.* (1986). This should then reveal what will no doubt be a close connection between the region of parameter space treated in the above model and integral and fractional quantum Hall effects. Already, a re-entrant phase diagram is emerging from experimental studies (Jiang *et al.*, 1990; Buhmann *et al.*, 1991) with a series of interspersed liquid and solid phases as ν increases. The "envelope" of these phases seems to be close to the melting line in Fig. A9.4.5. The host material may also have a strong influence on the phase diagram of the two-dimensional electrons. For instance, Kohler *et al.* (1986) have suggested that a phonon-mediated transverse charge-density wave state in a quantizing magnetic field may lead to Wigner crystallization over the whole range of magnetic quantization.

Finally, to summarize, the main conclusions from the model of Lea and March (1992) are:

(i) There are two possible instabilities: one arising directly from anharmonicity in the simple model presented; the other is a Kosterlitz–Thouless transition. As the physical parameters are varied, it is possible to find "crossover" between the instabilities.

(ii) The quantum limit $T = 0$ in zero field is clearly sensitive to (a) tunneling, which is not incorporated in the present model; it is related to Wigner oscillator wave-function overlap and (b) possible ring exchange, relating to magnetism, as discussed by March and Tosi (1980).

(iii) At the present stage of development of magnetic-field-assisted Wigner crystallization, the simple model presented here seems particularly useful. The main features of the phase diagram established experimentally by Andrei *et al.* (1988) and by Glattli *et al.* (1990) are compatible with the above predictions of Wigner crystallization in the high-field regime, except near $T = 0$ (see Fig. (A9.4.5) where more careful treatment of the low-lying excitations in the liquid and the crystal phases is clearly called for (for $B = 0$ see Ferraz *et al.*, 1978, 1979). Subsequent experiments on nonlinear electrical conductivity provide further strong evidence in support of pinned Wigner crystals over a substantial range of electric fields, followed eventually by "sliding" of the Wigner crystallites.

References

ABE, R. (1989). *J. Phys. Soc. Japan* **58:** 918.

ABOU-CHACRA, R., ANDERSON, P. W., and THOULESS, D. J. (1974). *J. Phys.* **C7:** 65.

ABRIKOSOV, A. A., GORKOV, L. P., and DZIALOSYNSKI, I. E. (1962). *Methods of Quantum Field Theory in Statistical Physics*, Moscow.

AHLRICHS, R., and SCHARF, P. (1987). *Adv. Chem. Phys.* **67:** 501

ALDER, B. J., CEPERLEY, D. M., and REYNOLDS, P. J. (1982). *J. Chem. Phys.* **86:** 1200.

ALLAN, N. L., WEST, C. G., COOPER, D. L., GROUT, P. J., and MARCH, N. H. (1985). *J. Chem. Phys.* **83:** 4562.

ALLOUL, H., and DELLOUVE, C. (1987). *Phys. Rev. Lett.* **59:** 578.

ALONSO, J. A., and MARCH, N. H. (1989). *Electrons in Metals and Alloys,* New York: Academic.

AMUSIA, N. YA., CHEREPKOV, N. A., JANEV, R. K., and ZIVANOVIC, DJ. (1974). *J. Phys.* **137:** 1435.

ANDERSON, J. B. (1975). *J. Chem. Phys.* **63:** 1499.

ANDERSON, J. B. (1976). *J. Chem. Phys.* **65:** 4121.

ANDERSON, J. B. (1980). *J. Chem. Phys.* **73:** 3897.

ANDERSON, M. A., and CAVE, R. J. (1991). *Chem. Phys.* **T54:** 1.

ANDERSON, P. W. (1958). *Phys. Rev.* **109:** 1492.

ANDERSON, P. W. (1961). *Phys. Rev.* **124:** 141.

ANDERSON, P. W. (1968). *Phys. Rev. Lett.* **21:** 13.

ANDERSON, P. W. (1969). *Phys. Rev.* **181:** 25.

ANDERSON, P. W. (1973). *Mater. Res. Bull.* **8:** 153.

ANDERSON, P. W. (1987). *Science* **235:** 1196.

ANDERSON, P. W. (1990a). *Phys. Rev. Lett.* **64:** 1839.

ANDERSON, P. W. (1990b). *Phys. Rev. Lett.* **65:** 2306.

ANDERSON, P. W. (1991a). *Phys. Rev. Lett.* **66:** 3226.

ANDERSON, P. W. (1991b). *Phys. Rev. Lett.* **67:** 2092.

ANDERSON, P. W. (1991c). *Phys. Rev. Lett.* **67:** 3844.

ANDERSON, P. W., BASKARON, G., and ZOU, Z. (1987). *Solid State Commun.* **63:** 973.

ANDREI, E. Y. (1992). Chapter in Butcher et al. (1992).

ANDREI, E. Y., DEVILLE, G., GLATTLI, D. C., WILLIAMS, F. I. B., PARIS, E., and ETIENNE, B. (1988). *Phys. Rev. Lett.* **60:** 2765.

AOKI, H., and KANIMURA, H. (1976). *J. Phys. Soc. Jpn.* **40:** 6.

ARAI, T., and YOKOYAMA, I. (1993). *Phys. Chem. Liquids* **26:** 143.

ARGYRES, P. (1967). *Phys. Rev.* **154:** 410.

AROVAS, D. P., SCHREIFFER, J. R., WILCHEK, F., and ZEE, A. (1985). *Nucl. Phys.* **B251:** 117.

ASCOUGH, J. A., and MARCH, N. H. (1990). *Phys. Chem. Liquids,* **21:** 251.

ASHCROFT, N. W., and LEKNER, J. (1966). *Phys. Rev.* **145:** 83.

Ashcroft, N. W., and Sraus, D. M. (1976). *Phys. Rev.* **B14:** 448.
Bachelet, G. B., Hamaan, D. R., and Schluter, M. (1982). *Phys. Rev.* **B26:** 4199.
Bachlechner, M. E., Macke, W., Miesenbock, H. M., and Schinner, A. (1991). *Physica* **B168:** 104.
Baeriswyl, D., and Maki, P. (1985). *Phys. Rev.* **B31:** 6633.
Baldereschi, A., and Tosatti, E. (1979). *Solid State Commun.* **29:** 131.
Baldereschi, A., and Tosatti, E. (1983). *Solid State Commun.* **44:** 1397.
Balseiro, C. A., and Castillo, H. E. (1992). *Phys. Rev. Lett.* **68:** 121.
Bamzai, A. S., and Deb, B. M. (1981). *Rev. Mod. Phys.* **53:** 95.
Banyard, K. E., and Ellis, D. J. (1975). *J. Phys.* **B8;** 2311.
Banyard, K. E., and Sanders, J. (1993). *J. Chem. Phys.* **99:** 5281.
Banyard, K. E., and Sanders, J. (1994). *J. Chem. Phys.* **101:** 3096.
Barnes, S. E. (1976). *J. Phys.* **F6:** 1375.
Bartel, L. C., and Jarrett, H. S. (1974). *Phys. Rev.* **B10:** 946.
Barth, U. von, and Hedin, L. (1972). *J. Phys.* **C5:** 1629.
Bartlett, R. J. (1981). *Ann. Rev. Phys. Chem.* **32:** 359.
Bartlett, R. J. (1989). *J. Phys. Chem.* **93:** 1697.
Baus, M. (1990). *J. Phys. Condens. Matter* **2:** 2111.
Beattie, A. M., Stoddart, J. C., and March, N. H. (1971). *Proc. R. Soc.* **A236:** 97.
Becke, A. D. (1988). *Phys. Rev.* **A38:** 3098.
Becke, A. D. (1992). *J. Chem. Phys.* **96:** 2155.
Becker, K. W., and Brenig, W. (1990). *Z. Phys.* **B79:** 195.
Becker, K. W., and Fulde, P. (1988). *Z. Phys.* **B72:** 423.
Becker, K. W., and Fulde, P. (1989). *J. Chem. Phys.* **91:** 4223.
Bednorz, J. G., and Müller, K. A. (1986). *Z. Phys.* **B64:** 189.
Bell, R. J., and Dean, P. (1970). *Disc. Faraday Soc.* **50:** 55.
Belyayev, A. M., Bobrov, V. B., and Trigger, S. A. (1989). *J. Phys. Cond. Matter* **1:** 9665.
Benson, J. M., and Byers Brown, W. (1970). *J. Chem. Phys.* **53:** 3880.
Bennett, M., and Inkson, J. C. (1977). *J. Phys.* **C10:** 987.
Bernholc, J., and Holzwarth, N. A. W. (1983). *Phys. Rev. Lett.* **50:** 1451.
Berrondo, M., Daudey, J. P., and Goscinski, O. (1979). *Chem. Phys. Lett.* **62:** 34.
Bersohn R. (1962). *J. Chem. Phys.* **36:** 3445.
Bethe, H. A. (1931). *Zeits für Phys.* **71:** 205.
Betsuyaka, M., and Yokota, I. (1986). *Phys. Rev.* **B33:** 6505.
Bhatia, A. B., and March, N. H. (1984). *J. Chem. Phys.* **80:** 2076.
Bhatia, A. B., Thornton, D., and March, N. H. (1974). *Phys. Chem. Liquids* **4:** 97.
Bishop, R. F. (1991). *Theor. Chem. Acta* **80:** 95.
Blackman, J. A., and Taguena, J. (eds.) (1991). *Disorder in Condensed Matter Physics,* Oxford: Clarendon Press.
Blankenbecler, R., Scalapino, D. J., and Sugar, R. L. (1981). *Phys. Rev.* **D24:** 2278.
Blazej, M., and March, N. H. (1993). *Phys. Rev. E.* **48:** 1782.
Bloch, F. (1930). *Z. Phys.* **56:** 706.
Bloch, F. (1933). *Z. Phys.* **81:** 363.
Bloch, F. (1934). *Helv. Phys. Acta* **7:** 385.
Bohm, D., and Pines, D. (1953). *Phys. Rev.* **92:** 609.
Bonner, J. C., and Fisher, M. E. (1964). *Phys. Rev.* **135A:** 640.
Bonsall, L., and Maradudin, A. A. (1977). *Phys. Rev.* **B15:** 1959.
Brandow, B. H. (1975). *J. Solid State Chem.* **12:** 397.
Brandow, B. H. (1977). *Adv. Phys.* **26:** 651.
Brinkman, W. F., and Rice, T. M. (1970). *Phys. Rev.* **B2:** 4302.
Brown, P. J. (1972). *Phil. Mag.* **26:** 1377.

Brush, S. G., Sahlin, H. L., and Teller, E. (1966). *J. Chem. Phys.* **45:** 2102.

Buhmann, H., Joss, W., von Klitzing, K., Kukuskim, I. V., Plaut, A. S., Martinez, G., Ploog, K., and Timofeev, V. B. (1991). *Phys. Rev. Lett.* **66:** 926.

Bush, I. J., Logan, D. E., and Madden, P. A., see Logan, D. E. (1991).

Bush, I. J., Logan, D. E., Madden, P. A., and Winn, M. D. (1989). *J. Phys. Condensed Matter* **1:** 2251.

Butcher, P. N., March, N. H., and Tosi, M. P. (eds.) (1992). *Low Dimensional Solids,* New York: Plenum.

Callaway, J., and Chatterjee, A. K. (1978). *J. Phys. Colloq. (France)* **39:** C-6, Pt 2: 772.

Callaway, J., and March, N. H. (1984). *Solid State Phys.* **38:** 136.

Campbell, D. K., DeGrand, T. A., and Mazumdar, S. (1984). *Phys. Rev. Lett.* **52:** 1717.

Campbell, D. K., Gammel, J. T., and Loh, E. Y. (1988a). *Synth. Met.* **27:** A9.

Campbell, D. K., Gammel, J. T., and Loh, E. Y. (1988b). *Phys. Rev.* **B38:** 12043.

Cann, N. M., Boyd, R. J., and Thakkar, A. J. (1993). *Int. J. Quantum Chem.* **48:** 1.

Car, R., and Parrinello, M. (1985). *Phys. Rev. Lett.* **55:** 2471.

Car, R., Tosatti, E., Baroni, S., and Lee Caprute (1981). *Phys. Rev.* **B24:** 985.

Care, C. M., and March, N. H. (1971). *J. Phys.* **C4:** L372.

Care, C. M., and March, N. H. (1975). *Adv. Phys.* **24:** 101.

Carr, W. J. (1961). *Phys. Rev.* **122:** 1437.

Carr, W. J., and Coldwell-Horsfall, R. A., and Fein, A. E. (1961). *Phys. Rev.* **124:** 747.

Castellani, C., Di Castro, C., Feinberg, D., and Ranninger, J. (1979). *Phys. Rev. Lett.* **43:** 1957.

Cave, R. J., and Davidson, E. R. (1988a). *J. Chem. Phys.* **88:** 5770.

Cave, R. J., and Davidson, E. R. (1988b). *J. Chem. Phys.* **89:** 6798.

Ceperley, D. M. (1978). *Phys. Rev.* **B18:** 3126.

Ceperley, D. M. (1981). In *Recent Progress in Many-Body Theories,* Zabolitsky, J. G. et al. (eds.), Berlin: Springer, p. 262.

Ceperley, D. M. (1983). *J. Comput. Phys.* **51:** 404.

Ceperley, D. M., and Alder, B. J. (1980). *Phys. Rev. Lett.* **45:** 567.

Ceperley, D. M., and Alder, B. J. (1984). *J. Chem. Phys.* **81:** 5833.

Ceperley, D. M., and Alder, B. J. (1987). *Phys. Rev.* **B36:** 2092.

Ceperley, D. M., and Kalos, M. H. (1979). In *Monte Carlo Methods in Statistical Physics,* Binder, K. (ed.), Berlin: Springer, p. 145.

Chandrasekhar, S. (1947). *Scientific Monthly* **64.**

Chang M., and Maki, K. (1983). *Phys. Rev.* **B27:** 1646.

Chapman, R. G., and March, N. H. (1988). *Phys. Rev.* **B38:** 792.

Cheranovskii, V. (1992). *Int. J. Quantum Chem.* **41:** 695.

Chihara, J. (1987). *J. Phys.* **F17:** 295.

Chui, S. T., Rice, T. M., and Varma, C. M. (1974). *Solid State Commun.* **15:** 155.

Chui, S. T., and Esfaniani, K. (1991). *Europhys. Lett.* **14:** 361.

Cioslowski, J., and Challacombe, M. (1992). *Theor. Chem. Acta* **83:** 185.

Cizek, J. (1966). *J. Chem. Phys.* **45:** 4256.

Cizek, J. (1969). *Adv. Chem. Phys.* **14:** 35.

Clementi, E. (1963). *J. Chem. Phys.* **38:** 2248.

Clementi, E., and Roetti, R. (1974). *Atom Data Nucl. Data* **14:** 177.

Coester, F., and Kümmel, H. (1960). *Nucl. Phys.* **17:** 477.

Cohen, L., and Frishberg, C. (1976). *Phys. Rev.* **A13:** 927.

Cohen, L., and Lee, C. (1985). *J. Math. Phys.* **26:** 3105.

Coldwell-Horsfall, R. A., and Maradudin, A. A. (1960). *J. Math. Phys.* **1:** 395.

Coldwell-Horsfall, R. A., and Maradudin, A. A. (1963). *J. Math. Phys.* **4:** 582.

Colle, R., and Salvetti, O. (1975). *Theor. Chim. Acta* **37:** 329

Colle, R., Montagnani, R., Riana, P., and Salvetti, O. (1978). *Theor. Chim. Acta* **49:** 37.

COLLE, R., MOSCARDO, F., PIANI, P., and SALVETTI, O. (1977). *Theor. Chim. Acta* **44:** 1.
CONNOLLY, J. W. D., and WILLIAMS, A. R. (1983). *Phys. Rev.* **B27:** 5169.
CONWELL, E. M. (1980). *Phys. Rev.* **B22:** 1761.
CONWELL, E. M., MIZES, H. A., and CHOI, H.-Y. (1991). *Synth. Met.* **41–43:** 3675.
CORDERO, N. A., ALONSO, J. A., LOPEZ, J. M., and MARCH, N. H. (1993). *Molec. Phys.* **79:** 393.
CORLESS, G. K., and MARCH, N. H. (1961). *Phil. Mag.* **6:** 1285.
COULSON, C. A., and FISCHER, I. (1949). *Phil. Mag.* **40:** 386.
COULSON, C. A., and NEILSON, A. H. (1951). *Proc. Phys. Soc.* **78:** 831.
COURANT, R., and HILBERT, D. (1953). *Methods of Mathematical Physics,* New York: Interscience, Vol. 1, Chapter 6.
COX, P. A. (1992). *Transition Metal Oxides,* Oxford: University Press.
CRUZ, L., and PHILLIPS, P. (1994). *Phys. Rev.* **B49:** 5149.
CUSACK, S., MARCH, N. H., PARRINELLO, M., and TOSI, M. P. (1976). *J. Phys.* **F6:** 749.
CYROT, M. (1972). *Phil. Mag.* **25:** 1031.
DALING, R., VAN HAERINGEN, W., and FARID, B. (1990). *Phys. Rev.* **B40:** 1659.
DALING, R., VAN HAERINGEN, W., and FARID, B. (1991). *Phys. Rev.* **44:** 2952.
DALING, R., VAN HAERINGEN, W., and FARID, B. (1992). *Phys. Rev.* **45:** 8970.
DANIEL, E., and VOSKO, S. H. (1960). *Phys. Rev.* **120:** 2041.
DAVIDSON, E. R. (1970). *Phys. Rev.* **A1:** 30.
DAWSON, K. A., and MARCH, N. H. (1983). *Phys. Lett.* **94A**: 434.
DAWSON, K. A., and MARCH, N. H. (1984). *J. Chem. Phys.* **81:** 5850.
DELLEY, B. (1982). *J. Chem. Phys.* **76:** 1949.
DELLEY, B., ELLIS, D. E., FREEMAN, A. J., BAERENDS, E. J., and POST, D. (1983a). *Phys. Rev.* **B27:** 2132.
DELLEY, B., FREEMAN, A. J., and ELLIS, D. E. (1983b). *Phys. Rev. Lett.* **50:** 488.
DELLEY, B., JARLBORG, T., FREEMAN, A. J., and ELLIS, D. E. (1983c). *J. Magn. Magn. Mater.* **3:** 100.
DEMARCO, R. R., ECONOMOU, E. N., and WHITE, C. T. (1978). *Phys. Rev.* **B18:** 3946.
DEVILLE, G. (1988). *J. Low. Temp. Phys.* **72:** 135.
DEVILLE, G. ET AL. (1984). *Phys. Rev. Lett.* **53:** 588.
DE WETTE, F. W. (1964). *Phys. Rev.* **A135:** 287.
DICKENS, P. G., and LINNETT, J. W. (1957). *Qu. Rev.* **11:** 291.
DIETERICH, K., and FULDE, P. (1987). *J. Chem. Phys.* **87:** 2976.
DIRAC, P. A. M. (1930). *Proc. Cambridge Phil. Soc.* **26:** 376.
DOBSON, P. J. (1976). *J. Phys.* **C11:** L295.
DOLG, M., FULDE, P., KUCHLE, W., NEUMANN, C. S., and STOLL, H. (1991). *J. Chem. Phys.* **94:** 3011.
DOREN, D. J., and HERSCHBACH, D. R. (1987). *J. Chem. Phys.* **87:** 433.
DOWKER J. S., and CHANS K. (1990). ICTP Trieste: Internal Report.
DREIZLER, R. M., and GROSS, E. K. U. (1990). *Density Functional Theory,* Berlin: Springer-Verlag.
DRICKAMER, H. G. (1965). *Solid State Physics* **17:** 1.
DRICKAMER, H. G., LYNCH, R. W., CLENDENEN, R. L., and PEREZ-ALBUERNE (1966). *Solid States Physics* **19:** 135.
DUBOIS, D. F. (1958). *Ann. Phys.* (N.Y.) **7:** 174, ibid **8:** 24.
DUPREE, R., KIRBY, D. J., FREYLAND, W., and WARREN, W. W. (1980). *Phys. Rev. Lett.* **45:** 130.
DUPREE, R., KIRBY, D. J., and WARREN, W. W. (1985). *Phys. Rev.* **B31:** 5597.
DURKAN, J., ELLIOTT, R. J., and MARCH, N. H. (1968). *Rev. Mod. Phys.* **40:** 812.
DURKAN, J., and MARCH, N. H. (1967). *J. Phys.* **C1:** 1118.
DZYALOSHINSKII, I. E. (1987). *Sov. Phys. JETP* **66:** 848.
ECONOMOU, E. N. (1983). *Green's Functions in Quantum Physics,* Berlin: Springer.
ECONOMOU, E. N., and COHEN, M. H. (1972). *Phys. Rev.* **B5:** 2931.
ECONOMOU, E. N., and POULOPOULOS, D. (1979). *Phys. Rev.* **B20:** 4756.
ECONOMOU, E. N., WHITE, C. T., and DE MARCO, R. R. (1978). *Phys. Rev.* **B18:** 3946.

EDWARDS, D. M., MANNING, S. M., and RETTKE-GROVER (1994). *Phil. Mag.* **B69:** 849.
EDWARDS, P. P., and SIENKO, M. J. (1978). *Phys. Rev.* **B17:** 2575.
EDWARDS, P. P., and SIENKO, M. J. (1981). *J. Am. Chem. Soc.* **103:** 2967.
EGAMI, T. (1979). *J. Appl. Phys.* **50:** 1564.
EGELSTAFF, P. A., MARCH, N. H., and MCGILL, N. C. (1974). *Can. J. Phys.* **52:** 1651.
EGOROV, S. A., and MARCH, N. H. (1994). *Phys. Chem. Liquids* **28:** 141.
EGOROV, S. A., and MARCH, N. H. (1995). *Phys. Chem. Liquids* (in press).
EL-HANNAY, U., BRENNERT, G. F., and WARREN, W. W. (1983). *Phys. Rev. Lett.* **50:** 540.
ELLIOTT, R. J., and KLEPPMANN, O. (1975). *J. Phys.* **C8:** 2737.
ELLIS, D. E. (1982). *Phys. Rev.* **B26:** 636.
EMERY, V. J. (1976). *Phys. Rev. Lett.* **37:** 107.
EMERY, V. J. (1987). *Phys. Rev. Lett.* **58:** 2794.
EMERY, V. J. (1989). *MRS Bull. (USA)* **14:** 67.
ENGEL, G. E., and FARID, B. (1992). *Phys. Rev.* **B46:** 15812.
ENGEL, G. E., and FARID, B. (1993). *Phys. Rev.* **B47:** 1593.
ENGEL, G. E., FARID, B., DALING, R., and VAN HAERINGEN, W. (1991a). *Phys. Rev.* **B44:** 13349.
ENGEL, G. E., FARID, B., NEX, C. M. M., and MARCH, N. H. (1991b). *Phys. Rev.* **B44:** 13356.
ENGEL, G., FARID, B., and DALING, R. (1994), *Phil. Mag.* **B69:** 901.
EVANGELOU, S. N., and EDWARDS, D. M. (1983). *J. Phys.* **C16:** 2121.
FALICOV, L. M., and HARRIS, R. A. (1969). *J. Chem. Phys.* **51:** 3153.
FALICOV, L. M., and KIMBALL, J. C. (1969). *Phys. Rev. Lett.* **22:** 997.
FANO, U. (1976). *Physics Today* **29:** 32.
FARID, B., GODBY, R. W., and NEEDS, R. J. (1990). *Physics of Semiconductors,* Singapore: World Scientific, p. 1759; Eds: Anastassakis, E. M. and Joannopoulos, J. D.
FARID, B., ENGEL, G. E., DALING, R., and VAN HAERINGEN, W. (1994). *Phil. Mag.* **B69:** 901.
FARID, B., HEINE, V., ENGEL, G. E., and ROBERTSON, I. J. (1993). *Phys. Rev.* **B48:** 11602.
FAZEKAS, P., and ANDERSON, P. W. (1974). *Phil. Mag.* **30:** 423.
FENG, S. (1994). *Phys. Rev.* **B49:** 2368.
FENTON, E. W. (1958). *Phys. Rev. Lett.* **21:** 1427.
FERMI, E. (1928). *Z. Phys.* **48:** 73.
FERRAZ, A., GROUT, P. J., and MARCH, N. H. (1978). *Phys. Lett.* **66A:** 155.
FERRAZ, A., and MARCH, N. H. (1979). *Phys. Chem. Liquids* **9:** 121.
FERRAZ, A., and MARCH, N. H. (1980). *Solid State Commun.* **36:** 977.
FERRAZ, A., MARCH, N. H., and FLORES, F. (1984). *J. Phys. Chem. Solids* **45:** 627.
FERRAZ, A., MARCH, N. H., and SUZUKI, M. (1978). *Phys. Chem. Liquids* **8:** 153.
FERRAZ, A., MARCH, N. H., and SUZUKI, M. (1979). *Phys. Chem. Liquids* **9:** 59.
FINK, J., VOM FELDE, A., and SPRÖSSERPREU, J. (1989). *Phys. Rev.* **B40:** 10181.
FIORENTINI, V., and BALDERESCHI, A. (1992). *J. Phys. Condensed Matter* **4:** 5967.
FOLDY, L. L. (1971). *Phys. Rev.* **B17:** 4889.
FOO, E.-N., and HOPFIELD, J. J. (1968). *Phys. Rev.* **173:** 635.
FRAGA, S., and RANSIL, B. J. (1961). *J. Chem. Phys.* **31:** 727.
FRANZ, G., FREYLAND, W., and HENSEL, F. (1980). *J. Phys. (Paris) Coll.* **41:** C8–70.
FRANZ, J. (1984). *Phys. Rev.* **B29:** 1565.
FRANZ, J. R. (1986). *Phys. Rev. Lett.* **57:** 889.
FREEMAN, A. J. (1983). *J. Magn. & Magn. Mater.* **35:** 31.
FREYLAND, W. (1979). *Phys. Rev.* **B20:** 5104.
FREYLAND, W. (1980). *J. Phys. (Paris) Coll.* **41:** C8–74.
FREYLAND, W. (1981). *Commun. Solid State Phys.* **10:** 1.
FREYLAND, W., and HENSEL, F. (1985). In *The Metallic and Nonmetallic States of Matter,* Edwards, P. O., and ROO, C. N. R. (eds.), London: Taylor and Francis, p. 93.

FRIEDEL, J. (1958). *Suppl. Nuovo Cimento* **7:** 287.
FRIEDEL, J., and SAYERS, C. M. (1977a). *J. Phys. (Paris)* **38:** L263.
FRIEDEL, J., and SAYERS, C. M. (1977b). *J. Phys. (Paris)* **38:** 697.
FRIEDEL, J., and SAYERS, C. M. (1978). *J. Phys. (Paris)* **39:** L59.
FRIEDMAN, L. R., and TUNSTALL, D. P. (1978). *The Metal–Nonmetal Transition in Disordered Systems,* Edinburgh: Scottish Universities Summer School in Physics.
GADRE, S. R., TOSHIKATSU, K., and CHAKRAVORTY, S. J. (1987). *Phys. Rev.* **A36:** 4155.
GALLI, G., and PARRINELLO, M. (1991). In *Computer Simulations in Materials Science,* Meyer, M., and Pontikis, V. (eds.), Dordrecht: Kluwer, p. 282.
GANN, V. V. (1980). *J. Nucl. Mater.* **90:** 144.
GASKELL, T. (1958). *Proc. Phys. Soc.* **72:** 685.
GASKELL, T. (1962). *Proc. Phys. Soc.* **80:** 1091.
GASPARI, G. D., and GYORFFY, B. L. (1972). *Phys. Rev. Lett.* **28:** 801.
GEBHARD, F., and VOLLHARDT, D. (1987). *Phys. Rev. Lett.* **59:** 1472.
GELDART, D. J. W., and RASOLT, M. (1976). *Phys. Rev.* **B13:** 1477.
GELDART, D. J. W., and TAYLOR, R. (1970). *Can. J. Phys.* **48:** 155, 167.
GELDART, D. J. W., and VOSKO, S. H. (1966).
GELL MANN, M., and BRUECKNER, K. A. (1957). *Phys. Rev.* **106:** 364.
GERJUOY, E. (1965). *J. Math. Phys.* **6:** 993.
GHOSH, S. K., and DEB, B. M. (1982). *Phys. Rep.* **92:** 1.
GHOSH, S. K., and SAMANTA, A. (1991). *J. Chem. Phys.* **94:** 517.
GILBERT, T. L. (1954). *Molecular Orbitals,* Pullman, B., and Löwdin, P. O. (eds.), New York: Academic.
GILLON, B., BECKER, P., and ELLINGER, U. (1983). *Molecular Phys.* **48:** 763.
GIRVIN, S. M., and JONSON, M. (1980). *Phys. Rev.* **B22:** 3583.
GIRVIN, S. H., MACDONALD, A. H., and PLATZMAN, P. M. (1985). *Phys. Rev. Lett.* **54:** 581.
GIRVIN, S. M., MACDONALD, A. H., and PLATZMAN, P. M. (1986). *Phys. Rev.* **B33:** 2481.
GLATTLI, D. C. (1990). *Surf. Sci.* **229:** 344.
GODANITZ, R. J., and ALDRICHS, R. (1988). *Chem. Phys. Lett.* **143:** 413.
GODBY, R. W., SCHLÜTER, M., and SHAM, L. J. (1988). *Phys. Rev.* **B37:** 10159.
GOGOLIN, A. A. (1988). *Phys. Rev.* **5:** 269.
GOODGAME, M. M., and GODDARD, W. A. (1981). *Phys. Rev. Lett.* **48:** 135.
GOULD, M. D., and PALDUS, J. (1990). *J. Chem. Phys.* **92:** 7394.
GOULD, M. D., PALDUS, J., and CHANDLER, G. S. (1990). *J. Chem. Phys.* **93:** 4142.
GRASSI, A., LOMBARDO, G., PUCCI, R., and MARCH, N. H. (1994). *Molecular Physics* **81:** 1265.
GREENFIELD, A., WELLENDORF, J., and WISER, N. (1971). *Phys. Rev.* **A4:** 1607.
GROS, G., JOYNT, R., and RICE, T. M. (1987a). *Phys. Rev.* **B36:** 381.
GROS, G., RICE, T. M., and JOYNT, R. (1987b). *Z. Phys.* **B68:** 425.
GROSS, E. K. U., and KOHN, W. (1985). *Phys. Rev. Lett.* **55:** 2850.
GRUNER, G. (1983). *Comm. Solid State Phys.* **10:** 183.
GUINEA, F., and ZIMANLYI, G. (1993). *Phys. Rev.* **B47:** 501.
GUNNARSSON, O. (1976). *J. Phys.* **F6:** 587.
GUTZWILLER, M. C. (1963). *Phys. Rev. Lett.* **10:** 159.
GUTZWILLER, M. C. (1964). *Phys. Rev.* **A134:** 923.
GUTZWILLER, M. C. (1965). *Phys. Rev.* **A137:** 1726.
GYGI, F., and BALDERESCHI, A. (1989). *Phys. Rev. Lett.* **62:** 2160.
HALPERIN, B. ET AL. (1984). *Phys. Rev. Lett.* **51:** 2302.
HALPERIN, B. ET AL. (1986). *Sci. Am.* **254:** 40.
HANSEN, J. P. (1972). *Phys. Lett.* **A41:** 213.
HANSEN, J. P. (1973). *Phys. Rev.* **A8,** 3096: see also Pollock, E. L. and Hansen, J. P. (1973) *Phys. Rev.* **A8:** 3110.

HARBOLA, M. K., and SAHNI, V. (1989). *Phys. Rev. Lett.* **62:** 489.
HASEGAWA, H. (1979). *Solid State Commun.* **31:** 597.
HASEGAWA, J. (1979). *J. Phys. Soc. Jpn.* **46:** 1504.
HASEGAWA, J. (1980a). *Solid State Phys.* **35:** 1.
HASEGAWA, J. (1980b). *J. Phys. Soc. Jpn.* **49:** 963.
HAWKE, P. S., ET AL. (1978). *Phys. Rev. Lett.* **41:** 994.
HEBBORN, J. E., and MARCH, N. H. (1970). *Adv. Phys.* **19:** 175.
HEDIN, L. (1965). *Phys. Rev.* **139:** A796.
HEDIN, L., and LUNDQVIST, S. (1969). *Solid State Phys.* **23:** 1.
HEINE, V., SAMSON, J. R., and NEX, C. M. M. (1981). *J. Phys.* **F11:** 2645.
HEITLER, W., and LONDON, F. (1927). *Zeits fü Physik* **44:** 455.
HENSEL, F., and UCHTMANN, H. (1989). *Ann. Rev. Phys. Chem.* **40:** 61.
HENSEL, F. (1990). *J. Non-Cryst. Solids* **117:** 441.
HENSEL, F., WINTER, R., and BODENSTEINER, T. (1989). *High Pressure Res.* **1:** 1.
HERMAN, F., and MARCH, N. H. (1984). *Solid State Commun.* **50:** 725.
HERMAN, F., VAN DYKE, J. P., and ORTENBURGER, I. B. (1969). *Phys. Rev. Lett.* **22:** 807.
HERRICK, D. R., and SINANOGLU, O. (1975). *Phys. Rev.* **A11:** 97.
HERRICK, D. R., and STILLINGER, F. H. (1966). *J. Chem. Phys.* **45:** 3623.
HERRICK, D. R., and STILLINGER, F. H. (1975). *Phys. Rev.* **A11:** 42.
HERSCHBACH, D. (1986). *J. Chem. Phys.* **84:** 838.
HILLER, J., SUCHER, J., and FEINBERG, G. (1978). *Phys. Rev.* **A18:** 2399.
HIRAKAWA, K. ET AL. (1985). *J. Phys. Soc. (Japan)* **54:** 3526.
HIRSCH, J. E. (1983). *Phys. Rev.* **B28:** 4059; *Phys. Rev. Lett.* **51:** 296.
HIRSCH, J. E. (1985). *Phys. Rev. Lett.* **54:** 1317.
HIRSCH, J. E., and SCALAPINO, D. J. (1983). *Phys. Rev. Lett.* **50:** 1169.
HIRSCH, J. E., and TANG, S. (1989). *Phys. Rev. Lett.* **62:** 591.
HLUBINA, R. (1994). *Phys. Rev.* **B50:** 8252.
HODGES, C. H. (1973). *Can. J. Phys.* **51:** 1428.
HOFFMANN-OSTENHOF, M., and HOFFMANN-OSTENHOF, T. (1977). *Phys. Rev.* **A16:** 1782.
HOFFMANN, M. R., and SIMONS, J. (1988). *J. Chem. Phys.* **88:** 993.
HOHENBERG, P. C., and KOHN, W. (1964). *Phys. Rev.* **136:** B864.
HOHL, D., NATOLI, V., CEPERLEY, D. M., and MARTIN, R. M. (1993). *Phys. Rev. Lett.* **71:** 541.
HOLAS, A. (1993). *ICTP Trieste:* lecture notes, and private communication.
HOLAS, A., and MARCH, N. H. (1987). *Phys. Chem. Liquids* **17:** 215.
HOLAS, A., and MARCH, N. H. (1991). *Phys. Lett.* **157:** 160.
HOLAS, A., and MARCH, N. H. (1994). *J. Molecular Structure (Theochem)* (in press).
HOLAS, A., and MARCH, N. H. (1995). *Phys. Rev.* **A51:** 2040: also *Paris DFT Conf* (to appear).
HOLDEN, A. J., and YOU, M. V. (1982). *J. Phys.* **F12:** 195.
HOLLISTER, C., and SINANOGLU, O. (1965). *J. Amer. Chem. Soc.* **88:** 13.
HORSCH, P. (1981). *Trans. Am. Nucl. Soc.* **34:** 442.
HORSCH, P., and FULDE, P. (1979). *Z. Phys.* **B36:** 23.
HORSCH, P., and KAPLAN, T. (1983). *J. Phys.* **C10:** L1203.
HOUGHTON, A., and MARSTEN, J. B. (1993). *Phys. Rev.* **B48:** 7790.
HOUSTON, J. E., and RYE, R. R. (1983). *Comm. Solid State Phys.* **10:** 233.
HUANG, C., MORIARTY, J. A., and SHER, A. (1976). *Phys. Rev.* **B1:** 2539.
HUANG, K. (1948). *Proc. Phys. Soc.* **60:** 161.
HUBBARD, J. (1957). *Proc. Roy. Soc.* **A240:** 539, ibid **A243:** 336.
HUBBARD, J. (1959). *Phys. Rev. Lett.* **3:** 77.
HUBBARD, J. (1963). *Proc. R. Soc.* **A276:** 238.
HUBBARD, J. (1964). *J. Phys. C.* **2:** 1222.
HUBBARD, J. (1969). *Phys. Rev.* **B19:** 2626.

HUBBARD, J. (1979a). *Phys. Rev.* **B19:** 2626.
HUBBARD, J. (1979b). *Phys. Rev.* **B20:** 4584.
HUBBARD, J. (1981a). *Electron Correlation and Magnetism in Narrow-Band Systems,* Moriya, T. (ed.), Berlin: Springer.
HUBBARD, J. (1981b). *Phys. Rev.* **23:** 5974.
HUNTER, G., and MARCH, N. H. (1989). *Int. J. Quantum Chem.* **35:** 649.
HYBERTSON, M. S., and LOUIE, S. G. (1986). *Phys. Rev.* **B34:** 5390, ibid **B32:** 7005.
HYLLERAAS, E. (1930). *Z. Phys.* **65:** 209.
ICHIMARU, S. (1982). *Rev. Mod. Phys.* **54:** 1017.
INKSON, J. (1972). *J. Phys. C.* **5:** 2599.
INKSON, J., and BENNETT, M. (1977). *J. Phys. C.* **10:** 987.
IOFFE, L. B., KALMEYER, V., and WEIGMANN, P. B. (1991). *PHYS. REV.* **B43:** 1219.
ISIHARA, A. (1989). *Solid State Phys.* **42:** 271.
JAIN, J. K. (1989). *Phys. Rev. Lett.* **63:** 199.
JAIN, S. C. (1991). *Phys. Rev.* **B41:** 7653.
JANCOVICI, B. (1977). *J. Stat. Phys.* **17:** 357.
JANKOWSKI, K., and MALINOWSKI, P. (1993). *Int. J. Quantum Chem.* **48:** 59.
JOHANNSSON, B., and BERGGREN, K. E. (1969). *Phys. Rev.* **181:** 855.
JOHNSON, M. D., and CANRIGHT, G. S. (1990). *Phys. Rev.* **B41:** 6870.
JOHNSON, M. D., and CANRIGHT, G. S. (1991). *Comm. Solid State Phys.* **15:** 77.
JOHNSON, M. D., and MARCH, N. H. (1963). *Phys. Lett.* **3:** 313.
JOHNSON, M. W., MARCH, N. H., PERROT, F., and RAY, A. K. (1994). *Phil. Mag.* **B69:** 965.
JOHNSON, R. A. (1989). *Phys. Rev. B.* **39:** 12554.
JONES, W., and MARCH, N. H. (1986). *Theoretical Solid-State Physics, Volumes 1 and 2,* New York: Dover.
JOO, J., PRIGODIN, V. N., MIN, Y. G., MACDIARMID, A. G., and EPSTEIN, A. J. (1994). *Phys. Rev.* **B50:** 12226.
JORGENSEN, J. D., SCHUTTLER, H.-B., HINKS, D. G., CAPONE, D. W., ZHANG, K., and BRODSKY, M. B. (1987). *Phys. Rev. Lett.* **58:** 1024.
KAIS, S., SUNG, S. M., and HERSCHBACH, D. R. (1993). *J. Chem. Phys.* **99:** 5184.
KAJZAR, F., and FRIEDEL, J. (1987). *Phys. Rev.* **B35:** 9614.
KALMAN, G., KEMPA, K., and MINELLA, M. (1991). *Phys. Rev.* **B43:** 14238.
KALOS, M. H., LEVESQUE, D., and VERKET, L. *PHYS. REV.* **A9:** 2178.
KALOS, M. H. (1984). NATO ASI Ser., Ser C, 19.
KAMIMURA, H. (1980). *Phil. Mag.* **B42:** 763.
KAMIMURA, H., and KANEHISA, B. (1972). *Solid State Commun.* **28:** 127.
KANAMORI, J. (1963). *Prog. Theor. Phys.* **30:** 275.
KANE, E. O. (1971). *Phys. Rev.* **B4:** 1910.
KANE, E. O. (1972). *Phys. Rev.* **B5:** 1493.
KAPLAN, T. A., HORSCH, P., and BORYSOWICZ, J. (1987). *Phys. Rev.* **B35:** 1877.
KAPLAN, T. A., HORSCH, P., and FULDE, P. (1982). *Phys. Rev. Lett.* **49:** 889.
KATO, T. (1957). *Commun. Pure. Appl. Math.* **10:** 151.
KAWAKAMI, N., and OKIJI, A. (1981). *Phys. Lett.* **86A:** 483.
KHAN, M. A., and CALLAWAY, J. (1980). *Phys. Lett.* **A76:** 441.
KIEL, B., STOLLHOFF, G., WEIGEL, C., FULDE, P., and STOLL, H. (1982). *Z. Phys.* **B46:** 1.
KILIC, S. (1970). *Fizika (Yugoslavia)* **2:** 105.
KILIC, S. (1972). *Fizika (Yugoslavia)* **4:** 195.
KIM, Y. H., and HEEGER, A. J. (1987). *Phys. Rev.* **B36:** 7252.
KIM, Y. H., and HEEGER, A. J. (1989). *Phys. Rev.* **B40:** 8393.
KIMBALL, J. C. (1973). *Phys. Rev.* **A7:** 1648.
KIMBALL, J. C. (1975). *J. Phys.* **A8:** 1513.

KIRZNITS, D. A. (1957). *Sov. Phys. JETP* **5:** 64.
KITTEL, C. (1963). *Quantum Theory of Solids,* New York: Wiley, p. 407.
KIVELSON, S., and HEEGER, A. J. (1987). *Synth. Met.* **17:** 183.
KIVELSON, S., and HEEGER, A. J. (1988). *Synth. Met.* **22:** 371.
KIVELSON, S., and HEIM, D. E. (1982). *Phys. Rev.* **B26:** 4278.
KIVELSON, S., SU, W.-P., SCHRIEFFER, J. R., and HEEGER, A. J. (1988). *Phys. Rev. Lett.* **58:** 1899.
KLEIN, D. J., ALEXANDER, S. A., SEITZ, W. A., SCHMALZ, T. G., and HITE, G. E. (1986). *Theor. Chim. Acta.* **69:** 393.
KLEIN, D. J., MARCH, N. H., and ALEXANDER, S. A. (1995). (to appear).
KLEIN, D. J., and PICKETT, H. M. (1976). *J. Chem. Phys.* **64:** 4811.
KLEIN, D. J., and SEITZ, W. A. (1980). *Int. J. Quantum Chem. Symp. 13,* **293:** (1974) *Phys. Rev.* **B10:** 3317.
KLEPPMANN, O., and ELLIOTT, R. J. (1975). *J. Phys.* **C8:** 2747.
KNOPP, K. (1945). *The Theory of Functions* (New York: Dover) Pt 1, pp. 92–111: 1947, ibid Part II, pp. 93–118.
KOBAYASKI, H., OHASHI, Y., MARUMO, F., and SAITO, Y. (1970). *Acta Cryst.* **B26:** 459.
KOHANOFF, J., and HANSEN, J.-P. (1995). *Phys. Rev. Lett.* **74:** 626.
KOHLER H. ET AL. (1986). *J. Phys.* **C19:** 5215.
KOHN, W., and SHAM, L. J. (1965). *Phys. Rev.* **140:** A1133.
KOHNO, H. and YAMADA, K. (1991). *Prog. Theor. Phys.* **85:** 13.
KOONIN, S. E., SUGIYAMA, G., and FRIEDRICH, H. (1982). In *Time-Dependent Hartree–Fock and Beyond,* Goeke, K., and Reinhard, P. G. (eds.), Berlin: Springer-Verlag, p. 214.
KOSTERLITZ, J. M., and THOULESS, D. J. (1973). *J. Phys.* **C6:** 1181.
KOSTERLITZ, J. M., and THOULESS, D. J. (1978). *Prog. Low. Temp. Phys.* **B7:** 371.
KOTLIAR, G., and RUCKENSTEIN, A. E. (1986). *Phys. Rev. Lett.* **57:** 1362.
KRISHNA-MURTHY, H. R., WILKINS, J. W., and WILSON, K. G. (1980). *Phys. Rev.* **B21:** 1003, 1044.
KRÖTSCHECK, E. (1995). In *Proc. Valencia Conf.* New York: Nova, in press.
KRYASCHKO, E. S., and LUDENA, E. (1990). *Energy density functional theory of many-electron systems* (Dordrecht: Kluwer).
KUBO, R. (1962). *J. Phys. Soc. Jpn.* **17:** 1100.
KUGLER, A. A. (1969). *Ann. Phys.* **53:** 133.
KÜMMEL, H., LUHRMANN, K. H., and ZABOLITSKY, M. (1978). *Phys. Rev.* **C36:** 1.
LAIDIG, W. D., and BARTLETT, R. J. (1984). *Chem. Phys. Lett.* **104:** 424.
LAIDIG, W. D., SAXE, P., and BARTLETT, R. J. (1987). *J. Chem. Phys.* **86:** 887.
LAM, P. K., and GIRVIN, S. M. (1984). *Phys. Rev.* **B30:** 473.
LAMING, G. J., NAGY, A., HANDY, N. C., and MARCH, N. H. (1994). *Molecular Phys.* **81:** 1497.
LANGLINAIS, J., and CALLAWAY, J. (1972). *Phys. Rev.* **B5:** 124.
LANGRETH, D. C., and MEHL, M. J. (1981). *Phys. Rev. Lett.* **47:** 446.
LANGRETH, D. C., and MEHL, M. J. (1983). *Phys. Rev.* **B28:** 1809.
LANGRETH, D. C., and PERDEW, J. P. (1975). *Solid State Commun.* **17:** 1425.
LANGRETH, D. C., and PERDEW, J. P. (1977). *Phys. Rev.* **B15:** 2884.
LANGRETH, D. C., and PERDEW, J. P. (1979). *Solid State Commun.* **31:** 567.
LANNOO, M., SCHLÜTER, M., and SHAM, L. J. (1985). *Phys. Rev.* **B32:** 3890.
LANTTO, L. J., and SIEMENS, P. J. (1979). *Nucl. Phys.* **A317:** 55.
LARSSON, S. (1981). *Chem. Phys. Lett.* **77:** 176.
LAUGHLIN, R. B. (1983). *Phys. Rev. Lett.* **50:** 1395.
LAUGHLIN, R. B. (1988). *Phys. Rev. Lett.* **60:** 2577.
LAWLEY, K. P. (1987). *Adv. Chem. Phys.* **67:** 69; *Ab Initio Methods in Quantum Chemistry.*
LAYZER, A. J. (1963). *Phys. Rev.* **129:** 897.
LAYZER, D. (1959). *Ann. Phys.* **8:** 271.
LEA, M. J., and MARCH, N. H. (1989). *Int. J. Quantum Chem. Symp.* **23:** 717.

LEA, M. J., and MARCH, N. H. (1991) : see Blackman and Taguena.
LEA, M. J., and MARCH, N. H. (1992). *J. Phys. Condensed Matter* **3:** 3493.
LEA, M. J., MARCH, N. H., and SUNG, W. (1992). *J. Phys. Condensed Matter* **4:** 5263.
LEA, M. J. ET AL. (1991). *Phys. Chem. Liquids.* **23:** 115.
LEE H. C. (1990). *Mod. Phys. Lett.* **B4:** 863.
LEE, C., and GHOSH, S. K. (1986). *Phys. Rev.* **A33:** 3506.
LEE, C., YANG, W., and PARR, R. G. (1988). *Phys. Rev.* **B37:** 785.
LEE, C., and GHOSH, S. K. (1986). *Phys. Rev.* **A33:** 3506.
LENNARD-JONES, J. E. (1937). *Proc. Roy. Soc.* **A158:** 280.
LEVESQUE, D., WEIS, J. J., and REATTO, L. (1985). *Phys. Rev.* **A33:** 3451.
LEVY, M. (1979). *Proc. Natl. Acad. Sci.* **76:** 6062.
LEVY, M., and GÖRLING, A. (1994). *Phil. Mag.* **B69:** 763.
LIDIARD, A. B. (1951). *Proc. Phys. Soc.* **A64:** 490.
LIDIARD, A. B., MARCH, N. H., and DONOVAN, B. (1956). *Proc. Phys. Soc.* **A68:** 112.
LIEB, E. H., and WU, F. Y. (1968). *Phys. Rev. Lett.* **20:** 1445.
LINDGREN, I., and LUNDQVIST, S. (1980). *Nobel Symposium* (Göteborg).
LINDGREN, I., and MORRISON, J. (1982). *Atomic Many-Body Theory,* Berlin: Springer, p. 376.
LIPPINARI, E., STRINGARI, S., and TAKAYANAGI, K. (1994). *J. Phys. Condensed Matter* **6:** 2025.
LOGAN, D. E. (1991). *J. Chem. Phys.* **94:** 628.
LOGAN, D. E., and SIRINGO, F. (1992). *J. Phys. Condensed Matter* **4:** 3695.
LOGAN, D. E., and WINN, M. D. (1988). *J. Phys.* **C21:** 5773.
LOGAN, D. E., and WOLYNES, P. G. (1986). *J. Chem. Phys.* **85:** 937.
LOUBEYRE P. (1982). *Phys. Lett.* **80A:** 181.
LOWDE, R. D., MOON, R. M., PAGONIS, B., PERRY, C. H., SOKOLOFF, J. B., VAUGHAN-WATKINS, WILTSHIRE, M. C. K., and CRANGLE, J. (1983). *J. Phys.* **F13:** 249.
LÖWDIN, P. O. (1955). *Phys. Rev.* **80:** 110.
LUNDQVIST, B. I. (1967). *Phys. Cond. Mat.* **6:** 193 and 206.
LUNDQVIST, S. (1983). in Lundqvist and March (1983).
LUNDQVIST, S., and MARCH, N. H. (eds.) (1983). *Theory of the Inhomogeneous Electron Gas,* New York: Plenum.
LUTHER, A. (1994). *Phys. Rev. B* **50:** 11446.
LUTTINGER, J. M. (1961). *Phys. Rev.* **121:** 942.
LUTTINGER, J. M. (1963). *J. Math. Phys.* **4:** 1154.
MA, S. K. (1985). *Statistical Mechanics,* Singapore: World Scientific.
MA, S. K., and BRUECKNER, K. A. (1968). *Phys. Rev.* **165:** 18.
MCWEENY, R. (1989). *Methods of Molecular Quantum Mechanics,* New York: Academic, p. 130, 379.
MCFEELY, F. R., KOWALCZK, S. P., LEY, L., and SHIRLEY, D. A. (1974). *Solid State Commun.* **15:** 1051.
MAHAN, G. D. (1990). *Many-Particle Physics, 2nd Ed.* New York: Plenum.
MAHAN, G. D. (1994). *Comments on Condensed Matter Physics* **16:** 333.
MANSFIELD, R. (1971). *J. Phys.* **C4:** 2084.
MAO H., and HEMLEY, R. J. (1989). *Science.* **244:** 1462.
MAO H., and HEMLEY, R. J. (1994). *Rev. Mod. Phys.* **66:** 671.
MAO H., JEPHCOAT, A. P., HEMLEY, R. J., FINGER, L. W., ZHA, C. S., HAZEN, R. M., and COX, D. E. (1988). *Science* **239:** 1131.
MARCH, N. H. (1957). *Adv. Phys.* **6:** 1.
MARCH. N. H. (1958). *Phys. Rev.* **110:** 604.
MARCH, N. H. (1972). *Phys. Lett.* **39A:** 150.
MARCH, N. H. (1975a). *Self-Consistent Fields in Atoms,* Oxford: Pergamon.
MARCH, N. H. (1975b). *Phil. Mag.* **32:** 497.
MARCH, N. H. (1981). *Mathematical Methods in Superfluids,* Edinburgh: Oliver and Boyd.

March, N. H. (1982). *NATO Adv. Study Inst.,* New York: Plenum. Ed. Devreese, J. T.
March, N. H. (1983). *Phys. Lett.* **A84:** 319.
March, N. H. (1985a). *Phys. Lett.* **108A:** 368.
March, N. H. (1985b). *Phys. Lett.* **A113:** 66.
March, N. H. (1986). *Phys. Lett.* **A113:** 476.
March, N. H. (1987). *Chemical Physics of Liquids,* New York: Gordon and Breach.
March, N. H. (1988). *Phys. Rev.* **A37:** 4526.
March, N. H. (1987). *Phys. Rev.* **A36:** 5077.
March, N. H. (1990). *J. Molecular Structure (Theochem)* **188:** 261.
March, N. H. (1990). *Liquid Metals: Concepts and Theory,* Cambridge: University Press.
March, N. H. (1991). *J. Phys.* **B24:** 4123: *Phys. Chem. Liquids* **22:** 191.
March, N. H. (1992). *J. Math. Chem.* **4:** 271.
March, N. H. (1993). *Phys. Chem. Liquids.* **25:** 65.
March, N. H. (1993). *J. Phys. Condensed Matter Suppl.* **34B:** B149.
March, N. H., and Bader, R. F. W. (1980). *Phys. Lett.* **78A:** 242.
March, N. H., and Cizek, J. (1987). *Int. J. Quantum Chem.* **33:** 301.
March, N. H., and Donovan, B. (1954). *Proc. Phys. Soc.* **A67:** 500.
March, N. H., Gidopoulos, N., Theophilou, A. K., and Sung, W. (1993). *Phys. Chem. Liquids* **26:** 135.
March, N. H., and Mucci, J. F. (1983). *J. Chem. Phys.* **78:** 6178.
March, N. H., and Mucci, J. F. (1992). *Chemical Physics of Free Molecules,* New York: Plenum.
March, N. H., and Murray, A. M. (1960a). *Phys. Rev.* **120:** 830.
March, N. H., and Murray, A. M. (1960b). *Proc. R. Soc.* **A256:** 300.
March, N. H., and Nagy, A. (1992). *J. Molecular Structure (Theochem).* **129:** 1992.
March, N. H., and Paranjape, B. V. (1995). *J. Phys. Chem. Solids* (in press).
March, N. H., and Parrinello, M. (1982). *Collective Effects in Solids and Liquids,* Bristol: Adam Hilger.
March, N. H., and Pucci, R. (1982). *J. Chem. Phys.* **75:** 497.
March, N. H., and Sampanthar S. (1964). *Acta Physica Hungarica* **8:** 502.
March, N. H., and Stoddart, J. C. (1968). *Rep. Prog. Phys.* **31:** 533.
March, N. H., Suzuki, M., and Parrinello, M. (1979). *Phys. Rev.* **B19:** 2027.
March, N. H., and Tosi, M. P. (1972). *Proc. R. Soc.* **A330:** 373.
March, N. H., and Tosi, M. P. (1973). *Phil. Mag.* **28:** 91.
March, N. H., and Tosi, M. P. (1985). *J. Phys. A.* **18:** L 643.
March, N. H., and Tosi, M. P. (1985a). *Phys. Chem. Liquids* **14:** 303.
March, N. H., and Tosi, M. P. (1985b). *Atomic Dynamics in Liquids,* New York: Dover.
March, N. H., and Tosi, M. P. (1973). *Ann. Phys.* (N.Y.) **81:** 414.
March, N. H., and Tosi, M. P. (1995). *Phil. Mag.* (in press).
March, N. H., and Tosi, M. P. (1984). *Coulomb Liquids,* New York: Academic.
March, N. H., and Tosi, M. P. (1995). *Adv. Phys.* (to appear).
March, N. H., and Wind, P. (1992). *Mol. Physics* **77:** 791.
March, N. H., Young, W. H., and Sampanthar, S. (1995). *The Many-Body Problem in Quantum Mechanics,* New York: Dover.
Matsubara T., and Toyozawa, Y. (1961). *Prog. Theor. Phys.* **26:** 739.
Matthiess, L. F. (1961). *Phys. Rev.* **123:** 1209.
Matthiess, L. F. (1987). *Phys. Rev. Lett.* **58:** 1028.
Mattis, D. C. (1987). *Phys. Rev.* **B36:** 745.
Mazumdar, S. (1987). *Phys. Rev.* **B36:** 7190.
Mazumdar, S., Lin, H. Q., and Campbell, D. K. (1991). *Synth. Met.* **41–43:** 4047.
Mentch, F., and Anderson, J. B. (1981). *J. Chem. Phys.* **74:** 6307.
Metzner, G. W., and Vollhardt, D. (1987). *Phys. Rev. Lett.* **59:** 121.

MEYER, W. (1971). *Int. J. Quantum Chem. Symp.* **5:** 341.
MIGLIO, L., TOSI, M. P., and MARCH, N. H. (1981). *Surf. Sci.* **111:** 119.
MILLS R. L., ET AL. (1979). *J. Appl. Phys.* **49:** 5502.
MILNE, E. A. (1927). *Proc. Camb. Phil. Soc.* **23:** 794.
MILOVANOVIC, M., SACHDEV, S., and BHATT, R. N. (1989). *Phys. Rev. Lett.* **63:** 82.
MIZES, H. A., and CONWELL, E. M. (1993). *Phys. Rev. Lett.* **70:** 1505.
MORAES, F., CLEN, J., CHUNG, T.-C., and HEEGER, A. J. (1985). *Synth. Met.* **11:** 271.
MOREO, A., SCALAPINO, D. J., SUGAR, R. L., WHITE, S. R., and BICKERS, N. E. (1990). *Phys. Rev.* **B41:** 2313.
MORI, H. (1970). *Phys. Rev.* **B43:** 5474.
MOSKOWITZ, J. W., and KALOS, M. H. (1981). *Int. J. Quantum Chem.* **20:** 1107.
MOSKOWITZ, J. W., and SCHMIDT, K. E. (1986). *J. Chem. Phys.* **85:** 2868.
MOSKOWITZ, J. W., SCHMIDT, K. E., LEE, M. A., and KALOS, M. H. (1982a). *J. Chem. Phys.* **76:** 1064.
MOSKOWITZ, J. W., SCHMIDT, K. E., LEE, M. A., and KALOS, M. H. (1982b). *J. Chem. Phys.* **77:** 349.
MOTT, N. F. (1966). *Adv. Phys.* **13:** 325.
MOTT, N. F. (1974). *Metal-Insulator Transitions,* London: Taylor and Francis.
MOTT, N. F., and KAVEH, M. (1985). *Adv. Phys.* **34:** 329.
MUCCI, J. F., and MARCH, N. H. (1983). *J. Chem. Phys.* **78:** 6178.
MURRAY, C. W., HANDY, N. C., and LAMING, G. J. (1993). *Mol. Physics* **78:** 997.
MURRAY, C. W., LAMING, G. J., HANDY, N. C., and AMOS, R. D. (1992). *Chem. Phys. Lett.* **199:** 551.
MURRELL, J. N. (1971). *Electronic Spectra of Organic Molecules*, London: Chapman and Hall.
NAGARA, H., NAGATA, Y., and NAKAMURA, T. (1987). *Phys. Rev.* **A36:** 1859.
NAGAOKA, Y. (1966). *Phys. Rev.* **147:** 392.
NAGASAWA, H., and MURAYAMA, S. (1980). *J. Magn. Magn. Mater.* **15–18:** 93.
NAGY, A., and MARCH, N. H. (1989). *Phys. Rev.* **A39:** 5512.
NATOLI, V., MARTIN, R. M., and CEPERLEY, D. M. (1993). *Phys. Rev. Lett.* **70:** 1952.
NEGELE, J. W., and ORLAND, H. (1988). *Quantum Many-Particle Systems,* Redwood City, CA: Addison-Wesley, p. 332.
NEILSON, D., SWIERKOWSKI, H., SJOLANDER, A., and SZYMANSKI, A. (1991). *Phys. Rev.* **B44:** 6291.
NELDER, J. A., and MEAD, R. (1965). *Comput. J.* **7:** 308.
NELSON, D. R., and HALPERIN, B. (1979). *Phys. Rev.* **B19:** 2457.
NERSESYAN, Y., LUTHER, A., and KUSMARTSEV, M. (1993). *Phys. Rev.* **B12:** 3908.
NIKLASSON, G., SJOLANDER, A., and SINGWI, K. S. (1975). *Phys. Rev.* **B11:** 113.
NOGA, J., and BARTLETT, R. J. (1987). *J. Chem. Phys.* **86:** 7041.
NOGA, J., and BARTLETT, R. J. (1988). *J. Chem. Phys.* **89:** 3401 (E).
NOOIGEN, M., and SNIJDERS, J. G. (1993). *Int. J. Quantum Chem.* **48:** 15.
NOZIÈRES, P., and PINES, D. (1958). *Nuovo Cimento* **9:** 470.
OGAWA, T., KANADA, K., and MATSUBARA, T. (1975). *Prog. Theor. Phys.* **53:** 614.
OLES, A. M., PFIRSCH, F., and FULDE, P. (1986). *J. Chem. Phys.* **85:** 5183.
ORTIZ, G., and BALLONE, P. (1994). *Phys. Rev.* **B50:** 1391.
OVERHAUSER, A. W. (1985). *Int. J. Quantum Chem.* **S14:** and private communication.
PAALANEN, M. A., GRAEBNER, J., BHATT, R. N., and SACHDEV, S. (1988). *Phys. Rev. Lett.* **61:** 597.
PAINTER, G. S. (1981). *Phys. Rev.* **B24:** 4264.
PALDUS, J., CIZEK, J., and SHAVITT, I. (1972). *Phys. Rev.* **A5:** 50.
PARRINELLO, M., and MARCH, N. H. (1976). *J. Phys.* **C9:** L147.
PAULING, L. (1949). *The Nature of the Chemical Bond,* Ithaca: Cornell University Press.
PEIERLS, R. E. (1956). *Quantum Theory of Solids,* Oxford: University Press.
PERDEW, J. P., and LEVY, M. (1983). *Phys. Rev. Lett.* **51:** 1884.
PERDEW, J. P., and ZUNGER, A. (1981). *Phys. Rev.* **B23:** 5048.
PERERA, A., WATTS, J. D., and BARTLETT, R. J. (1994). *J. Chem. Phys.* **100:** 1425.

PERRIN, R., TAYLOR, R., and MARCH, N. H. (1975). *J. Phys. F.* **5:** 1490.
PERROT, F., and MARCH, N. H. (1990a). *Phys. Rev.* **A41:** 4521.
PERROT, F., and MARCH, N. H. (1990b). *Phys. Rev.* **A42:** 4884.
PEUCKERT, V. (1976). *J. Phys.* **C9:** 4173.
PICKETT, W. E. (1986). *Comm. Solid State Phys. (GB)* **12:** 57.
PICKETT, W. E., and WANG, C. S. (1984). *Phys. Rev.* **B30:** 4719.
PINES, D., and NOZIERES, P. (1960). *Physica* **26:** S103.
PIPPARD, A. B. (1966). *Classical Thermodynamics,* Cambridge: University Press.
PLATZMAN, P. M., and EISENBERGER (1974). *AIP CONF. PROC. (USA)* NO. **24:** 394.
PLAUT, A. S. (1990). *Phys. Rev.* **B42:** 5744.
POPLE, J. A., KRISHNAN, R., SCHLESEL, H. B., and BINKLEY, J. S. (1978). *Int. J. Quantum Chem.* **14:** 545.
POUGET, J. P., and COMES, R. (1989). *Charge Density Waves in Solids,* Gorkov, L. P., and Gruner, G. (eds.), Amsterdam: North-Holland, p. 85.
POUGET, J. P., KHANNA, S. K., DENOYER, F., COMES, R., GARITO, A. F., and HEEGER, A. J. (1976). *Phys. Rev. Lett.* **35:** 445.
PRIGODIN, V. N., and ROTH, S. (1993). *Synth. Met.* **53:** 237.
PUCCI, R., and MARCH, N. H. (1987). *Phys. Rev.* **A35:** 4428.
PUCCI, R., SIRINGO, F., and MARCH, N. H. (1988). *Phys. Rev.* **B38:** 9517.
QUINN, J. J., and FERRELL, R. A. (1958). *Phys. Rev.* **112:** 812.
RAMAKRISHNAN, T. V. (1985) : see Freyland and Hensel (1985).
RAMIREZ, R., FALICOV, L. M., and KIMBALL, J. C. (1970). *Phys. Rev.* **B2:** 3383.
REATTO, L. (1988). *Phil. Mag.* **A58:** 37.
REATTO, L., LEVESQUE, D., and WEIS, J. J. (1986). *Phys. Rev.* **A33:** 3451.
REHMUS, P., and BERRY, R. S. (1979). *Chem. Phys.* **38:** 257.
REHMUS, P., KELLMAN, M. E., and BERRY, R. S. (1978a). *Chem. Phys.* **31:** 239.
REHMUS, P., ROOTHAAN, C. C. J., and BERRY, R. S. (1978b). *Chem. Phys. Lett.* **58:** 321.
RESTA, R. (1977). *Phys. Rev.* **B16:** 2717.
REYNOLDS, P. J., CEPERLEY, D. M., ALDER, B. J., and LESTER, W. A. (1982). *J. Chem. Phys.* **77:** 559.
RICE, T. M. (1965). *Ann. Phys.* **31:** 100.
RICE, T. M., and JOYNT, G. (1987). *Int. Conf. Valence Fluctuations,* Bangalore.
RICE, T. M., UEDA, K., OTT, H. R., and RUDIGIER, A. (1985). *Phys. Rev.* **B31:** 594.
RITSKO, J. J., MELE, E. J., HEEGER, A. J., MACDIARMID, A. G., and OZAKI, M. (1980). *Phys. Rev. Lett.* **44:** 1351.
ROSE, J. H. (1980). *Phys. Rev.* **B23:** 552.
ROSENBERG, B. J., and SHAVITT, I. (1975). *J. Chem. Phys.* **63:** 2162.
ROSS, M. (1985). *Rep. Prog. Phys.* **48:** 1.
ROTH, L. M. (1974). *J. Phys. (Paris)* **35:** C4-317.
ROTH, L. M. (1976). *J. Phys.* **F6:** 2267.
ROTH, L. M. (1978). In *Transition Metals,* London: Inst. Phys., p. 473.
ROUSSEAU, J. S., STODDART, J. C., and MARCH, N. H. (1972). *Kyoto Conf. Liquid Metals:* (1971) *J. Phys.* **C5:** L175.
RUNGE, E., and GROSS, E. K. U. (1984). *Phys. Rev. Lett.* **52:** 997.
RUTTINK, P. J. A., VANLANTHE, J. H., ZWAANS, R., and GROENBOOM, G. C. (1991). *J. Chem. Phys.* **94:** 7212.
SAITO, G., and KAGASHIMA, S. (1990). *The Physics and Chemistry of Organic Superconductors,* Berlin: Springer.
SAMPSON, J. B., and SEITZ, F. (1940). *Phys. Rev.* **70:** 806.
SANCHEZ, J. M., and DE FONTAINE, D. (1981). *Phys. Rev.* **B25:** 1759.
SAVIN, A., STOLL, H., and PREUSS, H. (1986). *Theor. Chim. Acta* **70:** 407.
SAYERS, C. M. (1977). *J. Phys.* **F7:** 1157.

SAYERS, C. M. (1982). *Z. Phys.* **B46:** 131.
SAYERS, C. M., and KAJZAR, F. (1981). *J. Phys.* **F11:** 1055.
SCALAPINO, D. J., and SUGAR, R. L. (1981). *Phys. Rev.* **B24:** 4295.
SCHINNER, A. (1987). *Phys. Chem. Liquids.*
SCHMELTZER, D. (1992). *Phys. Rev. B.* **45:** 3168.
SCHMELTZER, D. (1993). *Phys. Rev. B.* **49:** .6944
SCHMIDT, K., KRAEFT, W. D., and MARCH, N. H. (1991). *Phys. Chem. Liquids* **24:** 103.
SCHMIDT, K. E., and KALOS, M. H. (1984). In *Application of the Monte Carlo Method,* Binder, K. (ed.), Berlin: Springer-Verlag.
SCHORK, T., and FULDE, P. (1992). *J. Chem. Phys.* **97:** 9195.
SCHULZ, H. J. (1987). *Europhys. Lett.* **4:** 609.
SCHULTZ, H. G. (1991). *Int. J. Mod. Phys.* **B5:** 57.
SCHWINGER, J. (1985). *Phys. Rev.* **A32:** 26.
SEITZ, F., and TURNBULL, D. (1970). *Solid State Phys.* **24.**
SENATORE, G., and MARCH, N. H. (1985). *J. Chem. Phys.* **83:** 1232.
SENATORE, G., and MARCH, N. H. (1994). *Rev. Mod. Phys.* **66:** 445.
SENATORE, G., and PASTORE, G. (1990). *Phys. Rev. Lett.* **64:** 303.
SEWELL, G. L. (1986). *Collective Phenomena,* Oxford: University Press.
SHAM, L. J., and SCHLÜTER, M. (1983). *Phys. Rev. Lett.* **51:** 1888.
SHAM, L. J. (1973). *Phys. Rev. Lett.* **31:** 631.
SHULTZ, L. (1991). *J. Non-Cryst. Solids* **130:** 273.
SIGAGALAS, M. M., and PAPACONSTANTOPOULOS, D. A. (1994). *Phys. Rev.* **B50:** 7255.
SILVERA, I. F. *Phys. Rev.* **B43:** 10191.
SIMCOX, L., and MARCH, N. H. (1962). *Proc. Phys. Soc.* **80:** 830.
SINANOGLU, O., and BRUECKNER, K. A. (1970). *J. Am. Chem. Soc.* **88:** 13.
SINGWI, K. S., and TOSI, M. P. (1981). *Solid State Phys.* **36:** 177.
SINGWI, K. S., TOSI, M. P., LAND, R. H., and SJOLANDER, A. (1968). *Phys. Rev.* **176:** 589.
SIRINGO, F. (1995). In *Disordered Structures,* Srivastava, S. K., and MARCH, N. H. (eds.), Singapore: World Scientific.
SIRINGO, F., LEA, M. J., and MARCH, N. H. (1991). *Phys. Chem. Liquids.* **23:** 115.
SIRINGO, F., and LOGAN, D. (1992). *J. Phys.* **3:** 4747.
SIRINGO, F., PUCCI, R., and MARCH, N. H. (1988a). *Phys. Rev.* **B37:** 2491.
SIRINGO, F., PUCCI, R., and MARCH, N. H. (1988b). *Phys. Rev.* **B38:** 9517: 9567.
SIRINGO, F., PUCCI, R., and MARCH, N. H. (1989). *High Pressure Research* **8:** 1.
SLATER, J. C. (1951). *Phys. Rev.* **81:** 385.
SLATTERY, W. L., and DE WITT, H. (1980). *Phys. Rev.* **A26:** 2255.
SOMERFORD, D. (1971). *J. Phys. C.* **4:** 1570.
SORELLA, S. (1989). Ph.D thesis (ISAS: Trieste).
SPALEK, J. ET AL. (1983), private communication.
STAFSTRÖM, S. (1991). *Phys. Rev.* **B43:** 9158.
STIEBLING, J., and RAETHER, H. (1978). *Phys. Rev. Lett.* **40:** 1293.
STISHOV, S. ET AL. (1976). *Phys. Lett.* **59A:** 148.
STODDART, J. C. (1975). *J. Phys.* **C8:** 3391.
STODDART, J. C., BEATTIE, A. M., and MARCH, N. H. (1971). *Int. J. Quantum Chem. Symp.* **4:** 35.
STODDART, J. C., and MARCH, N. H. (1971). *Ann. Phys.* **64:** 174.
STODDART, J. C., WIID, D., and MARCH, N. H. (1972). *Int. J. Quantum Chem. Symp.* **5:** 745.
STODDART, J. C., ORTENBURGER, I. B., and MARCH, N. H. (1974). *Nuovo Cimento.* **23B:** 15.
STONER, E. C. (1938). *Proc. R. Soc.* **A154:** 656.
STONER, E. C. (1939). *Proc. R. Soc.* **A169:** 339.
STORMER, H. ET AL. (1980). *Phys. Rev.* **B21:** 1589.

STOLLHOFF, G., and BOHNEN, K. P. (1988). *Phys. Rev.* **B37:** 4678.
STOLLHOFF, G., and FULDE, P. (1977). *Z. Phys.* **B26:** 257.
STOLLHOFF, G., and FULDE, P. (1978). *Z. Phys.* **B29:** 231.
STOLLHOFF, G., and FULDE, P. (1980). *J. Chem. Phys.* **73:** 4548.
STRATONOVICH, R. D. (1958). *Sov. Phys. Dokl.* **2:** 416.
STRATT, R. M.. (1990). *Ann. Rev. Phys. Chem.* **41:** 175.
STRATT, R. M., and XU, B. C. (1989). *Phys. Rev. Lett.* **62:** 1675.
STRINGARI, S. (1995). In *Proc. Valencia Conf.,* New York: Nova (in press).
STURM, K. (1982). *Adv. Phys.* **31:** 1.
STURM, K. (1983). *Solid State Commun.* **48:** 29
STURM, K., and OLIVEIRA, L. E. (1981). *Phys. Rev.* **B24:** 3054.
STURM, K., ZAREMBA, E., and NUROH, K. (1990). *Phys. Rev.* **B42:** 6973.
SU, W. P., SCHRIEFFER, J. R., and HEEGER, A. J. (1980). *Phys. Rev.* **B22:** 2099, 171.
SUGIYAMA, G., and KOONIN, S. E. (1986). *Ann. Phys.* **168:** 1.
SZABO, N. (1972). *J. Phys.* **C5:** L241.
SZALAY, P. G., and BARTLETT, R. J. (1994). *J. Chem. Phys.* **101:** 4936.
TAKAHASHI, M. (1977). *J. Phys.* **C10:** 1289.
TAKAYAMA, H., LIN-LIU, Y. R., and MAKI, K. (1980). *Phys. Rev.* **B21:** 2388.
TAKEMURA, K. MINOMURA, S., SHIMOMURA, O., FUJII, Y., and AXE, J. D. (1982). *Phys. Rev.* **B26:** 998.
TAMAKI, S. (1987). *Can. J. Phys.* **65:** 286.
TANG, S., and HIRSCH, J. E. (1988). *Phys. Rev.* **B37:** 9546.
TAO, Y. K. (1990). *Physica C* **165:** 13.
TAUT, M. (1993). *Phys. Rev.* **A48:** 3561.
TAUT, M., and STURM, K. (1992). *Solid State Commun.* **82:** 295.
TELLER, E. (1962). *Rev. Mod. Phys.* **34:** 627.
TEN SELDAM, C. (1961). *Proc. Phys. Soc.* **79:** 810.
THOMAS, L. H. (1926). *Proc. Camb. Phil. Soc.* **23:** 542.
THOULESS, D. J. (1974). *Phys. Rep.* **13C:** 93.
THOULESS, D. J. (1977). *Phys. Rev. Lett.* **39:** 1167.
TOET, S. (1987). *MSc Thesis,* Eindhoven Univ. of Tech., Eindhoven, The Netherlands.
TOMONAGA, S. (1950). *Progr. Theor. Phys.* **5:** 544.
TORRANCE, J. B. (1978). *Phys. Rev.* **B17:** 3099.
TOSI, M. P. (1994). *Scuola Normale Superiore, Pisa:* lecture notes and private communication.
TRIGGER, S. (1976). *Phys. Lett.* **56A:** 325.
TUAN, D. F.-T. (1969). *J. Chem. Phys.* **50:** 2740.
TYUTYULKOV, N., KANEV, I., and CASTANO, O. (1980). *Theor. Chim. Acta* **55:** 207.
UNG, K. C., MAZUMDAR, S., and CAMPBELL, D. K. (1992). *Solid State Commun.* **85:** 917.
UNG, K. C., MAZUMDAR, S., and TOUSSAINT, D. (1994). *Phys. Rev. Lett.* **73:** 2603.
URSELL, H. D. (1927). *Proc. Camb. Phil. Soc.* **23:** 685.
USAMI, K., and MORIYA, T. (1980). J. Magn. Magn. Mater. **20:** 171.
UTSUMI, K., and ICHIMARU, S. (1981). *Phys. Rev.* **B23:** 3291.
VAN HAERINGEN, W., FARID, B., and LENSTRA, D. (1987). *Phys. Ser.* **T19:** 282.
VAN KAMPEN, N. G. (1981). in *Chaotic Behavior in Quantum Systems.* Natoasi Series B. Vol. 120, p. 309.
VAN STRAATEN, J., and SILVERA, I. F. (1988). *Phys. Rev.* **B37:** 1989.
VISSER, R. J. J., OOSTRA, S., VETTIER, C., and VOIRON, J. (1983). *Phys. Rev.* **B28:** 2074.
VOLLHARDT, D. (1984). *Rev. Mod. Phys.* **56:** 99.
VOLLHARDT, D., and WÖLFLE, P. (1980). *Phys. Rev.* **B22:** 4666.
VOM FELDE, A., SHROSSER-PROU, J., and FINK, J. (1989). *Phys. Rev.* **B40:** 10181.
VOSKO, S. H., WILK, L., and NUSAIR, M. (1980). *Can. J. Phys.* **58:** 1200.

WALTER, J. P., and COHEN, M. L. (1972). *Phys. Rev.* **B5:** 3101.
WANG, C. S., and KLEIN, B. M. (1981). *Phys. Rev.* **B24:** 3393.
WARD, J. C. (1950). *Phys. Rev.* **78:** 182.
WANG, W. P., PARR, R. G., MURPHY, D. R., and HENDERSON, G. A. (1976). *Phys. Lett.* **43:** 409.
WARREN, W. W. (1984). *Phys. Rev.* **B29:** 7012.
WARREN, W. W. (1987). In *Amorphous and Liquid Metals,* Lucher, E., Fritsch, G., and Jacucci, G. (eds.), *NATO Adv. Study Inst. Series E18,* Dordrecht: Martinus Nijhoff.
WARREN, W. W. (1993). *J. Phys. Condensed Matter Suppl.* **34B5:** B211.
WARREN, W. W., BRENNERT, G. F., and EL-HANANY, U. (1989). *Phys. Rev.* **B39:** 4038.
WATTS, J. D., and BARTLETT, R. J. (1994a). *J. Chem. Phys.* **101:** 409.
WATTS, J. D., and BARTLETT, R. J. (1994b). *J. Chem. Phys.* **101:** 3073.
WEGER, M., and FAY, D. (1986). *Phys. Rev.* **B34:** 5939.
WEN, X. G. (1990). *Phys. Rev.* **B42:** 6623.
WENDIN, G.. (1982). *AIP Conf. Proc. (USA)* no. **94:** 495.
WHITE, C. T., and ECONOMOU, E. N. (1978). *Phys. Rev.* **B18:** 3946.
WHITE, R. J., and BYERS BROWN, W. (1970). *J. Chem. Phys.* **53:** 3869.
WHITE, S. R., SCALAPINO, D. J., SUGAR, R. L., LOH, E. Y., GUBERNATIS, J. E., and SCALETTAR, R. T. (1989). *Phys. Rev.* **B40:** 506.
WIEGMANN, P. B. (1980). *Phys. Lett.* **80A:** 163.
WIGNER, E. P. (1934). *Phys. Rev.* **46:** 1002.
WIGNER, E. P. (1938). *Trans. Faraday Soc.* **34:** 678.
WIGNER, E. P., and HUNTINGTON, H. B. (1935). *J. Chem. Phys.* **3:** 764.
WILCZEK, F. (1990). *Fractional Statistics and Anyon Superconductivity,* Singapore: World Scientific.
WILSON, S. (1981). *Theoretical Chemistry,* Vol. 4, London: Royal Society of Chemistry, p. 1.
WILSON, S. (1988). *Electron Correlation in Molecules* (University Press: Oxford).
WINN, M. D., and LOGAN, D. E. (1989). *J. Phys. Condensed Matter* **1:** 1753.
WINTER, R., BODENSTEINER, T., and HENSEL, F. (1989). *High Pressure Res.* **1:** 1.
WINTER, R., and HENSEL, F. (1989). *Phys. Chem. Liquids* **20:** 1.
WINTER, R., PILGRIM, W. C., and HENSEL, F. (1994). *J. Phys. Condensed Matter* **6:** A245.
WIO, H. S. (1983). *Ann. Nucl. Energy* **11:** 425.
WOLFF, P. A. (1961). *Phys. Rev.* **124:** 1030.
WU, C. SUN, X., and NASU, K. (1987). *Phys. Rev. Lett.* **59:** 831.
WULMAN, H., and KUMEI S. (1973). *Phys. Rev.* **A9:** 2306.
XU, B. C., and STRATT, R. M. (1989). *J. Chem. Phys.* **91:** 5613.
XU, H., HANSEN, J.-P., and CHANDLER, D. (1994). *Europhys. Lett.* **26:** 419.
YAMADA, T., KUNITOMI, N., NAKAI, Y., COX, D. E., and SHIRANE, G. (1970). *J. Phys. Soc. Jpn.* **28:** 615.
YAMADA, Y. ET AL. (1985). *J. Phys. Soc. Jpn.* **53:** 3634.
YOKOYAMA, H., and SHIBA, H. (1987a). *J. Phys. Soc. Jpn.* **56:** 1490.
YOKOYAMA, H., and SHIBA, H. (1987b). *J. Phys. Soc. Jpn.* **56:** 3570.
YONEZAWA, F., WATABE, NAKAMURA, M., and ISHIDA, Y. (1974). *Phys. Rev.* **B6:** 2322.
YOU, M. V., and HEINE, V. (1982). *J. Phys.* **F12:** 177.
YOU, M. V., HEINE, V., HOLDEN, A. J., and LIN-CHUNG, P. J. (1980). *Phys. Rev. Lett.* **44:** 1282.
YOUNG, W. H., and MARCH, N. H. (1961). *Proc. R. Soc. A.* **256:** 62.
YOUNG, A. P., ET AL. (1989). *J. Phys. Cond. Matter* **18:** 2997.
ZANGWILL, A., and SOVEN, P. (1980). *J. Vac. Sci. and Technol.* **17:** 159.
ZENER, C. (1952). in *Imperfections in Nearly Perfect Crystals*: Ed. Shockley, W. (New York: Wiley).
ZHANG, F. C., and RICE, T. M. *Phys. Rev.* **B41:** 2557.
ZIMAN, J. M. (1961). *Phil. Mag.* **6:** 1013.
ZIMAN, J. M. (1964). *Adv. Phys.* **13:** 89.
ZUBAREV, D. N. (1960). *Usp. Fiz. Nauk.* **71:** 71.

Index